Energy: The
Next Twenty Years

Energy: The Next Twenty Years

**Report by A Study Group
Sponsored By The Ford Foundation And
Administered By Resources For The Future**

Hans H. Landsberg, Chairman

Kenneth J. Arrow
Francis M. Bator
Kenneth W. Dam
Robert W. Fri
Edward R. Fried
Richard L. Garwin
S. William Gouse
William W. Hogan
Harry Perry

George W. Rathjens
Larry E. Ruff
John C. Sawhill
Thomas C. Schelling
Robert Stobaugh
Theodore B. Taylor
Grant P. Thompson
James L. Whittenberger
M. Gordon Wolman

Ballinger Publishing Company • Cambridge, Massachusetts
A Subsidiary of Harper & Row, Publishers, Inc.

International Standard Book Number: 0-88410-092-8 (Hb)
0-88410-094-4 (Pb)

Library of Congress Catalog Card Number: 79-5226

Printed in the United States of America

Library of Congress Cataloging in Publication Data
Main entry under title:

Energy, the next twenty years.

   Includes bibliographical references and index.
   1. Energy policy—United States. 2. Power resources—United States. I. Landsberg, Hans H. II. Arrow, Kenneth Joseph, 1921- III. Ford, Foundation. IV. Resources for the Future.
HD9502.U52E56   333.7   79-5226
ISBN 0-88410-092-8
ISBN 0-88410-094-4 pbk.

Contents

111563

Part V
Improving the Process 509

List of Tables

List of Figures

Foreword

This study was commissioned by the Ford Foundation in 1978, and it is an honor to write its foreword as I leave the Foundation's presidency.

This is the third major independent study of energy problems that we have commissioned in the last eight years, and like its two predecessors, it appears at a timely moment. The central message delivered by the authors of the first study, *A Time to Choose*, was that this country could and should get along with less energy than historic patterns of growth suggested. That message was right and timely in 1974. The central message of the group that did the second study, *Nuclear Power Issues and Choices*, was that over the next generation we could and should handle nuclear power in ways that would reduce the dangers of nuclear proliferation, without giving up light water reactors for the present and other nuclear options for the future. That message was right and timely in 1976-1977. The central message of the present report is that energy—expensive today—is likely to be more expensive tomorrow and that society as a whole will gain from a resolute effort to make the price that the user pays for energy, and for saving energy, reflect its true value. And I myself think that the message will prove right and timely in 1979.

The members of this study group, many of them distinguished economists, have given heavy emphasis to the role of market forces, and I am glad that they have, partly because it was a valid criticism of *A Time to Choose* that it did not sufficiently consider those forces. The study group has also had the advantage of close experience of what government has and has not been able to do in the years since energy policy became a major issue. And while this study, like its predecessors, is addressed mainly to the American audience and to policy questions that can be decided within the United States, the group has made a major effort to take full and fair account of the differing perspectives from which other countries address the matter.

This report makes a particularly valuable contribution by its insistence on distinguishing real from unreal dangers. There is a real danger that there will be repeated special shocks like the one that was set off this year by changes in Iran. But the notion that we are headed toward a sudden worldwide energy "gap" created around 1985 by some inexorable excess of fixed demand over fixed supply—that is unreal. The study group is wary of easy rhetoric about "crisis," and while I myself believe that there are persistent problems quite grave enough to deserve the general title of energy crisis, I believe just as strongly that neither understanding nor the prospect of constructive action is advanced by foolish prophecy of an "energy gap." That helps as little as the false target of "energy independence" set by earlier leaders. Indeed, the single most important element in the continuing crisis of energy policy may be the continuing failure in our national understanding of it.

In addition to its insistence on respect for basic economic realities, this report contains much else that can advance understanding. It gives special emphasis to the growth of our dependence on oil imports and argues that there is high economic as well as political value in reducing that growth. It provides a cool assessment that may help to put limits on our hot arguments over the efficiency and equity of controlling and decontrolling oil and gas prices. It gives unusually knowledgeable accounts of the complexities that confound simplistic assessments, favorable or unfavorable, of coal, nuclear energy, and solar power. It gives trenchant analyses of the economics of conservation and environmental protection, with special attention to the ways of regulating air quality. In all this it draws on a vast body of technical data without forcing the reader, most of the time, beyond the limits of what can be understood with a first year knowledge of economics.

In its Overview the study group makes a number of specific policy recommendations. The most important of these—on decontrol of oil and gas prices and on public utility pricing—are politically controversial, so it is worth emphasizing that the Ford Foundation as an institution takes no position on any specific recommendation. We do believe that the demonstrated quality of its members, its intense collective effort of analysis, and its careful use of the best available evidence give the study group a powerful claim to be heard.

Let me end by offering our warm thanks to all members of the study group; to its small but excellent staff; to Resources for the Future, which administered the project; and above all to the group's chairman, Hans H. Landsberg. Mr. Landsberg's patience, persistence, and fairness, combined with his own extraordinary knowledge of the subject, are what have enabled the group to offer powerfully argued conclusions at a time when they can make a difference.

McGeorge Bundy

Preface

In the face of a flood of books, articles, speeches, and conferences on the subject of energy, produced by politicians, economists, physicists, ecologists, and even mystery writers, it is difficult to believe that anything new remains to be said. What, then, is the excuse for yet one more publication?

To find the answer, we must look back to mid-1975 when sharply colliding judgments on the desirable future course of nuclear energy persuaded the Ford Foundation to fund an attempt to raise the quality of discussion and to lower its temperature. The Nuclear Energy Policy Study (NEPS) group—twenty independent experts drawn from diverse professional and occupational backgrounds—was put together and a year later, in March 1977, issued a report, *Nuclear Power Issues and Choices.* The study was successful: it did improve the tone and sharpened the content of the nuclear debate; with facts and analysis it illuminated public policies and provided a reference point for the continuing examination of complex issues many of which have still not been resolved.

During the summer and fall of 1977 another energy source began to loom large—coal. Available in abundance, and not only in the United States, but beset by problems in production, in transportation, and in consumption, this time-honored energy source, ready for a revival, was bound to trigger controversy and policy debates and, perhaps, to encounter serious roadblocks. Accordingly, the Foundation, early in 1978, funded another study group, this one to assess the future of coal. But it recognized that it could not deal with coal in a vacuum. A comprehensive framework was required that compared coal to other energy sources in a consistent and comprehensive manner. The time frame of the study was to be the next two decades. This was seen as the "transition"—not a transition to a new stable condition, but to a set of new tendencies: from cheaper, abundant, reliable energy to costlier, scarcer, and more chancy energy supply; from principal reliance on conventional oil and gas to reliance on coal, nuclear, unconventional oil and gas, and renewable sources; from taking energy for granted to paying attention to it. The Foundation also specified that the study group complete its work within a year and a half, a demand designed to forestall the slippage to which committees are prone, to preserve the freshness of whatever the group might say, and to terminate the deliberations before the group ran out of steam.

To deal with its open-ended task within a tight schedule, the group had to be selective in its analysis. Some topics would have to be treated intensely, others cursorily or not at all. Little or nothing is said about geothermal energy, ocean oil spillage, the hazards of bringing liquid natural gas to coastal cities, the pros and cons of oil company divestiture, and many other matters.

The group felt strongly about one thing from the outset: a major aspect of the problem of energy is the way people think about it. Energy is crisis. Energy is all bad. Or all good. The "right" energy policy helps banish all evils—inflation, social injustice, international tension, environmental insults. What we hope to have done, at a minimum, is to help readers to think about the energy problem in a more sensible way than is possible if they rely solely on headlines, brief paragraphs written by columnists and spoken by television commentators, or partisan literature on all sides of the problem. If, in addition, decisionmakers find familiar principles well articulated or new ones persuasively put forward, that is an outcome we would obviously welcome. But we have no exaggerated notion of the significance of any single piece of writing in the massive outpouring of literature or that inspiration, innovation, and solution are there for the grasping.

A few words about the group's modus operandi. Beginning in late February 1978 we met once a month for two days; extended workshops were held in August 1978 and in January 1979. All of the report was written by the members of the group, drawing in part on background papers commissioned from outsiders. Drafts were discussed repeatedly in our monthly meetings and rewritten to capture the group's judgments. Thus, all drafts started as single viewpoints and slowly took on the coloration of the group. We ended up with a manner of "collective responsibility," though it must be stressed that not everyone agrees with each sentence that eventually survived the discussion, modification, and editing.

Readers may wonder how, confronted by so many complex and frequently divisive issues, a group like ours could have achieved unanimity. The simple answer is that throughout our deliberations, substantial differences emerged in judgment and in emphasis, on the exact wording of a recommendation, and on the need to include or ignore a specific issue. It is fair to say that each member of the group would probably have written a somewhat different report. Yet in the end, no member felt sufficiently critical of the thrust of the report to withhold consent. At the same time, all realized that this broad consensus leaves everyone free to continue to express his own views, if they conflict with what is found in these pages. What we achieved was a consensus on major issues and on the usefulness of the report rather than unanimity on all of its details.

Can any group, no matter what the background of its members, comprehend the manifold ramifications of the energy problem? We assumed not and so tried diligently to identify our ignorances and fill the gaps. In the early meetings we invited and listened to people from government, industry, and interest groups, such as in the consumer and environmental areas. We commissioned background papers by experts, several of which will be published in a companion volume, entitled *Selected Studies on Energy: Background Papers for Energy: The Next Twenty Years.* Later, we had our drafts read by outsiders and invited their criticism. In early 1979 a subpanel met in Paris with a number of Europeans to hear their reaction to what was then an interim version of the report's overview. At various stages we talked to legislators and their staffs. They encouraged us to speak up and speak out and not to be unduly restrained by second guessing the political feasibility of what we considered a desirable course. In some instances we have recommended policies where hard choices need to be made; in others we have limited ourselves to describing the issues involved and the instruments needed to make the choice.

Our focus and primary audience are quite clearly American. But we have tried to be conscious of positions, views, problems, and biases of other countries. We hope the report reflects our interest in and concern for their problems and will be useful and intelligible outside the United States.

As to the nature of the product, we have not tried to cast the various contributions into a common style to make it a "book" in the best sense. Time alone made this infeasible. Thus, the chapters differ in format, style, length, and degree of technicality. In an enterprise of this kind—halfway between a collection of essays and a once through narrative—a point of diminishing returns is reached in seeking conformity and symmetry. The one exception is the overview, which was begun early in our work and underwent continuing revision almost to the day it went to the printer. Taking our time with it enabled us to integrate rather than to assemble. This we believed important, as the overview is both part of the full report and intended to be read independently.

During the course of our deliberations, the energy world refused to stand still. Late in 1978 Iran went through a political upheaval, and oil production and exports first stopped, then resumed only slowly; reverberations continue, and the long-run outlook remains clouded. In March 1979 the worst nuclear accident so far recorded in the United States occurred at Three Mile Island in Pennsylvania and cast a long shadow over the future of nuclear energy. In May gasoline lines reappeared for the first time since embargo days, rapidly escalating oil prices led to rising clamor "to do something," and even Project Independence was having a degree of rebirth. There is little doubt that 1979 will be recorded as a year of perturbation in energy matters.

Any book dealing with issues and policy choices spanning a period of years or decades ahead, and especially one written on energy in 1979, courts the risk of seeming out of touch with whatever might be the crisis topic at the time of publication. But keeping readers abreast of day-to-day developments is the job of the press. Books must focus on longer range issues that may look a little different from the perspective of one month or year to another but that are not essentially altered by the headlines of the day. Moreover, as we stress throughout this report, policy is best not made in an atmosphere of crisis, and most of our findings and recommendations do not focus on any specific crisis and its management, a factor that may detract from the report's topicality but add to its life expectancy. Above all, government and the public alike are inescapably preoccupied with today's crisis, at the expense of the vitally important task of thinking about and fashioning policies for the longer run. We have aimed above all at contributing to the latter endeavor.

Nonetheless, when all is said and done, I believe that, had we begun our deliberations now rather than in late February of 1978, we would still be unlikely to have escaped wholly the mood of mid-1979. If I read the study group correctly, we would have put stronger stress and have done or promoted additional analysis on at least three subject areas.

One, given the widening gap between the world and the domestic price of oil, we would have emphasized even more strongly the need for not just the act of price decontrol but its finality. Second, we would have stressed more strongly the need for pushing ahead on alternatives to imported oil. In doing so, we would have had to give a little on our preference for letting market forces do the job, on the grounds that speed is important, partly for noneconomic reasons. Without dissent, special measures to promote conservation would have been endorsed, and with misgiving and with some dissent, a more determined though not more massive effort to move ahead on a supply alternatives program, including but not limited to synthetic fuel plants, such as adumbrated in Chapters 6 and 15. Finally, we would have dealt more fully with measures to manage short-term perturbations, although we never saw contingency planning at the microlevel as part of our task, and even though I harbor a strong suspicion that the crisis that erupts is, as often as not, like Three Mile Island, one not foreseen.

It remains for the chairman only to put down in writing his thanks to all participants. They range from those in the Ford Foundation most intimately involved: McGeorge Bundy, Marshall Robinson, and Allan Pulsipher; to Charles Hitch, president of RFF at the time; to the members of the study group itself; and to the members of our small "secretariat" at RFF: Fisher Howe who, assisted by Linda Perrotta, managed the logistics end of the project and saw to it that both people and documents went to the right place at the right time so that our meetings could be productive; Richard Sclove who not only kept terse and clear minutes of our meetings, but who played an essential role in reviewing, checking, questioning, and filling in of the chapters; my administrative assistant, Helen-Marie Streich; and Carolyn Cummings-Saxton who added to our research capacity in times of stress. Douglas Bohi, Joel Darmstadter, and Milton Russell of RFF, were ready with help whenever asked. Our editors, David Aiken, Ruth Haas, and Herbert Morton, did their best to turn a dozen and a half different prose styles into a semblance of orderliness and to reduce the volume of written material to manageable proportions. Only those who handled what we gave them can appreciate the magnitude of the task.

Finally thanks are due to a large number of persons who helped the Study Group in a variety of ways—by making presentations or otherwise joining in discussions at meetings of the Group, by writing background papers or brief statements, by filling gaps, and by written or oral review of portions of the report. Several of the background papers commissioned by the Group are being made available in a companion volume entitled *Selected Studies on Energy: Background Papers for Energy: The Next Twenty Years.* Its contents are listed in an Appendix. Other experts from government, industry, universities, and other fields, who helped at various stages of the undertaking are: John Ashworth, Jack F. Bennett, W. Michael Blumenthal, Clarence J. Brown, Douglas M. Costle, Richard N. Cooper, Robert Curry, Ray Downy, Rex Dupont, John Ecklund, A. Denny Ellerman, Maurice Ernst, Anthony C. Fisher, Joseph L. Fisher, Jerome G. Gaves, Michael Haltzel, C. Howard Hardesty, Edwin E. Herricks, Henry M. Jackson, Joseph P. Kalt, Juanita M. Kreps, Marc K. Landy, Kai N. Lee, John H. Lichtblau, Mike McCormack, Gordon J. McDonald, Bruce K. MacLaury, Richard R. Nelson, Alyce Newberg, John F. O'Leary, Frank Potter, Paul R. Portney, Penelope ReVelle, Philip R. Sharp, Don Smith, David A. Stockman, Karl Swenson, Piero Telesio, Lee C. White, Arthur W. Wright.

In addition, the following attended an international panel meeting in Paris at which an early draft of the Overview and the papers dealing with energy matters in countries other than the United States were reviewed: Salah Al-Shaikhly, Robert Belgrave, Jane Carter, Umberto Colombo, Thierry de Montbrial, Pierre Desprairies, Wolfgang Hager, Jack E. Hartshorn, Milton Klein, Ulf Lantzke, Jacques Lesourne, Amory B. Lovins, Bill Martin, Hanns Maull, Cesare Merlini, Zuhayr Mikdashi, Francisco R. Parra, Edith T. Penrose, Ian M.H. Smart, Roger Vaurs, Leonard Williams.

It is understood that none of the persons listed above take responsibility for the Study Group's findings and recommendations, though they may well find their viewpoints reflected here and there in the text.

May 31, 1979   Hans H. Landsberg

Energy: The Next Twenty Years
Study Group

LIST OF MEMBERS

Hans H. Landsberg (Chairman), Director, Center for Energy Policy Research, Resources for the Future

Kenneth J. Arrow, James Bryant Conant University Professor, Harvard University

Francis M. Bator, Professor of Political Economy, John F. Kennedy School of Government, Harvard University

Kenneth W. Dam, Harold J. and Marion F. Green Professor, University of Chicago Law School

Robert W. Fri, President, Energy Transition Corporation

Edward R. Fried, U.S. Executive Director, World Bank

Richard L. Garwin, IBM Fellow and Science Advisor to the Director of Research, Thomas J. Watson Research Center, IBM Corporation; and Professor of Public Policy, John F. Kennedy School of Government, Harvard University

S. William Gouse, Chief Scientist, The MITRE Corporation

William W. Hogan, Professor of Political Economy, John F. Kennedy School of Government, Harvard University

Harry Perry, Consultant, Resources for the Future

George W. Rathjens, Professor of Political Science, Massachusetts Institute of Technology*

*In January 1979, Professor Rathjens accepted a part-time appointment as Deputy U.S. Representative for Nonproliferation, with particular responsibility for U.S. participation in the International Nuclear Fuel Cycle Evaluation. His contribution to the study was substantially completed prior to that time.

Larry E. Ruff, W.R. Grace and Company

John C. Sawhill, President and Professor of Economics, New York University

Thomas C. Schelling, Lucius N. Littauer Professor of Political Economy, John F. Kennedy School of Government, Harvard University

Robert Stobaugh, Professor of Business Administration and Director of the Energy Project, Harvard University Graduate School of Business Administration

Theodore B. Taylor, Visiting Lecturer (Half-Time), Department of Mechanical and Aerospace Engineering, Princeton University

Grant P. Thompson, Senior Associate, The Conservation Foundation

James L. Whittenberger, James Stevens Simmons Professor of Public Health, Harvard University School of Public Health

M. Gordon Wolman, Chairman and Professor of Geography, Department of Geography and Environmental Engineering, Johns Hopkins University

Overview

1

INTRODUCTION

More than half a decade has passed since the oil crisis of 1973-1974 signaled a new era in U.S. and world history. The effort to develop a satisfactory policy response to what was once characterized as the "moral equivalent of war" has stretched out so long that weariness rather than vigor characterizes the national debate. Even the passage in 1978 of legislation designed to establish a national energy policy neither defined nor resolved the problems clearly: energy pricing and anti-inflation policies are at odds; energy and environmental objectives seem irreconcilable; there is no ready strategic petroleum reserve; a national consensus that solar energy is a good thing has yet to result in significant resource commitments, while support for nuclear energy, yesterday's hope for tomorrow, is eroding; and coal is marking time. Meanwhile, the slow, steady increase in the number of barrels of oil imported, and intermittent rude shocks such as the Three Mile Island nuclear accident and the summer gasoline scramble, provide reminders that much needs to be done.

To be sure, there have been encouraging developments in the past five years. Economic growth has resumed. Energy consumption per unit of output has declined steadily. It is estimated that half of U.S. homeowners have added insulation to their houses. Improved gasoline mileage for new automobiles is slowly but steadily raising the average performance of the U.S. auto fleet. Industry consumed 6 percent less energy in 1978 than in 1973, while producing 12

1

percent more output. These developments demonstrate what some have questioned: energy is an economic good, not unlike others; firms and individuals can, given the proper motivation, find ways of satisfying their needs with less of it.

Nonetheless, despite the progress that has been made, and even allowing most charitably for the delays that are characteristic of a democratic society and for the real conflicts of values and interests that make the energy problem more difficult than most, it is hard to escape the conclusion that the U.S. policy response has been seriously inadequate. Individual shocks have produced hurried and often poorly considered responses. Between shocks—which is most of the time—the sense of urgency has receded and basic, continuing problems have been only timidly addressed. The emphasis has been on the quick fix, the politically easy measure that hurts no constituents and produces more barrels of pork than of oil. Indeed, most of the positive developments in energy have occurred as much in spite of as because of any specific energy policy measures.

We are convinced that this situation is unsatisfactory if not downright dangerous for the United States and for the world. We are also convinced that the situation can be improved only by looking beyond the headlines of the day and putting energy in its proper time perspective. In energy matters, current decisions about location, investments, and designs have such a strong influence on the long-run future and are so strongly influenced by expectations about that future that short-term and long-term policies cannot be made independently. A coherent view of the problems and of their solutions over time must be developed, lest goals and programs for the short term frustrate those for the long term and vice versa.

In order to put energy into a more useful time perspective in this study, we have used twenty years as our time horizon. A period much shorter would be inadequate to include the effects of some policy actions that are being or should be taken now and would introduce too many constraints on what is possible. A period much longer would require speculation about unknowable matters and would make it too easy to dismiss real problems and interim solutions with the argument that over the long run everything adjusts to everything else. Of course, the shape of the future beyond this twenty year horizon is a central concern to our analysis, even though we do not deal with it directly. But the objective of today's policies should be to leave to the citizens of the next century a world with economic vitality, environmental health, and social and technological options sufficient to solve the pressing issues of their day, not to describe or prescribe their world in detail.

Clear thinking about energy problems requires an international perspective. Energy resources are a major element in international economic and political affairs, and energy-related events elsewhere can have direct consequences for the United States. Our study has been focused on energy policy rather than foreign policy and has been conducted by and primarily for citizens of the United States. We have not tried to analyze foreign policy issues or to suggest energy policies for other countries, whose situations may be quite different. But we have earnestly sought views from energy specialists outside the United States and have tried to be aware of how U.S. policies might affect others. And we view our recommendations for energy policy as appropriate for an internationally responsible United States.

With this twenty year, internationally conscious perspective, our study group has tried to identify some of the fundamental realities that, in our view, define the energy problem. Unlike the myriad of energy supply and demand forecasts in the literature—a myriad to which we have resisted adding—these realities, once stated and explained, are neither numerous, difficult to comprehend, nor surprising. But because we are persuaded that these realities are vital to a clear understanding of energy problems and because they have so strongly influenced our policy recommendations, we summarize them here and explain them at some length in the next section of this Overview.

Reality One: The world is not running out of energy

The energy resources of the United States and the world are huge, at prices not much more than about double those that prevail today. Use of these energy resources may be constrained by political or environmental factors, but the world is not "running out" of energy. With proper policy and planning and a willingness to pay the costs, energy can be produced to meet any reasonable projection of demand, without "gaps" or physical shortages.

Reality Two: Middle East oil holds great risks, but is so valuable that the world will remain dependent on it for a long time

The world is critically dependent on oil from the politically unstable Middle East, increasing the probability that otherwise minor events will result in major economic disruption or even war; the United States is vulnerable to such events because of its own oil imports and because of the even greater import dependence of some of its friends. This world dependence is due to the scarcity and geographic concentration of easy to produce oil and to the high costs of alternatives—facts that cannot be much changed by attacking the oil "cartel." Efforts to ease the world oil supply-demand situation can reduce this dependence, but only slowly and at high cost.

Reality Three: Higher energy costs cannot be avoided, but can be contained by letting prices rise to reflect them

Higher energy costs are a reflection of physical facts: the easy sources are about gone, while the plentiful sources are expensive to use safely. The higher costs need not have severe effects on economic welfare or lifestyles if they are properly managed; but it is a dangerous misconception to believe that government can somehow provide dependable, clean, and plentiful energy cheaply. The transition from lower to higher energy costs will be easier overall if prices rise to reflect the economic realities; but finding the political will and the technical means to deal with the income distribution and inflationary effects is difficult.

Reality Four: Environmental effects of energy use are serious and hard to manage

Some energy activities pose serious threats to human health and to the environment. The need to reduce those threats will be a major cause of rising energy costs and may even limit the extent to which some particular energy resources will be used. But there is a high degree of uncertainty surrounding the mechanisms and extent of damage and risks, and the costs of reducing the threats can depend critically on how the threats are defined and managed. It is important that environmental objectives be defined carefully and pursued efficiently, so that they can be achieved as fully as possible in the long run.

Reality Five: Conservation is an essential "source" of energy in large quantities

Both in the short and the long run, energy conservation is often the cleanest, quickest, and cheapest way to react to the inevitable higher energy costs. Over our twenty year period, conservation will inevitably become one of the most important energy "sources" quantitatively. Because effective conservation involves the decisions of millions of diverse individuals, with a few notable exceptions it cannot realistically be mandated or managed centrally, but requires that information and incentives be provided to energy users who make their own adjustments.

Reality Six: Serious shocks and surprises are certain to occur

Because the energy system is a complex combination of technology and society, the future is certain to contain serious shocks, most probably involving short-term supply interruptions and price instability in world oil markets. Preparation for such shocks is perhaps the most important (and neglected) function of energy policy. There are also sure to be surprises both pleasant and unpleasant regarding new supply and conservation technologies, so that the long-term outcome cannot reliably be predicted; a wide range of diverse options must be maintained precisely because we do not know which ones will ultimately prove to be most acceptable.

Reality Seven: Sound R&D policy is essential, but there is no simple "technical fix"
New technologies of energy production and conservation will be a major part of the best response to higher energy costs, and government policies toward R&D (and toward other things, such as energy pricing) will be a major influence in determining which technologies are developed and applied and when. But there is no single technical solution nor much likelihood that technology in general will be able to reverse the trend toward higher energy costs.

These seven realities are developed and analyzed further in the next section of this Overview, drawing on the material presented in the balance of this report. Our analysis of these realities and of the policy options available for dealing with them convinces us that U.S. energy policy should be developed around the following principal themes:

- Use market forces to price and allocate energy efficiently, because doing so will simplify most other aspects of energy policy;
- Develop ready contingency programs for a sudden interruption in oil imports, because such a cutoff is almost certain to happen somewhere, sometime, with potentially disastrous consequences;
- Recognize that economic and political factors in world oil markets make the cost of imported oil greater than measured by its price; any explicit measures to reduce imports can be economical if the per-barrel cost of achieving the reduction is not too high:
- Encourage energy production—including its equivalent, energy conservation—anywhere in the world, because world energy and economic systems are so interdependent that developments in one part of the globe affect everyone;
- Change some of the procedures and programs now used to deal with environmental conflicts, because the important substance of

the debates is too often lost in the maze of the process, and the costs of mismanagement are too high to be ignored;
- Use government research and development policies to define and develop a wide range of information and options, but not to push premature adoption of one or another technology.

The specific policies that we recommend for developing these themes are detailed in the balance of this report and are summarized in the final section of this Overview.

We believe our recommendations, taken together, provide a sound framework for facing an energy future that will be difficult but manageable. The fundamental changes we anticipate in the world energy situation over the next twenty years—such as a relative decline in oil and gas use, rising energy costs and prices, and increased efficiency in energy use—will occur whether the United States handles its energy policy wisely or foolishly. But U.S. energy policy will be a major factor determining whether the changes will be relatively smooth and easy or will result in unnecessary disruption, cost, or worse. If the required energy adjustments have to be forced by events that have not been anticipated and planned for, the next twenty years could see a series of perpetual crisis-oriented actions serving primarily to delay required long-term adjustments; this could result in an energy future much worse than the energy present. While we think that we understand the principal realities of the energy future and have some suggestions to make for dealing with them, until we see more evidence that policy is moving in the directions that we (among many others) think is necessary, we cannot be truly optimistic about energy in the next twenty years.

SEVEN REALITIES OF THE ENERGY SITUATION

Energy pervades a modern industrial society in ways that make it difficult to decide what is an energy problem and what is a problem of income distribution, environment, or international politics. Energy touches each of these (and many other) aspects of life, playing a major role in some and a minor role in others. As a result, when "energy" is selected as the central organizing theme, the "energy problem" looks different to different observers, depending upon perspective. Each of these perspectives can help identify real problems that must be addressed with specific policies. But for thinking about energy itself, the many perspectives need to be integrated into an overall view.

We are persuaded that one of the most fruitful ways to begin the process of integration is to consider energy as an economic variable. General economic concepts—such as the inevitability of trade-offs and the value of decentralized mechanisms in managing complex situations and contending interests—help in developing a coherent view of energy problems. This is not to say that the issues surrounding energy policy are solely economic, but rather that economic concepts provide an extremely useful organizing framework for dealing with this complex set of problems.

Our study group, diverse though it is in viewpoint and training, has its own implicit values, biases, and blinders, which have influenced our analysis and conclusions. Those who see energy as essentially a problem of technology or lifestyle or environmental quality will not be convinced that our framework of analysis is adequate. But those who are looking for a way to bring some order to a complex and confusing subject may find our approach of some value, even if they disagree with our specific recommendations. For this reason, we offer our views in two parts: first we expand upon the statement of energy realities mentioned briefly in the preceding section, and then we turn to specific policy recommendations. The balance of this report presents the analyses, background information, and detailed recommendations developed during our study.

Reality One: The world is not running out of energy

The potential energy resources of the United States and the world are so large that the ultimate physical exhaustion of energy in general is hardly a matter for present concern. In the long run of half a century or more, some of the more speculative forms of energy production, such as nuclear fusion and new applications of solar energy, will in all likelihood be capable of supplying essentially unlimited amounts of energy at costs that will probably be high, but easily manageable with the income levels of that time. Even if these ultimate solutions do not become available for a century or more, there are many energy sources that are somewhat more expensive than those of the recent past, but are known to be workable and available in quantities large enough to last through the next century and beyond—coal, the nuclear breeder, geothermal power, and known forms of solar energy. And finally there are enormous "supplies" of energy waiting to be tapped in the form of more efficient uses of energy and in the form of substitutions of ingenuity and information for energy.

For this reason, it is incorrect and misleading to define the long-run energy problem in terms of a gap, shortfall, or shortage, as though there were some natural definition of energy needs and some physical supply limits preventing these needs from being met. The energy is there to be had, at a cost, in virtually whatever quantities it may be demanded; and demand itself is a variable, depending on cost and availability of energy. While physical gaps or shortages may occur due to poor policy or inadequate planning, present or past, these should not be attributed to any general "running out of energy." It is not a physical lack of resources that accounts for people waiting in line for gasoline in mid-1979. Indeed, even a hundred years from now there can be plenty of fuel around to power private automobiles—if people want to pay enough for it and if policy allows technology and resources to be developed in a timely way.

As a result of poor policy and planning before and since 1973–1974, it is conceivable that the world could deplete certain sources of conventional energy (primarily low cost fossil fuels) before the long-term, virtually inexhaustible options are available in quantity. Even here, however, physical exhaustion is not the problem. Proved reserves of oil and gas—that is, identified deposits that can be produced economically under present conditions of price and technology, a category often likened to the industry's on-the-shelf inventory—in the United States will last about ten years at current production rates, while remaining discoverable and recoverable conventional oil and gas resources could allow production at present annual rates for an additional two to three decades. Coal can (at a cost, and with adequate time to make the required investments) do anything that oil and gas can do; U.S. coal reserves could support an energy demand of 100 quads per year for fifty years. (The United States is now using not quite 80 quads per year; a quad is a unit of energy equal to one quadrillion British thermal units—10^{15} Btus.) Additional estimated coal resources beyond those identified as reserves could last 300 years at this rate of consumption. Thus, aggregate oil, gas, and coal resources in the United States are more than adequate to meet energy needs until sources such as shale oil, tar sands, heavy oils, peat, and various widely abundant forms of gas can be developed; with these sources, U.S. fossil fuel resources stretch far beyond any reasonable planning horizon. Demand for these resources, and improved methods for producing and using them, will be developed in response to rising costs of traditional sources.

For the world as a whole, the situation is more uncertain, but the general picture is the same: conventional oil and gas resources are adequate to meet the energy demands of a growing world economy for several decades, and coal, shale, tar sands, and so forth can add a century or more to this estimate. In the case of the world supply of these fuels, the estimates of the recoverable quantities can be expected to increase with time, since large parts of the world are virtually unexplored for their resources.

Geographic distribution of high grade energy resources is uneven. The Middle East, the Soviet Union, and North America have the greatest concentration of oil and gas; the latter two plus China are the richest in coal. Other parts of the world appear to have fewer resources, although Africa and South America have been relatively little explored and may hold surprises. But to some extent, the problem of geographic concentration of energy resources will become less important as the high grade resources are used up; higher prices will make it economically practical to mine lower quality resources that according to accepted geologic principles are likely to be more widely distributed throughout the world.

Physical abundance does not mean that all or any of these resources will be, should be, or can be used without limit. They will all be costly to extract, use, and sometimes to transport, and even with careful control strategies there will remain serious potential threats to human health, aesthetics, and the physical and biological environment. The need to go further and deeper to get energy and the higher environmental and social costs associated with the lower grade energy resources will tend to offset improvements in the technology of extraction and use that are sure to occur and that have traditionally resulted in declining real energy costs. Although it is not certain how these countervailing tendencies will balance out, the best bet is that real energy costs will rise. And since energy is only one input into a dynamic, changing economic system, the world will adjust to the higher costs by using less energy than it otherwise would, even though the energy resources are physically there to be used.

There are other reasons to believe that some energy forms may never be used in the quantities that are physically possible. Nuclear fission power, for example, may be limited more by concerns about operating safety, waste disposal, and nuclear weapons proliferation than by resource availability. Or further scientific investigation may confirm the worst fears about energy systems that depend on burning fossil fuels—that the carbon dioxide they create adversely alters world climate. These possibilities persuade us that the world must try hard to find long-run alternatives such as solar energy and energy conservation, so that our grandchildren will have the option of not living in a world containing thousands of nuclear reactors, dozens of nuclear-fuel-processing plants, and huge complexes processing coal and its less desirable cousins. With luck and successful policy, the world may be able to have a prosperous and satisfying future without digging very far into its huge endowment of dirty and potentially dangerous fossil and nuclear fuels.

This recital of the general facts about resource abundance is not meant to suggest that there are no problems. Bad luck and bad policy may combine to inhibit the timely emergence of new sources and technologies. And even the best of policies cannot guarantee that energy costs or environmental disruption will not become very painful, at least temporarily while adjustments are being made. But the current energy problem does not lie in physical resource limits. It lies in a too heavy dependence on one or two sources of energy, in an unwillingness to face up to change, and in a fear unjustified by the facts that the costs of change will be unmanageable. It is to these issues that we now direct our attention.

Reality Two: Middle East oil holds great risks, but is so valuable that the world will remain dependent on it for a long time

The most vexing short-term energy problem for the United States and for the world is that the remaining supplies of low cost oil—that is, oil cheap to find and extract—are scarce and concentrated in a few, relatively unstable parts of the globe. While the world is moving away from these energy sources to others more widely distributed and more abundant, the high prices of oil strain the world's economies, and interruptions in the supply of oil could shatter world peace and stability. Problems that are minor by comparison—such as pricing energy efficiently, controlling strip mining, and redistributing income among winners and losers—take up most of the time and emotion in domestic policy debates, but pale in significance when compared to the truly difficult problems associated with world dependence on Middle East oil.

The United States now gets about half of its oil and a quarter of its total energy from imports, while Western Europe relies on oil imports to provide essentially all of its oil (the United Kingdom and Norway are the main exceptions) and about one-half of its energy, and Japan imports all its petroleum, representing over three-quarters of its energy. About 80 percent of the crude oil in world trade is produced by the countries that are members of the Organization of Petroleum Exporting Countries (OPEC); and about three-fourths of OPEC oil, or 60 percent of the oil in world trade, is produced in the Middle East. The Soviet Union has been meeting its own oil needs and those of Eastern Europe and even exporting a little oil to the West, but may cease doing so in the near future if (as some expect) oil production in the USSR is unable to keep pace with internal demands. Since 1973-1974, rising world income has offset the demand-reducing effects of the higher world oil prices, with the result that demand for oil in the world market, and hence for OPEC oil, has continued to grow. Further large increases are forecast for the 1980s.

The most serious problem associated with this high demand for imported oil in the United States and elsewhere is the dependence it creates on a small and unstable part of the world for a large share of a vital commodity. Because of this dependence, events in far-off places that interrupt the flow of oil can imperil the economies of the importing countries. While this problem is serious for the United States, it is far more serious for others, who would be devastated by a sustained interruption in oil imports. Many of the things that the United States might do to reduce the threat of an interruption in oil supplies are not really matters of energy policy as such and hence lie outside the scope of this study; we leave it to others to suggest ways that the United States should respond to Soviet-Cuban activities in critical areas, deal with the new regime in Iran and the old one in Saudi Arabia, prevent war between Israel and the Arabs, position the fleet to defend the sea lanes, and so forth. We limit our discussion and recommendations to U.S. *energy* policy.

The most obvious way that energy policy might try to reduce vulnerability to events in the world oil market is to reduce the level of oil imports. But there is no simple relationship between the import level on a day-to-day basis and the probability or impact of sudden interruptions in the flow of imports. A lower level of U.S. oil imports would ease pressure on world oil supplies and, perhaps, prices and is worth encouraging for that reason. However, to what extent this would make the Middle East political situation any less volatile or make use of the "oil weapon" any less tempting is not at all certain. Furthermore, incrementally reducing the level of oil imports by such measures as converting some oil-burning power plants to coal, improving automobile efficiency, encouraging solar home heating, building a few synthetic fuel plants, and stimulating marginal oil production would not necessarily make it any easier to accommodate a sudden decrease in the flow of the remaining imports; it may be harder to adjust to the sudden loss of a million barrels per day of imports after the easy import-reducing steps have been taken.

For example, lines at gasoline pumps seem invariably to produce political pressure to "do something," such as launching a massive synthetic fuels program. But installing the capacity to produce even 1 million barrels per day of synthetic liquid fuels—about one-eighth of current U.S. oil imports and of current U.S. domestic oil production—would require some $40 billion and ten years lead time. Because of the high capital cost involved, the synthetic fuels plants would probably be operated at maximum capacity continuously, providing little or no cushion against sudden import interruptions. A much better case can be made for beginning to install significant synthetic fuels capacity, in order to reduce pressure on world oil prices in the late 1980s, to learn what is involved in case such plants become necessary in the 1990s, or perhaps even to demonstrate that the United States is at last "serious" about energy problems. But it is a mistake to think that such a program could significantly reduce the vulnerability of the United States to unpredictable interruptions in the world oil market any time soon.

Even complete elimination of oil imports (which would be prohibitively costly within most of our twenty year period) would not fully insulate the United States from the effects of disruptions in the world oil market, because other countries will long remain dependent on oil imports. The United States could not stand idly by while its allies and trading partners were faced with a choice between economic collapse or military action. It may be, in fact, that the world oil market would be more prone to shocks and less able to adjust to them if the United States were to withdraw from it completely; a better long-term strategy may be for the United States to remain a major importer with a direct interest in maintaining the flow of oil into the world market, but with the capacity to reduce temporarily its own imports (or even to become an exporter) if necessary.

In any case, there is no doubt that the United States and others will continue importing a large share of their oil for most or all of our twenty year period and that all will be affected by shocks anywhere in the world oil market. Thus, the importing countries must work together to prepare for the kinds of interruptions in supply that are certain to occur, but at times, in places, and for reasons that are unpredictable. The energy policy options for doing so are discussed more fully under Reality Six.

Another serious problem caused by oil imports is their high cost, paid primarily to OPEC countries. In the United States, oil imports cost $40 billion in 1978 and represented nearly one-fourth of the cost of all imports. But these costs, large as they are, apparently are still less than the costs of doing without the imports. The fact that the payments go to foreigners is a complicating factor, but does not change the basic nature of the problem: the costs of domestic oil must be paid by using economic resources to sink holes in the ground, while the costs of imported oil must be paid by sending U.S.-produced goods and services abroad, now or in the future. The high international payments cause some problems that are potentially serious but that can be handled by increasing current exports and importing foreign capital to make productive investments in the U.S. economy. Even countries with oil imports proportionally much larger than those of the United States, such as Germany and Japan, have been able to manage their import bills. As long as imported oil is cheaper than domestic substitutes for it or than doing without it, imports make economic sense.

That the oil-importing countries can pay high prices for oil without collapsing economically and that they do pay these prices rather than get by without the oil does not prove that it is economically efficient or fair that the price be as high as it is. It may well be true, as is generally believed, that the world oil price since 1973-1974 has been set by a greedy cartel at levels that are far too high in some senses and that U.S. policy might be able to get prices down. Certainly oil prices are far above the cost of extracting oil in the productive oil fields of the Middle East, where development and lifting costs are often less than a dollar per barrel; and oil prices are clearly too high for the best short-run interests of the importing countries. But it is not so obvious that world oil prices since 1973-1974 have been grossly too high for the long-run efficiency of the world economy.

The true economic value of a scarce and depleting resource has little to do with the cost of extracting it or with what the buyers would like its price to be. Rather, the economic value depends on how long the resource is expected to last at projected usage rates, the cost of substituting other sources or doing without it when it is gone, and the productivity of investments elsewhere in the economy. The current and projected world oil prices are too high only if total world demand at lower real prices could be fully satisfied from low cost oil sources until economical supply and conservation alternatives make low cost oil unnecessary. In light of the scarcity of low cost oil resources in the world, the apparently insatiable world appetite for oil even at high prices, and the slow progress being made toward alternative supply and conservation measures, it is not at all clear just how much too high world oil prices have been since 1973-1974.

No one can be sure how high current and expected oil prices should be to provide the proper production, conservation, and substitution incentives to wean the world gradually away from low cost oil by the time it runs out. But market forces, given an opportunity to work, tend to price scarce and depleting resources to reflect their real, long-run scarcity value, at least as seen by the owners of the resource. An individual resource owner, whether a monopolist or a perfect competitor, will not sell now unless the price he gets reflects the expected future value; oil in the ground may indeed be as good as money in the bank—especially if the owner does not have much faith in either the money or the bank! When extraction costs are small compared to the real scarcity value of the resource, it may not make much difference whether the resource owners are acting competitively or monopolistically, because either way prices will be set high enough to stretch the life of the resource until alternatives become available. Prices may depend much more on the resource owners' individual views of the value of the current revenues relative to future revenues.

For these reasons, we think it is unproductive and distracting to blame the world's energy problems on the existence of OPEC and particularly to look for solutions in its destruction. Political events, a rapidly tightening world oil market, and changing perceptions about the energy future combined in 1973–1974 to give all oil-exporting countries, whether members of OPEC or not, a unique historical opportunity. Suddenly, oil was priced to reflect the long-term interests of those countries who had it to export, not the short-term interests of the international oil companies and their home countries. Whether or not the resulting prices are too high for the long-run efficiency of the world economy, there is not much prospect that the historical change from a consumer-dominated to a producer-dominated market will be reversed.

If OPEC really is a collusive cartel in the traditional sense and is effectively raising prices far above efficient competitive levels, the world can work and hope for its collapse. Immediately after 1973–1974, many experts were saying that OPEC would soon suffer the fate of most producer cartels, as its members individually began trying to take advantage of the too high prices by expanding output surreptitiously. But this has not happened, and we do not think that it is likely to happen. Even without periodic meetings and the other trappings of a collusive cartel, the individual oil-producing countries would now produce their oil at rates and sell it at prices reflecting its long-term value to them individually: Saudi Arabia would still be faced with the fact that its financial investments abroad are yielding essentially no real return after correcting for inflation and currency depreciation; Iran would still have the social problems resulting from trying to modernize rapidly; Mexico and Norway (not now members of OPEC) would still develop their oil resources deliberately, to make sure that they do not squander their national assets. Replacing OPEC with a noncommunicating group of self-interested competitors might now have very little effect on world oil production and prices.

In this situation, there are only a limited number of ways to try to limit world oil price increases. One is to persuade the governments of key oil-exporting countries that their interests lie in increasing production and holding prices down in the short run; military or diplomatic pressure or support of a particular government against internal or external threats might be the means of persuasion. However, this approach, even if it could work, might not be wise in the long run: forcing exporting countries to keep oil prices low so that consuming countries could continue importing large volumes of oil cheaply and painlessly would be shortsighted if the wells will run dry before new sources are developed.

Another way is to convince the oil producers that they have better investment opportunities than leaving oil in the ground. In large, heavily populated countries (Mexico, Iran, Venezuela) these alternative investments would probably be internal development projects; in less populous countries (Saudi Arabia, Kuwait) they would probably be financial or real foreign investments. Either way, developments since 1973–1974 suggest that it will not be easy to demonstrate (and may not even be true) that there are better investments than oil in the ground at current and projected prices. Nonetheless, there may be real opportunities for mutually beneficial international agreements regarding oil prices and investment of oil revenues—if the oil importers are willing to open up their internal economies to foreign investment on a large scale.

Finally, there is the approach suggested by ordinary market analysis: the best way to keep prices down in the market is to make do with less and to find other suppliers. Whether OPEC is more nearly a powerful cartel or a weak association of vigorous competitors, the forces of supply and demand are at work. A relatively weak demand at the prevailing price (a condition that prevailed from 1974 into 1978) makes it less likely that the official OPEC prices will increase; instead, price shaving may occur, as it did. On the other hand, when an important supplier such as Iran withdraws from the market, spot prices rise sharply; if the reduction in supply begins to look permanent, long-term prices rise as well. Other events with less direct immediate impact, but with implications for oil supply and demand in the long run, will also influence the world oil price:

projections of a world oil "crunch" in 1985 increase pressure for higher prices now; setbacks for the nuclear power program increase pressures for higher oil prices; new oil discoveries, or credible programs to augment supply or reduce demand, increase pressures for lower prices now and later. No monopoly or greedy cartel is necessary to explain such developments: the markets for soybeans, housing, or engineers exhibit the same behavior.

The world oil market is very complicated and not completely explainable in terms of simple market theories. Nonetheless, the United States may be able to help slow the increases in world oil prices by such measures as encouraging energy (and especially oil) conservation domestically; increasing production of energy in all forms from the large United States resource base; helping the developing countries in their energy' production and conservation programs; and coordinating research, development, and technology transfer programs internationally in ways that will help ease the world energy supply-demand balance. In an integrated energy and economic world, economical supply expansion and demand reduction anywhere helps everyone.

The problem with this general and by now quite conventional list of measures to "encourage," "stimulate," and "push" is that it does not answer the question of "how hard" beyond the level resulting from ordinary economic forces and high prices for imported oil. All manner of inefficient programs and expensive white elephant projects can be justified by a general mobilization to reduce oil imports, unless some clear criteria are developed for evaluating and selecting policies and projects. While we have no simple answer to the question of how hard to push on the world energy supply-demand situation, we do think that the decision criteria should reflect the fact that the real problem is oil in the world market. This suggests paying a premium for energy production or conservation that reduces demand for world oil, and the appropriate premium might be determined by estimating the extent to which reducing imports by one barrel may reduce the world oil price and, thus, the cost of all imported barrels.

The last point can be illustrated with a few numbers. Suppose Congress wanted imports in 1985 reduced from a projected level of, say, 12 million barrels per day to 11 million barrels per day and were considering adopting to this end a set of tax incentives and subsidies for alternative production or conservation. How large should the incentive be? One way to answer this question is to estimate how much the 1985 oil price would be reduced by the reduction in imports. If reducing oil imports by 1 million barrels per day would lower the 1985 world price from, say, $20 to $19.50 per barrel, the 1985 U.S. oil import bill would drop from $20 × 12 or $240 million to $19.5 × 11 or $214.5 million per day, a difference of $25.5 million per day. In a very real sense, the incremental million barrels of oil would cost $25.50 each if they were purchased, not the $20 per barrel world market price; it would be worth paying a cost of up to $5.50 per barrel above the world oil price in order to increase oil production or decrease oil consumption in the United States (or anywhere else, for that matter), taking into account only the savings to the United States. On the other hand, if it could be shown that decreasing imports would have no effect on the price of imports, then the value of reducing imports would not exceed the import price, and positive policies to discourage imports, beyond letting domestic prices rise to reflect import prices, would not be called for, at least not on purely economic grounds.

We do not know the quantitative relationship between United States oil imports and the world oil price over time and hence do not know how much more than the world price it would be worth to reduce United States imports. We are convinced, however, that higher import levels increase the tightness of the world market and hence make it more likely that price increases will occur and will stick. Thus, we think that imported oil should be regarded as probably having a higher cost than the world price, both because of the direct price effects discussed above and because of other adverse effects of higher oil imports, such as vulnerability to interruption and distortions due to balance of payments problems. On strict economic efficiency grounds, this argues for a surcharge on imported oil, with domestic prices allowed to rise accordingly; at the very least, it suggests that a policy of keeping domestic oil prices below the price of imports or a "windfall profits" tax that taxes new domestic oil while leaving imports unrestricted are of questionable economic wisdom.

In summary, there are great risks associated with reliance on imported Middle East oil, because it can be interrupted or increased in price without warning. This problem cannot be eliminated or even significantly reduced soon, and hence the importing countries must develop ways of responding to shocks in the world oil market. Trying to break up OPEC is not a particularly productive approach; it would do little to reduce the unreliability problem, might have little effect on the pricing and output decisions of the individual oil-exporting countries, and is unlikely to succeed in any case. Middle East oil is too valuable to do without; but market forces may help contain future price increases if the United States acts to expand supply and reduce demand in the world market.

Reality Three: Higher energy costs cannot be avoided, but can be contained by letting prices rise to reflect them

The quarter century ending in the early 1970s was unprecedented as an era of abundant, safe, secure, and cheap energy—or so it seemed at the time. Until late in that period, the discovery of large productive oil and gas reserves served to keep world oil prices relatively low and the price of natural gas at little more than the cost of transporting it. The emerging age of nuclear power promised electricity "too cheap to meter," while large conventional power plants brought improvements in efficiency and lower electricity prices. It was generally believed that synthetic fuels and hydrocarbons from distant places would cost only slightly more than conventional oil and gas. And there seemed to be no end in sight to this era of cheap, abundant energy.

With the benefit of hindsight, it can be seen that the received wisdom of that earlier time was wrong and that the world is now left with a legacy of planning and thinking that reflects yesterday's misconceptions more than today's realities. One cannot exclude the possibility of a technological breakthrough that will produce large amounts of inexpensive energy, but it certainly cannot be counted on during the next twenty years and is unlikely even beyond that period. The potential health and environmental effects of energy activities will be costly to control and to live with. Concerns about nuclear power or global climate could lead to very high costs of nuclear or fossil fuel or even prohibitions. Most of the signs point to higher energy costs for all countries, with most of the possible surprises being unhappy ones.

The sudden change in the world outlook for energy is commonly attributed to the emergence of OPEC as a visible and active force. It is our view, discussed above, that OPEC has been as much the bearer as the creator of the bad news about energy costs. Only in the late 1960s and early 1970s did the world begin to recognize just how scarce and valuable low cost oil, and especially Middle East oil, really is, and OPEC's flowering was largely the result of this sudden recognition. Oil prices are high and OPEC is important because energy in general is scarce or costly and is becoming more so, not the other way around.

The future of energy costs and world oil prices over our twenty year period is essentially unpredictable, and hence policy must be made recognizing the wide range of possible outcomes and must be flexible enough to accommodate developments as they occur. For what it is worth, we have used for our own internal discussion purposes a plausible scenario in which the real world oil price (by which we mean the inflation-adjusted, long-term contract price of the Saudi "marker crude") increases in fits and starts over the next twenty years, about doubling from mid-1979 to $30 to $40 per barrel (1979 dollars) by the year 2000; this is an average real rate of increase of 3 to 4 percent per year. This is only one of many possible scenarios, however; the world oil price in the year 2000 could well be anywhere from $20 to $50 per barrel (1979 dollars).

Whatever the world price is by the end of the century, it will not increase smoothly. It is more likely to remain constant or even decline in real terms for a few years (as it did from 1974 to 1978), then increase 25 percent (up to 35 percent, including inflation) or more in one year; even an increase as great as 50 percent in real terms over a short period cannot be ruled out, followed by a period of stability. The timing of these price surges cannot reliably be predicted by estimating when growing demand will outstrip supply capacity, creating a "gap" or "crunch" in the world market; such predictions ignore the ability of market forces to raise prices and/or expand capacity in anticipation of shortages, now that the oil-producing countries control their resources. Random political or economic events, such as the revolution in Iran, will probably be the principal factors determining the timing of price surges.

As the real world oil price increases, other energy sources and conservation measures will be of increasing importance, and average energy costs will increase at about the same rate (but from a lower base). The more oil prices increase, the more important other energy sources will become, providing an automatic damper on the overall energy cost increases. Eventually, costly but abundant sources will replace most of the low cost but increasingly scarce ones. Barring adverse developments of such severity that further use of one or the other energy source would become economically or politically infeasible, average real energy costs will probably begin leveling off early in the next century at about double today's level. Oil prices could, however, peak at an even higher level and then decline if low cost oil nears exhaustion before substitutes are ready in large quantities.

The economic impact of real energy cost increases of this magnitude will not be pleasant. Higher real costs mean that real incomes are lower than they otherwise would be, even if the economy adjusts without unemployment and recession; the same amount of effort will simply yield less end products and services. But the magnitude of these unavoidable higher costs should be put into perspective. In 1978 the U.S. gross national product (GNP) was $2.1 trillion (1978 dollars), and primary energy consumption was about 78 quads. If all this energy had been in the form of imported oil costing about $2 per million Btu, primary energy costs would have been 7.4 percent of the GNP. In fact, most energy costs far less than $2 so that primary energy costs were only about 5 percent of the GNP. If average real primary energy costs were to double from their 1978 level, while energy use grew in proportion with gross economic activity, then (apart from the macroeconomic problems discussed below) the real GNP would be about 5 percent less than it would be if average real energy costs stayed at 1978 levels. Similar relationships exist for other industrialized countries; and in developing countries, where commercial energy is a smaller share of total costs, the reduction in income caused by higher energy costs would be correspondingly smaller.

A reduction of 5 percent in the GNP of the United States is a huge cost. If the doubling in real energy costs occurred between 1978 and 1990, the absolute loss in income would be perhaps $150 billion in 1990 alone, equivalent to a cut in the average annual economic growth rate over the dozen years of about 0.4 percentage points—for example, from 3.5 to 3.1 percent. But a reduction in economic growth of this magnitude is not a serious threat to the health of an economy, even in developing countries. Jobs could continue to be created, albeit at lower real wages than otherwise, to absorb a growing labor force. Furthermore, the cost impact can be reduced by responding to higher energy costs in ways other than simply paying more for the same amount—that is, by energy conservation. There is no denying that, at best, higher energy costs will hurt; but if their impact is managed without adding self-inflicted wounds, the injury need not be seriously disabling.

One form of additional damage that is often difficult to avoid when energy costs increase is macroeconomic disruption—inflation, unemployment, and recession. Especially if energy costs increase suddenly, as they did when world oil prices quadrupled in 1973-1974, serious macroeconomic problems arise and add to the direct cost impact. But shocks as large and sudden as those of 1973-1974 are unlikely in the future, and even these are being weathered. That experience is worth reviewing.

When world oil prices suddenly quadrupled in 1973-1974, energy users in the importing countries could not immediately reduce their oil purchases and hence had to decrease their purchases of domestic goods and services in order to pay the higher oil import bill; countries with inadequate foreign exchange had to scramble to find credit. The oil exporters, on the other hand, were unable to increase their purchases and investments in the importing countries quickly, so they could do little except accumulate large amounts of cash, securities, short-term notes, and other liquid assets. At the same time, oil price increases showed up in consumer price indexes throughout the world, exacerbating fears that an already troublesome inflation was getting worse. The fear of fueling inflation made officials in importing countries reluctant to use expansionary monetary and fiscal policies to offset the depressing effects of the large outflows of money and decreases in domestic demand. The results were extraordinarily high rates of unemployment and inflation simultaneously, accompanied by accumulating debts and payment imbalances in international financial markets.

This disruption in national and international economic affairs was costly and potentially dangerous. Real income levels are still lower than they would have been if oil prices had remained low, at least in part because of a substantial period of lost output that can never be regained. Inflationary forces that existed before 1973–1974 persist, heightened by demands for "catch up" wage and price increases. Some countries have accumulated large external debts and are having trouble meeting their payment obligations. Another sudden doubling of world oil prices would be more difficult to absorb; and the oil price increases of 1979 are causing further problems. But on the whole, the damage has been contained: economic growth has resumed, even in the developing countries, and international financial institutions have managed to finance (and, where necessary, refinance) the deficits of developed and developing countries alike. The economic system has been shown to be remarkably resilient. With careful macroeconomic management, it should be able to handle future energy-related strains of the magnitude we think likely.

Successful management of higher energy costs also requires encouragement and coordination of a vast number of particularized, individual adjustments in order to minimize the overall adverse effects. When energy policy tries to prevent energy prices from rising in response to or in anticipation of higher energy costs, the things government must do directly to try to accomplish these microeconomic adjustments increase greatly in number and complexity: supplies and conservation must be stimulated by subsidies, regulations, exhortation, and gimmicks of various kinds; shortages must be allocated by complex bureaucratic systems; long-range investment and technology development must be stimulated by government; "profi-teers," "hoarders," and other citizens doing what comes naturally must be pursued and punished. Logic suggests and experience dem-onstrates that programs of this kind become increasingly inefficient, complicated, and unfair with time and are ultimately abandoned, but not before the distortions and delays they introduce raise costs—and, in the long run, prices—above what they would otherwise be.

The attempt to hold down energy prices in the United States is particularly insidious now, because of the interaction with oil im-ports. Domestic energy price controls decrease the supply of and increase the demand for energy, and particularly oil. Ordinarily such controls would result in shortages, rationing, black markets, and other signs that the policy is failing, leading to removal of the con-trols. But under the present conditions, the gap in the domestic oil market can be filled simply by importing more oil; and under present policy, refiners who do import oil are entitled to collect money from refiners who use cheaper domestic oil, in effect taxing domestic oil production and subsidizing oil imports. That the real economic costs to the United States would be less in the short run and in the long run if domestic prices were allowed to rise is not obvious to many consumers, voters, and politicians. Domestic prices are therefore kept controlled, real energy costs are higher than they need be, oil imports rise, and OPEC is blamed when the increased demand helps push world oil prices upward.

This problem, and many others, could be reduced by relying more on the market and less on price-setting and price-regulating bureau-cracies. This is not to say that markets can do everything, that they need no improvement or supervision, or that there is no role for ac-tive government policy. There are many areas where "letting the mar-ket work" is not enough: much research and development would not be undertaken in a pure market system, no natural market exists (al-though artificial markets can be created) for environmental harms, and various other social goals such as an equitable income distribu-tion need to be advanced in a way that the market acting alone will not do. Even in most of these areas, however, the other policy mea-sures that must be taken will generally be easier and more effective if energy markets are pricing energy at economically efficient levels.

Moreover, there are important parts of the energy system itself where markets may not function adequately. Public electric and gas utilities, in particular, are regulated as natural monopolies, subject to control of the prices they charge and the service they offer; the problem for public policy in this area is to encourage methods of regulation (and, where feasible, deregulation) that will cause the prices paid by customers to reflect the value of the energy they use. Outside the United States, the energy system consists of a collection of multinational oil companies, nationalized oil companies of importing countries, and governments of oil-exporting countries, fitting no simple definition of a market. On the other hand, within the United States, the energy markets (excluding the public utilities) seem to be workably competitive. There are large, integrated firms, but with no single one dominant. The many vigorous independents, competing resources, independent resource owners, and potential entrants make up a market as competitive, dynamic, and efficient as most in the real world.

Most advocates of energy price controls do not argue that markets are unable to price energy efficiently, but that market prices, however efficient, will cause hardships and inequities. It is important to remember, however, that controlling prices does not reduce the real costs that must be paid by someone, somehow. At best, the controls redistribute the costs in some more equitable fashion; at worst, they increase the costs by discouraging cost-reducing measures and then distribute the higher costs in some arbitrary but disguised fashion. For example, a consumer forced to heat his or her home with expensive electricity because natural gas hookups are frozen in the area reaps only costs from price controls that encourage the waste of natural gas elsewhere. Factory workers laid off because price-controlled fuel to industry is curtailed are paying costs that may be much higher than the costs of producing more fuel. Drivers spending hours in line for gasoline might gladly pay higher gasoline prices in exchange for shorter lines. Changing the form in which society pays higher energy costs by controlling prices does not necessarily reduce the resulting hardships and inequities.

To the extent that controlling energy prices can transfer more of the costs to those better able to pay them, the resulting improvement in income distribution may be felt to be more important than the losses in economic efficiency. However, the proportion of total expenditures used to purchase energy directly and indirectly declines only little with income level, and hence keeping energy prices low to everybody saves more in absolute terms and only somewhat less proportionately for a richer than for a poorer consumer. And poorer people are not necessarily better able to adjust to the various nonprice devices that must be used to equate supply and demand if prices do not. They cannot fly or rent a car when the gasoline pumps run dry, and they are more likely to lose their jobs when fuel is unavailable. Keeping energy prices artificially low is not an efficient or effective way preferentially to help the poor.

Another argument for energy price controls is that they reduce the transfer of income from energy consumers in general to energy producers. Such a transfer may be felt to be undesirable either because it is unfair or unearned or because, deserved or not, it is seen as adversely affecting the income distribution objectives of society. Again, however, it is not certain who ultimately gains and who loses, and by how much, if prices are controlled at low levels. Texas oil barons and stockholders in Exxon may be more conspicuous beneficiaries of energy price increases; but Appalachian coal miners, makers of storm windows, and academic energy experts all reaped windfalls when world oil prices rose. Even higher oil company incomes lead to higher corporate and individual income tax payments, higher severance taxes, increases in the values of pension funds, and increased endowments of nonprofit institutions. On the other hand, if energy prices are controlled, lobbyists, accountants, and lawyers skilled at using complicated regulatory procedures for the benefit of wealthy clients become wealthier; inefficient oil refiners otherwise unable to compete get rich through special provisions of the regulations; and firms and individuals who correctly anticipated higher energy costs and acted accordingly in their energy conservation and production plans find their foresight unrewarded.

Not only is there great uncertainty about how large the windfalls from a sudden price increase are and where they ultimately lie, but there is danger in establishing the precedent that successful gamblers will not be allowed to collect or to keep their winnings. It is not easy to distinguish careful, calculated risk-taking from blind luck. Many investments that will most likely turn out badly or only "so-so" are undertaken because there is a small probability that they will pay off handsomely. Eliminating this up side potential by making windfalls fair game for the price controllers and the tax collectors will not encourage the kind of far-sighted risk taking that is essential to the solution of energy problems.

Nevertheless, there is little doubt that the quadrupling of world oil prices in 1973–1974 was a special case, and hence public policies to reduce the windfalls resulting from energy investments made prior to 1973 would not do great damage to short-run efficiency, long-term incentives, or society's sense of fairness. This need not be done by keeping prices low, however, since selective taxes can capture the windfalls. In principle, such a windfall tax might apply to pre-1973 investments in a broad range of things—coal mines, Toyota franchises, degrees in petroleum engineering. In practice, it would be applied to some categories of "old" oil.

For energy-related investments made after 1973–1974, and especially for investments yet to be made, the situation is quite different. There has been and will continue to be great uncertainty about the future of world oil prices, as well as about the technical success of many ventures, such as deep drilling, technological research, and enhanced oil recovery. Investors may want to risk or may have already risked their money on such a venture, on the off chance that it will be a technical bonanza or that the world oil price will go to $50 per barrel soon; domestic oil prices have long been scheduled for decontrol in 1981, with no windfall tax attached, and hence such investors have had some reason to think that they might be able to keep their winnings if they are right, just as they will surely bear the losses if they are wrong. Such gambles will increase the probability that the world oil price does *not* go to $50 soon and will help to contain the damage if it does so. Changing the rules after the fact, or even announcing ahead of time that, say, 50 percent of all gains due to price increases will be taxed away, does not help encourage risky or high cost ventures.

A final reason often given for trying to control energy prices is the need to check inflation. Higher real energy costs mean that energy prices, to reach efficient relative levels, must increase faster than the overall price level. In principle, a decline in other prices and in money incomes could raise energy prices relatively but leave the general price level unchanged. But since prices and money incomes in a modern economy do not easily decline, higher relative energy prices tend to increase the general price level—that is, they add to inflation.

The management of inflation is an important and difficult problem at the end of the 1970s, and energy policy should not make it worse without good reason. Wage and price decisions are inevitably affected by perceptions of rising energy costs. A sudden, one time increase in energy prices would echo in those decisions for years as workers and firms made vain attempts to catch up; a slower but more persistent increase in real energy prices would be a chronic source of pressure for cost of living adjustments; and the hidden costs of price controls would put upward pressure on prices. It is not clear which of these evils is least bad for the management of inflation. But energy pricing policy must recognize its impact on inflation and may have to compromise accordingly.

In summary, high and rising energy costs are probably inevitable over our twenty year period, for reasons much more fundamental than the existence of OPEC. Inflation and recession problems of the magnitude caused by the events of 1973–1974 should not recur unless there is another price shock of comparable severity—which we regard as unlikely, now that the historical shift from consumer to producer control of world oil resources is about complete. Letting domestic energy prices rise to reflect real costs and scarcities is the most efficient way to contain the inevitable cost increases; the income distribution effects of higher energy costs can be handled by direct assistance to specific groups and by limited taxes on windfalls without making costly energy appear cheap to everyone. Higher energy costs will hurt, but if properly managed, need not seriously threaten the health of the world's economies.

Reality Four: Environmental effects of energy use are serious and hard to manage

Many energy-related activities damage or threaten to damage human health, the natural environment, or the earth's biologic and geophysical systems. Some of these activities involve environmental effects that are of such potential magnitude and are so difficult to control as a technical matter that they may ultimately constrain the use of some energy sources; the aforementioned possibility of an adverse climatic effect of carbon dioxide from fossil fuel burning is the principal example in this category, to which other analysts might add radioactive waste disposal. Another category includes potentially serious environmental effects that are controllable as a technical matter, but are now being so badly managed that both energy and environmental values may suffer unnecessarily in the long run; air pollution from fuel burning is probably the most important example here. Problems in either category hold the potential for changing the energy future substantially.

There are also environmental problems related to energy that are localized or relatively easy to solve as a technical matter and that will probably be solved (albeit with difficulty) within existing laws and policies. The reclamation of mined land, for example, in most instances is not difficult or costly compared to the value of the resource extracted, and mining can be prohibited entirely in the most sensitive areas without significantly increasing costs to the economy. The problems associated with boom towns or coal trains affect relatively few people and are not ordinarily life or death matters. Even the problems of water availability for and water pollution from energy activities need not, in our view, be insurmountable obstacles to energy development in general, even if they do thwart some particular projects. This is not to say that any of these problems are trivial or that all of them will be handled satisfactorily. Indeed, it may well be that failure to resolve conflicts over seemingly mundane matters such as who should pay for relocating railroad tracks could seriously interfere with the solution to national energy problems. But on the whole, solution of these problems primarily requires persistent effort and negotiation on a case-by-case basis; if the economic gains to be made by resolving the conflicts are great enough, the conflicts will be resolved—though not necessarily quickly.

Our analysis of the various environmental effects of energy production and use and of the technical and economic feasibility of controlling them leads us to a cautiously optimistic, conditional conclusion: given careful and flexible management, energy can be produced and consumed in the United States at levels we think likely over the next twenty years, without undue harm to human health, natural systems, or aesthetic values in general. The most important qualifications to this conclusion involve the carbon dioxide and air pollution problems referred to above. The carbon dioxide problem could begin acting as a constraint on fossil fuel use near the end of the century, but its principal implications for now are that a focused research program must be vigorously pursued and that nonfossil options must be developed and maintained for the long run. And the air pollution problem can be managed satisfactorily (although not necessarily without some degradation in air quality in some areas) with the right policies. It is the difficulty of managing these problems, rather than the substantive nature of any one of them, that we view as crucial for energy policy and that we discuss here.

The difficulty of managing energy-environmental conflicts can be reduced in part by more and better scientific and economic information. Presumably, it always helps to know more; and we, along with everyone else, urge that more research be done and more information be gathered. But the most important and most neglected features of environmental management are the inherent, even irreducible, scientific uncertainties and the need for societal judgment when basic values conflict. More and better technical information and careful, logical analysis are essential in such situations but cannot settle the basic issues. There are important gains to be made in environmental policy by improving the social processes for decisionmaking and managing in the face of great uncertainty and value conflicts, as well as by improving scientific knowledge.

The uncertainties surrounding environmental problems come from the basic complexity and the dynamic nature of natural and social systems. Both ecology and economics remind us that when dealing with complex evolutionary systems, it is not possible to identify simple cause and effect relationships or to predict with confidence the ultimate effects of action or inaction. For example, while there has long been persuasive evidence that human health is impaired by breathing polluted air, science has not been able to determine which pollutants, in what combinations, at what exposures, cause which health effects; in fact, it becomes increasingly clear that such simple questions have no simple answers. Similarly, the ultimate effects of carbon dioxide on climate depend on complex biological, geophysical, and social interactions that would not be fully understood or repeatable even if the experiment were conducted on the earth itself. If action must be delayed until science has all the answers, it may be too late.

Complexity and uncertainty are inherent also in the social aspects of environmental management. Decisions about which sources to control, how, when, and with what technology can have large effects on the costs of accomplishing an objective, and minor differences in the objective itself can result in major differences in the costs of accomplishing it. In the long run, both the costs and the effectiveness of control measures will be influenced by the details of investment, research and development, shifts in processes and locations, and other factors that cannot be predicted or controlled precisely. If policy encourages these factors to develop over time in ways that make it easier and less costly to control effects, both environmental and energy-economic values can benefit in the long run, even if one or the other is compromised temporarily or locally; conversely, policy can discourage the right overall, long-run changes by excessive concentration on narrow, short-run objectives.

Faced with such complexities and uncertainties, society must deal with two related but distinct problems: basic judgments must be made about what level of costs and risks should be incurred in one area of life in order to reduce costs and risks elsewhere; and social processes and institutions must be established for implementing and continually modifying these judgments in the light of new information and changing circumstances. Failure either to make the required social judgments or to provide mechanisms for implementing and modifying them efficiently can result in environmental management programs that are ineffective and too costly.

We stress these central facts of environmental problems and policy for one simple reason: U.S. environmental policy has consistently ignored them. Basic societal questions are too often avoided or left unresolved in the open political processes where they should be handled and are pushed into administrative and legal channels where narrow issues of fact and procedure act as proxies for the underlying basic conflicts. In other areas, precise objectives, detailed schedules, and specific solutions, which might better be determined on a decentralized or even on a market basis, are spelled out in legislation and regulations as though they, and not some underlying, broader objective, represented the social consensus. Since these detailed goals are defined as though there were no uncertainty or scope for flexibility, and since the implementation programs are incapable of dealing efficiently with the millions of details and uncertainties, the goals are periodically thrown out and replaced with new, even more detailed absolutes. Instead of producing steady but flexible programs for making progress toward general social objectives openly agreed upon, these environmental policies generate costly and interminable conflict and indecision over details and lead to disappointing progress.

There are two areas in particular where these problems threaten both energy and generally accepted environmental objectives. One concerns private intervention in governmental decision processes—traditionally affecting nuclear power plant licensing and siting, but increasingly affecting many other energy decisions as well. Too often, the procedures originally intended to broaden participation in governmental decisions are used to delay and frustrate the implementation of social judgments. Broad participation in governmental processes is something to be encouraged; but it must be made more constructive and less a source of obstruction. For example, rule-making proceedings can rely more on notice and comment and less on adjudicatory hearing; the extent to which generic issues can be raised in specific cases can be limited; and Congress itself can be more explicit about its intentions, so that administrators and the courts are able to refer to clear public policy objectives in reaching their decisions. At first glance, the Three Mile Island accident would seem to dim the chances for reform in this area; but the reevaluation that is now going on also offers an opportunity to seek fundamental changes that both speed up and improve the process of making nuclear power and other decisions.

The other area in which conflicts between energy and the environment are seriously mishandled is air pollution control policy. The underlying assumption of the U.S. Clean Air Act (CAA) is that science can find certain levels of air pollution that are "safe," that air everywhere must meet that standard by a certain date, and that regulatory processes can dictate the detailed responses of each of thousands of emission sources. But the effects of air pollution are too complex to be captured in a rigid number: air is only more or less harmful, depending on the concentration and duration of exposures; the combinations in which pollutants are inhaled; and the size, health, age distribution, and smoking habits of the population exposed. Likewise, the rate at which progress should be made toward cleaner air and the ultimate "best" level of air quality depend strongly on the costs involved, as well as on such characteristics of a region as population and the aesthetic value of clean air. And the technical and economic factors involved in pollution control are too complex and dynamic to be handled by purely administrative means. The present policy ignores the constantly changing and poorly understood relationships among technology, scientific information, economic cost, and social desires; a complex, variable world cannot be reduced to simple absolutes and uniform regulations. The debates over setting the ambient air quality standards, the deadlines, and the technological rules and then changing them every few years result in delay, cost, and uncertainty while doing little to improve air quality.

The rigid standards and deadlines were put into the CAA for what were regarded as good administrative reasons: experience and logic suggest that regulatory programs become ineffective when forced to recognize the kind of complexity, uncertainty, and ambiguity inherent in air pollution problems. The costs of controlling pollution and even the costs imposed on society by the pollution that it is not worth controlling should be included in the prices of energy and other things. It is absolutely necessary that social judgments be made about how much to spend to reduce pollution, how fast to make progress overall, how far to go ultimately. But it is neither necessary nor practical to use the processes of social decision to make all the detailed judgments about when, where, and how; and yet that is just what the current air pollution policy tries to do. Both energy and environmental values would benefit by adoption of alternative policy approaches that would leave to government what it should be doing—setting the basic objectives or providing the proper incentives—while leaving to the private sector the job of figuring out the details. We offer some specific suggestions for such a policy in Recommendation Four.

Reality Five: Conservation is an essential "source" of energy in large quantities

Energy is never used to the extent that it could be as a physical matter or would be if it were free—it always has been and always will be conserved, because there is a cost to getting and using it. In order to estimate how much energy is being "not used" in any actual or hypothetical situation, it is necessary to estimate how much more energy would be used in some other hypothetical situation; the difference can be called "energy conservation." Since there is no unique way to define what energy use would be in the hypothetical case with "no" conservation, there can be no unique definition or measure of energy conservation. It is whatever it is defined to be for the purpose at hand.

For certain analytical purposes, however, energy conservation can be thought of as an incremental source of energy, similar to new sources of oil, coal, or nuclear power, and can be measured by the extent to which certain energy-conserving actions or events reduce the use of these other sources. The energy "produced" by conservation has a number of advantages over energy produced in the conventional sense. It is almost always cleaner than other forms of energy, and those disturbances to the environment and human health that are associated with it, such as less ventilation and hence less healthy air inside well-insulated buildings, are generally localized and easily manageable. It does not draw directly on foreign resources and hence creates fewer international security, trade, and payments problems—though foreign products that use energy more efficiently or come from areas with lower energy costs may be imported in larger quantities as part of domestic energy conservation. Some forms of conservation can be adopted with shorter lead times than can other energy sources; and some can be purchased in small increments by making a series of small investments, rather than by making a single large commitment as is required for a mine, a coal gasifier, or a solar satellite. Most energy conservation investments carry relatively low technical risk; there is little doubt that they work as promised and will continue doing so for a long time.

As with any energy source, some forms of energy conservation are cheaper than others, so that unit costs increase as the source is called upon to supply more. There are some energy conservation measures that are virtually costless once people become energy use conscious and change simple habits, such as shutting off lights in empty rooms. There are measures, such as insulating buildings and installing heat recovery equipment in factories, that cost something but that, up to a point, are cheaper than the energy they save. As energy prices increase, people will find ways to do much the same things they have always done, substituting other economic factors for energy; and they will find that there are different, less energy-using things they can do to try to satisfy their basic needs and desires as far as possible.

There is nothing inherently good about these energy-conserving activities, just as there is nothing inherently good about pumping oil from the ground. Energy conservation, like oil production, is an economic activity that can be pushed so far that it ceases being cheap and easy and becomes costly and inconvenient—but it may still be preferable on economic grounds to the alternatives. Even reductions in energy use that are very painful and economically disruptive, such as those resulting from a sudden interruption in oil imports, may be cheaper than the alternatives—the most economical "strategic reserve" may be a margin of energy consumption that can be quickly eliminated by emergency measures. Over time, higher energy costs may result in energy-conserving shifts in lifestyles and investment patterns leading to lower measured GNP than would be obtained from a strategy of forced investment in energy production. Given the deficiencies of GNP as a measure of economic welfare, however, the future with lower energy use may well be economically preferable despite its lower measured GNP. Whether or not the label "energy conservation" is applied to cuts in energy use involving significant changes in human activities, the principle remains the same: energy is an economic variable, and using less of it as it becomes more costly is the economically rational thing to do.

With all the advantages that energy conservation possesses and with the long history of low energy costs in the United States, it is not surprising that conservation now appears to be quantitatively one of the most important incremental energy sources, both for responding to short-term interruptions in other energy supplies and for reducing the demand for other energy forms in the long run. In the event of, say, a sudden 20 percent decrease in oil imports to the United States (amounting to a decrease of about 5 percent of total U.S. energy use), energy conservation would today be the only "source" capable of expanding much in response. Sudden reductions in energy use on this scale would be expensive and disruptive and perhaps should not be called "conservation," but they would have to do the job because nothing else could.

Over a period of years or decades, energy conservation can become quantitatively the most important energy source of all. For example, prior to the events of 1973–1974, most forecasters expected the United States to be consuming over 90 quads of energy by 1980. It now looks as if 1980 energy consumption will be only about 80 quads, partly because forecasts of economic activity have been lowered and partly because energy is being used more efficiently. But on any reasonable definition, well over half of the difference of 10 quads should be attributed to energy conservation; of the other domestic energy sources, only coal and nuclear have increased significantly since 1971, by about 2 quads each.

Even more dramatically, before the changed perceptions about energy, forecasts of U.S. energy consumption in the year 2000 were generally in the range of 130 to 175 quads, compared to more recent forecasts of only 90 to 120 quads. Again, some of the reduction is due to reduced forecasts of income. But if most of the difference of about 30–40 quads is interpreted as the energy "supplied" by conservation, then the increase in conservation over the next twenty years becomes an energy source of the same quantitative importance as coal, petroleum (domestic and imported), or nuclear by the year 2000.

The bulk of the energy conservation that has occurred and will occur is the natural and economically efficient response to higher energy costs, as these costs are reflected either in prices paid by energy users or in shortages, curtailments, and prohibitions of various kinds. Either way, it will become increasingly costly to get energy, and individual and business purchasers will respond by using less. Because effective and efficient energy conservation is the result of millions of individual decisions, each one influenced by highly particularized circumstances and preferences and made by purchasers with limited resources to devote to exploring and evaluating options, there is an economic argument for teaching and educational programs directed to the individual energy buyer. Appliance labeling, audits of homes for their energy efficiency, energy extension services, development of "house doctor" programs to provide information and services, and provision of direct training to small industries can help make consumers aware of the energy effects of the decisions they are making.

Such information and technical assistance programs will be more effective if it is in the interests of individual energy users to save energy. If the prices consumers pay for energy reflect the underlying cost of the energy being consumed, they will find it in their own interest to conserve energy without exhortation or coercion by the government. Furthermore, prices that reflect higher costs will provide a powerful incentive to inventors to develop and to entrepreneurs to promote energy conservation technologies that are not economic at lower, controlled prices. Only when individuals and enterprises can capture the benefits of their inventiveness and effort can new technologies be expected to emerge and to be applied.

It is not always easy to make prices to consumers fully reflect the economic value of energy, particularly since almost half of the energy consumed in this country is sold to its final users by regulated companies. In periods of inflation or rising relative energy costs, current regulatory practices prevent prices from rising enough to reflect fully the true economic value of the energy sold, encouraging more consumption than is efficient and tending to discourage use of products (such as insulation or solar hot water heaters) that compete with energy from the utility. This suggests that relying on the market alone will not result in as much energy conservation as is economically warranted. In view of the high economic and other value of energy conservation, some more direct policy actions can be justified.

For example, the federal automobile efficiency standards have no doubt forced the industry to move toward smaller, lighter cars more quickly than the controlled price of gasoline alone would have done. These regulations, because they specify fleet averages and not individual model limits and because they provide for payment of a moderate penalty if the fleet averages are slightly high, provide a reasonable degree of choice and flexibility. They also cause certain distortions: trucks above 3 tons (which include a variety of personal transportation vehicles) are exempted; manufacturers who produce or import a limited model line can be unfairly and uneconomically advantaged or disadvantaged. They do not discourage driving or speed the turnover of the existing auto fleet as effectively as higher gasoline prices would. And the precise level of the targets in future years should be examined carefully to be sure that the cost per unit of energy saved does not become unreasonable. All in all, a tax on fuel inefficiency might be preferable to the present regulations; but the regulations have pushed in the right direction and at more or less the right rate.

Similarly, regulation of energy consumption in new buildings can play a useful supplemental role. The market consisting of equipment makers, realtors, builders, contractors, renters, and buyers is a complex and, in many ways, imperfect one in which information and incentives are often poor. Almost half the dwelling units that will exist in the year 2000 have yet to be built, and it is important that those units be built to reflect the high and rising costs of energy over their lifetimes. Government standards can help, if they are based upon a careful examination of the life cycle costs of various conservation actions in different parts of the country in order to make sure that the standards satisfy a reasonable cost effectiveness test. However, the conflicts that have surrounded the setting of such standards under present law indicate that it is not easy to decide just what is and what is not cost-effective in the highly diverse and individualized area of designing and constructing homes and other buildings.

When the energy prices paid by consumers do not reflect the true social cost of the energy used, it is also economically justified, in principle, to subsidize energy conservation measures. Although it is very difficult to implement efficiently a subsidy program for something as diffuse as energy conservation, the existence of such a program (e.g., tax credits for home insulation or for energy-saving investments in industry) can serve to stimulate and maintain interest in looking for ways to invest to save energy and can result in some actions that, while cost-effective for the society as a whole, would not be so for an energy consumer paying uneconomically low prices.

In summary, energy conservation is one of the most important "sources" of energy, which will be used to substitute for other forms of energy as they become more costly and scarce in the next twenty years and beyond. Increased use of this source is a trend to be welcomed, even encouraged, by explicit policy, not fought. Although the bulk of the conservation will be the result of normal economic forces and individual, self-interested actions, energy policy has an important role in reducing and offsetting some important market imperfections and stimulating use of this energy source to its full economic potential.

Reality Six: Serious shocks and surprises are certain to occur

Energy policy must deal with a complex mixture of poorly understood physical realities, unstable social institutions, and unpredictable human nature. The only certainty is that nothing is certain, that the best laid plans will go awry. There are some events with potentially serious consequences that are almost certain to happen—but there is no telling just where, when, or why. There are a few suggested events or discoveries that probably will not happen, but that will drastically change the course of things if they do. And of course, the real surprises will be things nobody has yet thought of. But shocks and surprises will occur; and there is no excuse for being shocked by or unprepared to deal with the consequences when they do.

As we note elsewhere, the most serious and likely energy shocks over the next twenty years will be associated with the world oil market and particularly with the oil-exporting countries in the Middle East. The probability and effects of disruptions in the world market might be reduced in various ways, perhaps including reducing oil imports below what they otherwise would be. But whatever is done, the United States and others will import a large share of their oil for most or all of our twenty year period and will be affected by shocks anywhere in the world oil market. The importing countries must work together to prepare for the kinds of interruptions in supply that are certain to occur—but at times, in places, and for reasons that are unpredictable.

A reduction in the volume of oil flowing into the world market will cause prices on the spot market to leap upward; even if long-term contract prices are honored for the oil exports that continue, by the time the oil makes its way through the system to consumers, it will tend to be priced at its spot scarcity value. Conversely, a sudden increase in the price at which oil is available, even if unaccompanied by explicit limitations on quantity, will result in less oil being purchased. In principle, there is little difference between a reduction in quantity supplied and an increase in price—although given the difficulty of reducing oil use quickly, a small reduction in quantity supplied can be equivalent to a large increase in price. But similar analytic and policy approaches are appropriate in either case.

When the supply of oil is suddenly interrupted or the price is increased on the world market, the economic value of the oil is increased. Price controls may reduce the extent to which domestic prices rise, but do not solve the real problem, which is the need to decrease use and increase production of energy in general and of oil in particular. Also, reserve stocks (both those set aside for just such an emergency and those held for ordinary commercial purposes) can be drawn upon until the temporary situation passes or to help ease the transition to a new, persistent condition.

No single action, such as driving automobiles less or drawing upon a strategic reserve, can be relied upon if an interruption in supplies is very large for very long. If prices to both buyers and sellers of energy are allowed to increase freely in an emergency, a wide variety of constructive individual actions will be taken to prepare and to respond: standby production capacity, alternative fuel capability, and more storage capacity will be installed and used by energy producers and consumers—and even by the dreaded speculators and hoarders—and will then be used where it is most valuable in an emergency; conservation plans will be readied and then used when the time comes; market signals will tell refiners whether they should maximize gasoline or fuel oil output and will tell distributors when gasoline is more valuable in California and when diesel oil is critical to farmers. The largely market-directed energy system has long dealt with such complex problems in the United States and is at least as capable as are governmental processes of dealing with the details of emergency situations, where specialized knowledge, quick responses, and ability to take into account individual situations are essential.

The confusion and disruption in domestic oil markets in mid-1979 provide some indication of the extent to which the United States is unprepared to deal with interruptions and perturbations in world oil markets and largely because domestic markets are not allowed to operate freely. When events in Iran reduced the flow of oil into the world market, the U.S. Department of Energy (DOE) urged oil-importing companies not to bid in the spot market for incremental supplies. Domestic oil producers subject to price controls had little incentive to increase or accelerate production from high cost sources. Industrial plants with access to oil at low, controlled prices and fearful of the regulatory mess they would find themselves in if they tried anything new resisted DOE's urging that they switch to natural gas—despite the fact that as a result of partial price decontrol less than a year ealier, the natural gas "crisis" had become the natural gas "glut." So neither supply nor demand adjusted much, shortages developed, and oil had to be allocated by government.

DOE issued erratic and conflicting instructions to refiners and distributors about what to produce and to whom to sell it. Rigid allocation formulas based on fuel usage in some past period sent gasoline to Tennessee where it was not sold, while Californians stood in line; after the governor of California called on the president, lines shrank in California but lengthened in New York. Suddenly, the District of Columbia became the "gas line capital" of the nation, and a puzzled and angry Congress tried in vain to learn from federal officials what had prompted this development. Officials debated whether service stations should be ordered to stay open or to close on Sundays, while making it illegal to charge higher prices as incentives for providing services in off hours or in high demand locations. The biggest "price gougers" according to the pricing regulations sometimes had the cheapest gas and the longest lines. People wasted hours in lines, disrupted traffic, cancelled long-planned vacations, and vented their wrath on oil companies, station owners, the government, and each other.

Eventually, DOE reversed itself and encouraged importers to bid for supplies in the world spot market. It even granted subsidies, in the form of additional "entitlements," to importers of fuel oil, on the grounds that the U.S. market was not receiving its proper share of supplies. In the end, the pressure of U.S. demand on world oil prices was greater, and the total costs of the internal adjustments were higher, than they would have been if domestic energy markets had been allowed to operate freely. Owners of domestic energy resources made less money, and consumers as a group probably paid fewer dollars for energy *directly* than they would have otherwise. But this experience with a shortfall of a few percent in oil supply should raise doubts about the generally accepted notion that the government should intervene massively in the market whenever an emergency arises.

Nonetheless, in the event of a reduction in oil supplies and upward pressure on prices, the insistence that government do something will be intense—especially if, as in 1979, the market is already so regulated that it cannot respond adequately. Emergency programs developed in advance, when clear and careful thinking is possible, are likely to be better than crisis-motivated responses. These programs can in some cases rely on market forces to deal with the details of the situation, using government to establish the overall objectives, to deal with special cases of hardship, and to modify government rules that may be inappropriate under the emergency conditions. An example is "white market" gasoline rationing, in which ration coupons are distributed in quantities and to recipients determined by government, but can then be bought and sold in a national market so that gasoline goes to individuals and regions willing to pay the most for it. Such a market can help protect social or regional economic interests without severe distortions. Similarly, strict air pollution regulations and timetables might be relaxed temporarily, in exchange for payment of high fees or taxes; such a policy would recognize the increased relative value of energy during the emergency, while minimizing the demand for special exemptions. Hard as it is to get agreement on such measures, the issues should not be ducked until it is too late for careful consideration.

Whatever policy measures are taken to deal with a sudden decrease in supply (and increase in the domestic value) of oil, they will be costly and disruptive at best. Faced with the prospect of having to use these measures, countries will naturally consider exertion of diplomatic or even military pressure to try to keep the oil flowing.

The ability to increase domestic oil supply temporarily could be of great value in reducing the temptation to overreact if oil imports are interrupted. As discussed previously, it is doubtful that a synthetic fuels industry would be flexible enough to provide this sudden supply expansion ability. Shut-in domestic oil wells might help. But in order to be able to provide for sudden increases in domestic oil supply, a strategic petroleum reserve or stockpile is essential for dampening shocks to the economic system and in providing an interval during which efforts can be made to mediate disputes or to fashion countermeasures; it becomes less necessary to respond quickly to every oil supply restriction as if it were a serious threat to the economy.

A strategic stockpile may be held either privately or publicly, so long as its existence is mandatory and its release is subject to government approval. For example, the right to import oil could be made dependent upon demonstrating that some proportion of the amount imported is being held under bond in ready reserve. Private stock-pilers would have every incentive to find low cost ways to hold the required reserves; storage space could be purchased in a government-operated salt dome project or from private parties with underutilized tankage, private salt dome facilities, or the like. At least some of these facilities would be cheaper and more quickly available than the government facilities now being built, and they would naturally be dispersed around the country near where the oil would actually be used.

Private stockpiling so organized would (1) automatically place the cost of this security measure on those who occasioned it—the consumers of imported oil, to whom storage costs would be passed forward—(2) discourage imports, and (3) encourage domestic energy production. Similar results would be obtained from a stockpile held by the government and funded by fees levied on the imported oil.

The appropriate ultimate size for a strategic stockpile depends on the probability that it will be needed, its potential for reducing damage in the event of a supply interruption, and the degree to which the nation is willing to accept risks. A stockpile is not cheap to build and to hold; but the largest proportion of the cost is the annual capital cost of the investment in the stored oil, which may well be offset by subsequent increases in the value of the oil. Since oil imports are unlikely to be cut off completely, and since domestic energy production should be increased and consumption decreased as part of the response to a supply interruption, a stockpile equal in size to one-fourth or one-third of annual imports would provide a large measure of protection. Estimates are, for example, that the United States lost less than 100 million barrels of imports in the first four months of 1979 due to the problems in Iran; this amounts to less than 3 percent of one year's imports, and a substantial difference in the strain imposed could have been achieved with replacement of even one-fourth of that amount by withdrawals from strategic storage.

In addition to acquiring a stockpile, it must be decided under what conditions and how rapidly it is to be drawn down, and how the oil is to be allocated. If problems develop in the world oil market, it will be difficult to determine just how serious the problem is and to decide whether to use the reserve early to forestall damage or to save it for later when even worse problems may develop. Furthermore, since there will often be no way to know initially whether a change in the world oil market (such as a reduction in Iranian exports) is a temporary aberration or a new permanent condition, use of the strategic reserve should be integrated with other emergency measures in ways that lead naturally to the right long-term adjustments, if these become necessary. A strategic stockpile can be counterproductive if it is used in an unpredictable, destabilizing fashion or if it serves to delay adjustments too long. Carefully thinking out of contingency plans and decision options before an emergency develops is likely to result in better management of a strategic reserve than waiting until the crisis strikes.

Most analyses of supply crises focus on quantity measures of the "shortfall," which is the difference between some guess about what imports would be if there were no crisis and some guess about what imports are now. The imprecision of this notion can lead to a great deal of sterile debate and delay. For example, the Congress and DOE argued long and hard first about what the shortfall due to Iran would be and then about what it had been; and a major weakness of the International Energy Agency agreements on oil sharing is that the parties may never agree on which countries have "lost" how much oil in an emergency. Furthermore, a quantity measure is not directly relevant to estimating the economic impact of a shortfall, since it provides little information on the cost of adjustment. These considerations suggest that decisions about when and how to implement emergency measures, either internationally or domestically, should be based on something other than hypothetical measures of the quantity shortfall. In Recommendation Five we outline some ways in which price-based policies might be used to manage a strategic petroleum reserve.

Instability of world oil markets is not the only potential source of shocks and surprises in the energy future. New information or discoveries about environmental impacts from burning coal or new events and information further eroding public confidence in nuclear power could prevent one or both of these energy sources from becoming as important as they otherwise might. New sources of gas, conservation, or solar energy may hold pleasant surprises, providing enough economical and clean energy to limit the extent to which less desirable sources must be used. All of these possibilities argue that a wide range of options should be actively explored in order to maximize the probability of pleasant surprises, to learn about the unpleasant truths as soon as possible, and to be able to respond to the unpredictable future as it unfolds.

No doubt, with bad luck and bad policy, there could be enough unpleasant shocks and surprises to make the energy future very bleak indeed. Our reading of the situation suggests to us that good policy can make some of its own good luck, so that the nervousness and pain occasioned by sudden and unexpected shocks and changes in direction can be kept within reasonable limits. Unfortunately, the actions of U.S. energy policymakers—both in the Congress and in the executive branch and especially in combination—over the past five years give us little confidence that the right policies are likely to be put in place. The world is lucky that nothing worse than the revolution in Iran has occurred to disrupt world oil markets—yet.

Reality Seven: Sound R&D is essential, but there is no simple "technical fix"

Americans are natural optimists about the ability of science to solve human problems. The familiar statement, "If we can get a man on the moon, surely we can . . .," reflects a deep faith in technology. But some important features of energy problems limit the extent to which technology will be *the* solution. Technology will be an important, even critical, part of the solution, of course; but there is not likely to be a simple "technical fix"—no single device or process that the technological establishment can develop in a crash program, leaving everybody else to go about business as usual.

The basic energy problem is that low cost energy such as oil is more scarce and abundant energy such as coal is more costly than was generally believed as little as ten years ago and will probably get more so over the next twenty years and beyond. In principle, technology could change this situation; there is no law of physics or economics that says that technology will never make solar energy or fusion power cheaper than oil energy was ten years ago, the way it made oil cheaper than coal and wood in the past. And there are some possible breakthroughs, particularly in solar energy, that could dramatically change the cost outlook. But on the whole, energy from futuristic sources such as fusion show little promise of reversing the trend toward higher energy costs. The abundant fossil, nuclear, and hydropower resources of the earth can be exploited now—it is just costly to do so safely and cleanly, and there is not much prospect that technology will change this fact dramatically. Technology will be able to help slow the rate of cost increase as lower quality energy sources are increasingly drawn upon, but it is not likely to solve the problem of higher energy costs.

Since technology in itself offers no complete "fix," society will have to make extensive changes in the way it gets and uses energy. Successful adjustment over the next twenty years and beyond will require many new and improved technologies for energy production and conservation and for protection of the environment. But new methods and machines will not be enough. Also essential will be improved knowledge of the health, ecological, and geophysical impacts of energy-related pollutants; better information about the earth's resource base and natural systems; and more understanding of the interactions among energy use, economic growth, and social change. Thus, technological progress in many areas simultaneously, and improved understanding of basic natural and social processes, must both be central objectives of energy policy.

The production of basic technological, scientific, and social knowledge is one of those activities that unaided markets cannot be expected to stimulate adequately, and hence government support is clearly called for. Even where the market would eventually get technology developed and applied, there may be social benefits exceeding private benefits that justify pushing some energy technologies faster than the market would ordinarily do. But if policies toward science and technology are to be effective, two special characteristics of energy technology must be kept in mind—the role of the private sector in energy technology and the significance of relative costs.

In the United States, the private sector is the source and eventual user of most of the scientific knowledge and technological advances in important parts of the economy, including those involving the use and production of energy. While some research results are used by government directly in its own programs—such as knowledge about the health effects of pollutants or about what motivates people to conserve energy—most of the results of energy-related research, even that sponsored by government, must be used by the private sector if they are to be used at all. In this respect, energy research differs critically from other government-sponsored research. While government military and space programs are the users of most government-sponsored science and technology, private utilities, individuals, and firms will decide whether, when, and how they will use knowledge generated by the energy R&D program. A "successful" research program producing nice gadgets nobody in the private sector wants is no success at all.

Because the energy problem is primarily one of cost, a successful energy R&D program is one that lowers cost (including external costs due to environmental damages and import dependence); a project that simply adds to the long list of known ways to get expensive energy is of little value. If prices do not fully reflect costs and benefits, then the government may see relative costs somewhat differently than the private sector does. But the fact that the government rather than the private sector puts up the R&D money does not mean that fundamentally different criteria should be applied: unless money spent now will, with reasonable probability, result in energy costs being lower than they otherwise would be in the not too distant future, it ought not be spent, whether it is taxpayers' or stockholders' money. By the same reasoning, if there are several projects that can be undertaken, those likely to yield the greatest reduction in energy costs should be given priority.

These two characteristics of energy R&D—the role of the private sector and the importance of relative costs—suggest that the most successful government R&D programs of the past are not appropriate models for an energy R&D program. Both the Manhattan Project (which developed the atomic bomb) and Project Apollo (which put men on the moon) had a single customer for their output—the U.S. government. In these projects, cost was no object, either in the development program itself or in the final product. There was no substitute product on the shelf that would be used if the new product turned out to be 10 percent more costly. If the devices worked technically, the project was a success. But energy technology R&D will be successful only if many individual users, each facing a different set of particular circumstances, decide that the new technology has economic advantages over a wide array of alternatives available in the market.

For every new technology that succeeds in the private sector, many fail; new energy technology will—and should—exhibit the same high failure rate. But when research is sponsored by government, the discipline of the private sector is often lacking. The market stimulates development of many diverse ideas on a small scale, but only when someone is willing to bet his own money on the idea; then, before really big money is spent, all but a few possibilities are weeded out in a competitive environment in which real technical and economic factors are all important. Government administrators, on the other hand, have difficulty deciding which new ideas are worth developing even on a small scale and hence tend to select only a few "safe" ones; then, once a technology or project develops a constituency that wants quick results (or simply construction jobs in the home state), it is too often pushed to the expensive large-scale stage when technical or economic factors do not warrant it. The combination of too much expensive construction and not enough inexpensive thinking results in a lot of money being spent with disappointing results.

For these reasons, private sector involvement in energy R&D is essential, even when there are good reasons for government funding. If energy pricing policy provides the proper signals and incentives, the private sector can ordinarily be expected to take a proven scientific concept, explore and develop its potential, estimate its economic viability, and move it into the marketplace at about the right time and rate. In certain special cases involving extraordinary social costs or risks, government action may be necessary to establish financing or risk-sharing arrangements that simulate market incentives. But, in most important cases, the fact that there are risks or large costs involved should not prevent a promising idea from being applied: the private sector knows how to assess and spread market and technological risks and to assemble capital, if the basic concept is promising. If the government provides adequate support for basic research, education, and information gathering, the private sector can ordinarily be trusted with the job of putting the results of science to productive use.

If the private sector decides that the economic payoff from a technology is so far in the future or so unlikely that it does not want to go ahead now with its own money, that is prima facie evidence that the time or technology is not yet right. Like all such prima facie evidence, it may be rebutted if closer scrutiny reveals other reasons for the private sector's reluctance to proceed or special reasons for government support. But as a rule, instead of pushing ahead to perfect a specific and economically questionable technology that the private sector will not support, it is preferable to go back to the R&D stage, explore further a number of alternative ideas, and wait until there is more real interest from those who will have to put up the money and use the technology in the end. The bypassed technology is not lost thereby, but remains available, presumably in improved form, in case nothing better turns up or conditions change.

This line of reasoning is the basis for the general rule that the scale up of technology to near commercial size should not ordinarily be undertaken until the private sector is willing to take on a large share of the financial risk. There are some arguments for making exceptions to this rule. For example, research on fusion requires very large facilities simply in order to investigate the fundamental physical principles involved; but even here, there is a large jump between proving the concept technically and demonstrating it commercially, and the private sector can help take that jump with some of its own resources.

More difficult problems are posed when technology may be useful in providing insurance against conceivable but unlikely adverse events. For example, there is some small probability that the United States will have to launch a crash program to replace imported oil completely by, say, 1995. Given the long lead times involved in such an effort, perhaps a number of synthetic liquid fuel plants using different processes and resources should be built in the early 1980s, so that the crash program—if needed—would have design, construction, and operating experience to draw upon. But no private firm, on its own, will build a large plant that, except under highly unlikely conditions, will be a sure loser. Thus, if the insurance is valuable enough, society as a whole should support the early plants.

The government may also want to subsidize the production of

certain kinds of energy that are more valuable than their relative market price would indicate, because they have less adverse environmental impact, or they reduce oil imports, or price controls distort market signals. Putting some large demonstration plants in the R&D budget is one way to provide such a subsidy. Calling something R&D when it is not, however, just because it is easier to justify a subsidy that way, is seldom good public policy; it is likely to result in little useful R&D (because only safe plants will satisfy the real objective of increasing production) and little contribution to supply (because too few plants can be built, and each will have to appear to be a research project).

The insurance and the supply subsidy arguments combined suggest that the government might want to stimulate construction of some prototype plants, probably producing liquid fuels as a substitute for imported oil, earlier than the private sector is likely to do on its own. There are some real dangers in such a move, because white elephants and serious disappointments can result. These dangers can be reduced, if not eliminated, by limiting the number and in some instances the size of plants, buying only the most economical insurance available, and using procurement policies that put some of the risk on the private sector. Specific suggestions along these lines are made in the recommendations section of this Overview.

Although the bulk of this discussion has dealt with energy supply technologies, energy conservation technologies, and nontechnological research are equally important parts of an effective energy R&D program. The recommendations section outlines some of our specific ideas in these areas.

NINE RECOMMENDATIONS FOR ENERGY POLICY

The preceding discussion of some of the central realities of world energy problems and of the lessons we draw from them contains many general and implicit policy recommendations. We now offer some more specific ideas about what we think should be done about energy problems. The policy recommendations we make here do not represent a complete program or even the only correct path to follow; but we feel that they are sensible approaches to the major issues that must be on any reasonable energy policy agenda. They are presented below in abbreviated fashion; more background information and detailed analysis are contained in the balance of this report.

In line with our overall theme that, whatever else is done, market forces should be relied upon as much as possible in implementing energy policy, our first two recommendations concern energy pricing: regulation of prices paid to producers of energy, particularly fossil fuel, where the issue is the social and economic effects of higher prices rather than any technical inability of markets to function efficiently; and the problem of pricing energy efficiently to consumers, where a pervasive pattern of utility price regulation has long existed and the policy problem is how, not whether, to regulate. Our second group of recommendations deals with science and technology policy, environmental management, and contingency planning and crisis management; here our suggestions concern ways to improve the efficiency and effectiveness of government programs themselves. Finally, we make recommendations concerning four specific energy sources, because special government programs do or should exist for them: nuclear fission, coal, conservation, and solar energy.

Recommendation One: Decontrol Oil and Gas Producer Prices

We see no technical or economic reason not to move toward total decontrol of prices paid to domestic producers. The primary energy-producing industries are workably competitive domestically, and the market—if it is allowed to operate—can be expected to produce, conserve, and allocate energy with reasonable efficiency and without causing major inequities or hardships. Price decontrol would, to some extent, stimulate domestic production of oil and gas and is worthwhile for that reason. But we do not think that the major benefits of decontrol will come from whatever short-term surge in oil and gas production may result. The more important benefits are long term and reach throughout the economy; decontrol would have constructive effects on patterns of energy consumption, on choices between imported oil and domestic energy sources, and on long-run investment choices in both energy production and consumption. It is an essential step toward putting U.S. energy policy on a sound economic basis for the long run. There are, of course, strong objections to decontrol, primarily regarding its effects on income distribution and inflation. These have been described and discussed under Reality Three.

Decontrol Domestic Oil Prices. We recommend that the complicated and inefficient system of crude oil price controls, complete with its tax on domestic production and subsidy for imports, be phased out as quickly as possible. A wellhead tax on "old" oil would be advisable as a means of capturing windfalls; but truly new oil—

that is, oil produced as a result of actions taken after a specified date—should be freed from price controls and "windfall profit" taxes. We see little wrong with President Carter's proposal of April 1979 to phase out price controls beginning June 1, 1979, and to tax "old" oil, and we recommend most strongly that Congress not reverse this action; but the "windfall profits" tax on new oil, as proposed by the president, should be omitted unless it is absolutely necessary to gain political support for the rest of the package or to minimize the probability that price controls will be reimposed at some future date. It is essential that investors in energy production know that in the long run, they will be paid at least as much for oil that they produce domestically as they and others will be paid for oil that they import.

We recognize that there may be value in reducing oil imports even below the level resulting from producer price decontrol, in order to lower pressure and perhaps prices in world oil markets. This consideration, plus the fact that those who use imported oil should bear the costs of the insecurity that comes with it, argue for imposing a surcharge on oil imports and letting domestic prices rise to reflect the import price plus the surcharge. However, we do not recommend going so far at this time. Letting domestic prices go to the world level is enough of a step in the right direction and even that continues to face serious political opposition. We suggest decontrol now, deferring for later consideration the possibility that prices even above the world oil price should be imposed.

Deregulate Natural Gas Wellhead Prices. The Natural Gas Policy Act of 1978 has eased many of the most serious problems caused by the earlier practice of using public utility regulatory concepts to set wellhead prices on interstate gas. Deregulation of most "new" gas is promised in 1985 or 1987 at the latest, and incentive pricing concepts are introduced. The system of price regulation established by the Act is unnecessarily complicated: it requires too many categories of gas and difficult well-by-well determinations, its definitions are unclear and confusing, and the rate of allowed price increases may be too low in light of recent world oil price movements. Amendments to simplify and speed up the decontrol process would be desirable. But the higher prices and promised deregulation are encouraging signs. We hesitate to recommend reopening the entire issue once again on the chance that a modified law would be better; but we certainly recommend that wellhead gas prices be decontrolled as quickly as allowed under the Natural Gas Policy Act.

Recommendation Two: Make Utility Prices to Consumers Better Reflect Real Costs

Natural gas and electricity are generally provided to consumers by public utilities, which have long been regulated as "natural monopolies." Traditionally, the goal of regulation has been to set prices no higher than necessary to cover actual costs incurred and to provide all the service demanded at those prices. The recent rapid increases in the costs of primary energy and of construction have made the cost of replacing or expanding utility facilities much higher than historical average costs. This confronts regulators with a dilemma. On the one hand, pricing utility services at the low average or historical costs gives consumers inadequate incentive to conserve or to use less costly forms of energy and tends to encourage investment in utility facilities that cost more than they are worth. On the other hand, pricing services at the higher replacement costs that exist today would result in high bills to the customers and profits to the companies higher than they have traditionally been permitted to earn.

Apply Marginal Cost Pricing Principles. It is easy enough to state the economic principles that could resolve this dilemma: customers should be charged the full marginal cost of the electricity or natural gas they use; utilities should invest in such a way that their profits are maximized given this price structure; and if the resulting profits are thought to be too high, the excess profits should be distributed in such a way that neither consumer nor investor incentives are changed as a result. This principle is not easy to apply even conceptually, and in practice, only crude approximations to the ideal are possible given the technical problems of defining marginal costs and of metering usage continuously. Nonetheless, we recommend that utility pricing be based as much as possible on these marginal cost principles.

The most obvious obstacle to applying the marginal cost principle is that consumers do not like higher utility bills, even if they are told that any excess profit to the utility will be taxed away and used for their general benefit. But workable approximations to true marginal cost pricing can be developed that do not necessarily raise the average consumer's utility bill. Generally, these involve providing each consumer with a certain amount of low-priced service and then charging higher prices for service above that amount and less during periods of low system demand. Such pricing systems provide an economic incentive to cut back consumption and shift to off-peak periods, but need not raise the average bill. There are some difficult problems involved in deciding where the dividing line between low-priced and high-priced service should fall, and atypical consumers may end up paying more or less than their "fair share," in some sense. But on the whole, those who use more of the higher cost energy will pay more, while those who use less of it will pay less—and that is a large part of what marginal cost pricing is all about.

Limit "Rolled In" Pricing of Energy. Marginal cost pricing also helps to prevent uneconomic utility investments. When prices to consumers are based on average or historical costs that are below the cost of incremental sources, utilities can satisfy growing demand (or replace depreciated capacity or depleted reserves) only by producing or purchasing energy that costs more than consumers will pay for it. In effect, when the high costs of new sources are "rolled in" with the low costs of old sources to compute the average cost on which price is based, the old sources are being taxed to subsidize the new ones. At its best, this keeps existing transportation and distribution systems utilized at higher levels than they might otherwise be, thereby spreading their fixed costs over more units and helping keep down average costs and hence consumer prices. However, at its worst, rolled in pricing encourages utilities (whose profits are based on invested capital) to pursue capital-intensive projects that produce energy at unit costs higher than the costs of available non-capital-intensive alternatives and far above what consumers would be willing to pay for the incremental energy.

Rolled in pricing is now widely used by utilities. We recommend that the practice be phased out where it exists and not be extended to other energy sources that are so costly relative to alternatives that they clearly would not be used without rolled in pricing. However, like the general marginal cost pricing principle of which this is just a special case, this is easier said than done. Thus, it is also important that public utility regulators insist that utilities actively seek out and acquire the next cheapest source of energy (including such nontraditional sources as industrial cogeneration, solar energy in its various forms, and even conservation) as they expand or replace energy supplies.

Price Backup Capacity Economically. Another important reform for utility pricing concerns the pricing of backup utility services for customers who generate some or all of their own power with, for example, windmills or industrial cogenerators. Utilities correctly observe that such customers should contribute to the cost of the central station facilities that are held in reserve for backup in case of severe weather, maintenance problems, or unexpectedly large demands. The principles of marginal cost pricing can guide regulators here as well. Too often utilities set uneconomically high standby charges as a way of discouraging competition from their customers. Fixed monthly charges, plus a per unit price for energy bought from or sold to the utility, might allow economic, decentralized energy sources to operate. As the number and the diversity of customer-owned sources attached to the system increase, the central station backup capacity dedicated to each customer, and hence the fixed monthly charge, should decrease. The economic role of utilities might well change over time because of such pricing reforms, with benefits in system reliability and lower total cost to society.

Treat Business and Personal Energy Use the Same.   It is important to note that marginal cost pricing does not mean that the high cost of incremental supplies should be loaded onto industry so that households are protected from higher prices. Even less does it mean that prices should reflect higher costs only when the higher prices do not induce conservation or fuel switching. The point of marginal cost pricing is to induce everyone, including households, to use energy in any form only to the extent that it is truly economical. The "incremental pricing" provisions of the Natural Gas Policy Act of 1978 badly misapply marginal cost pricing concepts; they are basically intended to minimize the gas bills of households at the expense of industry, without reducing use of natural gas. Because the price of domestic oil is now controlled to artificially low levels in the United States, there is some logic in encouraging industry to stay with gas by juggling gas prices, as the NGPA tries to do; but it results in artificially low gas prices and creates an administrative nightmare. We recommend decontrolling producer prices and pricing energy near marginal cost to all users, rather than creating favored classes of energy users.

Recommendation Three: Use Science and Technology to Generate and Define Basic Options, While Relying Primarily on the Private Sector to Develop and Deploy Technology

The market cannot be expected to provide the right amount or kinds of basic scientific research or information gathering, but can ordinarily move ideas from the lab to the marketplace at about the right rate. We think it is important, therefore, that government policies concentrate on encouraging programs to enhance basic knowledge and to provide a wide range of competing technological concepts that can be evaluated and, when warranted, picked up and applied by the private sector. In particular, we recommend the following.

Increase emphasis on nonhardware research. Basic research in physics, chemistry, health sciences, and earth sciences directed toward understanding the processes of combustion, the impacts of pollutants on health, the effects of carbon dioxide on climate, and so on is essential. Social science research to identify and explore policy options and their impacts and general information programs informing people what individual options they have and what the costs and benefits might be are as important as the hard science and engineering research. Legal and institutional research looking at barriers and incentives to private actions can assist in making better governmental decisions. This sort of research and education is particularly valuable where energy conservation is concerned, because conservation depends so strongly on individual actions by informed and motivated consumers. Programs of this kind are cheap when compared to the cost of one commercial-sized energy supply demonstration plant. The prospect of earning an occasional "Golden Fleece" award and opposition from industries opposed to better information in the hands of the public should not dissuade policymakers from encouraging this sort of activity.

Develop Competing Processes and Generic Technologies. Where technology and hardware are concerned, government programs should stress exploration of many competitive concepts at small scale and development of generic processes such as energy storage and low temperature heat engines that are potentially applicable to many specific technologies and energy sources. A wide variety of competitive ideas generated through parallel approaches to the same technological end is both a useful spur to government program managers and a safeguard against premature selection of an intriguing but ultimately noneconomic concept; competition is likely to produce better ideas and, in the end, better technologies. Similarly, pursuit of generic technologies safeguards against getting locked into any specific system or resource. Happily, pursuing a number of ideas to roughly the working model stage is relatively cheap.

Rely on Private Sector Selection and Management. Even when there is good reason for government funding of energy science and technology, there may not be good reason for government selection and management of the work to be done. Since the bulk of energy research is intended for use by entities outside of government, it is logical that individual scientists and firms in the private sector should have a major hand in selecting the research projects and technologies. A number of devices, such as advisory committees, institutional research grants, and larger tax credits for industrial research grants to universities, are available and should be used more extensively for this purpose.

When it comes to demonstration and deployment of technology, the private sector should be given primary responsibility for choice, design, construction, and operation. For example, if the objective is to demonstrate or simply to subsidize the production of synthetic liquid fuels from fossil fuels, the government could simply ask for bids or offer to purchase the product, leaving it to private parties to choose the technology, the resource, the scale, the location, the price, and so forth. By accepting bids or buying product only in quantities, at prices, and with other characteristics consistent with program guidelines, the government could accomplish its objectives with minimal direct involvement, limited perhaps to loan guarantees if private funding is less than forthcoming. If the private sector does not bid or offer to sell at prices the government is willing to pay, that is prima facie evidence that the costs or technical risks may be higher than was thought and, as all such evidence, subject to scrutiny; but unless there are reasons for withdrawing or sweetening the offer, it should be left standing to provide a goal for private developers.

Marketlike mechanisms of this sort should also be used when the objective is not R&D so much as simply subsidizing supply. Because they make the costs explicit, put an upper limit on them, and let the private sector bear some of the risks and rewards, such measures are definitely preferable to regulatory devices or cost-plus procurement. For example, requiring all refiners to use synthetic crudes for, say, 5 percent of their throughput hides the costs and puts no limit on them; and simply constructing a synthetic fuel facility at government expense provides no incentive for the private sector to be careful in designing, constructing, or operating the facility. Here as elsewhere, the market can be used to advantage to further the goals of energy policy.

Pursue Large-scale Demonstrations Selectively and With Great Care. A commitment to demonstration or commercialization of any technology, especially at a large scale, is a pivotal decision that slows development of new ideas and runs the risk that the selected concept will not prove viable in the private sector. We believe that the commitment to government-sponsored demonstration or deployment projects should be the exception, not the rule. The rule should be to price energy properly, to encourage environmentally sound energy projects whether for supply or conservation, and to expect the private sector to read the signals and take the risks. Exceptions to this rule should be selected and managed with great care. Some possible exceptions, and our thoughts on each, are presented below.

Coal Utilization Technology. The use of coal is limited by the difficulty of burning it cleanly. Technology to enable utilities to use coal serves a clear need, and government should risk erring on the side of demonstrating too many rather than too few possible methods. Improved pollution control devices, solvent-refined coal solids, fluidized bed boilers, and combined cycle generating plants are all worth encouraging. Improved technology would serve a similarly clear need in industry. Current government programs for industrial technology, such as those for small fluidized bed boilers and small coal gasifiers, seem to be on target, and technologies such as co-generation and low and medium Btu gasification are sufficiently advanced that industry needs little technical help in deploying them when they are economical. However, further small industry demonstration programs might be of significant value at low cost.

High Btu Gas From Coal. High Btu gas from coal can, it is estimated, be produced at costs on the order of $5 to $7 per million Btu, and no technology now on the horizon will lower this cost significantly—although financing gimmicks of various kinds might allow a plant to sell gas at a lower price and still report an accounting profit. Gas from many conventional and unconventional sources is or could become available in uncertain but probably large quantities at costs quite a bit lower—on the order of $3 to $5 per million Btu—and even this gas might not have a market for years in competition with other sources of energy, including low cost natural gas. Thus, there is no economic justification for pushing ahead with large-scale, commercial coal-to-pipeline gas projects at this time. However, as insurance against the possibility that notions of natural gas availability are exaggerated, or that world oil prices increase more rapidly than we think likely, we recommend construction of pilot plants, on the scale of 300 to 600 tons of coal per day, for one or two of the best new technologies in this field.

Supplementing Liquid Fuels. Because we view oil imports as central to energy problems, we think increasing domestic liquid fuels production is essential, though not a useful way to cope with temporary perturbations, as discussed under Reality Two. Enhanced oil recovery, which is probably the best supplemental source of domestic oil over the next few decades, deserves aggressive support, especially as past achievements seem to lag badly behind potentials; this, however, may be one of those problems better solved by pricing oil correctly and letting the private sector operate. Shale oil requires little technology development, but environmental problems continue to loom large. Coal liquefaction is expensive and likely to remain so; further small-scale work on many approaches should be stimulated, and foreign efforts should be closely followed. This appears to us to be a situation in which the government should offer to purchase some synthetic liquid fuels at a premium price (perhaps $25 per barrel, increasing at least with inflation), obtaining diversity of sources but leaving it to the private sector to determine whether shale oil, even after overcoming the environmental drawbacks to its use (if this is possible), is cheaper than liquids from coal or tar sands (or garbage or corn), and how the costs of the best of these compare with the price of imported oil.

Breeder Reactors. The U.S. breeder reactor program is a billion dollar program stuck somewhere between research and demonstration: it is too large to be a proper research program, and if it is a demonstration program, agreement seems to be lacking on what is to be demonstrated. Our judgment is that breeders will not need to be deployed on a commercial scale before 2010 or so, if then, so that the United States can postpone any decision about construction of a large (but still not commercial scale) breeder until the mid-1980s. The intervening years should be spent evaluating the economic, safety, environmental, and proliferation characteristics of alternative breeder designs, in cooperation with the utility industry, but also with the involvement of engineers from the vendor firms and the national laboratories, who may be less conservative in outlook and design philosophy. The United States should go ahead with a major breeder demonstration program only if it has been clearly established that programs in France or elsewhere will not provide a satisfactory basis for commercialization and licensing of breeders in the United States or if there is good reason to believe that a United States breeder would be sufficiently better or different to provide a real alternative.

Recommendation Four: Adopt a Different Approach to Air Pollution Control

The conflicts between energy and environmental values are real and will be difficult to resolve under the best of conditions. However, the present concepts and programs of pollution control make these conflicts even more severe, disruptive, and costly than they need to be and threaten to damage both energy and the environment in the long run. Therefore, rather than recommending minor changes in existing procedures and particular standards within the existing programs, we recommend moving toward a different philosophy of pollution control, but in a way that builds upon existing programs. We take the particularly troublesome case of air pollution control as an example, but the principles apply to other areas of environmental concern as well.

The underlying assumptions of the U.S. Clean Air Act are that science can define achievable levels of air pollution that are "safe," that air everywhere must meet that standard by a certain date, and that regulatory processes must decide the details of the response required by each of thousands of pollution sources. As discussed previously, this approach is neither scientifically nor administratively sound. We recommend that a different approach be incorporated into the Clean Air Act through a process along the following lines:

1. Research on the scientific facts underlying pollution should seek to define the general relationships among pollution, human health, and other values that are protected by cleaning the air, not to find "no damage" levels. Careful measurements, research on atmospheric transport and chemistry, models of dispersion, and studies of health effects should be emphasized as a search for facts and relationships, not as a defensive hunt for support of current standards and regulations.

2. Rigid deadlines and standards should be deemphasized, but overall progress should be speeded up. Unreasonable goals and deadlines waste time and resources in sterile conflict rather than making progress toward cleaner air. The goal of air pollution control policy should not be to accomplish specific air quality standards by specific deadlines, but rather to make continuous progress over time toward the general objective of cleaner air, giving priority to those areas where the value of cleaner air is greater for health, aesthetic, or ecological reasons. Local or temporary degradation can occur within an overall picture of improving air quality. The important questions concern where the nation as a whole will be in ten or twenty years, not whether an arbitrary air quality standard will be met everywhere in 1983.

3. Air pollution control policy should concentrate on providing incentives for making progress toward cleaner air in a way that is cost-effective over time. Emission charges, marketable discharge permits, and similar marketlike devices should be used; in the balance of this report we make some particular suggestions for incorporating such devices into existing programs dealing with new source perfor-

mance, with regions where air quality standards have not been attained, and with prevention of significant deterioration.

In making such recommendations, we are not suggesting that too much money is now being spent on air pollution control or that the dollar value of clean air can be determined through some technocratic cost-benefit calculation. We recognize that the value of clean air can be determined only by collective, political processes, in which estimates of the dollar value of health effects and other damages caused by pollution will be only one (albeit an important) factor. But we do not think that the present system, in which absolute goals are set without regard to the costs of accomplishing them by what purports to be a purely "scientific" process, is the best way to make social judgments and trade-offs openly and carefully. Nor do we think that traditional regulatory processes can cope adequately with the complex, dynamic, and long-term economic problems involved in controlling pollution.

For these reasons, we recommend starting with the programs now in place, but beginning to move toward a rather different policy philosophy, in which social judgments would be made about whether to apply more or less pressure toward cleaner air, leaving market processes to determine the details. Some small steps in this direction are now being taken in the form of the "emission trade-off policy" for areas that have not met air quality standards. We are convinced that in the long run, widespread and imaginative use of such devices will result in significantly cleaner air at lower cost than will the present scientifically questionable and administratively cumbersome policy. Similar marketlike devices could be useful in managing other difficult environmental problems.

Recommendation Five: Prepare For Disruption in World Oil Markets

There have occurred and there will occur again disruptions in world oil markets, causing sudden loss of major suppliers or price increases, temporarily or indefinitely, for reasons that cannot be predicted or controlled. The United States will, at least throughout our twenty year period, be vulnerable to such shocks, if not directly then through the impact on friends and allies in the world, and must work with others to prepare for such disruptions. We recommend the following steps.

Develop an Effective Oil Stockpile Program. Poor policy guidance and program management of the strategic petroleum reserve program cost the United States the opportunity to develop a significant oil stockpile while the world oil markets were slack in 1976–1978. Now, in the aftermath of events in Iran, it will be much more costly to buy oil and put it back into the ground. Oil is more expensive now than it was, and additional purchases for the stockpile while market conditions are tight add to pressures to increase prices further. Nevertheless, a significant strategic reserve is so important as a deterrent to potential interruptions, as a cushion for cool thinking in crises, and as a means of smoothing price shocks that purchases of oil for the stockpile should be vigorously pursued. Market conditions should be a constraint only if they are likely to affect price critically.

Strategic oil reserves need not and should not be held exclusively and perhaps not even primarily by the government. The private sector should be encouraged to build and hold stockpiles, by requiring importers to hold reserves, by contracting with refiners and others, and by letting it be known that stockpilers will be allowed to profit from higher prices if the stockpiles are called upon. Such measures could increase the amount of oil stored and improve the efficiency of the stockpile program.

Use Market Forces to Manage Stockpiles and Crises. A large, readily available oil stockpile must be used correctly if it is to be a major stabilizing factor in the U.S. and world energy situation. Because price is the economically relevant measure of the tightness of the world oil market, we recommend that decision rules based primarily on market variables play a central role in deciding when and how to use the strategic reserve. For example, oil could be sold from the stockpile to anyone who wants it, at a price (say) 50 percent above the landed price of imports averaged over the preceding twelve months. Such a policy (perhaps raising the percentage premium as the stockpile is drawn down so that it is not depleted too quickly) would tend to put a ceiling on panic buying in the spot market, would help to smooth out sudden price jumps, and would put the reserve oil into normal commercial channels with a minimum of bureaucratic delay and confusion. Then, when the crisis ended or the surge in prices slowed down, reliance on the stockpile would decline naturally and smoothly.

Emergency measures for expanding oil supplies and reducing energy demand must be evaluated and prepared beforehand, so that they can be used in conjunction with the strategic reserve; these will both stretch the useful life of the stockpile and begin the process of adjustment that will be necessary if the "temporary" situation turns out to be permanent. We recommend that market forces be utilized to the maximum possible extent in implementing these measures.

If a significant strategic reserve is in place, triggered by prices as suggested above, the best policy is probably simply to let prices rise until demand and supply are in balance. The strategic reserve will put a limit on the immediate price increase; and if the disruption lasts so long that the reserve begins to be depleted, it is no longer a temporary emergency but a new permanent condition that should be reflected in higher energy prices in general.

If there is no strategic petroleum reserve acting to limit the short-run price increases, political opposition to letting the market solve the problem will be greater; even here, however, it may be the best solution, if combined with measures to ease hardships in extreme cases. Temporary taxes on energy use, combined with distribution of the proceeds in a way that is perceived to be fair, can reduce the extent to which "hoarders" and "profiteers" benefit from the higher prices paid by consumers—but it will be at the cost of discouraging the kind of anticipatory stockpiling and investing, and extra energy production, that could help reduce the damage. If rationing or allocation systems are to be used, they should work on market principles. For example, ration coupons should be legally saleable, so that the user faces a high cost when he decides to buy or not to sell each coupon and so that those with the most valuable uses for the rationed commodity get it. Any of these measures—letting the market work, taxes, or white market rationing—which allow decentralized responses to individual circumstances would be preferable to the rigid, centralized allocation programs superimposed on an already over-controlled market.

Work on Cooperative International Strategies. A shock in world oil markets affects every country, and it is to the advantage of all to spread the remaining supplies to minimize harm. The United States has entered into commitments through the International Energy Agency that address these issues in a limited way. Even these limited international arrangements are not fully recognized or appreciated by the U.S. public and perhaps even the Congress; and recent developments suggest that the administration may not be as diligent as it should be in this matter. We recommend, therefore, that U.S. energy policymakers act to strengthen the IEA processes and to impress on the public the existence and importance of U.S. obligations. The world political and economic systems are too interdependent for the United States to try to go it alone in the event of a serious disruption in world oil markets.

Diversification of energy sources worldwide would serve many purposes, among them making it easier for the world market to accommodate disruption in any one place. We recommend that such diversification be a major objective of U.S. policy, even when it involves such measures as providing energy technology to countries who are not necessarily friendly to the United States.

Recommendation Six: Continue Efforts to Reduce the Problems Associated With Nuclear Power

No energy source is without problems. Each is burdened with its own set of short- and long-run drawbacks of high cost, potential calamities, or limited applicability. Eliminating any source from the menu of options at this time merely increases the probability that problems will arise in the future for which there is no good solution. Thus, despite continuing concerns about nuclear weapons proliferation, reactor safety, and radioactive waste disposal, we recommend that nuclear power not be excluded as an energy option either in the United States or abroad and either in the short run or in the long run. Regions of the world that lack significant fossil fuel reserves are especially prone to favor nuclear power and for good reasons.

The relative attractiveness of nuclear power (including breeder reactors) and the extent to which it is ultimately used will depend critically on what is learned about the climatic and other effects of fossil fuel use and on the rate at which solar and other nonfossil energy sources (including conservation) become economically viable on a large scale. There is no way to predict all these factors, and hence we do not know when or whether nuclear fission will contribute a large share to total energy supplies. But nuclear energy should continue to be developed as a long-term option. Furthermore, it should be applied in its various forms where and when it is economical and accepted by the public. Some particular policy directions that we recommend are outlined below and detailed in the balance of this report.

Adopt a Positive Antiproliferation Stance. The problem of nuclear weapons proliferation, to the extent that it is exacerbated by civilian nuclear power, is related primarily to facilities for the enrichment of uranium and the separation, shipment, and subsequent storage of plutonium from spent reactor fuel or from breeder reactor "blankets." The United States, the USSR, China, Great Britain, France, India, Germany, the Netherlands, and Japan all have or soon will have facilities for enriching uranium and/or separating plutonium; the first five of these countries have already produced nuclear weapons, while the last three could but have agreed not to, have little motivation for doing so, and have accepted international safeguarding of their nuclear facilities. The existence or addition of enrichment and reprocessing facilities in these countries (with the possible exception of India) does not significantly worsen the nuclear weapons proliferation problem. However, further spread of such facilities to other countries, such as Brazil, Pakistan, or South Africa, could worsen the proliferation problem and should be discouraged.

Our analysis of nuclear power convinces us that there is no persuasive technical or economic reason for the United States to reprocess fuel from commercial reactors or to move to commercial breeder reactors within this century and probably for a decade beyond. Each forecast or event that scales down estimates of future nuclear capacity or increases estimates of uranium reserves reinforces this conclusion. However, other countries—especially those with limited domestic energy sources—see things differently. There is consequently movement abroad toward reprocessing and breeders. Current U.S. policy is to discourage this trend, primarily (as a result of the Nuclear Non-Proliferation Act of 1978) by denying U.S. nuclear technology and services to countries that do not accept specified conditions on nuclear fuel cycle activities.

While we agree with the objective, the approach has, in our view, become counterproductive. It is a source of friction between the United States and others, and it reinforces doubts about the reliability of U.S. commitments and of foreign sources of nuclear technology and services in general. Countries simply go elsewhere for what they need and have more incentive to develop domestic capabilities that cannot be so easily interrupted by U.S. action. Thus, we recommend that U.S. nonproliferation efforts take a less heavy-handed and more positive approach, offering economic incentives and supply assurances to dissuade others from premature or otherwise undesirable enrichment, reprocessing, and breeder development. In particular, we recommend that the present "go slow" policy on reactor fuel reprocessing and breeder commercialization be continued within the United States; but in the world market the United States should continue to remove restrictions (except the requirement for international safeguards) to international trade in nuclear technology and services, so that countries not now possessing enrichment and reprocessing capability have less reason to obtain it.

Continue to Define and Improve Long-Run Nuclear Options.   It is desirable that nuclear fission be maintained and improved as an energy option for the next century, even if rising costs and concerns about safety slow its growth and limit its use in the short run. This objective can be advanced most successfully by pursuing a broad program of research, information gathering, component development, and systems analysis, rather than prematurely pushing specific designs to the expensive commercial stage. For example, breeder concepts in addition to the liquid metal fast breeder reactor (LMFBR) should be explored, including designs with lower breeder gains but with other desirable attributes; waste disposal techniques must be developed, with particular emphasis on detailed characterization of specific candidate sites; and estimates of uranium resources (including costs and lead times) must be improved, for the United States and the world.

The problem of nuclear reactor operating safety has taken on new importance in light of the accident at Three Mile Island. Detailed recommendations on our part dealing with training, monitoring, instrumentation, and contingency planning would be premature at the time of writing, since various investigations with access to the full range of information are still underway. Nonetheless, we expect these investigations to result in recommendations and actions with major impact on the design and operation of reactors, on the economics of nuclear power, and perhaps even on the institutional setting in which nuclear power will ultimately operate. The Three Mile Island accident has certainly set back nuclear power in the short run and has reemphasized the importance of developing non-nuclear alternatives for the long run, but has not, in our view, diminished the importance of maintaining and improving the nuclear option for the long run.

Recommendation Seven: Work to Improve the Acceptability of Coal

Conservation apart, for most of our twenty year period, coal is the only energy source that can increase its absolute contribution rapidly and economically, in this country and in other important parts of the world. Increased use of coal does not depend on finding it—there is no secret about where it is—but on making its use more acceptable without pricing it out of the market. The following general recommendations for helping to do so are limited here to only three issues, but it should be noted that our recommendations on oil and gas price decontrol, air pollution, and science and technology have a strong bearing on coal use.

Facilitate Coal Use in Industry. Forecasts of rapidly rising coal use in the United States generally assume rapid expansion of coal consumption by electric utilities, but especially by industry. Yet there are major obstacles. Weaning industry away from gas and oil, perhaps even more so than reducing automobile fuel usage or improving residential heating efficiency, will be a slow and costly process. Specifically, it is unlikely that coal can push significantly into medium- and small-scale industrial applications unless progress is made on improved pollution control technologies and improved combustion technologies; it may be, in fact, that such applications of coal are economically impossible under current rigid environmental regulations. Simply assuming that an expansion in industrial coal use will occur can result only in disappointment and failure to adjust public policies.

The policies we recommend for facilitating industrial use of coal are not those that use regulatory procedures to try to force industrial plants to convert to coal from oil and gas. The many diverse circumstances in which coal might be applied in industry cannot reasonably be investigated and evaluated by regulatory bureaucracies. As experience with these forced coal conversion programs demonstrates, coal will be used by industry when and where it is the most economical thing to do, considering all the hassle and inconvenience involved in coal use, and not much otherwise.

There are a wide range of things government might do to encourage industrial coal use, including R&D on improved technology and improvement of coal transportation and distribution systems. The most important involve making air pollution control programs more flexible and letting prices of oil and gas rise to levels reflecting real costs and scarcity; our recommendations in these areas are discussed elsewhere. Here we are primarily recommending that focused, positive, and realistic programs, and not wishful thinking or regulatory mazes, be developed to stimulate industrial use of coal.

Resume Federal Coal Leasing. The moratorium on leasing of federal coal lands that has been in effect since May 1971 should be terminated quickly. Although large acreages and tonnages—17 billion tons or so—are now under lease, it is not known how much of this is efficiently minable, especially in the West. Size, location, and other characteristics make much of the leased land less desirable economically and environmentally than new tracts could be. Even if half the tonnage under lease is efficiently minable (and the fraction may be less), this tonnage would support a net increase in annual production of only about 500 million tons, less than is forecast in optimistic appraisals of coal's role by 1985.

The objectives of federal coal leasing should be to make low cost coal available in efficient blocks, to obtain a fair royalty for the public treasury, and to protect the public interest in sound environmental practices. The bidding system that now protects the public interest by including acceptable mining and land restoration conditions, stipulating performance requirements, and giving the leasing agency the right of refusal of any bid should be used in leasing new acreage. If demand for coal develops as rapidly as some anticipate, the new acreage will be important in controlling costs; if these anticipations turn out to be wrong, mining will not take place, and nothing will be lost.

Learn as Much as Possible as Soon as Possible about the Carbon Dioxide Problem. The sooner it is learned just what the relationships are among coal use, CO_2 buildup, and global climate, the sooner the long-term potential for coal and other fossil fuels can be defined. The need is not just for "more research," but for focused programs with the objective of defining the principal uncertainties, identifying the important "lead indicators," and determining the critical decision points for coal over the next several decades. Responsibility for such a program should be placed unequivocally on an established independent government body, coupled with a periodic reporting requirement. Since the problem is a worldwide one, a parallel responsibility needs to be placed in the international field, where care must be taken not to allow the issue to be considered as merely an interesting scientific curiosity. Generous and reliable funding will be vital, but money alone will not produce the answers.

Recommendation Eight: Vigorously Pursue Conservation as an Economical Energy Source

Specific energy-using processes will be made more energy-efficient and basic human needs will be satisfied in less energy-intensive ways as a natural and economic result of higher energy costs. The most important thing energy policy can do to stimulate efficient development of the conservation "source" is to let the prices of other energy forms rise to reflect their real economic value. But, as we have noted elsewhere in this Overview, the process of getting prices to the right levels is neither swift nor certain. Although shortages, allocation systems, and direct regulations will also provide incentives for conservation, as long as energy prices are not at levels reflecting the full costs (including the costs of environmental damages, reliance on insecure foreign sources, etc.), there are strong arguments for policies that stimulate conservation directly; and even if prices are about right, there are some policy actions that can improve the workings of the market. We recommend the following:

Temporarily Subsidize Conservation Investments. Although consumers may find it hard to accept, the fact is that energy prices to consumers are now too low to reflect the true economic cost of energy. Furthermore, there is inadequate public recognition that energy costs will almost surely become and remain even higher. As a result, there is insufficient incentive for individuals to take action now that will yield energy savings of value to the society as a whole in the long run. Thus, we recommend that the government subsidize conservation investments until other sources are efficiently priced. It is difficult to find ways to provide such incentives efficiently and particularly to avoid "windfall profits" to consumers as a result of subsidizing the many actions that would be taken in response to higher energy prices even without a subsidy. Nonetheless, for the next few years policy should try to err on the side of too much incentive rather than too little. The long-term administrative problems can be minimized by limiting the duration of the subsidy programs. The existing federal income tax credit for energy-saving measures taken by homeowners is a limited step in the right direction.

Use Federal Facilities as a Conservation Proving Ground. The government should ideally have a long time horizon for planning and for calculating costs and return on its investments. But often government budget processes encourage minimizing the front end costs that are part of the budget for a particular year or agency, rather than weighing future costs that are in another budget. Valuable legislative direction now exists for encouraging government agencies to use life-cycle-costing techniques in purchases. This legislation should be administered so that government provides a market for energy-conserving technology. But the federal government, particularly, should go further. It should aggressively try energy conservation ideas in its own activities, with the expectation that some of them will not work out. Those that are successful should be widely publicized.

Increase Nonhardware Conservation Research. Our reading of the research budget of the Department of Energy suggests that too little effort is going into conservation research generally and that nonhardware research is particularly slighted. Much of energy conservation is not a matter of inventing a new gadget, but rather of discovering ways to remove barriers or provide incentives, so that people will be able to make choices of less energy-intensive methods of benefit to them and the nation. We recommend that more research on such barriers and incentives be undertaken. The total amount of money spent on such research can never be large compared with the cost of massive hardware research in supply technologies; but money intelligently spent here may repay the investment manyfold.

Aggressively Demonstrate and "Market" Conservation to Consumers. Energy is used directly by consumers in many ways. Only if individuals understand the options available to them will they be able to make efficient decisions. There is no well-developed industry promoting and selling conservation; and under the best of circumstances, a private market in conservation information and equipment will consist of many small firms of uncertain competence and reliability. We recommend that federal, state, and local governments aggressively "market" conservation information, especially to households, consumers, and small commercial establishments, but also to large commercial operations and industry. The marketing should provide information concerning specific applications, preferably put before the potential energy consumer at the time a decision is at hand. Appliance and product labeling, hot lines for advice, mandatory disclosure of energy use of buildings, energy consumption "extension" services, and other actions that focus attention on energy use should be pursued.

Recommendation Nine: Remove Impediments to Use of Solar Energy

We believe that, in the coming decades, solar energy should and will become one of the most important energy sources and that it is essential as a long-term option. Public opinion appears strongly to favor the use of solar energy; but this potential public demand for solar energy is at least partly suppressed by policies that prevent prices of fossil fuels and nuclear energy from reflecting their true marginal costs. Here as elsewhere, our most important recommendation is that all energy be priced at its true economic value, so that solar energy can find its appropriate role within an overall energy policy framework that relies primarily on market forces to determine when, where, and how particular energy sources are used. Because existing energy prices now put solar energy (like conservation) at a particular disadvantage in the marketplace and because of the value of the solar option in the long run, we recommend that special policies be developed to encourage use of solar energy, the first of which, not here repeated, is the same as proposed in Recommendation Eight: Temporary subsidization of investments. Others are listed below.

Develop New Mechanisms for Financing Solar Energy Investments. Solar energy systems, even those that are economic today, are often at a disadvantage in the marketplace because they involve unconventional methods of purchasing energy. Customarily, utilities purchase capital facilities with borrowed and stockholder or government funds (and sometimes at subsidized rates of interest) and then pay for those facilities with the revenue obtained by selling energy to consumers over many years. Utility customers have low front end costs and little to worry about except paying their monthly bills. With on-site solar energy systems, such as home heating systems, the consumer must finance the capital investment in the home mortgage, with a special loan, or from personal resources. The high interest rates on individual loans and the task of arranging the financing tend to discourage such investments even when, from the point of view of society as a whole, they are more economical than further investments by the utility. Thus, we recommend that the federal government develop policies that make capital more readily available for decentralized solar energy investments, such as encouraging public utilities to own solar devices used by their customers, guaranteeing bank loans, or establishing special purpose lending institutions.

Emphasize Cost Reduction, New System Concepts in R&D. Solar energy will be used sooner and more extensively the less costly it is. Inventing and demonstrating new ways to obtain solar energy that is expensive are not much help. We recommend that government R&D programs not spend large sums demonstrating devices, such as solar power towers or power-generating windmills, that involve known technology and have little prospect of becoming significantly cheaper. Instead, the emphasis should be on identifying the basic reasons for present high capital or operating costs of solar systems and then focusing R&D programs on ways to reduce these costs. New construction and collection materials and methods; energy storage devices; and new concepts for collecting, distributing, and using solar heat (or cold, such as ice formed in the winter) should all be explored seriously so that many options for using solar energy will be available.

Actively "Market" and Plan for Solar Energy. Solar energy is in many ways similar to energy conservation, both institutionally and technically, and hence most of the recommendations made earlier for marketing energy conservation apply to solar energy as well— provide information and technical assistance, use government as a proving ground, and so forth. In addition, some actions taken now have the potential for making solar energy much easier or harder to utilize in the future, when the technology has improved—for example, the orientation of streets and buildings and the density of development will influence the feasibility of installing solar systems in the future. We recommend that land-use-planning processes include access to solar energy as one of the many (often conflicting) desirable objectives to be considered in urban design.

Energy and the Economy

Introduction

The long-term energy problem is primarily one of adjusting to rising costs. All the fuels—gas, oil, and coal—will become progressively more expensive because of higher extraction costs and the higher costs of cleaning them up for use or cleaning up after them. The rise in costs has been accelerated and dramatized by the effectiveness of OPEC over the past half dozen years, but the problem predates OPEC and is at least somewhat independent of it. Indeed, one of the best kept secrets about energy is that the problem began to raise concern before 1973. In any event, the problem is not that world demand will someday overtake world supply. Rather, the problem is the difficulty of adjustment to high and rising costs.

Not all fuels will show the same increases in cost, but because of substitutability among them, there is always a strong sympathetic effect as a rise in one—say, the price of oil or gas—raises the price of another—say, coal, as well as other substances. Demand slowly shifts from fuels showing the greater cost increases to those showing lesser cost increases.

Although it seems highly likely that the general trend of fuel prices will be upward—with oil prices possibly more than doubling by the end of the century and the average prices of all fuels possibly increasing as much as 100 percent—there is still great uncertainty about the costs at which different fuel resources are likely to become economical to extract and consume. It is not out of the question that oil and gas resources, especially gas, will be discovered on a large enough scale within the coming decade to keep fuel prices from rising at all. There are also great uncertainties about technologies for extracting liquids from shale or coal and uncertainties about the environmental costs (especially the costs of protecting against hazards to human health) of increasing the use of fossil fuels. But even sizable petroleum reserves are still being discovered, recently in Mexico but before that in the North Sea, the north slope of Alaska, and other places. Yet on the whole, surprises are more likely to be unpleasant ones.

Whatever increase in the price of oil and of all energy that one assumes, between now and the year 2000, it is hard to know whether to expect a steady rise in the relative cost of oil and other fuels or abrupt increases reminiscent of 1973-1974 or 1978-1979. And it will make a difference. Sudden price jumps are disruptive. They will be politically conspicuous, both abroad and in the United States, appearing to demand regulatory measures and the abandonment of the pricing mechanism. Abrupt oil price increases are harder to accommodate in the balance of payments. They raise inflationary problems in an aggravated form. Furthermore, a period of abrupt large increases rather than steady price increases will entail a greater average rate of price increase over the next twenty years. Thus, prominent among the various uncertainties is the question of whether the long-term price increase will be steady or erratic.

Two other dimensions bedevil the analyses of energy policy. One is the matter of differential impact. Because fuels tend to be visible, homogeneous, and a conspicuous item of consumption, when gasoline prices or heating fuel prices or gas or electric utility prices go up, people are aware of it. Everybody knows the price of gasoline, and most people know the price they pay for heating fuel—and they seem to care. But Americans are affected differently by energy price increases, depending on the region in which they live, income level, and urban or rural status. Energy producers and policymakers seem to enjoy little public trust; public opinion polls show that a majority of the people think that an energy shortage is a great hoax or conspiracy. Oil companies are unpopular and easy targets of charges that they are engaged in some kind of skullduggery. Energy policy often appears to contain some kind of challenge to one's lifestyle, partly because there is an apparent notion—exaggerated but by no means quixotic—that the automobile is central to the American way of life and must not be challenged. The congressional abhorrence of a gasoline tax seems greater even than the abhorrence of the sales tax. And now, even more than a couple of years ago, there is an apparent unwillingness of the public to trust government officials, even the president of the United States, on matters relating to fuel prices.

Second, for the United States, as for other countries, imported oil has become a dramatically large component in the balance of international payments. That adds a dimension of difficulty to energy policy and enhances some problem of economic management whenever oil prices suddenly increase. There is, furthermore, a cumulative problem. If oil-exporting countries are slow to expand their own imports and accumulate large liquid balances in dollars and other currencies, their various portfolio adjustments can have potentially startling effects on the capital account balance of payments and on exchange rates. In the past couple of years the main oil-exporting countries have displayed a much greater capability to absorb imports than was thought likely three or four years ago, so that the problem of accumulating large volatile liquid asset balances has not materialized in the manner that appeared so worrisome recently. Nevertheless, the fact that so much oil is imported makes it a degree more difficult to manage than the impact of indigenous supplies.

The foregoing is a brief synopsis of the subject matter contained in this part. Chapter 2 delves into the relationship between energy use and the growth of the economy and finds that there is more flexibility than we used to believe, both here and abroad. Surveying the many projections that have been made, the chapter puts energy use in the year 2000 at no less than 100 quads and no more than 120, compared with about 80 in 1979. It cautions that the speed of adjustment to lower energy growth will depend heavily on the push it receives from higher energy prices.

Chapter 3 investigates a specific aspect of future demand—the potential for energy conservation, price-induced and otherwise. Conservation is one of those things that everyone favors, but few people define in the same way. In Chapter 3 we try to be specific. We stress again the crucial role that prices play in motivating people and enterprises to use energy efficiently, and we remind the reader that a unit of energy conserved can often be the equivalent of a unit of energy produced. In addition, we devote much attention to the appropriate role of government and to the institutional setting under which conservation can be expected to flourish, including such, at first glance, arcane matters as utility rate regulation. Current conservation programs are extensively described and evaluated.

Chapter 4 lays out the problems associated with keeping events in the energy sector from disturbing the performance of the economy: price levels, aggregate output, and employment can all be adversely influenced. Some costs need to be paid simply because more real resources have to be expended to obtain energy; other costs arise because economic policy is unable to take offsetting measures. Especially in a strongly inflationary period, with substantial unemployment and an unfavorable trade balance, the "right" policy is hard to come by, and it is difficult to assess the trade-offs that are worth making.

Since short-term crises are apt to aggravate the long-term problem, the chapter next takes up means of lessening the disruptive impact of such events, including what at one time was widely considered to be the most threatening aspect of the energy problem, the large imbalances created in the international payments pattern. It notes that up to the present, individual nations have managed to cope with both the surpluses in the hands of the oil-exporting countries and the deficits accruing to some of the oil importers. On close inspection, it turns out, in fact, that other elements such as general export competitiveness and rates of inflation influence performance as much as the degree of oil import dependence or perhaps more. Nonetheless, whether and how well the oil importers would withstand another siege similar to that of the immediate postembargo period of 1973–1974 and how much comfort we may draw from the fact that we have survived it remain questions to which there is no clear answer, though reduced oil imports would no doubt ease management of the problem.

Part I next moves to an analysis of the costs of price regulation and the distributional effects of price decontrol. Since questions of equity have so greatly bedeviled the energy debate, it is important to sort out both the underlying principles and the magnitude of the problem. Chapter 5 concludes that there are several categories of regulatory costs. First are inefficiencies. Controls stimulate demand beyond what the true value of oil would call for, depress domestic supply below it; they subsidize imports and tax domestic output. Other costs, probably much larger but not lending themselves well to measurement, arise from administrative complexity, the evasions of the system, the discouragement of expensive exploration, and so on. All in all, the chapter concludes that measurable costs of regulation are significant, but that they become large enough to suggest their abolition only when hard-to-quantify costs and distortions are taken into account.

As for the shifts in wealth and income that deregulation would entail, the chapter argues that the equity aspects of decontrol tend to be exaggerated. Much of the distributional problem affects crude oil sellers versus refiners rather than consumers. Consumers in the aggregate would not be deprived of amounts large enough to make deregulation inadvisable and the income shift from poor to rich is less pronounced and neat than usually assumed. In any event, those least able to bear additional costs should be helped, preferably through supplying financial assistance rather than energy, though political realities may make the latter solution inevitable.

Chapter 6 discusses the most pressing part of the energy problem—the high volume and cost of imported oil. It demonstrates the advantages that would flow from reducing oil imports and argues for various means of achieving reduction, first among which is price decontrol. Should this measure bring disappointing results, import fees are the next step in a sequence of increasingly direct policies. Caution is advised against any system of direct limitations through setting tonnage or dollar quotas.

Portions of the five chapters that make up Part I are necessarily technical and may not be easy reading for the noneconomist. In the instance of Chapter 2 it was found advisable to present in the companion volume of background papers, to be published as *Selected Studies in Energy: Background Papers for Energy: The Next Twenty Years*, a complete exposition of the reasoning and calculations.

The Future Demand for Energy

The development of energy policy and the evaluation of alternative energy futures depend in large measure on our common understanding of the diverse components and prospective change of energy demand. The purpose of this chapter is to chart the uncertain evolution of energy use. First we sketch the current pattern of energy consumption to illustrate its variety and to underscore the potential for adaptation during the newly perceived era of higher energy prices. The focus then turns to a review of the many forecasts of energy demand. Rather than construct still one additional projection, we apply a simple model to place the various forecasts in perspective, to isolate the key assumptions, and to relate the expected long-run adjustments to the observed short-run behavior in the energy system. Although we concentrate here on the aggregate demand for energy in the United States, we recognize that energy problems are both fuel-specific in character and international in scope; therefore, we summarize briefly the implications of the collateral data for other countries, and, in the next to last section, we illustrate the probable range of fuel mix changes over the next two decades. The final section relates this investigation of energy demand to the central conclusions of the overall study.

ENERGY DEMAND

Until recently, energy policymakers expected the aggregate demand for primary energy to grow steadily. Before 1970, there were abundant supplies of cheap energy. Markets for new energy-using technologies, such as air conditioning, were expanding rapidly, and the growth in energy demand closely paralleled the growth in the gross national product. Between 1950 and 1970 the U.S. economy grew at 3.2 percent per year, the demand for energy grew at 3.4 percent per year, and the year-to-year changes in both were closely related (see Figure 2-1). This close link between energy and the economy led to a widespread presumption that the growth in energy consumption was essential for the future growth and health of our economy.

Of course, this period was also a time of great change in the energy system. The relative importance of coal decreased sharply, with petroleum and natural gas taking its place: the fraction of total energy consumption coming from coal dropped by half during this period, while the proportion coming from natural gas increased by nearly the same amount. (To measure energy use in the aggregate, we add up the heat content of the constituent sources. The common denominator unit usually employed is the British thermal unit—Btu for short.) The form of energy use changed as well, with the energy

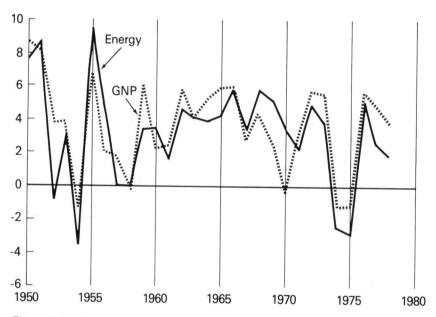

Figure 2-1. Changes in Primary Energy and GNP, 1950–1978.

Source: The data for GNP changes are from the *Economic Report of the President* (Washington, D.C.: Government Printing Office, January 1979). The energy data are from the Bureau of Mines for 1950–1974 and from the Department of Energy for 1974–1978.

going to electricity generation increasing from less than one-sixth to nearly one-fourth of all the primary energy used. These changes were of great importance to the energy industry, but passed unnoticed or were taken for granted by the average energy consumer.

By 1970, however, the energy system began to change its complexion in ways that affected our lives and our perceptions of the future. The regulation of natural gas prices became an increasingly controversial issue, and the first curtailments of natural gas consumption, intended to allocate limited supplies in the face of a growing demand, had begun. The domestic production of crude oil began to decline, with natural gas production soon to follow. The environmental problems associated with the burning of fossil fuels had moved to the forefront of national concern; for example, the regulation of powerplant emissions had become a major element of national policy. Finally, the events of 1973, with its oil embargo and the abrupt increase in the price of oil, created intensified concern with the economic consequences of energy problems and a fear that the future would be one of energy scarcity unlike the past of energy abundance.

This new awareness of the importance of energy has led to a re-evaluation of both the role of energy in our economy and the forecasts of the likely future demand for energy. At one extreme, there persists the traditional view that energy consumption is closely linked to the level of economic activity and that any expansion of the economy requires a proportional increase in energy supply. From this perspective, which, as noted, seems supported by U.S. data from the years before 1970, economic growth largely determines energy demand. An implication of this extreme view would be to treat energy demand as a requirement to be met, notwithstanding the sacrifice of substantial resources or the compromise of environmental goals, or both. At the opposite extreme, there is the view that increased use of energy is only one of many ways to accomplish our real goals and that in the interest of preserving our limited energy supplies for the future and improving the environment today, energy demand can be reduced without great sacrifices in welfare. The truth probably lies somewhere in between these extreme views, but the evaluation of the potential flexibility of energy use will have a great influence on the types of policies recommended for our energy future. Those who see energy demand as inflexible will favor dramatic programs to expand energy supply, even to the point of large-scale subsidization. Those who see more flexibility in the way energy is used will choose the energy conservation policies that could lead to sharp changes in our energy use patterns.

DIVERSITY OF ENERGY USES

Energy use is pervasive in our economy. Ultimately, energy demand stems from consumer demand for goods and services—either directly, as when we heat our homes or drive our cars, or indirectly, through the energy used to grow the food we eat or to make the clothes we wear. When we consider the indirect effects as well as the direct effects, we find that total energy use by consumers in different income classes is roughly proportional to their total expenditures. As shown in Figure 2-2, increasing income is associated with a decrease in the proportion of expenditures spent directly on energy but an

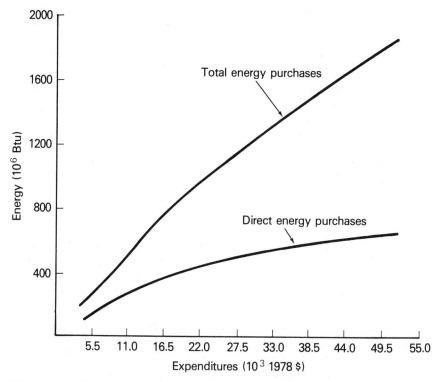

Figure 2-2. Energy Purchases versus Household Expenditures.

Source: The data are taken from R. Herendeen and J. Tanaka, "Energy Cost of Living," *Energy* Vol. 1, no. 2 (June 1976). The original data are in 1961 dollars and were converted here using the GNP deflator. Note that the Herendeen-Tanaka figures refer to household expediture classes. We assume that, by and large, the findings are also applicable to income classes, although some groups (e.g., retirees) have spending characteristics not related to current income.

increase in the proportion spent indirectly through the purchase of other goods and services. It is true that higher income groups have a slightly smaller proportion of all their expenditures in their energy budgets than the lower income groups, but the drop is less than many might believe. Rich and poor alike allocate nearly the same proportion of their expenditure to energy. Comprehensive energy taxes, therefore, will absorb a similar proportion of the expenditures of the rich as of the poor. Of course, as with any proportional tax, the absolute burden will weigh less heavily on the higher income groups.

Energy consumption is more varied across regions than across income classes. Differences in weather patterns, travel distances, and industrial composition can produce dramatic differences in the level and type of energy used in different parts of the country. In 1971 the average dollar of output required 124,000 Btus of total energy per year in Louisiana but only 25,000 Btus per year in Connecticut.[1] As one might expect, the type of energy used exhibits even greater geographic variability. Thus, most of the imported residual fuel oil is consumed in New England, where it was attractive for environmental reasons but where the increased price of oil imports in 1973 presented an especially acute problem.

Fuel mix characteristics and sectoral patterns of energy use further demonstrate the heterogeneous nature of energy consumption. As Table 2-1 shows, the industrial and electricity sectors virtually exhaust the consumption of coal, while the transportation sector dominates petroleum consumption. (Gasoline, in turn, accounts for over two-thirds of the petroleum used in transportation.)

The diversity of energy consumption patterns is even more pronounced if we look within the industrial sector. The range of energy intensity varies by a factor of ten across the industries in a ninety sector industrial classification. Representative examples of industries with high and low energy intensities are shown in Table 2-2. Clearly a change in the composition of final demand for goods and services could have a significant impact on the total demand for energy or the demand for any individual fuel.

Despite its statistical convenience, a concentration on equivalent heat content in measuring total energy demand disguises the diverse characteristics of the many forms of energy. Natural gas is a clean-burning fuel, but it is difficult to deliver except through a network of pipelines. Electricity is versatile and has a variety of specialized uses,

---

1. C. Starr and S. Field, "Energy Use Proficiency: The Validity of International Comparisons," *Energy Systems and Policy* 2, no. 2 (1978). The figures are in terms of real 1978 dollars.

Table 2-1.   Fuel Use by Sector, 1978 (10^{15} Btus).

	Residential- Commercial	Industrial	Transportation	Total Net Energy Demand	Electric Utility[a]	Total Primary Energy Demand
Coal	0.3	3.1	—	3.4	11.5	14.9
Petroleum	7.2	6.0	18.9	32.0	4.1	36.1
Natural Gas	8.4	8.4	—	16.8	3.2	20.0
Electricity	4.2	3.8	—	7.9	-7.9	—
Hydroelectric	—	—	—	—	3.0	3.0
Nuclear	—	—	—	—	3.0	3.0
Total	20.0	21.5	18.9	60.2	17.0	77.1

Source: Estimated based on data from the Department of Energy *Monthly Energy Review* (MER), March 1979. The growth rates implicit in the MER data are applied to the data according to the accounting conventions used in the OECD *Energy Balances of OECD Countries, 1974-76.* (Paris: OECD, 1978). The OECD accounting conventions differ from those used in the Department of Energy or the Bureau of Mines, although the total, primary energy figures are comparable. To maintain consistency with later comparisons across countries, the primary figures are used throughout. See the supporting paper by W. Hogan, "Dimensions of Energy Demand," *Selected Studies on Energy: Background Papers for Energy: The Next Twenty Years* (Cambridge, Mass.: Ballinger, forthcoming) for further details. The commercial sector use was approximately 40 percent of the total residential and commercial consumption.

[a]The entries for coal, petroleum, natural gas, hydroelectricity, and nuclear in this column represent primary energy inputs into electric generation. The negative entry for the electricity line represents electricity consumed by end use sectors. The sum of the column thus represents losses in electricity generation, transmission, and distribution.

but it is an expensive energy form for many purposes. Coal is an inexpensive fuel, but it is dirty when burned, and there are handling problems that restrict its use to large-scale facilities. Oil is a good transportation fuel, and it can be used for many other needs, but at current prices, it is too expensive for use in large boilers. Each of these fuels has been applied according to its specialized characteristics, and we have a complex balance of supply and demand as a result.

The variations we have observed over type and place suggest that the patterns of energy use may be varied over time as well. The long-run flexibility of energy use, which we shall see is the key to understanding the impacts of future energy scarcity, may show a similar high degree of diversity. As we change the prices of different energy forms and change these prices in different ways, we may expect the mix of activities in the economy and the demand for energy in any one of these activities to change in a great variety of ways. Given the diversity of energy uses, and looking to a future of higher energy prices, the historical link between the growth of energy and the

Table 2-2. Energy Intensities of Selected Industries (1967 input-output table).

Industry	*Primary Energy Intensity (1,000 Btus per dollar of GNP, in 1978 dollars)*
Water Transportation	117
Primary Iron and Steel	115
Chemical and Fertilizer Mining	103
Plastics	100
Air Transportation	99
Office and Computing Machines	19
Agricultural, Forestry, and Fishery Services	17
Radio and Television Broadcasting	14
Finance and Insurance	12
Communications	9
U.S. Average	38

Source: C. Bullard, B. Hannon, and R. Herendeen, *Energy Flow Through the United States Economy* (Urbana: University of Illinois Center for Advanced Computation, 1975). This study is a ninety sector aggregation of the 1967 U.S. input-output table converted to the flow of embodied energy. The data were updated from 1967 to 1978 dollars using the GNP deflator. The U.S. average is from Bureau of Mines data on energy, which differs by 10 percent from the data of Bullard et al.

growth of the economy is not convincing evidence that no changes in energy use patterns are possible. We have observed changes in the *composition* of energy demand historically; perhaps with higher prices there will be changes also in the *level* of energy demand.

FLEXIBILITY IN ENERGY DEMAND

Despite the strong link between energy and the economy suggested by Figure 2-1, there is evidence of a substantial potential flexibility in energy use per unit of output. This evidence is found through both the examination of the aggregate statistics of the past and the evaluation of the many energy-using technologies available now or likely to be in use in the future. Perhaps the most striking and most popular evidence of the inherent flexibility in energy demand is found in the comparison of the energy use patterns of different countries. For example, the now familiar contrast between Sweden and the United States, countries with similar standards of living but very different levels of energy demand, is prima facie evidence that the level of economic activity does not by itself determine the level of energy demand. Consider Figure 2-3, which is typical of the comparisons

Energy consumption (in quads)
per million dollars of GDP

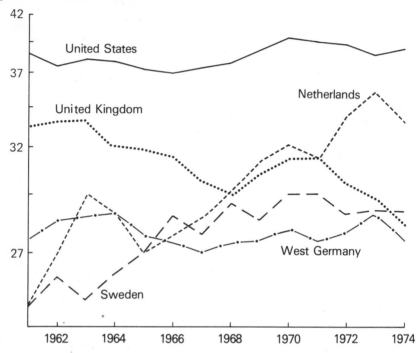

Figure 2-3. Energy-Output Ratios for Five Selected Countries, 1961–1974.

Source: From J. Darmstadter, J. Dunkerley, and J. Alterman, *How Industrial Societies Use Energy: A Comparative Analysis* (Baltimore: Johns Hopkins University Press for Resources for the Future, 1977). The data were converted to Btus per dollar of 1978 GDP by using 0.04 quadrillion (10^{15}) Btus per million tons of oil equivalent (see Darmstadter et al. p. 15) and the U.S. GNP deflator. See also L. Schipper and A. Lichtenberg, "Efficient Energy Use and Well Being: The Swedish Example," *Science* (December 1976). 194

that can be drawn with international data. The United States has a substantially higher energy use per dollar of output than other industrialized countries—more than one-third higher than Sweden. Such simple statistics are found at the core of many of the charges that the United States is profligate in its energy uses; furthermore, these data are often used to call for great changes in the level and composition of energy use without sacrificing in any of the more fundamental dimensions of economic welfare.

These international comparisons can be misleading, of course, because the countries differ in numerous ways that might affect the level of energy demand, and the experience in one country might not be transferable to another. This controversy, and the increased importance of energy, stimulated a careful examination of the international data, and the results indicate that the truth falls somewhere in between the possible extreme views.[2] On the average, approximately 40 percent of the difference in energy use between the United States and other countries can be attributed to the difference in the mix of activities in the different economies. Compared with other countries, the United States has a larger proportion of expenditures for energy-intensive activities, chiefly energy purchases for household operations and gasoline for private automobile transportation. The other 60 percent of the difference is attributed to other factors—principally differences in the efficiency of energy use found in different countries. It is especially this latter component of the lower energy use that might be transferable across countries. If we take Sweden, which uses 75 percent of the energy per dollar of gross domestic product as the United States, as the typical case, this would imply that the United States has the flexibility to improve its overall energy intensity by at least as much as 15 percent through efficiency improvements.

One explanation of the cause of these energy demand differences across countries is often found in the sharp differences in energy prices faced by consumers in the United States compared to consumers in other countries. The evidence is most striking in the prices paid for gasoline. Figure 2-4 displays the variation across many countries in the price paid for gasoline (with the variation in price due chiefly to differences in tax policies) and the quantity of gasoline used per dollar of economic output. The level of gasoline demand seems to be significantly affected by the price the consumer must pay. When the price is low, there is little incentive to conserve on gasoline, and consumption is high. But when the price is high, the consumer adjusts by using more efficient cars or switching to more efficient forms of transportation. A similar pattern exists in other sectors, although the more heterogeneous nature of the energy products makes the energy consumption profile more difficult to summarize with a single index.

An examination of automobile efficiencies reinforces the evidence for the existence of a substantial potential flexibility in energy use patterns. It is a commonplace that the average car in the United States auto fleet consumes much more gasoline per passenger mile

2. J. Darmstadter, J. Dunkerley, and J. Alterman, *How Industrial Societies Use Energy: A Comparative Analysis* (Baltimore: Johns Hopkins University Press for Resources for the Future, 1977).

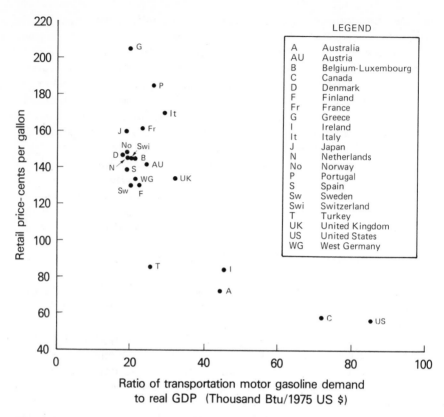

Figure 2-4.   Cross-country Comparison of Motor Gasoline Demand in 1975.

Source: Data are from the *Economic Report of the President* (Washington, D.C.: Government Printing Office, 1978). The conversions to 1975 U.S. dollars are apparently based on the then prevailing exchange rates, not purchasing power priorities as used elsewhere in this chapter. Price is not the only explanation for the consumption differences; for example, the highest consumption is in countries with longest travel distances.

than the counterpart in other countries. This fact, and the central importance of automobile use in determining the demand for petroleum, prompted the U.S. Congress to legislate strict standards for automobile efficiency in an effort to reduce the dependence on imported petroleum. Starting at less than 15 miles per gallon in 1975, the average new car efficiency is required to improve to 27.5 miles per gallon by 1985, and most analysts agree that this standard will be met. Although the existence of the efficiency standards may substantially weaken any potential further response to gasoline price increases,[3] there are indications that the technological conservation potential for the automobile is not exhausted by the current standards; as much as a 40 percent further increase in automobile efficiency is feasible and might be economically attractive.[4]

To take another example, a careful analysis of household refrigerator-freezer combinations indicates the potential for as much as a 50 percent reduction in energy use, depending on the initial cost increase that the consumer is willing to absorb[5] (see Figure 2-5). Of course, it is difficult to tell what will be required to induce consumers to adopt any particular technology, and there must be a gradual adjustment as old equipment is replaced, but the evidence is accumulating that there is substantial flexibility in energy use. The most comprehensive and credible attempt at an analysis of individual technologies envisions the possibility—over a forty-year time span—of as much as a 40 percent improvement in the energy intensity for the entire economy, depending on future increases in the price of energy to consumers.[6]

Chapter 3 of this volume takes this subject further, examining individual energy-using technologies and the conservation problems and opportunities inherent in past legislation and existing institutions. This selective inquiry into the flexibility of energy use in a few specific applications suggests that the demand for energy can often be reduced without substantially impairing productivity for many processes that use energy as one among many inputs. A section of Chapter 4 summarizes the economic role of energy in the aggregate and indicates that, in a climate of rising energy prices, the ability to adjust energy use patterns should permit the growth in energy demand to be substantially decoupled from the growth in the economy. In the next section, we anticipate these themes by further quantifying the flexibility in energy demand and examining the implications, in the large view, for energy demand growth over the next two decades.

3. J. Sweeney, "U.S. Gasoline Demand: An Economic Analysis of the EPCA New Car Efficiency Standard," in R.S. Pindyck, ed., *Advances in the Economics of Energy and Resources* (Greenwich, Conn.: JAI Press, 1979).

4. CONAES Demand and Conservation Panel, "U.S. Energy Demand: Some Low-Energy Futures," *Science* 200, no. 4338 (April 14, 1978): 142-52.

5. R. Hoskins and E. Hirst, *Energy and Cost Analysis of Residential Refrigerators* (Oak Ridge, Tenn.: Oak Ridge National Laboratory, January 1977). Reviewing a draft of this chapter S. Peck observed that moving from the least to the most energy-efficient combination saves 8,000 Btus per day. At $3 per million Btus, this is a savings of $10 per year achieved for an initial investment of $80 or a 12 percent rate of return on a ten-year life cycle.

6. CONAES Demand and Conservation Panel.

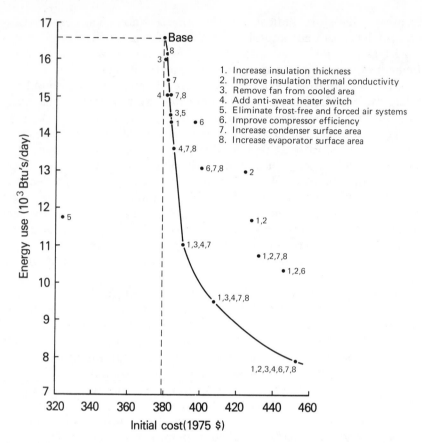

Figure 2-5.   Energy Use versus Retail Price for Various Design Changes for a 16 cu. ft. Top Freezer Refrigerator.

Source: R. Hoskins and E. Hirst, *Energy and Cost Analysis of Residential Refrigerators* (Oak Ridge, Tenn.: Oak Ridge National Laboratory, January 1977).

ESTIMATING SUBSTITUTION POTENTIAL

The starting point of our aggregate analysis of energy demand is to treat energy as an economic good; in particular, energy is assumed to be only one input to production processes, only one item used to meet the needs of users. This assumption is motivated by the specific examples of the trade-offs between energy and other goods—for example, between energy and capital investment in refrigerators—and from the basic common-sense judgment that our current demands for energy would not persist under higher price regimes. The profit-maximizing firm and the cost-conscious energy consumer will consider the prices of all inputs when making decisions about the mix of goods and services to be utilized. As the price of energy increases, we assume that energy users exploit the opportunities to substitute more energy-efficient machinery, to redesign production processes, to insulate homes, to install heat pumps, to switch to buses and subways—even to lower thermostat settings and to observe speed limits. The energy user makes these and many other substitutions to reduce the cost of meeting the more fundamental objective of producing a product or maintaining an acceptable standard of living.

The degree to which these adjustments will be made depends on many factors. These are, of course, technical constraints, but the actual substitutions may be motivated as much by convenience and personal taste as by the engineering characteristics of specific energy uses. For example, the existence of hand-operated can openers does not mean that cooks will (or should) abandon the electric counterpart. The measurement of the energy substitution potential, therefore, must be based on the observed behavior of energy users confronted with different prices and with the opportunity to make consumption choices. The result, if not the process, can be summarized by the proportional reduction in energy use, for a fixed level of activity, in response to a given percentage increase in the energy price. The ratio is the elasticity of energy demand (a convenient statistic that we will use extensively); hence, if the elasticity of demand is 0.5, a 10 percent increase in the price of energy produces a 5 percent decrease in energy demand, a decrease achieved by substituting a collection of other inputs and processes for the energy forgone.

The aggregate elasticity of energy demand will be useful in summarizing the substitution potential, but it is not useful for actually estimating the degree of potential substitution. As we have seen, energy is not a single commodity, and the full richness of the energy system must be considered when estimating the elasticity of energy demand. Only the most general linkage can be established between the abstraction of an aggregate elasticity of total energy demand for all uses and the disaggregated estimates of the many adjustments, made by different users, confronted with different prices, for different energy products.

One of the important ways in which the links between the aggregate elasticity and the elasticities for individual sectors or fuels are obscured involves the diversity of prices that must be considered. The price to the user is the relevant price in determining the substitution among the many energy and nonenergy inputs for production or consumption. We often speak of the price of energy in terms of the price of primary energy or, more narrowly, in terms of the price of imported oil. But no consumer actually purchases energy at the price of production, and the prices to consumers do not vary in constant proportion with the price of primary energy. Hence the same absolute price change and the same demand response will lead to different estimates of the price elasticity depending on the point of measurement selected. For example, in 1972 the price of oil at the wellhead accounted for slightly more than one-fifth of the cost of gasoline at the pump. Hence, assuming a straight pass-through of increased crude oil costs and no change in refinery operations, the elasticity of demand measured at the pump would be five times the elasticity measured at the wellhead.

Despite, or perhaps because of, these difficulties of measurement and interpretation, there have been many different attempts to examine the data and to estimate the elasticities of energy demand. Many factors, such as weather, population, and economic activity, must be controlled to isolate the effects of price alone, and the short-run effects must be separated from the long-run adjustments that may take many years to complete. In addition, the data are not perfect, and there is no guarantee that the changes of the past, over one range of prices, will be indicative of the changes in the future, over a very different range of prices. The results are not without controversy, therefore, but some patterns do emerge.

One survey of a number of different studies compared the long-run elasticities of demand for different consuming sectors, as shown in Table 2-3. These figures indicate a substantial elasticity of energy demand—that is, a significant flexibility to substitute for energy use. For example, the lowest estimate of the elasticity of demand for

Table 2-3. Estimates of Long-run Elasticity of Energy Demand Delivered to the Consumer.

Sector	Range of Studies	Most Likely Range
Residential	0.28 to 1.10	0.7 to 1.1
Industrial	0.49 to 0.90	0.8 to 0.9
Transportation (Motor gasoline)	0.22 to 1.3	0.7 to 1.0

Source: R. Pindyck, "The Characteristics of Energy Demand," in J. Sawhill, ed., Energy Conservation and Public Policy (Englewood Cliffs, N.J.: Prentice-Hall, 1979). The survey by Pindyck covered several detailed econometric studies. The range of estimates was narrowed to the most likely range based on Pindyck's qualitative discussion of the results.

motor gasoline is consistent with the data in Figure 2-4 where consumption and prices were compared across countries.

These aggregate elasticities are roughly consistent with the potential reductions in energy use found in detailed technological analyses. The implication is that a doubling in the real cost of energy delivered to the consumer could produce up to a 50 percent reduction in the intensity of energy use. Of course, this would require more than a doubling of the cost of primary energy, but this represents a substantial flexibility in the long-run demand for energy. And these aggregate figures disguise even greater adjustments in the demands for individual fuels; the increase in the price of any individual fuel will cause a shift away from that fuel, in addition to a reduction in the total energy demand. The evaluation of this more volatile shift in the mix of fuel demands requires a more complete specification of the effects of the changes in energy prices; these fuel mix adjustments are discussed in more detail in a later section.

DEMAND FORECASTS

To forecast energy consumption, analysts must simultaneously determine the cost and the availability of alternative forms of energy supply and evaluate a host of other factors that will determine the patterns of energy consumption. Our way of life, the design of our cities, the level and composition of economic activity, the stock of energy-using equipment, the changing nature of energy-using technologies—all will contribute to determining the level of future energy demand. The complete details are required for many analytical purposes, such as evaluating the potential market for some new energy-using technologies, but more forecasts are based on one or another model that approximates the energy system by submerging many of the details in an aggregate analysis.

Many such forecasts of energy demand have been prepared, often through the use of simplifying assumptions that treat the major uncertainties by examining a number of possible energy futures. Table 2-4 shows a sampling of these forecasts.[7] There has been since 1973 a steady decrease in forecast demand as well as a gradual convergence of the estimates. There is a similar convergence in the associated estimates of oil and gas imports. Important differences remain, however, and the differences become more pronounced when con-

7. For those interested in a larger portfolio of forecasts, the International Energy Agency has compiled seventy-eight major forecasts prepared over the last ten years; J.R. Brodman and R.E. Hamilton, "A Comparison of Energy Projections to 1985" (Paris: IEA, January 1979).

Table 2-4. Selected Forecasts: Total Primary Energy Demand and Imports (10^{15} Btus).

Study	1985 Demand	1985 Imports	1990 Demand	1990 Imports	2000 Demand	2000 Imports
Project Independence,[a] 1974	102.9-109.1	6.6-24.8	—	—	—	—
FEA National Energy Outlook,[a] 1976	90.7-105.6	11.8-25.2	114.0-121.7	11.6-41.4	—	—
A Time to Choose:[a]						
Historical Growth	116.1	—	—	—	186.7	—
Technical Fix	91.3	—	—	—	124.0	—
Zero Energy Growth	88.1	—	—	—	100.0	—
USDI, Energy Through the Year 2000:[a]						
1972 Forecast	116.6	—	—	—	191.9	—
1975 Forecast	103.5	—	—	—	163.4	—
Shell:						
1976 Forecast[b]	96.2	24.0	110.0	22.0	—	—
1978 Forecast[c]	90.4	22.6	101.2	22.6	—	—
Exxon,[d] 1977	93.0	27.0	108.0	26.0	—	—
National Energy Plan,[e] 1977	97.0	23.0	—	—	—	—
CIA,[f] 1977	98.6	26.8	—	—	—	—
CONAES MGR,[g] 1977						
DESOM	—	—	112 -114	—	—	—
ETA	—	—	103.1-104.8	—	—	—

NORDHAUS	—	—	80.5-96.2	—	—
SRI			108.9		—
CRS,[a] 1977	91.2-98.4	19.7-38.9	104.5-113.4	20.5-45.2	94-136(2010)
CONAES DEMAND, 1978[h]	—				—
MOPPS, 1978[i]	94.6	15.3	—	—	117.3
EIA, 1978[j]	91.2-96.9	21.4-22.9	100.7-109.4	24.1-29.7	—

NORDHAUS — — 80.5-96.2 — — —
SRI 108.9
CRS,[a] 1977 91.2-98.4 19.7-38.9 104.5-113.4 20.5-45.2 94-136(2010) —
CONAES DEMAND, 1978[h] — — — — —
MOPPS, 1978[i] 94.6 15.3 — — 117.3 12.5
EIA, 1978[j] 91.2-96.9 21.4-22.9 100.7-109.4 24.1-29.7 —

[a]Extracted from *Project Interdependence: U.S. and World Energy Outlook Through 1990* (Congressional Research Service, Committee Print 95-33, November 1977), p. 125.

[b]From *National Energy Outlook*, Shell Oil Company, September 1976.

[c]From *National Energy Outlook*, Shell Oil Company, February 1978.

[d]From *Energy Outlook: 1977-1990* (New York: Exxon Corporation, August 1977).

[e]From *National Energy Plan*, White House, April 1977, p. 96.

[f]From *The International Energy Situation's Outlook to 1985* ER77-102404 (Washington, D.C.: Central Intelligence Agency, April 1977).

[g]From Modeling Resources Group, Committee on Nuclear and Alternative Energy Systems (CONAES), *Energy Modeling for an Uncertain Future* (Washington, D.C.: National Research Council, 1978) (for selected models).

[h]From CONAES Demand and Conservation Panel, "U.S. Energy Demand: Some Low-Energy Futures," *Science* 200, no. 4338 (April 14, 1978): 142-52. The scenarios related here are Scenarios III and IV, which range from no price increases to a doubling of prices by 2010. The panel also examined two cases where prices to consumers had quadrupled by 2010. In these cases, projected energy demand fell to a range of 60 to 75 10^{15} Btus.

[i]From U.S. Department of Energy, "Market-Oriented Program Planning Study" (draft; Washington, D.C.: DOE, 1978).

[j]From Energy Information Administration, U.S. Department of Energy, *Annual Report to Congress*, DOE/EIA-0036/2, Vol. 2, 1977 (Washington, D.C.: DOE, released April 1978).

sidering not only the level but the composition and rate of change of U.S. energy demand.

For the present study, there is no need to prepare yet another detailed assessment of the likely future composition and level of energy demand. As Table 2-4 shows, analysts have given this subject extensive attention. We do see a pressing need to distill a qualitative understanding of the character of future energy demand and the sensitivity to variations in key uncertain parameters and assumptions.

The specification of the growth in the population, changes in the labor force, and the aggregate output of the economy must be the starting point of any projection of energy demand levels. That is, notwithstanding the great potential flexibility of energy use, the growth of the economy will be a key factor in determining the demand for energy at any given level of energy prices. This direction of effect, from the economy to the energy sector, is particularly important during the short run: witness the experience since 1973. Energy demand grew by only 0.5 percent per year between 1973 and 1977, as compared to 3.1 percent per year for the years preceding 1973. This deviation from the historical growth rate was caused equally by a reduction in the use of energy per unit of gross national product and by a reduction in the growth rate of the economy. Without increases in energy prices, there might be only limited incentive to substitute other inputs to save energy, and the future growth of energy demand would be determined in large measure by the future growth of the economy. The importance of the economic growth assumptions can be seen if one were to examine the underlying documentation for the forecasts collected in Table 2-4; a great part of the differences between the energy demand forecasts can be traced to different assumptions about the level of future economic growth.

The projection of aggregate economic activity is difficult enough; the translation of economic forecasts to energy demand forecasts is complicated further by the changes in the composition of output, as we have seen. This issue has been addressed, for example, by Hudson and Jorgenson through the application of their energy-economic model, considering the role of energy and the impacts of the changing composition of economic output. This analysis indicates that an increase in energy prices could change the intensity of aggregate energy use through a shift in the composition of output, from the energy-intensive manufacturing sector and toward the less energy-intensive services and communications sectors.[8]

8. E. Hudson and D. Jorgenson, "The Economic Impact of Policies to Reduce U.S. Energy Growth," *Resources and Energy* 1, no. 3 (November 1978): 205-29.

The regulation of energy prices and energy-using technologies introduces additional uncertainties into any demand forecasts. The administrative control of prices prevents the consumer from recognizing the true scarcity of energy, which may distort and increase energy demand. Conversely, as we have seen in the case of the automobile, the regulation of energy use can reduce both energy demand and the future responsiveness to energy price changes. These countervailing effects of regulation must be recognized to avoid one type of double counting—forecasting a decrease of the same energy demand once because of higher prices and then again because of efficiency regulations.

In any case, the adjustments in energy demand in response to increases in the price of energy will not occur instantly. In the long run we may have a great deal of flexibility to substitute labor, materials, or new processes for energy, but in the short run the energy intensity of the economy is more narrowly restricted. Consumers are not likely to replace or retrofit all energy-using devices as soon as they see higher energy prices. The more likely response is a gradual turnover of capital stock, with a slow adaptation to the optimum, long-run level of energy demand. In fact, the rate of adjustment to the long-run levels of energy use is probably slow enough for the price changes that have already occurred to have an impact on energy demand for the next twenty years—that is, the energy system was not in long-run equilibrium in 1978. To the extent that improvements in energy efficiency occur primarily in conjunction with the purchase of new equipment, then the slowness of the turnover of the capital stock would guarantee that some residual part of the energy efficiency even at the turn of the twenty-first century will still be determined by the composition of the capital stock today.

These uncertainties, and the slow processes of adaptation in the energy system, imply that the next twenty years will be very much a period of adjustment in energy demand. Let us focus on those adjustments.

ENERGY DEMAND ADJUSTMENTS

For many years real energy prices declined and energy demand grew at a relatively constant rate, but in 1973 the sharp increase of oil prices marked the beginning of what may be a long period of change and adjustment in the level and composition of energy demand. Reinforced by the presence of environmental problems and complicated by a high rate of general inflation, these adjustments were still very much under way more than five years later. Any investigation of future energy demand must consider both the continuing adaptation to past changes—for example, in the price of energy—and the effects of future rates of increase of energy prices and economic activity.

Our exploration of these adjustments will be improved by refining our measure of aggregate energy consumption. We need a yardstick that can be related easily to the data and yet interpreted as an index of the final consumption level of the energy consumer. The difficulties of developing such an index are apparent. It must provide a consistent aggregation of the diverse energy products, ranging from the coal consumed in industry at prices as low as $0.80 per million Btus to the electricity consumed in the home at prices as high as $10 per million Btus delivered. It is not possible to specify a perfect index, but it is possible to construct a good approximation by carefully accounting for the different prices paid for different energy products.

The aggregation employed here is essentially the one developed by Pindyck[9] for an international comparison of energy demands. The concept and quantity employed for each year and sector are the same as the net energy consumed—that is, the equivalent heat content, delivered to the consumer, net of the energy lost in generating and distributing electricity. This follows the long practice of the Bureau of Mines as continued by the Department of Energy.[10] The prices are those paid by consumers (as opposed to wellhead or primary energy prices), aggregated across sectors, weighted according to the value shares of expenditures. One advantage of measuring at this point is that it is the best approximation to the point where energy choices are actually made by the consumer. A disadvantage is that the resulting energy demand total is not the same as the total primary energy demand that is the focal point of so many energy forecasts, such as those in Table 2-4. The energy losses in electricity

---

9. R. Pindyck, *The Structure of World Energy Demand*, (Cambridge, Mass.: Massachusetts Institute of Technology, September 1978). The data are taken primarily from the extensive data base assembled by Pindyck, and we used his translog aggregation equations to construct the energy price variable for each major sector. The aggregation across sectors was accomplished using a Cobb-Douglas price function with the 1972 value shares as weights. For further details, see the supporting paper, by W. Hogan, "Dimensions of Energy Demand," in *Selected Studies on Energy: Background Papers for Energy: The Next Twenty Years* (Cambridge, Mass.: Ballinger, forthcoming).

10. Energy Information Administration, *Annual Report to Congress*, DOE/EIA/0036/2 (Washington, D.C.: Government Printing Office, U.S. Department of Energy, 1978). Readers interested in a cogent discussion of alternative aggregation conventions will enjoy E.R. Berndt, "Aggregate Energy, Efficiency, and Productivity Measurement," *Annual Review of Energy*, 1978.

generation would have to be added to the net demand to obtain the equivalent primary energy demand.[11]

If the scale of economic activity is assumed not to affect the composition of energy demand—such as by precluding the possibility of major structural shifts in the economy—then the major uncertainty associated with economic forecasts can be avoided in the analysis by concentrating on the ratio between the delivered net energy demand and the gross domestic product (GDP), hereafter referred to as the net energy GDP ratio.[12] The data for the United States and selected industrialized nations are summarized in Figure 2-6, which can be compared with the similar data for primary energy shown in Figure 2-3; qualitatively, the two sets of data tell the same story. The ratio between net energy and GDP for the United States is higher than that for most other countries. However, it has dropped steadily, from 33,400 Btus per 1978 dollar of output in 1972 to 29,100 Btus per 1978 dollar of output in 1978. The data from the other countries do not show quite so regular a pattern, but there is a general, if slight, reduction in the overall intensity of energy use. Part of the explanation for this improvement in energy efficiency can be found in the trends in energy prices. The aggregate energy prices for the consumers in each of these countries are displayed in Figure 2-7. There has been a significant real increase since 1972 in the price of energy, although the increase is much less (even after adjusting for the big rise in overall inflation) than many would expect given the dramatic changes in the prices of some important energy components. An example of such changes is oil in the United States, which increased by 120 percent between 1972 and 1978. The difference can be explained by noting that oil is only one component of aggregate energy input; many products are still sold under long-term contracts at low prices. And the costs of such primary energy forms as oil, coal, and gas are only a part of the price of delivered energy. Hence, the nearly threefold multiplication of imported oil prices since 1972 has led to only a 30 percent increase in the delivered price of all energy in the United States.

11. In 1978, electricity accounted for less than 15 percent of the distributed net energy, but was growing slowly. With a 30 percent efficiency of generation, a 15 percent elasticity would imply a primary energy demand of 1.5 times the net energy demand.

12. The international comparisons are expressed in terms of the gross domestic product to remove the net factor income originating in overseas enterprises and investments, which is included in GNP. The difference between GNP and GDP is quantitatively small for the United States; in 1977 the difference amounted to less than 1 percent of GNP. For convenience, the U.S. figures for GNP are used as the estimate for U.S. GDP.

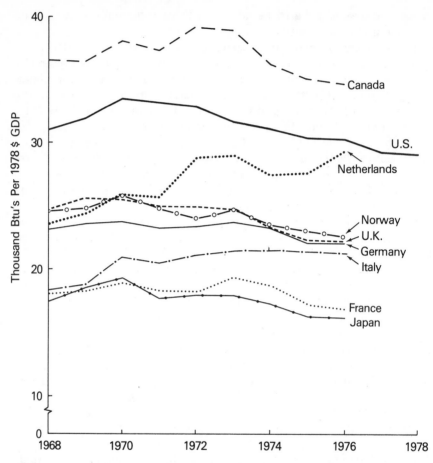

Figure 2-6.   Net Energy GDP Ratio: Selected Industrial Countries.

Source: Quantity data are from OECD *Energy Balances.* For further details, see the supporting paper by W. Hogan, "Dimensions of Energy Demand," in *Selected Studies on Energy: Background Papers for Energy: The Next Twenty Years* (Cambridge, Mass.: Ballinger, forth coming).

The uncertainties associated with future changes in energy prices can be illustrated by projecting the net energy GDP ratio using a number of different price assumptions. We examine primarily the data for the United States, from which we may draw analogies for the other industrialized nations. The construction of the aggregate price index involves several steps to integrate the prices of many fuels across many sectors, but the results can be summarized in terms of

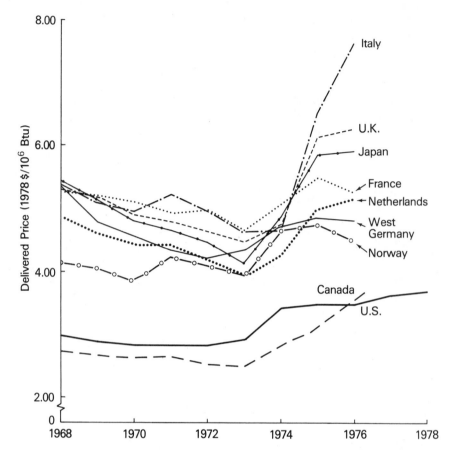

Figure 2-7. Price of Energy Delivered: Selected Industrial Countries.

Source: Price data are primarily from R. Pindyck, *The Structure of World Energy Demand* (Cambridge, Mass.: Massachusetts Institute of Technology, September 1978). See W. Hogan, "Dimensions of Energy Demand," in *Selected Studies on Energy: Background Papers for Energy: The Next Twenty Years* (Cambridge, Mass.: Ballinger, forthcoming), for a description of data sources and updates.

the relationship between the price of delivered energy and the price of primary energy. Let us assume the following:

1. The average price of coal (in 1978 dollars) gradually approaches $1.50 per million Btus, up from $.95 in 1978, and is then unaffected by the price of oil.
2. The price of gas is equated to the price of oil, gradually approaching $4.50.

3. The markups over the primary energy costs remain constant in real terms.

After all the long-run substitutions are made, the delivered price of energy will be as in Figure 2–8.

Let us also assume that the change in energy efficiency takes place only with the introduction of new energy-using equipment—for example, the efficiency of new cars increases, but auto owners do not drive less. The relationship over time between delivered energy prices

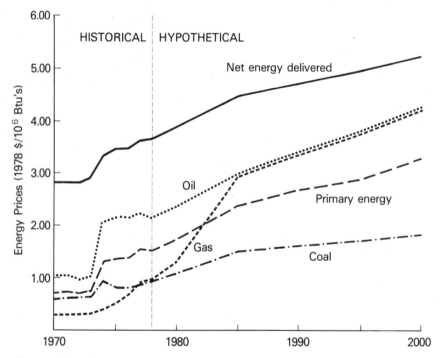

Figure 2–8. Illustration of Relation between Primary Energy Prices and Net Energy Delivered Prices, in the United States.

Source: The price data are from R. Pindyck, *The Structure of World Energy Demand* (Cambridge, Mass.: Massachusetts Institute of Technology, September 1978), and his aggregation equations are used to construct the sector long-term price indexes, assuming that the commercial sector pays the same prices as the residential sector. The sectors are combined in a Cobb-Douglas aggregation to produce the delivered price for all energy. See the supporting paper, "Dimensions of Energy Demand," by W. Hogan, in *Selected Studies on Energy: Background Papers for Energy: The Next Twenty Years* (Cambridge, Mass.: Ballinger, forthcoming).

and the net energy–GDP ratio can then be established, once we specify the long-run price elasticity and the rate of introduction of new equipment. The result yields a plot similar to a traditional demand curve and reflects a degree of price responsiveness that increases as we look further into the future.

Consider Figure 2-9, showing a family of demand curves based on

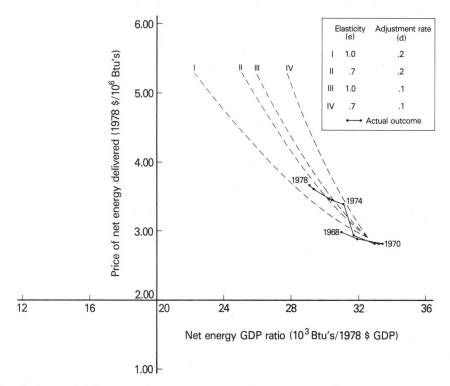

Figure 2-9. U.S. Net Energy Demand Curves for 1978: Selected Elasticities and Adjustment Rates.

Note: See accompanying text for explanation of this graph.

Source: The data for 1968-1978 are from R. Pindyck, *The Structure of World Energy Demand* (Cambridge, Mass.: Massachusetts Institute of Technology, September 1978), and the OECD. The demand curves are derived from a model with a Koyck lag adjustment process for the energy-GDP ratio. It is assumed that the prices change along the whole price path, between 1969 and 1978, by scaling the actual price path to match the proportioned change in 1978. For further details see the supporting paper, by W. Hogan, "Dimensions of Energy Demand," in *Selected Studies on Energy: Background Papers for Energy: The Next Twenty Years* (Cambridge, Mass.: Ballinger, forthcoming).

data for the United States. Four curves are shown, one for each combination of two possible long-run elasticities (e) and two possible rates of introduction of new equipment (d). The price elasticities were selected to be compatible with the range of estimates from Table 2-3 but to examine both high and low values ($e = 0.7; e = 1.0$). Recall that these are the elasticities in terms of delivered, not primary, energy. The selection of the rate of introduction of new energy-using equipment is more problematical; two values are considered here—$d = 0.10$ and $d = 0.20$. An introduction rate of 10 percent—that is, $d = 0.10$—implies that in any year, 10 percent of all energy-using equipment is newly installed and is assumed to be at the optimum energy efficiency at the price in that year. An economic growth rate of 3 percent per year and an average life of equipment of fifteen years would be consistent with a rate of introduction of new equipment of 0.10; if the average life were six years, the rate would be 0.20.

The demand curve in Figure 2-9 for each combination of demand elasticity and introduction rate is based on a forecast from 1969 through 1978 and can be compared with the actual outcomes (represented by the solid line) for the intervening years. For example, in 1978 the ratio between net energy and GDP was 29.1, down from the 1969-1970 average of 32.7, and the predicted value for the high elasticity, slow adjustment case ($e = 1.0; d = 0.10$) is 29.9.

At least two lessons can be drawn from Figure 2-9. First, there has been a notable reduction in the net energy GDP ratio for the United States, consistent with our expectations given the rise in energy prices. Second, however, it is too early to choose confidently, on the basis of these data alone, between the cases of a low price elasticity with a quick adjustment and a high price elasticity with a slow adjustment. (Of course, even the low price elasticity is higher than some of the estimates summarized in Table 2-3.) This short-term evidence, despite its many limitations, supports the view gleaned from Table 2-3 and the studies of the long-run substitutability of energy. There is a substantial potential to change the intensity of energy use, and we are going through a period of adjustment to new patterns of energy consumption; further reductions in energy intensity can be expected, even if the final outcome in terms of the level of demand is in doubt.

If we focus on these two cases, high and low elasticity, we can use the same method to examine the likely range of energy demand over the next twenty years. By the year 2000, the adjustment to the recent price increases may be nearer completion, but further increases in prices are widely anticipated; hence, we may see a process of continuous, if gradual, change with far different future net energy GDP ratios. Figure 2-10 summarizes the two demand curves for the year 2000 based on the parameters for the two cases that best fit the data through 1978. (Again, the solid line shows actual recent data.) Both curves imply that there could be a substantial further reduction in the demand for energy. For example, using our earlier assumptions, as in Figure 2-8, if the price of oil increased to $4.50 per million Btus, the price of net delivered energy in the aggregate would rise to approximately $5.50 per million Btus. According to Figure 2-9, the corresponding ratio between net energy and GDP could be anywhere from 19 to 22; but in either case, it would be more than 25 percent below the level of 1978. This would be a substantial reduction in the energy intensity of the economy, consistent with the lower forecasts in Table 2-4.

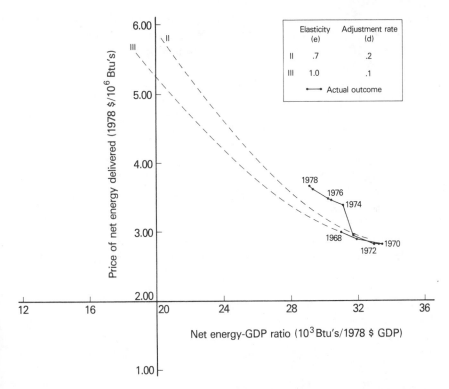

Figure 2-10. U.S. Net Energy Demand Curves for 2000: Selected Elasticities and Adjustment Rates.

Source: The method applied for Figure 2-8 was applied for a forecast through the year 2000; see note for Figure 2-9.

These figures can be converted to total primary energy demand, for comparison with other studies, by specifying (1) the degree of electrification (to account for the 70 percent of primary energy lost during conversion and distribution of electricity) and (2) the growth in the economy. In 1978, electricity accounted for less than 15 percent of the delivered energy. If there were no major increase in the degree of electrification, then the ratio of primary energy to delivered energy will be 1.35.

If the economy grows at 3 percent per year, total output will be $4 trillion ($10^{12}$) (in 1978 dollars) by the year 2000. Under these conditions, a net energy–GDP ratio of 22 implies a total primary energy demand in the year 2000 of 119 quadrillion (10^{15}) Btus (22 × 1.35 × 4 = 119). If the growth in the economy were only 2 percent per year, or $3.3 trillion by 2000, then the primary energy demand figure drops to 98 quadrillion Btus. These figures can be compared with the 1978 primary energy demand of 77 quadrillion Btus. Other projections, consistent with the range of demand curves in Figure 2-10, can be prepared in the same manner.

The implications of a sensitivity analysis for the United States, based on Figure 2-10, are that some combination of high economic growth rates, low energy prices, and an increase in the degree of electrification would be required to produce a primary energy demand in excess of 120 quadrillion Btus in the year 2000. Conversely, low economic growth, high energy prices, or a decrease in electrification would tend to yield primary energy demands for the year 2000 less than 100 quadrillion Btus.

INTERNATIONAL ENERGY DEMAND

If we extend this sensitivity analysis to energy demand in other industrialized countries, we would obtain similar results. Higher energy prices should reduce energy demand. The reduction may be smaller than for the United States because consumers in most other countries already pay higher energy prices than in the United States. This will be especially true assuming all future increases in energy prices come only from increases in primary energy costs. Because primary energy costs are a smaller proportion of delivered energy prices in other industrialized countries than in the United States, the relative increase in delivered prices, when oil prices increase, will be correspondingly less in those countries. Reinforcing this lesser price effect, the economic growth rates in the other industrialized countries are likely to be higher than the U.S. growth rate. This has been true historically; between 1961 and 1970 the U.S. economy grew at

4 percent per year in real terms, and the remainder of the OECD countries grew at 5.5 percent per year.[13] Although output in these other economies should grow more slowly in the future than in the past, their share in world economic activity and world energy demand should be increasing. But with increasing energy prices, we would expect energy demand growth to be much slower in the future than in the past.

At the higher end of our range of sensitivity tests—that is, an elasticity of one—energy prices rising only at the rate of economic growth imply consistency in the level of long-run energy demand. Of course, we do not expect delivered energy prices to increase this rapidly; there are abundant, but expensive, sources of energy supplies that should place a long-run ceiling on energy prices.

The ability to adjust to higher energy prices does not mean that higher prices are to be desired: there are real costs to be incurred when energy becomes more expensive. It only means that the flexibility in the energy system provides many means to accommodate the higher costs, minimizing the economic effects. But most of the evidence for this flexibility is drawn from analyses of data from the industrialized nations; we may be less sanguine about the outlook for the developing countries, at least for those developing countries that do not export energy. The structures of the economies in the developing countries are different from those of the industrialized nations. Accepting a single estimate for the elasticity of energy demand in all OECD countries is problematical: it would take a heroic leap to apply the same elasticity to the developing countries, and many observers argue that the elasticities are lower. For example, the limited investigation by Pindyck, who provided our estimates of the elasticities for OECD countries, indicates that the price elasticity for gasoline demand is only 0.3 to 0.5 for selected developing countries.[14]

An even more serious problem that complicates the projection of energy demand for the developing countries relates to the effect of economic growth. For the United States, we assumed implicitly that a doubling of the level of economic activity would double the demand for energy, assuming that energy prices did not change. In fact, the increase in energy demand might be somewhat less for several reasons—for example, the increase in output might come from the service industries, which are less energy-intensive. We did not pursue this argument; it would only tend to reduce further our

13. OECD *The National Accounts of OECD Countries 1974*, vol. 1 (Paris: OECD, 1976).
14. Pindyck.

relatively low estimates for the likely range of energy demand. But for the developing economies, the situation is likely to be reversed. Much of the current energy consumption in these countries is not part of the traditional energy trade, and this energy use does not appear in the statistics; wood and dung are not counted in our compendiums of quads. As the developing economies grow, and grow rapidly relative to the economies of the rest of the world, most of the demand for energy will be for the oil, gas, coal, and electricity of the industrialized world, especially for the oil. This signifies that energy demand in developing countries would be growing at rates higher than their rate of economic growth.

When we also recognize that developing economies should be growing rapidly over the next twenty years, we can conclude that the growing demand for energy in these countries will place a disproportionate demand on the world energy market, as well as on their own balance of payments. The magnitude of this problem remains highly uncertain, however. Energy availability will continue to be a problem of the first importance to the developing countries, but estimates of the feedback effects on the world energy system, particularly in terms of oil supplies and prices, are the subject of much debate. Recent projections of the demand for commercial energy for the year 2000 in the developing countries range from as low as 50 quadrillion Btus to as high as 100 quadrillion Btus, compared to the 19 quadrillion Btus consumed in 1972.[15] By any account, this is a rapid rate of growth; at the upper range of the forecasts, the LDC demands could have a significant effect on the world energy market. Ensuring the production of this energy, and financing its purchase, remain as major items on the agenda for energy policy in the international arena.

INTERFUEL SUBSTITUTION

The concentration on the analysis of aggregate energy demand is an important first step, particularly if the aggregate demand turns out to be lower than previously expected. In this event, the pressure to develop expensive new supply technologies or stiff conservation measures may not be great, and there is time to test alternatives, conduct the basic research and development, and collect information as we go.

The simple demand curve, relating net energy GDP ratios to

15. See the supporting paper by D. Bakke, "Energy in Developing Countries," in *Selected Studies on Energy: Background Papers for Energy: The Next Twenty Years* (Cambridge, Mass.: Ballinger, forthcoming).

energy prices, is useful for sensitivity studies to establish the likely range of energy demand, but it is not adequate for developing projections of the demand for individual fuels. From the perspective of U.S. national policy, the most pressing problems stem from the demand for oil and/or gas, more directly, for imported oil and gas. A reduction in the aggregate demand for energy may be beneficial for many reasons—improving the environment, for example. But the extent to which it reduces the demand for imported oil depends in part on the degree to which it is possible to substitute abundant sources of supply, such as coal, for the imported oil. The opportunities for this interfuel substitution are not revealed through the use of the aggregate demand curve; a more disaggregated model is required for this purpose. Such models are available (at the cost of increased complexity) for studying alternative energy futures.

Most econometric demand models are constructed at a level of detail that captures the major fuel substitution possibilities. These interfuel substitution opportunities, such as substituting electricity for oil in heating, are even greater than those for saving all energy, as in the use of more efficient insulation to reduce heat loss. The potential sensitivity of demand composition to changes in the prices of individual fuels can be indicated by examining the estimates of own-price elasticities. The own-price elasticity is the percentage reduction in the demand for any one fuel if its price rises by 1 percent and the prices of all other fuels remain constant. Consider the case of electricity, for example, in Table 2-5. This is a metasurvey—a survey of surveys of econometric estimates of the own-price elasticity of electricity, usually measured as delivered to the consumer. Although there is a substantial range of uncertainty in the estimates of this one parameter, the elasticity is large, and it will be important in determining the role that electricity might play in the nation's energy future.

Table 2-5. Surveys of Own-Price Elasticity for Electricity (long run).

	Residential Sector	*Industrial Sector*
Taylor[a] (8 studies)	0.90 to 2.0	1.25 to 1.94
MRG[b] (3 studies)	0.78 to 1.66	0.69 to 1.03
Pindyck[c] (4 studies)	0.3 to 1.2	0.5 to 0.92

[a]L.D. Taylor, "The Demand for Electricity: A Survey," *The Bell Journal of Economics* 6, no. 1 (Spring 1975).

[b]Modeling Resources Group, CONAES, *Energy Modeling for an Uncertain Future* (Washington, D.C.: National Research Council, 1978).

[c]R. Pindyck, "The Characteristics of Energy Demand," in J. Sawhill ed. *Energy Conservation and Public Policy* (Englewood Cliffs, N.J.: Prentice-Hall, 1979).

The impacts of these large substitutions can cut both ways, increasing or decreasing the demand for electricity. With small increases in oil and gas prices and no change in electricity prices, there may be a continuing shift away from fossil fuels and toward electricity. But if electricity prices increase and oil and gas prices do not, demand for electricity will drop. The sensitivity of fuel mix and total demand is illustrated in the more detailed studies conducted by the Demand Panel of the CONAES study.[16] This work used both bottom-up engineering and top-down econometric analyses to explore the sensitivities of demand to dramatic changes in energy prices. The CONAES sensitivity tests were limited to one set of demand elasticities, which are consistent with the higher estimates considered here, and the results are similar. For example, the electricity demands for the year 2010 double if electricity prices are cut in half while other prices are held constant. And this large potential for substitution was found in both the engineering and the econometric analyses. The foregoing remarks apply largely to the stage at which energy is delivered to the final user. As energy prices change relative to each other, there are also—and even greater—opportunities to change the mix of fuels used in producing electricity.

Two Disaggregated Models

Our simple demand curve could be extended, of course, to investigate the role of electricity, first by introducing two forms of energy: electric and nonelectric. But there are many other details for which we should adjust, including supply conditions and the differential effects on the prices of electric and nonelectric energy. To illustrate the range of variation that might exist for the composition of energy demand, we have compared two models. The first model was used by the Energy Information Administration (EIA) in preparing the *Annual Report to Congress.*[17] The second is the U.S. component of the OECD demand model developed by Griffin for the National Science Foundation.[18]

The EIA demand model is an econometric model developed by using a cross section of data for regions in the United States. In this model, the aggregate elasticities of delivered energy are below the low end of the range of figures used in the previous sensitivity tests. For example, the model uses a long-run price elasticity of 0.5 for the residential sector and 0.3 for the industrial sector. Of course, the

16. CONAES Demand and Conservation Panel.
17. EIA.
18. J. Griffin, "An International Analysis of Demand Elasticities Between Fuel Types" (report to the National Science Foundation, 1977).

Table 2-6. EIA Energy Price Assumptions (1978 dollars per million Btus).

	Scenario C			*Scenario F*	
	1975	*1985*	*1990*	*1985*	*1990*
Industrial Coal	1.07	1.94	2.03	1.96	2.10
Industrial Gas	1.30	2.36	2.96	2.54	3.24
Residential and					
Commercial Gas	1.86	3.12	3.80	3.83	4.38
Residual Oil	2.30	3.06	3.23	3.66	4.58
Gasoline	5.52	6.09	6.28	6.68	7.74
Industrial Electricity	6.78	9.50	9.98	9.01	10.94
Residential Electricity	11.23	12.11	12.45	12.80	11.45

Source: EIA, *Annual Report to Congress,* DOE/EIA-0036/2 (Washington, D.C.: DOE, 1978). Adjusted to 1978 dollars using the GNP deflator.

interfuel substitution elasticities are higher, often greater than 1. The Griffin model, in contrast, was estimated using a cross section of data from different countries, much like the Pindyck study mentioned above. The aggregate elasticities in this model are near the high end of the range of figures used in the previous sensitivity tests; it uses a price elasticity of 0.8 to 1.0. Again, the interfuel substitution elasticities are greater than the aggregate price elasticity.

These two models, therefore, have been selected to bracket the range of sensitivity tests that we discussed earlier. In this way, we can examine the implications of changing prices for the changes in the demand for individual fuels in the context of assumptions governing our previous aggregate demand analysis for individual fuels. Two scenarios are reported here for each model by exploiting the extensive work done in the preparation of the EIA *Annual Report to Congress.* Two of the scenarios reported by the EIA correspond roughly to low and high price assumptions for the range of energy products (see Table 2-6). The EIA's Scenario C is a medium GNP growth, medium oil price case. The EIA's Scenario F is a medium GNP, high oil price case. The results of these two scenarios are taken from the EIA's report.[19]

Using the assumptions in the EIA report (in which the assumptions are spelled out with unusual care), Griffin's model was run for Scenarios C and F, exclusive of the exogenous conservation savings assumed by the EIA.[20] The version of the Griffin model used has

19. EIA.
20. S. Granville, "Runs of the Griffin Model Under the DOE Price Assumptions" (memo, Stanford University, July 28, 1978). This memo details the simulation of the Griffin model. The EIA conservation figures of 1.97 quads in 1985 and 3.2 quads in 1990 were not subtracted from the Griffin model.

actual data only through 1973; hence, it is at a disadvantage relative to the EIA analysis, which begins in 1975. In an attempt to restore some consistency, we have scaled the results from the Griffin model so that the forecast from 1973 predicts 1975 demands exactly.[21]

The two scenarios involve very detailed price paths for a number of energy products. We use the price assumptions of the scenarios to exploit the richness of the models by permitting the prices of different fuels to change at different rates. The price assumptions for selected energy products are reported in Table 2-6, which is drawn from the EIA report. Both models were run with these prices as inputs. These varying rates of price change create the incentive for interfuel substitutions.

The results of the four model simulations, adjusted as described above, are shown in Figure 2-11. The differences here are substantial, reflecting the differences in the elasticities embedded in the two models. In comparing forecasts, it seems more important to agree on the model than on the price assumptions. For example, the delivered-net-energy-GNP ratios for 1985 in Scenarios C and F are 27.1 and 24.5 for the EIA model and 20.3 and 19.5 for the Griffin model[22] — a greater variation across models than across scenarios. Most of the difference in the results of the two models is in the projections of oil and coal consumption. The greatest difference in the detail is found in the forecasts of electricity generation, with the growth rate from Griffin's model between one and two percentage points below the growth rate from the EIA model.

The details of the two models tend to confirm the results of the sensitivity analysis with the simple demand curve. Prices and rates of

21. We note in passing that the estimates from the Griffin model are below actual primary energy demand by 5 percent for 1975 and by 13 percent for 1977. This may be attributed to the lag parameters, which imply a very rapid adjustment to high energy prices. We are not aware of any similar test with the EIA forecast.

22. The assumed GNP level is $2,722 billion (1978 dollars) in 1985 and $3,094 billion in 1990. The implied delivered-net-energy-GNP ratios in Figure 2-11 are as follows:

| | Scenario | | | |
| | 1985 | | 1990 | |
	C	F	C	F
EIA	27.1	24.5	28.3	26.0
Griffin	20.3	19.5	19.2	17.9

These values can be compared with the sensitivity tests with the simple demand curve. Note that most of the growth in primary energy demand projected with the EIA model comes from a rapid increase in electrification.

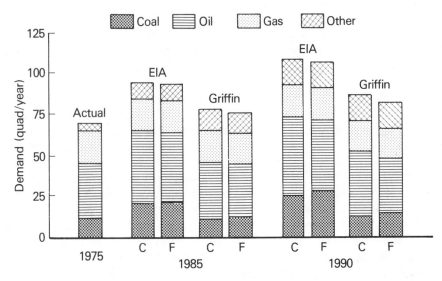

Figure 2-11. Composition of Primary Energy Demand (Scenarios C and F).
Source: The data for the graph appear in the supporting background paper by
W. Hogan (see fn. 9 above for reference).

price change will be important in determining energy demands during
the next decade, but substantial uncertainty remains in the measurement of the responses to higher prices. However, with all but the
most pessimistic assumptions, the results suggest that energy demand
may not grow as rapidly as indicated by the higher end of the range
of forecasts in Table 2-4. The gradual response to higher energy
prices, and lowered expectations for economic growth, may reduce
some of the urgency for the development of new energy sources.
Furthermore, if the price of oil increases more rapidly than the prices
of other energy products, the substitution away from oil, in addition
to the reduction of total energy demand, will reduce the demand for
oil imports as more abundant fuels replace the use of oil.

CONCLUSIONS

We should not rely on a single forecast of energy demand. Energy
markets are too complex, energy products are too diverse, and the
unknowns are too many to be reduced confidently to one description of future events. The growth of the economy, policies designed
to affect demand, the resolution of important environmental debates,
and the evolution of energy prices will have dramatic effects on the
level and composition of future energy demands. The uncertainties

associated with these critical elements are great and can never be fully resolved until after the fact. But careful analysis can help us understand the implications of the uncertainties and to delimit the range of future demand levels or the likely responses to changes in key conditions.

The shock of the dramatic changes in world oil markets that began in 1973 stimulated many studies of the options available for changing the growth in energy demand. Our examination of the results of these studies suggests that there are substantial opportunities to substitute insulation, new equipment, new production process, or changes in consumption patterns to reduce the consumption of energy. The evidence is found in both the examination of specific technologies—for example, the 45 percent improvement in efficiency now mandated for automobiles—and in the aggregate statistical data, where delivered price elasticities as high as unity imply that a doubling of the delivered price of energy could improve energy efficiency by 50 percent.

In view of the evidence of many opportunities for reducing energy demand, it is not necessary to produce a precise forecast based on a single view of the future. The long-run economic impacts of higher energy prices may be much smaller than might be expected, at first glance, for a commodity as important as energy. If the effect of higher energy prices and reduced energy use is to substitute other materials, equipment, and processes in place of energy rather than to curtail otherwise productive activities, then the economic cost of higher energy prices can be a relatively small proportion of the economy. And the best decisions today, as outlined elsewhere in this volume, do not depend critically on the demand outcomes within the range that appears to be most probable.

The process of adjusting while we move toward these long-run substitutions may be another matter. The flexibility of the economy and our ability to change the mix of energy demands are substantially less in the short run. In the immediate future, the demands for energy will be determined largely by the composition and design of our stock of energy-consuming equipment, which can be modified only slowly. All of the adjustments to the recent changes in prices may not be completed by the turn of the century. And the process of change, as we reshape the energy system, can have major impacts on large and important sectors of our economy and our society—even though the costs to the losers may only slightly exceed the gains of the winners, so that the aggregate change may look small.

The challenge, then, is to manage the adjustments. We have many opportunities to mismanage if we become blinded by extreme views that ignore the potential flexibility in the system or see a single panacea in energy conservation. We need to evaluate the costs and benefits of each energy use. The best way to achieve this evaluation is to make sure that the energy consumer has the proper information about the relative scarcity of each form of energy, chiefly through the price system. In this way, the consumer can compare the scarcity-reflecting price with the benefits of the energy use.

If prices reflect the true scarcity of energy supply—including the many effects on resource exhaustion, the environment, and national security—and if the consumer pays these prices, then the level of energy demand must be, in a certain sense, correct. It is not so important to forecast the correct level of energy demand as it is to ensure that the system gives the correct signals to the consumer. The projections of very high energy demands are often the starting point for an analysis of impending crisis in the energy system. But if the difficulty of supplying the high demands is translated into higher energy prices, the flexibility to change the uses of energy will lead to a reduction in demand until the costs of further energy consumption just balance the benefits. This is the theory of the market system. When left alone, the market system for energy has worked reasonably well in the past, and it can be made to work well again in the future.

The higher price of energy is neither a crisis nor the essence of the energy problem. If we can manage the adjustment to a new era of energy scarcity, the higher prices may be the solution to the energy problem. A sensible energy-pricing policy can help us manage the many adjustments that must be made by providing the incentives to make the substitutions that are possible. With these higher energy prices, future energy demand levels may be much lower than might be expected from the historical trend, a trend established during a period of decreasing energy prices. Conversely, if prices to consumers are held artificially low, we can make the problems of import dependence and supply allocation even more serious than the "crisis" situation we find ourselves in today.

Energy Conservation: Opportunities and Obstacles

For many people energy conservation carries, wrongly we believe, a connotation of deprivation, curtailment, and reduction in amenities. Some mistakenly use the term "mandatory conservation" in talking about such actions as closing service stations on Sundays to restrict driving, making the term a euphemism for an enforced denial of something people want as part of the good life. Yet with equal vigor, many treat the subject as if it held some special place in a pantheon of virtues; descriptions of America as a "wastrel society" serve as exhortations toward self-improvement.

We adopt quite a different viewpoint about energy conservation, a viewpoint guided by the economics of self-interest rather than either hardship or moralistic appeals. We mean by conservation those energy-saving investments, operating decisions, and changes in the goods and services that we buy and use that save money over the life of energy-consuming products. Money can be saved by substituting intelligence, prudence, maintenance, better equipment, or different equipment for purchased energy; the substitution should be made up to the point where the cost of not using the energy is equal to the cost of the energy saved.

This economic emphasis does not mean that we are oblivious to the many other positive values that may induce energy conservation. Actual changes in desires of consumers indicating shifts in preference away from energy-intensive goods and services toward simpler, less consuming lifestyles may be an important factor in spurring conservation. So may be a desire to preserve environmental values. These essentially noneconomic forces could well swamp the merely economic forces that we have chosen to discuss, and we cannot discount them as important determinants of energy consumption and efficiency in a society. But our deliberations have led us to the conclusion that such noneconomic motivations in society should not be imposed through the use of energy policy as such. Rather, if government wants to influence such choices at all, it should do so through more transparent measures. For example, creation of pedestrian-free zones in urban areas is more properly undertaken for reasons of making the city more pleasant to live in rather than for the indirect, if beneficial, consequences in energy conservation that may also result.

More than that, we have concluded that economic forces can have an important effect on those major energy-consuming sectors that respond only to efficiency arguments. We believe that energy consumption should not be the touchstone of determining behavior in the society, but rather that energy should be treated like other commodities and, in general, that preferences of consumers should be demonstrated through the marketplace. We also believe that there are substantial benefits in using market or marketlike mechanisms for assessing energy conservation goals and performance.

In principle, if energy were priced correctly—that is, at a price that equaled its marginal cost of production plus such factors as the cost of depletion, environmental damage, and national security—there would be no need for public policies dealing explicitly with energy conservation. The operation of the market would ensure that each person in a position to make a decision would purchase only enough energy to satisfy needs that could not be met more cheaply in some other fashion. Desires that took too much money to satisfy would be frustrated, so that greater desires could be satisfied. To be sure, there would be lags as the system readjusted itself to new higher prices, and some would suffer relatively temporary deprivations until capital equipment could be improved or replaced. Likewise, compassion and equity would dictate that the poor be given special assistance in meeting the costs newly imposed by the market. But if the system worked smoothly and efficiently, there would be very few cases where it would be worth the cost and effort required to crank up a bureaucratic system to regulate or subsidize.

Unfortunately, as described elsewhere in this study (see especially Chapters 2, 4, and 6), the United States is far from pricing energy correctly. Even though we have adopted the relatively distant perspective of twenty years for this study, movement toward economically rational pricing of energy may be even more glacial. Consider the fact that decontrol of natural gas will materialize in 1985 at the earliest and even that period can be extended. Beyond natural gas, the situation remains unsettled. The decontrol of oil prices continues to generate bitter fights, even though President Carter's proposal for decontrol by 1981 went into effect June 1, 1979.

Thus, because prices at the energy production end are likely to remain under government control well into the next twenty years, we must relate our assessment of energy conservation benefits and potentials to a discussion of energy conservation policies. Such a discussion of positive government regulations or subsidies does not imply retreat from our belief that market and marketlike systems hold the greatest promise for helping the country cope with energy issues. Regulations and other government actions do, however, need to be considered as a substitute for market forces where those forces do not otherwise exist.

There is another, more deeply embedded reason for discussing positive energy conservation strategies in a book that generally takes a strong market orientation. Table 2-1 indicates why. Nearly half the energy sold to ultimate consumers in this country is subject to price control because it is in the form of natural gas or electricity; where that energy is sold by a utility, control of price is virtually complete.[1] A homeowner who has watched utility bills rise rapidly since the early 1970s may find it difficult to take seriously the proposition that the energy is worth more than it costs. But unless the prices to electricity and natural gas consumers in some way reflect the replacement cost of the energy consumed, end users will be insulated from the full effects of rising energy costs on their energy consumption decisions.[2] The size of that protection can be enormous. In the Pacific Northwest, for example, power that must be generated by nuclear reactors to meet growth in load may cost as much as ten times the cost of power supplied by hydroelectric generators. Because of the large amount of energy affected by public regulations and the sometimes enormous gap between replacement costs and the prices actually billed to customers, a proposal for reliance on the market alone is simple minded and out of touch with political reality. As we shall see later, "deregulating" public utility prices is no simple matter, either in theory or in practice. Thus, the very existence of

---

1. Not quite all of the energy aggregated under natural gas and electricity figures is actually regulated by public bodies; notably, intrastate sales of natural gas were unregulated at the time the estimation of one-half was made. But the totals support the statement in the text as to orders of magnitude.

2. As used in this chapter, "replacement cost" is the same as "marginal cost" as defined by economists. In utility usage, "replacement cost," "incremental cost," and "marginal cost" have taken on specialized meanings somewhat unrelated to their customary definitions.

regulated utilities makes government policies an important part of an energy conservation chapter.

In this chapter, we first discuss the nature of energy conservation in somewhat more detail to make clear what we mean by the term and why we have chosen to take the economic view of the matter. Next we discuss some of the estimates that have been made for the amount of energy that could be conserved in this country without cutting into the standard of living that we enjoy. Next we describe in broad outline selected federal programs to encourage or mandate energy conservation. Where possible, we assess the strengths and weaknesses of those programs, attempting to draw general conclusions concerning energy programs that hold promise and those that do not and should be discarded. We then turn to a discussion of the public utility sector, because an understanding of how prices are and might be set by utilities helps set the limits of market-based strategies for many energy users. That discussion of utility pricing includes suggestions for reforming rate structures. Finally, we present a menu of possible energy conservation strategies that the nation should adopt to supplement the operation of the market.

The major conclusions of this chapter are as follows:

1. Energy conservation—as we have defined it—is a continuing source of energy that has positive private economic, environmental, and public economic benefits worth exploiting to their maximum. Although not all ideas that march under the banner of "conservation" are worth their cost, there is an inventory of good conservation ideas that should be explored. In some instances, regulations and subsidies can assist in that exploration.

2. The most important actions that can be taken to encourage conservation are to proceed with decontrol of oil and natural gas prices. The longer the delay in taking these actions, the longer we will continue to use energy inefficiently. There are enormous economic gains to be made from exploiting the conservation resource. Prices based on marginal costs are an opportunity to exploit these gains and should not be resisted.

3. There is a substantial difference between the marginal cost of oil and natural gas and the price paid by consumers of those products. That price difference will and should close, but until it does close, substantial opportunities for energy conservation will be lost needlessly. Therefore, we need positive public policies on energy conservation.

4. Regulations for enforcing energy conservation can be useful. The automobile efficiency standards, the Building Energy Performance

Standards (BEPS), and the appliance efficiency standards are (or can be) successful regulatory devices, even if it is difficult to be certain whether they are strictly economically justified at current levels. In any case, these regulations now exist and thus have passed the test of political acceptability. We see great promise for their success because they deal with discrete, energy-consuming assets where consumer preference alone may not be sufficient to bring about high energy efficiencies.

5. Direct regulation of energy consumption of machines used in industry and agriculture appears to be inappropriate because energy price changes alone can cause required efficiency improvements. However, a strong government-supported program of education (perhaps including testing and labeling where possible) should be adopted.

6. More research in the energy conservation field should be directed toward nonhardware issues than it has been in the past. Nonhardware research, which includes legal, institutional, behavioral, and other social science investigations, is relatively inexpensive to support, yet there are potentially high returns to society from such research. The government is the only source likely to support it in sufficient amounts.

7. Public utilities pose a particular problem for pricing energy at its marginal cost. Because of the difference in cost between peak supplies and off-peak supplies and because of the enormous difference in cost between energy supplied by older plants or by older gas wells and energy supplied by new generating plants and new wells, pricing utility-supplied energy at its marginal cost would result in large revenue overcollections. In spite of the difficulties, development of rate structures that reflect both the differences in cost that come from time of use and from source should be seriously pursued.

THE NATURE OF ENERGY CONSERVATION

We begin this section of the chapter by considering various ways of defining energy conservation. We shall argue that conservation amounts to little more than achieving economic efficiency in the use of a resource.

Perhaps the most simple way of determining whether we are conserving energy would be to compare the amounts of energy used at a given time with that used at a later time. Less energy used would equal energy conserved. But this approach does not stand scrutiny, or else keeping homes at 45°F would be conservation. In fact, it is simply curtailment. Under such an accounting system, barricading the superhighways to keep cars at home would be an acceptable conservation practice. But energy is not so precious that all other values in life must fall so that a single Btu can be saved. Instead, we need some more subtle measure. Several ingenious criteria that have been suggested as the proper methods for distinguishing genuine energy conservation from curtailment are discussed below.

The laws of thermodynamics tell us how far current engineering practices depart from the goals of theoretical efficiency. In an influential study, *Efficient Use of Energy: A Physics Perspective*,[3] a group assembled under the auspices of the American Physical Society (APS) suggested that the second law of thermodynamics was an appropriate test for measuring the efficiency of carrying out a given task. The lesson of the APS study is that the efficiency of a given device must be judged against the efficiency of a perfect device performing that task, even though that perfect device has not been or cannot be invented. Put another way, the APS study argues that measuring efficiency against the existing stock of capital goods understates the potential for conservation.[4]

Useful though the insights of the APS study may be, they expressly omit two important considerations. First, they ignore the question of cost. There is no way to tell in advance if the increase in energy efficiency that would be gained by approaching the ideal more closely would cost so much that the advance would not be worth it. Second, and equally crucial, the viewpoint neglects the fact that the same mix of human desires can be achieved in a variety of ways. For example, instead of changing from baseboard resistance heaters to an ultra high efficiency heat pump, it would be possible to deliver the same desired end product—a warm house—through caulking, insulation, and careful design. Exclusive concentration on the energy that is supplied directly to carry out a service tends to obscure the very important gains in real efficiency that can be derived from replacing energy by a capital input. This same substitution can take place between energy consumption and information (for example, thermo-

3. American Institute of Physics, *Efficient Use of Energy: The APS Studies on the Technical Aspects of the More Efficient Use of Energy*, AIP Conference Proceedings Series No. 25. (New York, 1975.)

4. Consider the use of home heating equipment powered by electricity. If we limit our inquiry to heating only by running a current through a wire to make it hot, it is hard to imagine how we could make that process more efficient than it already is. Thus, we might be tempted to call it 100 percent efficient use of electricity for heating. But if, instead of using electricity to heat a wire, we use it to drive a compressor motor in a heat pump, higher efficiencies result. The heat pump extracts heat from where it is (outside, even in cold weather) and puts it to where it is wanted (inside).

stats that control shading devices for the south and west sides of buildings; telephone shopping instead of driving); energy consumption and labor (whisks instead of electric beaters; teenage chores instead of trash compactors); and energy-consuming and non-energy-consuming activities (backpacking instead of trailbiking; mopeds instead of 8 cylinder cars).[5] It is not enough to look exclusively at the efficiency of carrying out a task in the way we now do it or even as we might do it with a different machine. To be more helpful, we should look at the service that is now provided and decide, first, whether we care enough about that service to continue desiring it at higher prices and, second, if we do care, how to procure it as efficiently as possible.

A more productive measure of conservation, in our view, is to consider the cost of saving energy. That is, we should compare the total cost of satisfying some human desire if conservation steps are taken with its cost if conservation steps are not taken; we would then learn what it costs to save a given amount of energy. If we properly discount future costs and benefits (so that the future is expressed in terms of the present) and make reasonable predictions concerning future energy costs, we can find some point at which the costs associated with a more expensive piece of equipment, a different operating style, or doing the task in some other way would equal the costs of the amount of energy to be saved. At that point, a movement toward either more energy consumption or more capital investment would be a decision entailing higher costs. Because money saved on one buying decision can be used to enjoy goods or services otherwise out of reach, the rational consumer, if properly informed, presumably would never take any other course.

To be sure, this bland description of the system for deciding what amount of energy conservation we consider "proper" hides a large number of difficult issues. First is the question of future energy prices. If any individual had perfect knowledge of the range and trends of future prices, he or she would not waste time optimizing purchases of refrigerators or cars: such time would be much more productively spent advising heads of state and corporation executives! It is far from trivial to accurately estimate energy prices even for next year. Price changes will be a feature of the energy future, and there is no way to avoid that uncertainty. Some broad

5. To be sure, one cannot ignore that the nature of the task may itself be dependent on how it is met. If the task is simply "mobility," five people crammed into a fuel-efficient subcompact car achieves what might otherwise involve a roomier, less fuel-efficient vehicle. Here the stated task should probably be "mobility amidst perceived comfort."

range of predictions is possible, and with the effect of discounting (see below), mistakes in judgment about price levels in the more distant future count less heavily in the decision in any case. If the other advantages of following an economic definition of energy conservation are sufficiently compelling, the inability to perform the task ideally becomes less important.

Second, there is wide disagreement concerning the proper discount rate that should be applied to express future energy savings and future costs (if any) in present dollars for the purpose of making comparisons. Discounting is an essential feature of any analytic scheme that tries to measure costs or benefits over time. Discount rates are a far from perfect tool for weighing how much we will care about some cost or benefit in the future. Nonetheless, they represent in a crude manner the fact that money invested today for energy conservation could instead be invested elsewhere to produce future benefits. The exact figure chosen for discount can be—and has been—the subject of hot debate; the principle should not be.[6]

Third, the economic criterion that we adopt implies that we know many things that are in fact beyond our knowledge. We do not have a very clear idea, for example, of how energy is really used in a home. Even to the extent that we can disaggregate information into functions—such as lighting, refrigerating, or heating—the operating patterns of a typical appliance, though knowable in principle, remain uncertain. How many times a day is the door of the average freezer opened and for how long? How large a problem is lack of weather-stripping? There is at least as much uncertainty about how a device will be used in the future. Present operating practices may be known, future ones guessed. For example, a manual defrost freezer might use half the amount of electricity as a frost-free model; but this performance is strongly tied to how often the manual model is defrosted, which in turn depends on the climate and how often the door is opened. That question cannot be answered for the future with any certainty. A childless working couple in a dry climate may defrost their refrigerator only occasionally; if that same family moves to a

---

6. We refrain from discussing here the important questions of the propriety of present generations making decisions that affect future generations and of weighing the importance of present decisions on those unborn either through the mechanism of the discount rate or by assumptions as to how they will value a resource. These difficult problems are not confined to the field of energy policy. See Talbot Page, *Conservation and Economic Efficiency: An Approach to Materials Policy* (Baltimore: Johns Hopkins University Press for Resources for the Future, 1977).

different climate and has children, they will need to defrost frequently but may not do it. Accurate testing and data collection can only go so far in answering these questions. Judgment, guessing, and luck will continue to be important ingredients in determining what purchase is best for a given buyer.

With all these difficulties and uncertainties associated with an economics-based definition of energy conservation, a reader might be tempted to ask why we continue to use the test. There are three reasons. First, even with the market imperfections in the real world, we are persuaded that a system that depends more on the private decisions of thousands of individuals and firms than on centralized decisionmaking is generally more responsive to the desires of a variety of persons, each of whom has a different mix of needs and ways to satisfy them. This observation is as true for energy conservation as it is for clothing. We believe that a system in which energy conservation depends on whether it makes economic good sense is much more adequately validated by the behavior of large numbers of people than any system that starts with an a priori judgement of the kind of lifestyle that others should adopt. To the extent that individuals must make guesses about the future in order to behave in their own economic self-interest, government regulators too must make those or similar guesses. Second, for all their imperfections as an analytic tool, the principles that we urge are well understood, and their limitations and biases have been explored both practically and theoretically. Finally, and persuasively, we have noted that unlike exhortations or pure regulations, price works. For example, in a time in which many Americans did not believe that an energy crisis existed or that, if it did, it was the result of conspiracy among the oil companies and in which polls revealed that more than half of the American public did not know that we imported any oil from abroad, home insulation sales soared. Price was talking to the consumer when administration policy, television programs, and newspaper articles had failed to convince. This observation has been replicated in many other sectors of the economy. There is sound pragmatic evidence that slowly, with complaint and suspicion, and—to be sure—suboptimally, economics does matter to individuals and firms in making decisions.

Even where, in later discussion, we support the use of regulations for certain sectors of the economy, we urge that those regulations receive a market test of efficiency. A good regulation, we believe, should mimic the action of the market, and departures from this general test should be explicit.

MODELING ENERGY CONSERVATION POTENTIALS

There is considerable evidence that energy conservation opportunities in American society are both large and relatively unexplored. Several studies have even suggested that we could enjoy the same standard of living and yet use 30–50 percent less energy than we do now. Conservative assessments disclose potentials that are dramatic when measured against the conventional wisdom of a few years ago. The director of the planning staff of the Electric Power Research Institute has written, for example:

> Energy management technologies which are currently available and appear to be economic could reduce total US fuel consumption by 28 to 46 percent by the year 2000, compared with an extrapolation of present usage patterns. About one-third of this savings, or 9 to 18 percent of total energy use, could be achieved by using electricity more efficiently, thereby reducing fuel needed by generating plants.[7]

Careful sectoral studies have calculated a potential for enormous amounts of energy conservation that satisfies our definition of being cost-effective. Eric Hirst of Oak Ridge National Laboratory has calculated that in the household sector alone, growth rates could be held to 0.4 percent per year between 1975 and 2000, assuming no lifestyle changes on the part of American households and no use of solar energy for any household functions.[8] Houses built according to specifications developed jointly by a builder and a manufacturer of insulation not only cost less to heat and cool, but are cheaper to build as well.[9] New automobiles, on a fleet average basis, now get more than 19 miles per gallon, compared with a preembargo average

7. R.L. Rudman, "Practical Savings Achievable with Efficient Energy Use," in Craig B. Smith, ed., *Efficient Electricity Use: A Reference Book on Energy Management for Engineers, Architects, Planners, and Managers*, 2nd ed. (New York: Pergamon Press, 1978), p. 3.

8. Eric Hirst, *Residential Energy Conservation Strategies*, Report No. ORNL/CON-2 (Oak Ridge, Tenn.: Oak Ridge National Laboratory, September 1976), p. 33. In a more recent study—Hirst, *Energy and Economic Benefits of Residential Energy Conservation RD&D*, Report No. ORNL/CON-22 (Oak Ridge, Tenn.: Oak Ridge National Laboratory, February 1978)—Hirst computes the net economic benefit to the nation from improved technologies for providing residential services such as space and water heating. Between 1977 and 2000, improvements in that equipment would cost consumers $2.6 billion more than they would otherwise spend, but household fuel bills would be reduced by $20.3 billion, leaving a net economic benefit of $17.7 billion.

9. See Owens-Corning Fiberglas Corporation, Insulation Operating Division, *Energy Saving Homes: The Arkansas Story*, (Toledo, Ohio: Owens-Corning Publication No. 4-BL-6958-C, June 1976). See also, "Fuel Costs Cut Sharply in Six New Virginia Homes," *Washington Post*, March 5, 1979, p. C-3, in which it is reported that certain houses using well-known energy-conserving construc-

of little better than 11 miles per gallon, and automobile manufacturers believe that the ultimate standard of 27.5 miles per gallon can be met. A Resources for the Future study projects an absolute decline in nationwide gasoline consumption by the year 2000.[10] Commercial and industrial success stories of modest energy conservation efforts paying off handsomely can be found in the pages of *Energy User News* on a regular basis.

This casual listing is suggestive of a trend, though hardly persuasive of its extent or depth. In principle, such quantification should be possible in a rough way. Essentially, one would build an economic model of the economy, then let prices rise in the model to their marginal costs. Running the model first with prices held to regulated levels would provide a base case. A second run (or better, series of runs) would give a range of plausible futures. In cases where prices to the end user were unlikely to reach marginal cost because of structural or other reasons in a given industry, the model might be modified to allow for regulatory changes or institutional practices that could be adopted to correct for market failures.

Such an economic model would be large and intricate, but is within the grasp of the science. The more difficult question is whether any model could capture the shifts in technologies and desires that higher energy prices would produce. Although simply assuming existing technologies is illuminating, the real world is far more complex. As we have seen earlier, conservation consists not only in using less energy in existing technology, but also in finding new ways to perform the same service or finding another service that relative price shifts now make more desirable. Models can only hint at the complexity of the trade-offs inherent in these shifts. In a good model, these trade-offs are included in part by assumed improvements in technologies or shifts, but the underlying data for making those assumptions are very poor indeed.

Chapter 2 addresses some of the aggregative econometric models that have tried to estimate future energy demand. These models clearly suggest that much lower total demands for energy can support a high standard of living in a healthy economy. Such modeling can only hint at the possible, not spell out details of managing—in the real world.

---

tion techniques cost owners approximately $2,000 more than homes not using those techniques, yet used 17 to 20 percent of the amount of fuel oil in a winter that conventional houses would use. The construction techniques included double insulation and double walls, south-facing exposures, double-glazed windows, and vestibules inside the front door.

10. S.H. Schurr, et al., *Energy in America's Future: The Choices Before Us* (Baltimore: Johns Hopkins University Press for Resources for the Future, forthcoming).

It is in supplementing the shortcomings of modeling that sectoral studies can be invaluable. Such studies look at each major energy-consuming part of an economy with the help of engineers and technically trained persons. For each sector, questions are asked about the possibility of saving energy in a cost-effective fashion. Although such engineering-based predictions also do not have any real way to predict new technologies or new consumer tastes, the combination of aggregative economic modeling and "bottom-up"engineering evaluation provides a powerful case that conservation potentials are large indeed. Recently, the National Academy of Sciences issued a detailed study by the Demand and Conservation Panel of the Committee on Nuclear and Alternative Energy Systems (CONAES), based on examination of many sectors of the American economy. In a 1978 article in *Science*, that panel concluded:

> in every sector of the economy major increases in energy efficiency can be made by using presently available technology, and even greater improvements can be made with technology now under development. The large discrepancies between present energy efficiencies and those that are thermodynamically obtainable, together with economic analysis, show that improvements in efficiency of 1 percent per year or more are sustainable over a number of decades in the United States.[11]

The CONAES panel's work emphasizes the fact that energy conservation actions have a continuing payoff. They do not simply provide a brief hiatus in a hopelessly exponential growth curve; instead, conservation can fundamentally change the shape and even the slope of that curve.

We have not conducted any independent examination of possible energy futures for America. But review of work prepared for our group and of the work of others has persuaded us that conservation will be an important ingredient in coping with the energy problem. If we fail to conserve, the transition can and probably will be very difficult and very expensive indeed. But if economic and regulatory means are intelligently combined, conservation can help reduce crisis to mere concern.

EXISTING CONSERVATION PROGRAMS

Congress has eneacted three waves of energy legislation: the 1975 Energy Policy and Conservation Act (EPCA),[12] the 1976 Energy

---

11. Demand and Conservation Panel of the Committee on Nuclear and Alternative Energy Systems, "U.S. Energy Demand: Some Low Energy Futures," *Science* 200, no. 4338 (April 14, 1978): 151.
    12. P.L. 94-163.

Conservation and Production Act (ECPA),[13] and the 1978 legislation known collectively as the National Energy Act.[14] Taken together, these acts—notwithstanding some worthwhile features—constitute a disjointed, ill-coordinated set of regulatory and educational programs to encourage energy conservation. This section briefly describes selected federal programs, discusses their strengths and weaknesses, and attempts to abstract from the programs generally fruitful approaches for further federal action.

The various pieces of legislation classify energy conservation actions primarily by the sector of the economy that they are designed to affect. Thus, household appliances are treated in one section, automobile efficiency standards in another, and voluntary industrial reporting in a third. For our purposes, we classify the actions according to the policy tool they use. Government has three basic methods of affecting the behavior of its citizens and of the economy. The first is education. By means of exhortations, detailed studies communicated to potential users, and hands-on training provided to contractors, government can urge individuals and firms to follow a publicly desirable course of action. Second, government can regulate. Detailed rules, enforced by the criminal law or by civil penalties, are designed to call forth certain kinds of behavior. Third, government can provide financial incentives through subsidies (including tax credits) to encourage desirable behavior. Both regulation and subsidization have their converse: government can refuse to regulate an economic activity, confining its actions to those necessary to develop and maintain a competitive market; and government can make activities more expensive by taxing them.

To a large extent, these powers of government become more effective in provoking the desired behavior as they move from education through regulation to subsidies or taxes. But they also become more difficult to enact. Laws that powerful interest groups dislike and fear may work are frequently parried by a plea for more public education. Nonetheless, each type of control has a role to play in public policy, and a limited government role may ultimately be better at provoking economically efficient behavior than a more pervasive regulation or subsidy.

The federal energy legislation passed so far has examples of educational strategies, direct regulations, and subsidies. In the following three sections, we discuss examples of these programs, their effects, and their prospects.

13. P.L. 94-385, August 14, 1976.
14. Consisting of the National Energy Conservation Policy Act, P.L. 95-619; the Public Utility Regulatory Policies Act, P.L. 95-617; the Powerplant and Industrial Fuel Use Act, P.L. 95-620; the Natural Gas Policy Act, P.L. 95-621; and the Energy Tax Act, P.L. 95-618, all effective November 9, 1978.

Educational Strategies

National legislation provides both for education of the public at large and for specific educational or informational programs to encourage energy-conserving behavior at a moment of decision. General educational programs include dissemination of publications on home energy conservation practices, a limited program to "market" energy conservation ideas to small businesses directly and through trade associations, and a few public service announcements. One program is the Energy Extension Service. The service, consciously modeled after the Cooperative Extension Service of the Department of Agriculture, is designed to give homeowners, businesses, and others advice about energy conservation and use. The effort began as an experiment in ten states and, as of 1979, was being expanded nationwide. Some states, such as Pennsylvania and Tennessee, run university-based programs that provide technical information to businesses and others located in the state. These answer line services have noted a large increase in the number of requests for energy conservation services. Likewise, the Department of Housing and Urban Development supports a toll free telephone service for answering questions about solar energy and for gathering and disseminating information on that energy source.

Although general information can create a climate for energy conservation, consumers are more likely to pay attention to advice that is offered when a purchasing decision is at hand. The prototype for such programs is the requirement that automobile manufacturers label the tested mileage ratings of their cars and include that information in advertising. The auto efficiency labels must contain information about the estimated annual fuel costs associated with the operation of the automobile and the range of fuel economy of comparable automobiles (whether or not manufactured by the maker of the labeled car). It is difficult to assess whether the labeling program itself has had an effect on consumer preferences for smaller cars. The downturn in purchases of larger cars that immediately followed the 1973-1974 oil embargo was soon reversed by renewed consumer interest in larger, more energy-intensive cars. That interest, however, may have been equivalent to the last fling of a bachelor about to settle down to married life: small cars are once again capturing a larger share of total sales. In any case, the effect of the labeling program has been obscured by the automobile efficiency standard program, discussed in more detail below.

A more comprehensive program of testing and labeling (and ultimately of mandatory standards) exists for household appliances. The EPCA establishes a mandatory program of testing and labeling products used in residences. The act specifically mentions thirteen consumer products, including refrigerators, freezers, dryers, water heaters, air conditioners (both room and central), and furnaces. The secretary of energy has the authority to add other products to the list on a finding that the appliance uses above a certain minimum amount of electricity or its equivalent per year and that inclusion in the group would carry out national goals.

Once a product is covered, the Department of Energy, in consultation with the National Bureau of Standards, devises testing procedures, and the Federal Trade Commission promulgates a label that must be prominently affixed to the appliance at the point of sale. The labels compare the particular appliance with others in its class and show a range of estimated annual energy costs for some products and an energy efficiency rating for others (including information for the consumer to consider if the particular consumer's own energy costs differ from the national averages). Again, as in the case of automobile labeling, it will be difficult to assess the effect of this informational program on appliances because Congress, in the National Energy Act, has made the appliance program subject to mandatory efficiency standards.

Assessment of Educational Programs. Although general educational programs are helpful for increasing awareness of energy conservation, there is considerable evidence that detailed, site-specific information is more useful to individuals and firms considering energy conservation investments. To a large extent, the ordinary advertising and marketing practices of those who stand to profit from conservation activities can be depended upon to convey information. However, conservation has two particular difficulties that justify government-sponsored programs. First, investments and changes in operations designed to conserve energy affect entire energy-using systems in such a way that the maximum benefits can only come from considering a cluster of actions and investments as a package and comparing it with other option clusters. For example, a homeowner who wants to install better insulation and draft-excluding door sills might find that he could get along with smaller heating and cooling equipment. Before going ahead, though, the homeowner should compare the option with, say, a passive solar design for the structure itself. Manufacturers of particular components are ordinarily poorly suited to give advice on such system packages. Second, many energy conservation actions do not require anyone to buy anything. More careful management of existing capital stock or substitutions in process or product have no natural profit-making constituency to give advice and information.

To be sure, a new cadre of professionals may be developing who give advice to individuals and firms on energy conservation. Many developers of large commercial buildings are hiring special energy conservation consultants. Several industrial firms that have themselves been successful in saving energy have begun to market their expertise. We can expect this private response to grow as energy prices rise. But this is no argument that government should not help impartial expert consulting services become available to individuals and firms.

We recommend that the federal government fund the development of training courses and experimental programs at the state and local level to provide professional advice at a cost-recovering fee to individuals and firms interested in energy conservation investments, product changes, or operating changes. We recommend that the services developed under such an experimental program be permitted to advertise, to solicit business, and to develop a market for similar services and that the federal government set a time after which its financial, marketing, and managerial services will terminate.

Research and Development

Although we deal more fully with research and development in Chapter 15, a few words on nonhardware R&D seems appropriate here. R&D provides the grist not only for energy-conserving hardware, but also for information pointing to overlooked opportunities for conservation.

In our examination of the menu of federal R&D in energy conservation, we are struck by the heavy emphasis on developing energy-conserving equipment. Better light bulbs and advanced engines receive their due, but scant attention goes to studying motivations to conserve or to legal and institutional research. Yet in our judgment, these are exactly the areas that are most productive for a further expansion of knowledge. Much of energy conservation is merely enlightened self-interest on the part of consumers large and small. Once informational and institutional barriers are overcome, market forces should provide the incentives for development of the hardware needed to meet a more costly energy future. In fiscal year 1979, the Department of Energy's estimated total budget authority was nearly $12.8 billion. Of this, about $1 billion was allocated to energy conservation programs. Thus, energy conservation represented about 8.1 percent of DOE's total budget; the corresponding figure in fiscal year 1978 was 6.3 percent. These figures can be compared to DOE's fiscal year 1979 budget authority for programs in energy supplies—$3.4 billion, or 27 percent of the total DOE budget. Although it is difficult to assess exactly the breakdown in this conservation budget among technology research, development and demonstration, nonhardware research, and programs to implement energy conservation for legislatively determined target groups (such as low income consumers), nonhardware research appears to be the smallest of the four. Probably substantially less than 4 percent of the total conservation budget is devoted to such research.[15] Put another way, less than 0.4 percent of the department's total budget was available for research on providing information that will lead to better conservation efforts in the framework provided by the marketplace.

We recognize that simple comparisons of percentages of budgets allocated to different categories have limited significance. Particularly in the case of nonhardware research, one would expect the budget to be much smaller because no purchases of large equipment are required to conduct such research. Thus, budget figures are only a surrogate—and perhaps a poor one—for the point we are making. Even so, the level of research as measured by the activity developed as well as by its budget figure appears inadequate in these areas. The low commitment of funds suggests a similar low commitment of interest to these important areas.

We recommend that the research budget of the Department of Energy be augmented to provide for a larger sum to be devoted to studies on ways to conserve energy other than by way of new hardware. Research that spans life cycle costing, development of new financing instruments, legal and institutional barriers to conservation, motivations for conserving energy, effects of various government and private actions, and other topics that will lead to a better understanding of how to conserve and how to assist consumers and firms to conserve are extremely high payoff areas that have been underfunded by the federal government.

Education and increase and diffusion of knowledge on energy conservation are not all that will be required from government in the next twenty years, but government does have a vital role as educator and supplier to citizens of information concerning energy conservation. Therefore, we strongly support programs that reach out to individuals and firms at the time that they are making decisions with informed, impartial, and accurate information, based on research. With this assistance, the price system can be largely relied upon to force the transition to a more conserving society.

Regulatory Strategies

Two of the labeling programs that began as voluntary programs to provide consumers with more product information were transformed

15. Richard Sclove, "The Department of Energy's Non-Hardware Energy Conservation Programs" (background paper done for study group, on file at Resources for the Future, 1755 Massachusetts Avenue, N.W., Washington, D.C. 20036).

by Congress into mandatory standard-setting programs before their effect could be assessed (in the case of automobiles) or the labels were even in place on the goods (in the case of household appliances). These standard-setting programs are briefly described below, together with a new program under development to mandate a national building code for energy conservation.

First, let us consider the automobile efficiency standards, which are the prototype of later standards. The standards are mandated under EPCA, Sections 301 et seq. Automobile manufacturers are required to improve the efficiency of ordinary passenger automobiles according to the schedule shown in Table 3-1. The standards under EPCA are related to "fleet average" fuel economy. This means that not every automobile must meet the fuel economy standards, but the average of all automobiles that a given manufacturer predicts it will make and sell in a given model year must meet the standard. No incentives are awarded for producing a fleet average better than the standards, but penalties are substantial for doing worse than the average. There has been continual discussion about the wisdom of amending this section to require that all automobiles meet the minimum standards. While Congress rejected a gas guzzler tax that penalized cars doing substantially worse than the average, it included in the Energy Tax Act of 1978[16] a tax imposed on a manufacturer whose fuel economy fails to meet efficiency standards. The tax in 1980 ranges from $200 per automobile for cars with a mileage rating of at least 14 but less than 15 miles per gallon up to $550 for cars with a mileage rating of less than 13 miles per gallon. Each year a new schedule of gas guzzler taxes is added, so that in 1986 a car with at least 21.5 but less than 22.5 miles per gallon pays $500, while a car with efficiency of less than 12.5 miles per gallon pays $3,850.[17] Our analysis indicates that manufacturers will probably react to such a gas guzzler tax by adjusting their marketing plans and prices so that the tax will have no effect beyond the efficiency standards.[18]

16. P.L. 95-618, §201.

17. The tax applies to automobiles used primarily on public roads and weighing 6,000 pounds or less. Nonpassenger vehicles, including vans and trucks, are an important exception to this gas guzzler tax. As noted later in the text, there is already considerable movement by consumers from automobiles (covered by mileage standards) to vans (not covered). The gas guzzler tax, as written, will tend to accelerate this trend.

18. See James L. Sweeney, "Energy Policies and Automobile Use of Gasoline," in *Selected Studies on Energy: Background Papers for Energy: The Next Twenty Years* (Cambridge, Mass.: Ballinger, forthcoming). Sweeney states, "For example, under current gasoline prices the imposition of a gas guzzler tax would leave fuel efficiency precisely at the standard unless that tax were large enough to cause the standards to be surpassed. In such a case the existence of the stan-

Table 3-1. Automobile Efficiency Standards.

Model Year	Average Fuel Economy Standard (miles per gallon)
1978	18.0
1979	19.0
1980	20.0
1981	a
1982	a
1983	a
1984	a
1985	27.5

[a]The Energy Policy and Conservation Act of 1975 states that for these model years, the secretary of transportation may set fuel economy standards "at a level which the Secretary determines (A) is the maximum feasible average fuel economy level, and (B) will result in steady progress toward meeting the average fuel economy standard established by or pursuant to this subsection for model year 1985." In fact, the secretary, by regulation announced in June 1977, set the figures at 22 miles per gallon for 1981; 24 miles per gallon for 1982; 26 miles per gallon for 1983; and 27 miles per gallon for 1984. The effect of this weighting of the efficiency standards more heavily toward the first of the controlled period will be to bring forth the benefits of the regulation faster. As this volume goes to press, the automobile companies have urged the House Subcommittee on Energy and Power to change the accelerated schedule established by the secretary.

The automobile efficiency standards appear so far to have achieved their stated goals. Judging from sales records, the more efficient and smaller cars that the standards have produced do not seem to have seriously inconvenienced the purchasing public. The value of the fuel saved has tended to soften the impact of rising automobile prices, and many of the most efficient automobiles are in the lower price ranges. The larger number of smaller cars has also made purchasers more understanding of small cars in general and may further help to increase their acceptability and sales. However, counterfactors have also been at work. Some people with smaller cars may actually be driving more, thus offsetting the goal of reducing fuel consumption. Also, until very recently, people have been purchasing the non-regulated, larger light trucks (pickups and vans) for personal use at an

dard would be irrelevant in determining mean efficiency. . . . In the broad inter-mediate range, changes in gasoline price *or moderate changes in most other policy instruments* will have no effect on new-car mean fuel efficiency. Automobile manufacturers will simply change pricing and marketing strategies to compensate for any policy shift so as to continue to just meet the standards. While such moderate policy changes can be expected to alter the relative prices of new cars of different fuel efficiencies, they will be totally ineffective in modifying new car mean efficiencies."

ever increasing rate. Because these vehicles are generally more durable than automobiles, the effect of their lower fuel efficiency will be felt for a long time to come. Finally, there is little evidence that the public has been adequately prepared for some of the less pleasant features of cars that must be built to comply with the increasingly severe regulatory standards. For example, although most people do not haul trailers, many cars are sold with provision for hauling them. Lower performance cars may not be able to do that. This may be a trade-off well worth making, because not many people must haul trailers frequently; that trade-off ought to be explained and sold to consumers openly.

Whatever the problems that may arise in the future for automobile standards, Congress is clearly persuaded that such direct regulatory action is effective, based on the evidence so far, and has extended the concept to other fields. In the National Energy Act, Congress required the secretary of energy to establish mandatory energy standards for consumer appliances, replacing previous statutory language making such standards merely permissive. Now, once a standard is set, no manufacturer may sell an appliance that does not comply. According to the statute, "Energy efficiency standards for each type (or class) of covered products . . . shall be designed to achieve the maximum improvement in energy efficiency which the Secretary determines is technologically feasible and economically justified."[19] The statute directs the secretary to consider the following factors to determine whether a standard is economically justified:

(1) the economic impact of the standard on the manufacturers and on the consumers of the products subject to such standard,
(2) the savings in operating costs throughout the estimated average life of the covered products in the type (or class), compared to any increase in the price of, or in the initial charges for, or maintenance expenses of, the covered products which are likely to result from the imposition of the standard,
(3) the total projected amount of energy savings likely to result directly from the imposition of the standard,
(4) any lessening of the utility or the performance of the covered products likely to result from the imposition of the standard,
(5) the impact of any lessening of competition determined in writing by the Attorney General that is likely to result from the imposition of the standard,
(6) the need of the Nation to conserve energy, and
(7) any other factors the Secretary considers relevant.[20]

19. National Energy Conservation Policy Act, P.L. 95-619, §422.
20. Id.

It is instructive to consider this list of criteria governing establishment of standards. If one believes that regulation should be invoked only to correct marketplace deficiencies, then regulations of economic behavior ought to force behavior that mimics the market as far as possible. Only item 2 in the list above appears to direct the secretary's attention to market considerations and there only in an imperfect way. But the presence of numerous other factors that the secretary is instructed to take into account blunts this simple economic test.

At the same time, the directives add up to a grasp that is too short. As we discuss later in this chapter, the price consumers pay for energy is not what is costs to provide it. Regulations and structure of the utility industry generally keep the price below the cost of replacing the energy. Regulations could correct for this market failure by directing the secretary to consider the marginal cost of energy in making the economic determination called for in item 2 above. Then at least the regulation would perform a useful service not performed by the market. But Congress, failing to take this step, limited itself to requiring that state utility regulators study rate reforms. No doubt the efficiency standards will force appliance manufacturers to be more conscious of that aspect of their engineering. But as drafted, they fail to direct the secretary to move toward a satsifactory substitute for market forces.

The third illustrative program is the development of the building code for energy performance. This program, called the Building Energy Performance Standards (BEPS), was established by EPCA in 1976.[21] The standards are to be performance based—that is, they should not specify how a particular building should be constructed, but rather should describe the desired energy performance of the building. At the time of writing, the standards are not available, even in proposed form. Nonetheless, their development has been available for public review; tracing this development illustrates how difficult it is to regulate as complex a system as a building.

The statute mandating BEPS requires that the standards be applied at the design stage and regulate the energy performance of a building. The government hired the American Institute of Architects Research Corporation (AIA/RC) to collect the data upon which standards would be developed. AIA/RC selected 1,661 buildings designed shortly after the oil embargo, on the somewhat dubious theory that the architects designing those buildings would have put the state of the art of energy conservation in their design stage. Using a proprietary computer program owned by the Edison Electric

21. P.L. 94-385, Title III, §301 et seq.

Institute, the buildings were "modeled" to calculate how many British thermal units per square foot the buildings would use per year. Unfortunately, the researchers did not go back to see whether the buildings as built performed as the computer predicted. As a cross-check on the computer simulations, a small sample of the original buildings were selected for redesign. Their original architects were given a three day course in energy conservation and told to redesign the building to save more energy. (The redesign instructions included the important limitation that the new building was not to cost any more than the original building. This instruction eliminated most capital improvements to make the redesigned buildings more energy-efficient, except that capital no longer required for large heating and cooling equipment was freed to be used elsewhere in the building.) Interestingly enough, the computer calculations indicated that four out of five of the redesigned buildings were more energy efficient than all but the top fifth of the designs in the original sample. This illustrates the power of paying attention to a problem and learning how to deal with it.

Economic calculations have been notably missing in the process. Cost figures for the changes were not made available, and little attention was paid to the question of money, either in the data-gathering phase or in the analysis. To the extent that economic analysis has been added to the standard development process relatively late in the game, it appears likely that the energy costs to be used are those actually paid by the homeowner to heat or cool a home. As we explain later in this chapter, those prices are typically below the cost to the nation of replacing the energy consumed—that is, they are below the marginal cost of the energy. Analysis of the regulations based solely on prices paid by the consumer will therefore understate the value to the nation of more energy-efficient buildings. Better design, more insulation, and more intelligent control systems must be bought at marginal costs while the fuel consumed is measured in average costs. The regulations, if promulgated in the fashion that seems most likely, will thus fail to push a traditionally sticky market toward economically more sensible behavior.

This failure to use the regulatory process for correcting deficiencies in the residential energy market is unfortunate because the housing market is almost a classic case in which intelligently conceived regulation has a place. Homebuyers do not generally think in terms of life cycle costing. Energy-conserving features are hard to evaluate, even in an older house, and almost impossible to judge in a new house where there is no history of utility bills to rely upon. Labeling would help consumers judge relative efficiencies of homes, but studies suggest that such labeling would be difficult to do accurately from a technical standpoint.[22] Therefore, a standard that was relatively easy to administer, that took into account the benefits of energy conservation to both the consumer and to the nation, and that permitted exceptions in cases where direct regulation was inappropriate[23] would have a great deal to commend it. It remains to be seen whether the BEPS will measure up to these tests.

These illustrations of regulatory standards for energy conservation are probably going to be joined by other actions taken by a Congress unwilling to wait for the slower, though more persistent, market forces to operate. For example, there has been discussion of making the voluntary reporting requirements for large industrial users the basis of a mandatory energy conservation program. Other examples of legislation seeking to correct market failures induced in part by price regulation may surface from time to time. Therefore, it is important to assess whether these programs have been a success and whether they should be emulated. Perhaps more important is to consider whether, even where energy is appropriately priced, some regulations may further enhance efficiency.

Assessment of Regulatory Programs. The major regulatory program that has had time to prove itself is the system of automobile efficiency standards. Yet even in this case, our analysis suggests that the undisputed early success of the standards may represent only the easy part of the path and that more difficult trade-offs lie ahead

---

22. See, for example, the series of careful experiments conducted over the past seven years at the Twin Rivers development near Princeton, New Jersey. That work suggests that identical townhouses can have actual energy consumption figures that vary by a factor of two. Moreover, the Twin Rivers studies show that there are many features of homes that have energy consequences discovered only after the home is lived in. For example, no computer simulation would have identified the heat loss associated with a wooden shaft surrounding a flue in a Twin Rivers townhouse; but that stack proved to be an important feature in venting warmed air into the attic where it was wasted. For an introduction to the important Twin Rivers study, see Robert H. Socolow, *The Twin Rivers Program on Energy Conservation in Housing: Highlights and Conclusions* (Springfield, Va.: National Technical Information Service, HCP/M4288-01, UC-95d, August 1978).

23. Flexibility in regulatory standards is often more easily achieved if the standard is expressed in terms of performance desired rather than the means that must be used to achieve it. For example, instead of telling automobile manufacturers that they must use lighter bumpers and smaller engines, the efficiency standards simply set a goal and let the manufacturers meet it in the most efficient method possible. Similarly, the BEPS are designed to set a certain energy efficiency goal, leaving designers free to meet it as they will. This approach generally gives more flexibility in administration and makes it possible to have fewer exceptions that must be granted by cumbersome administrative processes.

as standards are pushed to their limits. Nonetheless, their success is striking. According to background work commissioned for this study,[24] no other policy seriously proposed after the oil embargo would have had the effect that has been and will be realized by the efficiency standards. The analysis suggests that the standard may reduce long-run gasoline consumption by about 26 percent from what it would have been otherwise. New car efficiency is projected to increase by 47 percent, and vehicle miles by 8.8 percent. The net effect will be a drop in consumption of 26 percent. The gas guzzler tax (as originally proposed in the National Energy Plan) would have had only a modest additional effect in the long run—a reduction of some 1.3 percent in gasoline consumption. Taking another option, a gasoline tax that increased by 5 cents per year up to a maximum of 50 cents per gallon in 1985 would have affected consumption by only 6.8 percent in the long run. Thus, not only were such standards politically relatively easy to pass, but they have pushed fuel efficiencies in the correct (that is, economically efficient) direction.

Does this generally favorable experience give us any larger hope for regulation as a strategy to enforce energy conservation? The problems of using regulations for this purpose are well known: special cases are hard to deal with without making regulations so long and complicated that they benefit only lawyers; regulations often set standards that impose unacceptably large marginal costs of achieving the final increments; regulations often encourage greater efforts to beat them than to meet them; and regulations sometimes cause perverse results, so that, for example, automobiles are built to meet the mileage tests designed by EPA rather than to perform optimally on the highway.

But regulations also have some important positive effects. Where exact information is difficult or expensive to get for a consumer, regulations can help substitute for the "perfect knowledge" that a market economy presupposes. (Even though the regulator's knowledge may be no more perfect than the consumer's, at least in the aggregate, government is generally in a position to assemble somewhat better data than a consumer working alone.) Where energy considerations are a small part of any one individual's buying decision but a large part nationally, regulations may serve to limit the range of choices so that consumers are led to act in a way that will benefit them all. Regulations can serve where structural characteristics impede or frustrate operations of market forces. Finally, regulations may appear to be simpler to enforce where there are relatively few manufacturers of a product and many consumers.

24. See Sweeney.

Recognizing that regulatory strategies for energy conservation are one of the strategies that will be used by Congress and state legislatures during the period of transition, we urge that regulations be considered only after a careful examination of use of price and market systems. Further, marketlike strategies can often be used when no natural market exists for a desirable outcome.[25] Where regulations are used, they should be expressly based upon cost-benefit studies that take into account the marginal cost of energy to the nation, so that the regulation can serve to mimic a market. Finally, we believe that regulations characteristically outlast their usefulness and should ordinarily be subject to a review—one that includes reexamination of the cost-benefit ratio—no less frequently than every seven to ten years.

Carefully crafted regulatory strategies can serve to move the nation quickly on a path toward energy conservation while prices work their slower, more thorough adjustments, but the inflexibility of regulations, once they are in place, should make policymakers cautious about adopting them for every inviting target of opportunity.

Subsidies for Energy Conservation

Federal legislation now provides a few subsidies for energy conservation investments. Generally, the subsidies either offer tax credits for building improvements or make federally insured mortgage money more easily available for improvements that involve energy conservation. In addition, there is a relatively small program of outright grants to low income persons for energy conservation improvements in their homes.[26] The subsidies, taken together, are not large, at least when compared with direct or indirect subsidies for production of fossil fuels.

Our study group generally looks askance at subsidies. As a working rule, we believe that an investment in energy conservation—just like an investment in an oil shale facility, a central station generating plant, or a solar power satellite—should be forced to meet a market test. If the product is demanded by the market, it should be produced; if not, it has failed to be sufficiently attractive (when com-

25. For an example of the use of such a marketlike strategy for enforcing social goals, see Chapter 11.

26. Title II, part 2 of the National Energy Conservation Policy Act, P.L. 95-619, expands the weatherization program and coordinates three existing programs. Expenditures are limited to $800 per dwelling unit for materials; labor is generally donated or available under the Comprehensive Employment and Training Act (CETA). In rural areas where CETA labor cannot be found, an additional $700 may be expended per dwelling unit for labor. Funding levels were authorized to increase from $65 million to $130 million for fiscal year 1978.

pared to other consumer choices) to be afforded the use of public resources. Making conservation survive a market test is a powerful method for sorting good (that is, economically attractive) conservation ideas from bad ones. But notwithstanding this general rule, are there cases where energy conservation actions that are economically attractive should be made even more attractive by the use of subsidies? Three considerations lead us to the conclusion that energy conservation investments should be treated more generously in a contest for public support.

First, energy conservation can buy time for the nation in a swift way. Unlike other supply sources that may take a decade or two to bring on line, energy from conservation can in many cases be "produced" quickly and economically from millions of "decentralized sources." To be sure, at least a small part of this conservation energy comes initially not from genuine, economic conservation, but rather from curtailment. But changes in habit that are first perceived as being curtailment may lose that stigma. In most states, the 55 mile per hour speed limit is an example of changed preferences that reduce energy consumption. This ability to provide the equivalent of a quickly deliverable supply of new energy makes conservation especially valuable.

Second, energy conservation has economic advantages that are hard to find so well packaged in any other energy resource. By its nature, spending on energy conservation is mostly domestic, so that no trade deficits are involved (importing efficient foreign automobiles is an exception and perhaps a transient one). Inflation is generally reduced by energy conservation investments. (Of course, some of these advantages are diluted if an activity is so heavily subsidized as to compromise efficient use of resources.)

Third, the sources of energy conservation tend to be environmentally benign in their application and environmentally helpful in their effects. Although there has been some concern about indoor air quality in buildings that are sealed too tightly and research has been undertaken into the possible ill effects of fiberglass, those and other possible environmental effects of conservation are minor and manageable when compared with those associated with energy production, transportation, and use and even with imports. Moreover, to the extent that energy conservation programs reduce the need for conventional energy, the ill effects engendered by the supply-transport-use chain are eliminated—or at least delayed until a future day when improved technology might further mitigate their damaging effects on health and environment.

These three effects argue in favor of energy conservation as a general national strategy. But they are powerful arguments for favoring energy conservation investments specifically at a time when a large number of energy sources are competing for federal support. Conservation, as we have defined it, yields a greater return than many of those supply alternatives currently receiving support.

But if energy conservation investments make private economic sense for the person making them, should they be subsidized by the nation? Why reward virtue that carries its own reward? The answer has to do with the perversity of average cost pricing, already touched upon. As we shall see particularly in the case of price-regulated utilities, the price paid by the ultimate consumer is frequently less than the cost to the nation of the energy consumed. Power purchased during peak periods may be priced at only one-sixth or one-eighth of its actual cost. With this strong distortion in energy pricing, it is not surprising that energy conservation investments look relatively less attractive to the user of utility-supplied power than they actually are to the nation. Rate reform for public utilities should, therefore, be high on the national agenda. But the process will inevitably be slow in coming and painful. Until rates more nearly match marginal costs (at least for their final units of consumption), there is a strong argument for subsidizing those conservation opportunities that are attractive to the nation but unattractive to the individual consumer.

A second, less attractive reason to consider subsidy for energy conservation might be labeled the "another pig at the trough" theory. Virtually every conventional form of energy production and consumption receives some kind of subsidy, either directly or indirectly through the tax system. One study has estimated that the cumulated total of federal energy subsidies through 1976 approached $200 billion (in constant 1976 dollars).[27] The tax code is full of subsidies for production of depleting fuels and for their consumption but it remains relatively silent in rewarding conservation. To be sure, still additional subsidies are not a good idea and are difficult to remove once they have become part of a tax code. Perhaps subsidies of all kinds should be governed by an explicit phaseout or sunset provision. But if such a sunset provision is not adopted, it seems at best casual and at worst unprincipled to support subsidies for energy sources that are economically unwise and environmentally difficult, while turning away from subsidizing attractive energy forms such as conservation. It is difficult to be both principled and politically realistic in this area, but lack of principle may disadvantage conservation to the harm of the nation.

Finally, we believe that some energy conservation investments that are economically attractive to their owners are not made be-

27. Battelle Northwest Laboratory, *An Analysis of Federal Incentives Used to Stimulate Energy Production* (Richland, Wash.: Battelle, November 1978).

cause of inertia. There surely is room for small subsidies to act as "pump primers" for energy conservation. For example, some of the tax credits that have been granted for energy conservation investments may serve that purpose. They are not large enough to introduce fatal distortions into the market, yet they serve to spread the conservation ethic by drawing the attention of the public to the opportunity. Thus, we suggest as the general rule letting the market test whether a conservation activity should take place, but considering some subsidies as justified to correct market failures (in the case of utility-supplied fuels) or tolerable (as limited offsets to other subsidies).

PUBLIC UTILITY REGULATION AND ENERGY CONSERVATION

Natural gas and electricity each account for roughly 20 quads of domestic energy consumption. (Some 3 quads of that natural gas are used to generate electricity.) A great deal has been written in the press and elsewhere about the importance to the nation of the field price deregulation of natural gas that will take place under the National Energy Act, but few people note that at the other end of the pipeline—the local distribution company—little or no action has been taken to bring the full production costs of gas to the attention of its ultimate users. Instead, government has followed regulatory policies that result in allocations of gas, refusal of new hookups to potential users who would like to buy gas if it were offered for sale, and other devices that substitute for the allocation that would occur in a freely operating market.

Although electricity markets have not been disrupted by curtailments and allocation, price distortions at the consumer end prevail there as well. In this context, it is interesting that despite spiraling construction and operating costs, on average (expressed in 1972 dollars) the price to end users of residential electricity per kilowatt hour has gone only from 2.4 cents to 2.9 cents during the period 1973 to the third quarter of 1978. This amounts to only a 21 percent increase, less than half the rate of inflation during that period.[28]

The reason for the perverse behavior in these two important fuels is easy to discover: the distribution of natural gas and the distribution of electricity are treated as natural monopolies in this country, and even if price regulation of natural gas at the wellhead eventually disappears, gas pipeline and distribution utilities will continue to be

28. U.S. Department of Energy, Energy Information Administration, *Quarterly Energy Indicators* (Washington, D.C., January 15, 1979).

regulated. To the extent that energy conservation represents a fundamental part of the nation's energy policy, this regulation frustrates that policy.[29] Because of our conviction that price should be used as the basic means of encouraging energy conservation and measuring its success, an understanding of utility regulation and its peculiar impact on energy supply and demand is vital to any analysis of energy policy in the United States. This section describes briefly how utility rates are set. Next we describe some of the effects of the way that they are set. Finally, some suggestions for reforming those rates are described and urged for adoption.

The Art of Setting Utility Rates

In ordinarily competitive markets, price is not "set" in the usual sense of that term. Instead, through a process that matches the desire of the buyer for the lowest cost and the desire of the seller to obtain the highest price, a market price is arrived at, continually readjusting as conditions change. Quite a different concept has been followed in the case of public utilities. There, judges and legislatures observed early that certain services "affected with the public interest" could be provided to a community without costly duplication of expensive facilities by having only one supplier—that is, such services might be treated as though they were natural monopolies. A good part of the impetus for this treatment derived from a fear of the market power of those firms; regulation was also to protect the public.

In the United States, public bodies acknowledged these physical and economic facts in law by granting an exclusive monopoly territory to one company to provide a service. In exchange for the captive audience of consumers, the company gives up the right that a private enterprise customarily has to set its own rates and terms of service to its customers. The company has, instead, to propose rates and conditions of service that must be approved by a public control body. Although regulators seek to mimic the actions of a free market as if one existed, they have also explicitly adopted other goals. James Bonbright, author of a widely used text on public utility rate setting, has remarked, "to a substantial extent, sound ratemaking policy is a policy of reasonable compromise among partly conflicting objectives."[30] These objectives have included providing a fair rate of

29. See the broadcast remarks of President Carter to the nation, April 18, 1977, in which he stated: "the cornerstone of our policy is to reduce demand through conservation. . . . Conservation is the quickest, cheapest, most practical source of energy." Reprinted in Bureau of National Affairs, Inc., *Energy Users Report*, Reference File, p. 21:0665, November 17, 1977.
30. James C. Bonbright, *Principles of Public Utility Rates* (New York: Columbia University Press, 1961), p. viii.

return to the utility on its invested capital so that additional money will be attracted to the business as needed, disciplining the utility to ensure that it maintains efficient operations, controlling demand so that consumers don't overuse the product, and indirectly distributing income.

How are rates reflecting this natural monopoly operation actually determined? Prices for regulated utilities are generally set in lengthy and complex proceedings designed to establish a price level that will, in the aggregate, provide the utility a fair rate of return on its rate base. The "rate base" is calculated with reference to the original (current dollar) cost of the utility's existing plant, less the accumulated depreciation. Thus, the utility is told that it can expect to make a certain level of profits on its investments in capital equipment in addition to recovering its costs of actually conducting the business. This rate of return plus the expenses make up the total amount of money that must be collected from consumers for the service rendered.

Which consumer pays how much is decided in two steps. First, customers are grouped into broad categories (customer classes) that have similar usage and demand patterns. Different customer classes have different costs associated with serving them. For example, a large factory that buys many thousands of kilowatt hours of electricity per month at the voltage delivered from the high tension wires is cheap to serve: a single transmission line can carry a great deal of electricity; no transformers are required; only one meter is needed for a large sale of product. On the other hand, residential consumers have associated with their service expensive door-to-door meter reading, underground distribution lines, and relatively small sales per customer. For this reason, the rates that should be charged to different classes of customers vary for real cost-related reasons. By a complex process that has a larger component of judgment attached to it than many years of seasoning would make one suspect, the utility and the public control agency assign to each customer class responsibility for the equipment that members of that class use in common. At the end of this process, the total revenue that must be raised is thus allocated among the customer classes; traditionally, the three classes are residential, commercial, and industrial. Each of these three classes must then have rates set that will cover the amount of the total costs that have been assigned to it.

Thus, the final part of rate setting is determining the actual rates that will be charged within each customer class to ensure that the class provides its allocated share of the revenue. Here, there is considerable difference between the ways in which small and large users are treated. Recall that there are some components of servicing a customer that do not depend on how much electricity or gas he uses. The pipes or distribution lines, the meter reader's salary, the legal department, and the generating plant all cost the same amount of money no matter how much energy is purchased. On the other hand, the costs of supplying the energy do vary according to when the energy is consumed, for how long, and in what amounts. For residential customers, the energy component (kilowatt hours purchased or therms of gas consumed) and the fixed cost portion (constant costs) are usually rolled into one figure. To ensure that each customer pays a proper share of the fixed costs even if he uses only a small amount of energy, the utility prices the first units of consumption quite high or adds a separate customer charge to every bill; additional units of consumption more nearly approximate the variable cost of providing the energy. For larger industrial customers, the bill is typically divided into segments representing the energy purchased, the fixed costs associated with servicing the customer, and (frequently) a "demand charge" intended to compensate the utility for the cost of holding large amounts of capacity in reserve for occasional or peaking use by the industrial customer. (Put differently, the demand charge is assessed so that industrial customers have an economic incentive to use a relatively constant daily and annual amount of electricity or gas. Spikes and valleys in consumption by large users can impose enormous costs on a utility that must build capacity that sits idle except when a large demand occurs. The charge makes spikey customers help pay for that idle capacity or, better, encourages them to avoid making spikey demands.)

There is a final wrinkle in pricing that applies most dramatically in the case of electricity. The power company's actual costs of providing electricity to customers vary enormously according to the time of the day and the season of the year. During a hot, muggy summer afternoon in the East, residential users run air conditioners at the same time that stores, office buildings, and industries are making large demands. To meet this demand, the electric company must put on line older, less efficient generating plants that are kept only for times of large demand and may also have to add gas turbine generators that are cheap to purchase but have high operating and fuel costs. Conversely, in the middle of a winter night, the utility need run only its best baseload plants that use cheap fuel efficiently.

In spite of this variability in usage, the price charged to a customer at any one time to buy electricity has, with a few recent exceptions, been independent of the cost of producing it. Consequently, daytime electricity is too cheap; nighttime electricity is too expensive. A somewhat similar problem can occur for natural gas. If the utility has not stored enough gas, it must meet periods of peak demand by manufacturing gas from petroleum products and mixing this gas with the natural gas stream. Ideally, the decision whether to store gas against a peak day or to manufacture it should be made by comparing the marginal costs of undertaking either alternative and picking the cheaper alternative. Even though this is the way that the gas company decides which it will do, the customer is denied price information that would allow him to undertake a third alternative—not to consume the gas at all.

One can readily understand from a description of this process that any relationship between the price charged to the ultimate customer and the price that would be charged in a competitive situation is accidental. The process, as customarily carried out, makes no provision for the fact that the product (electricity or gas) costs different amounts to supply at different times of the day and year. It makes no provision for the fact that the prices charged the user of the service fail to reflect the costs incurred by the supplier to meet specific or new demands. Rather, these prices are governed by the costs of meeting the total demand, averaged more or less across all the existing units.

A final flaw in the economics of rate setting for regulated utilities is the fact that utilities have traditionally offered declining block rates. These rates have made it possible for large users of electricity or gas to pay less than small users for the final units of consumption. Justified in the past by the fact that scaling up the utility system produced savings for all customers, there are few if any places in the country where such declining block rates or other forms of volume discounts are justified today.

Effects of the Present Price Structure on Energy Conservation

Public utilities pay for their privilege of monopoly by agreeing to serve all customers. This agreement is both embodied in the terms of their franchise and embedded as a quasi-moral duty in the minds of utility executives. A utility's decision to add capacity to meet this legal and perceived moral obligation is generally based upon its projection of the demand for its service at regulated prices. Within this process, utilities do not compare the profitability of an investment in increased capacity with other possible investments that would provide the same services to society—for example, better insulation of customers' homes. Indeed, the system of utility price regulation and capacity expansion restricts a utility's ability to make such comparisons.

This system has generated increasing controversy in recent years, at least in part because of the increase in costs associated with additional capacity. As capacity expansion has necessitated large rate increases, the rate-setting process has attracted more participants. Much of the controversy has centered around utility rate structure, reflecting a common sense concern that declining block rates and volume discounts that encourage consumption should be reexamined in an era when reducing consumption costs less than increasing production. The issue has become more and more critical because of the recent gap between existing utility rates and the marginal costs confronting utilities. This gap has widened dramatically since about 1975 for three basic reasons:

1. Real construction costs have escalated so that the replacement cost of a utility's plant far exceeds its original cost. Because rates are based upon a fair rate of return on original cost, rates charged to users do not accurately reflect the costs of expanding capacity.
2. While the OPEC price action increased the market-clearing price for oil and gas, price regulation and the operation of long-term contracts have held oil and gas prices considerably below the estimated cost of an extra barrel of oil to the American economy (see Chapter 6).
3. Inflation itself can widen the gap between rates based on original costs and replacement costs by driving up the nominal cost of new capacity. But utilities are not allowed to account for inflation by writing up their assets to the level of replacement costs.

As a result of these trends, gas utilities that charge industrial users, say, $1.75 per thousand cubic feet (Mcf) are simultaneously adding some components of new supply at costs of between, say, $3.50 and $5.50 per Mcf, and some electric utilities that still charge industrial users a couple of pennies per kilowatt hour are busily adding capacity that ought to result in prices of perhaps 4 to 5 cents per kilowatt hour. Although it may mitigate the disruption associated with higher energy prices in the short term, this practice tends to produce a number of distortions in the economy.

First, the low rates encourage consumption by customers that can be supplied by the utility only at a higher replacement cost. Customers are discouraged from investing in energy conservation or non-utility substitute fuels, which may be cheaper than the costs of the new supplies a utility is adding.

Second, very expensive new sources of energy can be added to a system without any real market test. Because their costs are "rolled in" with cheaper, older sources of energy, no consumer is confronted with a price that accurately reflects these costs.

Third, customers' choices among utilities, both regionally and by fuel type, are distorted. The utilities with the lowest average rates have an advantage in competing for new hookups, while economic efficiency would dictate encouraging new hookups to utility systems with the lowest marginal costs.

For our purposes, the most important effects are the first two. An example may make them clearer. Let us suppose that there is a section of the country, much like the Pacific Northwest, where the bulk of the electric power is supplied by hydroelectric generation. Because of favorable living conditions, more people are moving into the area, and the demand for electricity to serve their needs is rising. Unfortunately, the sites to build additional hydroelectric generating facilities are now used up, so that new electricity must come from steam generating plants. The differences in the costs of supplying the new power, when compared with the old, are startling. The largely hydroelectric system supplies power at about 1.5 cents per kilowatt hour; the power to be generated by nuclear power plants, assuming that no other sources were available, would cost the consumer about 4 cents per kilowatt hour.

The service that the new customer desires is a warm home. The utility, under present conditions, can supply him that warmth in only one way—by building new capacity. If the price of electricity is 1.5 cents per kilowatt hour, the homeowner, examining the capital costs of providing himself with a warm home, will use electric baseboard heating. Looking at other alternatives, he notes that if electricity should go to 2 cents per kilowatt hour, it would be advantageous for him to insulate his attic heavily so that he would use less electricity in running the baseboard heaters. If the price of electricity were 3 cents per kilowatt hour, he would insulate, and he might install a heat pump, a piece of equipment whose higher cost would be more than justified by the energy saved.

Under present utility rate-setting practices, however, the utility would forecast the need for additional capacity, build a nuclear plant, and roll in the price of 4 cents per kilowatt hour from that plant with the price from the much larger hydroelectric base of 1.5 cents, so that the cost to the customers might only go to 1.6 cents or so. Customers would continue using baseboard electric heating and limit the amount of insulation in their attics. Some years later, when the system had continued to add more and more of the higher priced nuclear capacity to meet growing demands, the consumer might begin to appreciate that the marginal costs of new capacity had been hidden from him and that he chose unwisely.[31]

Obviously, there is a strong argument for some public policy measure that would correct the distortion and lead the prospective homeowner (or other category of user) to behave as if he or she were confronting the marginal costs that the system as a whole is experiencing. Using the above example, what are some conceivable policy choices?

1. Charge all electric users 4 cents per kilowatt hour for all their electricity.
2. Revise the rate structure and charge only new customers (those who join the system after a certain date or who add significant loads to the system after the set date) 4 cents per kilowatt hour. This system is called "vintaging" customers.
3. Develop a regulatory program to mandate insulation of roofs and to prohibit the sale and installation of electric baseboard heaters.
4. Allow the electric utility to help pay the costs of attic insulation and heat pumps and other devices that provide desired services more cheaply than building new generating capacity.

Unfortunately, only the first of these solutions really addresses the underlying economic problem of underpricing of the regulated energy source. Options 3 and 4 permit the "customer–utility system" unit to make only those energy conservation decisions that involve capital equipment. We might suppose, for example, that if the customer knew that electricity to heat the home would cost 4 cents per kilowatt hour, he or she might decide to set back the thermostat at night and use the money saved for something he or she valued more than a warm house at night. But the benefits to the system of that choice will be far less obvious to the homeowner if the higher marginal price does not call attention to the fact.

31. For some time to come, natural gas customers will behave in a similarly misinformed way. The Natural Gas Policy Act of 1978—a component of the National Energy Act—establishes a system that puts all of the higher costs of new natural gas first on industrial consumers, up to the point that the cost of gas causes industrial users to switch to alternative fuels. When all industrial facilities served by a pipeline reach the Btu equivalency of the cost of an alternative fuel, the higher rates will be shared by other customer classes as well. The effect of this provision is to hide marginal costs of natural gas from many customers and to encourage them to continue making unwise investments in gas-consuming equipment. (See P.L. 95–621, Title II.)

Option 2 gives the advantages of marginal cost pricing to new customers, but would be an administrative nightmare and might well be illegal in most states. New customers would be relatively easy to identify, although there would be substantial incentives to try to pretend that you were an old customer by slipping in under the old customer's name and not notifying the utility. For old customers increasing their usage dramatically (say by building a new factory next to the old one), there would be different rates applied to the two plants, or possibly, the new rates might then apply retroactively to the old plant as well (a positive discouragement to plant expansion in an energy-intensive industry). The political difficulties would also be immense. New customers would be a small group at first, but their numbers and concern would grow. It would be difficult to explain why it was fair that the new family on the block had to pay electricity bills more than double those of families that had bought a month earlier. Finally, most states, either by statute or in their constitutions, forbid rates that are "discriminatory." This has been interpreted to mean rates that are not justified by differences in costs of providing the service. No clear case could be made under the prevailing statutory tests of nondiscrimination that a vintaged, two part rate structure was fair.

As a final disadvantage, option 4 (letting the utility invest in conservation devices within the customer's own premises) has been limited severely by the provisions of the National Energy Act,[32] which prohibits utilities from making loans or supplying and installing energy conservation measures except in a number of limited cases:

• Utilities may install furnace efficiency modifications or clock thermostats or devices associated with load management techniques for the type of energy sold by the utility.
• Utilities may make loans not exceeding the greater of $300 or the cost of purchase and installation of the furnace modifications and load management techniques described above.

(The law contains a "grandfather clause" permitting continuation of existing programs, and the Department of Energy may grant exemptions to the absolute prohibitions if certain conditions are met.)

Option 1—pricing all electricity at its full marginal cost to all customers—is not without its disadvantages either. First, from a

32. P.L. 95-619, §216.

technical standpoint, marginal cost pricing for both gas and electricity implies that the rates charged should vary according to the condition of the system. During peak periods when more expensive gas supplies are being used or less efficient generators are turned on, the price should rise; during off-peak times, the price should fall. Few customers have meters that can record different rating periods. Thus, there would be the physical and capital costs of replacing customers' meters. It is likely, however, that the costs of "smart" meters will fall rapidly as the market for them expands; the problem is simply one of starting the ball rolling.

Second, changing to marginal cost pricing would mean that revenues would increase enormously for almost every utility. If we consider that massive income transfer unfair and that utilities should be limited to their traditional rate of return on investments as a measure of profit, then some method must be found to dispose of the newly acquired excess revenue. The most straightforward method would be simply to tax it away and return the proceeds to the general revenue fund, using that fund for some purpose that carried out general welfare objectives (such as helping low income consumers improve their energy-consuming homes and appliances). Alternatively, some method might be devised to return the overcollection to the electric and gas consumers by means of a credit to their bill. However, the credit would have to be independent of the amount of energy consumed to avoid defeating the object of the exercise— letting consumers know the marginal cost of the energy they consume. Possible rebate strategies include a flat, per meter rebate, a rebate on the first units of consumption (that are presumably used for necessities and are less elastic), or a rebate based on social policy grounds, helping individuals and lower income users while not rebating as much to higher income users or firms. (This last rebate strategy might again fall afoul of the prohibition on discriminatory rates, however.)

Any transition to a marginal cost regime would be enormously complicated by a fact that has nothing to do with conservation: charging customers of one utility the marginal costs of service may cause an undesirable switch to another fuel, unless we are confident that all fuels are priced at marginal cost. This lends a Catch-22 quality to efforts aimed at repricing gas, for example, to marginal cost. Unless electricity and oil are also priced at marginal cost levels, such pricing may lead to undesirable fuel switching. If price control keeps oil prices below replacement cost level, it is difficult to price gas and electricity correctly, lest this lead to a rise in oil imports. This problem is most acute with respect to industrial users who have substantial flexibility to switch between oil and gas, depending on relative prices.

The difficulties of the solutions that we tend to favor led the author of a background paper prepared for the study group to conclude:

> In general, the gap between marginal costs and current prices may be politically impossible to close. Instead, a combination of efforts to revise rate structure to provide more accurate price signals to consumers and various regulatory measures will probably have to be undertaken. Active utility involvement in energy conservation and alternative energy supply investments, appliance efficiency standards, building standards, and mandatory industrial efficiency standards may all be necessary to ensure that those conservation investments which are less costly than new supply will actually be made.[33]

This judgment may accurately reflect the political situation now. However, we believe that the benefits of making the prices paid by consumers of regulated utility services more nearly match the underlying costs are so great that considerable time, attention, and political skill should be devoted to the task. A number of statutory pressures will cause utilities to calculate their marginal costs of operation and expansion, and that information alone will help regulators shape rates that reflect long-run marginal costs. But the basic problem of limiting total revenue collections to a rate of return based on embedded historical costs will have to be confronted. Cautiously, and inevitably, with almost as many steps sidewards or backwards as in the correct direction, state commissions, whether acting on their own or prodded by federal law, will have to make certain that electricity and natural gas prices to consumers are set so that consumers see in their bills the marginal costs associated with higher consumption and the marginal savings associated with lower consumption. Overcollection of revenues is not, in our judgment, an insurmountable problem, since by law and regulation, the books and records of the utility companies are open for inspection and review by the regulators. A guiding principle to start with would be that the fruits of the overcollection should be given to the poor to help them meet the extra energy expenses that a marginal cost rate would entail.

33. Philip J. Mause, "Price Regulation and Energy Policy," in *Selected Studies on Energy: Background Studies for Energy: The Next Twenty Years* (Cambridge, Mass.: Ballinger, forthcoming).

The alternative to rate reform is continued distortion in the market and failure to exploit a large source of potential energy conservation. Likewise, solar energy and other forms of nonconventional energy forms will be discriminated against, and their market penetration slowed down. Economic energy conservation and a rising use of renewable resources should not be sacrificed to a short-term, illusory policy to keep prices below costs.

Economic Management of the Energy Problem

4

The long-run energy problem is best understood as a problem of rising costs that will have to be accommodated by rising prices. The pace and timing of the increases are uncertain. The rise may be gradual or may come in jumps. The outcome will depend on OPEC politics, on the availability of domestic substitutes for imported oil, and on demand.

Barring a persistent shutdown in supply by a major producing country, oil will reach about $30-$35 a barrel (in 1978 dollars) over the next fifteen or twenty years, a doubling of its real cost. What will be the consequences of such a doubling? Or, more precisely, how much will be lost in real income, consumption, and additions to wealth?

The answer comes in two parts. The first contains the terms of trade effect. This is the direct burden of having to pay OPEC a higher price. For each OPEC barrel we will have to trade larger claims on U.S. output and wealth. Simultaneously, as we shift to costlier domestic supplies of energy, there is the domestic productivity effect. It will take more resources to produce a unit of final domestic output. These we call the primary costs.

Second, an increase in the oil price, especially a large and sudden increase, can impose secondary costs—that is, they can exacerbate recession, inflation, and balance of payments difficulties and complicate the problem of fiscal and monetary management.

The task of this chapter is to sort out these primary and secondary domestic burdens, considering both the unavoidable costs of rising energy prices and the costs that may be avoided or reduced by economic management. A separate section discusses the international economic consequences of higher oil prices.

THE PRIMARY BURDEN OF INCREASING ENERGY COSTS

The contribution of primary energy to GNP can be indicated by two measures: (1) Properly priced at world market prices, the value of fuel used directly and indirectly in producing the GNP amounts to about 6 to 7 percent of the total GNP. (2) The value of imported fuel, the only component of energy supply whose real net cost to the United States could increase very quickly, amounts to about 2 percent of GNP.

These ratios are critical, because they set an upper limit on the aggregate primary burden imposed by increases in energy costs. Imagine, to take an extreme case, that OPEC doubles the real price of imported oil overnight. Such a large and sudden shock would pose acute problems. As the next section suggests, we would probably suffer large secondary costs. But in the short run, even before we had time to take advantage of substitution possibilities, the primary burden imposed on the United States would amount to an annual loss of 2 percent of potential GNP. In terms of productive capacity, the United States would be where it had been eight to twelve months before the increase; there would be an enormous loss in absolute amount, but hardly a catastrophe. Moreover, given time, we would begin to find substitutes for OPEC oil, and the relative burden would shrink.

If one is concerned with the near-term primary burden imposed on the United States by a large, rapid increase in the OPEC price, the relevant ratio is the 1.6 percent of U.S. productive capacity that is devoted to buying energy from abroad. Any increase in the price of domestically produced fuel would transfer real income from oil users to owners of domestic oil but would not increase the burden on the United States in the aggregate.

If, however, one is concerned about the cost of all energy doubling over, say, fifteen to twenty years, then the ratio that matters is the ratio of all energy costs to GNP, currently about 6 to 7 percent. Since energy use is growing somewhat more slowly than the GNP, that ratio is likely to be no greater than 5 to 6 percent in fifteen to twenty years. It follows that a doubling of the real price by then would entail a "burden" amounting to between 5 and 6 percent of

GNP, in the absence of any substitution or adjustment to the rise in price. With plausible allowance for adjustment, the annual primary loss would be less than 4 percent of potential GNP.

In twenty years, the GNP is likely to amount to $4 trillion ($10^{12}$) or more in 1978 prices, so a loss of 4 percent would be enormous, but in relative terms, once again, it would present no catastrophe. Twenty years from now we would find ourselves permanently about 2 years behind where we would have been if the real cost of energy had remained what it is today.

The precise size of the primary burden imposed by a doubling of the oil price will be sensitive to the adjustment responses of households and businesses. Given time for adjustment, and if we let the price system do its job, substitution will substantially reduce the primary burden. But even if we allow prices to mediate energy scarcity —and perhaps especially if we do—a large sudden oil price increase may impose what we have called secondary costs. The next section of this chapter explores these possible secondary consequences, particularly the connection between oil price shocks and inflation and recession.

THE SECONDARY BURDEN: INFLATION AND RECESSION

Between 1973 and 1975, while the price of imported oil increased fourfold, the United States suffered its worse recession since the 1930s, and inflation shot up to double digit figures. It is widely thought that the rise in the price of oil caused the setback to the economy, but there is confusion about the precise nature of the link. Were inflation and recession unavoidable consequences of the increase in the price of oil, or could they have been prevented by a different policy? Would another doubling of the price of oil have the same effects?

To answer these questions, we must explore the connections between increases in the price of oil and the economic choices facing the United States—choices concerning total output and income, employment, inflation, the balance of payments, and the distribution of income. From the perspective of policy, we must also examine the special problems posed by an energy price "shock" for fiscal and monetary management.

Neither recession nor inflation can be attributed mainly to rising energy costs. Those maladies have to do, rather, with the interaction among three things:

1. Disturbances to which the economy is subject as a matter of course, including changes in technology, in preferences of consumers and asset holders, and in government policy, as well as the vagaries of international politics and of the weather—in fact, any combination of events that gives rise to rapid shifts in demand or supply in particular markets that are of appreciable size relative to the economy as a whole;
2. Structural characteristics of modern industrial economies that condition the behavior of wage rates and prices in particular markets; and
3. Policy responses of government, as influenced by institutional constraints, imperfect information, uncertainty, conflicts of interest, and conflicts among goals.

Energy is relevant because an increase in its price constitutes a disturbance in the above sense. An increase in the price of imported oil works much like an excise tax levied by OPEC. The increase directly affects the price level, as measured by a weighted average of prices such as the official Consumer Price Index (CPI) or the broader GNP price index (the "GNP Deflator"). It also diverts spending from home-produced goods to imports, reducing total spending (aggregate demand) for U.S. output.

An increase in the price of domestically produced oil is like an excise tax levied by Americans on Americans. It, too, affects the price level and may reduce total spending for U.S. goods. (The cutback in spending by consumers of oil, whose real incomes would fall because of the higher oil price, is likely to exceed the increase in spending by producers of crude oil and land- or leaseholders whose incomes would rise.[1]).

The ultimate quantitative consequences for inflation and output will depend on the size and timing of the oil price increase; on the effect of the consequent price level increases on wage settlements; and on the fiscal, monetary, and regulatory response of government. We believe that real oil prices are headed higher, though the path is likely to be irregular. For purposes of analysis, however, we will explore

---

1. If the government taxes away the royalties, the net effect on aggregate demand will depend on the size and distribution of the associated rebate. (An increase in the price of domestic crude does not impose a primary, terms of trade cost on the United States. It does redistribute pretax income from oil consumers to crude producers, lease- and landholders, and so on.)

two special cases: (1) a persistent increase in real oil prices of 4 percent per year, which would result in a doubling in about eighteen years; and (2) a sudden, one time increase of 50 to 100 percent sometime in the next decade or two.

A Gradual Increase in Oil Prices

A continuing increase of 4 percent per year in the real price of oil would not in itself reduce aggregate demand enough to damage output and employment. The associated year-to-year shifts in net spending from home goods to imports would be just one more datum facing the fiscal-monetary managers, a sort they must routinely cope with. Moreover, the direct effect on the inflation rate, while it would persist and grow, would also be small. It would be small because the initial share of the GNP for imported oil (2 percent) and for all oil (4 percent) is small. To illustrate, a persistent real increase of 4 percent per year in all oil product prices would initially raise the annual inflation rate, as measured by the GNP price index, by about one-sixth of a percentage point (.04 × 4.0). Even after eighteen years and a doubling of the price of oil, the direct effect on the annual rate of increase in prices would be less than one-third of a percentage point. (Throughout, it would accurately reflect the transfer of income from oil consumers to OPEC and domestic crude producers.)[2]

Would such a slow rise in the price of oil cause any secondary damage? It could. If the slowly accelerating rise in the CPI were to provoke a catch up reaction in wage settlements, either inflation would accelerate more quickly or—if the government responded by tightening fiscal-monetary policy—income and employment would be lower. However, the extra damage, while uncertain in either case, is not likely to be large; even if it were, it would be a mistake to blame the oil price. If a small price disturbance can cause a lot of trouble—

2. The leverage exerted on the price level by an annual 4 percent increment in the oil price would grow, because the 4 percent would apply to a growing portion of GNP. Even with time for adjustment, the price elasticity of demand for oil is less than unity, and hence the share of oil in GNP would increase with the rise in its price. The direct result would be a gradual upward creep in the inflation rate. (The relevant "share" is the total value of oil used up in the economy in both intermediate and final use, divided by the GNP.)

The above arithmetic assumes that the oil price increase would not affect the exchange rate. If the incremental leakage into net imports were not balanced by capital inflow, the dollar exchange rate would decline, and that would give the price level a further upward push and, with a lag, cushion the decline in aggregate demand. However, a variety of other influences are likely to dominate the effect on the exchange rate; there is no reason to think that an increase in the oil price would in and of itself cause the dollar to decline in the exchange markets relative to other oil importer currencies. This issue is discussed further below.

that is, if wage settlements are ultrasensitive to acceleration in the CPI, making the price level incipiently unstable—we are in trouble in any case. An oil price rise would simply make the inflationary bias in our wage- and price-setting institutions a little more onerous.[3]

A Rapid Increase in Oil Prices

A large, rapid increase in the real price of oil, such as an increase of 30 to 50 or even 100 percent in a year, would be a different matter, even if it happened only once in a decade or two. Such a major disturbance is much more likely to inflict substantial secondary damage, on top of the direct, terms of trade cost.

For one thing, it would cause a large incipient drop in total spending for U.S. goods. Currently, a doubling of the price of oil in a period of one year would result in an initial reduction in annual spending by some $35 to $45 billion, even if the associated proceeds to domestic crude producers were either spent by them or taxed and rebated to consumers in general. Because the downward shift would be large and rapid, government money managers would find it hard to implement measures promptly for fiscal and monetary compensation. Moreover, there would be a danger of an inappropriate wage response unless an effective wage-price standard policy were already in place.

In order to offset the reduction in aggregate domestic demand, the government would have to cut tax rates, increase expenditures, or both and to speed up the growth in money supply, despite the transient increase in the inflation rate. In the absence of a wage reaction, such an expansionary policy would pose no dilemma. The difficulty would consist in making a diagnosis quickly and confidently of the

3. We would be in trouble in any case, because the economy is asymmetrically subject to upward price disturbances from many other sources besides the oil price, such as the weather, a commodity boom abroad, or a fall in productivity growth. In the absence of corrective policy, accelerating inflation would result.

The pace at which a CPI-induced wage response would speed up the inflation rate, assuming that fiscal-monetary policy tried to maintain output and employment, would depend on the institutional calendar time lag in the wage response and on the size of the response "coefficient." Until recently the lags have been generally quite long and the coefficient quite small. However, they have been changing for the worse, and with the inherited, predisturbance inflation close to double digit rates, there would be at least a small risk of an inflation acceleration scare.

With respect to the other horn of the dilemma, there is a great deal of uncertainty about how much extra permanent unemployment would be needed to prevent acceleration. We do know that slowing down inflation once it has speeded up would require a lot of transient unemployment, as well as perhaps some permanent increase. Nevertheless, when all is said, real trouble is likely to develop only very slowly, if at all, from such a gradual increase in the price of oil.

size and one shot character of the oil price increase and in avoiding delays while Congress and the executive branch argue about which expenditures to increase and whose taxes to cut.[4]

If there is no wage catch up, the secondary cost of a large, rapid, one shot increase in the price of oil—the extra cost on top of the terms of trade cost—would depend on the size and speed of the government's fiscal and monetary response. But because a doubling of oil prices would add about four percentage points to the cost of living, there would be pressure for wage and other cost of living adjustments. The government would therefore confront a dilemma. The acuteness of this dilemma would depend on the magnitude, timing, and persistence of the wage reaction, and on its sensitivity to extra unemployment.

To take a simple example, assume that wage boosts matched price increases percentage point for percentage point, but with a one year lag. In that case, if the oil price doubled in one year, the permanent underlying inflation rate would increase by about four percentage points. A fiscal and monetary policy designed to avoid a large secondary burden of lost income and extra unemployment—a nasty recession—would have to accommodate that four point increase in the pace of wage-price inflation. And the increase would persist long after the price of oil stopped rising.

Under our one shot assumption, the real oil price stops rising at the end of a twelve month transition period; thereafter, it remains at its new, higher level. Why then would the rate of inflation end up permanently higher? Refer to our example. Wage rates, after failing to respond during the transition year, begin rising at an annual rate that is four percentage points faster than before, due to a futile attempt by workers to offset the 4 percent loss in current real income. If there is no accompanying change in the growth rate of labor productivity (measured in home-produced goods), a four point jump in the rate of wage increase will result in a four point jump in the rate of increase of producers' labor costs per unit of output. To offset the resulting squeeze on profit margins, producers will respond by marking up their prices. As long as wages rise four percentage points faster than before, prices will rise about four percentage points faster than before, with zero gain in real wage rates.

4. It is possible that the larger oil price rise would make it easier to take compensatory action. The very size of the increase might make its one shot character more plausible, and the need for fiscal-monetary action, in the light of the 1973-1975 experience, would be dramatically clear. A lot would depend on the inherited rate of wage-price inflation. If it was 2-3 percent per annum, expansionary fiscal and monetary measures would be easier to take than if the pre-disturbance inflation rate was already at double digit rates.

Workers bid up wage rates, trying to evade the unavoidable loss in real income inflicted on them by the increase in the real price of oil. They will succeed only in pushing up the price level. It is like a dog chasing its own tail: wages chase other wages and prices, pushing up costs; prices respond to costs; and the government lets the money supply expand to keep money markets from tightening because the alternative of tight money (and tight fiscal policy) will compress real output and employment quickly and a lot, and it will decelerate inflation only slowly and a little. In general, the faster and more persistently wages respond to the CPI, the worse will be the effect on inflation.

Mainly because it does not really know how to "offset" just right, the government might prefer the other horn of the dilemma. If it does, it can try to use fiscal and monetary policy to create enough extra unemployment to offset the wage response and prevent a rise in the permanent inflation rate. But either way, a large, rapid increase in oil prices would impose an extra burden of inflation or unemployment, or some of each, on top of the direct cost of higher priced oil.

What Is the Likely Wage Reaction?

What is the wage reaction likely to be, especially in the absence of an effective policy setting standards for wages and prices?[5] Recent evidence is inconclusive. It suggests that the price shock of the mid-1970s had only a modest direct effect on wage rates. But that result is both incomplete and uncertain. Because wages and prices tend to move together, it is difficult to disentangle the effect of prices on wages from the momentum exercised by past wage rate increases.[6] Moreover, even if the effect was small, that fact would not yield much reassurance for the future. Cost of living adjustments (COLA clauses) are increasingly prevalent in wage contracts. More important, it seems unlikely that the response of wage rates to changes in consumer prices in a not very tight–not very loose labor market—and especially the response of the politically visible, pattern-influencing bargains—is governed by any predictable regularity that is invariant to recent inflation rates, to the timing and shape of the price disturbance, and to political circumstance. The uncertainty is intrinsic.[7]

5. And in the absence of significant reduction in real demand—that is, at the previous rate of unemployment and capacity utilization.

6. The evidence is incomplete because it is not incompatible with a large *indirect* price on wage effect. If a speedup in price inflation causes a jump in a few large, visible settlements, it might perturb the wage structure and set in motion a cumulative process of wage-chasing wages.

7. The issue concerns the nominal outcome of a set of interdependent, strategic bargains whose nominal results are not at all solidly anchored in the real econ-

Pessimists and optimists will view the fact of uncertainty in different ways. The pessimists will observe that a rapid one shot increase in oil prices might have a large and persistent effect on the underlying rate of wage-price inflation. The optimists will respond that that outcome is not inevitable. A policy of wage-price standards might significantly reduce the secondary damage—even an informal, one shot policy that deals with a sudden spurt in oil prices and that relies on public education about the one shot character of the price rise and about the futility in the aggregate of everybody trying to catch up. The objective of such a policy would be to help us absorb the primary reductions in real income and real wages, without either a persistent speeding up of inflation or the imposition on ourselves by a noncompensatory fiscal monetary policy of a significant additional cost in unproduced real income and extra unemployment.[8]

What Conventional Fiscal-Monetary Policy Can and Cannot Do

A crucial point in all this is that a fiscal and monetary policy that tries to suppress a catch up reaction in nominal wage rates by letting the decline in aggregate nominal demand run its course would have no direct anti-inflationary effect on wage rates and prices. It can affect inflation only to the extent that a reduction in aggregate spend-

---

omy, in contrast with real wage rates, which will move with average productivity.

If the situation is perceived as a strategic bargaining process, the monetary authorities are of course an implicit "participant." Indeed, in one view about expectation formation, they are the dominant participant. If only they would credibly commit themselves to a hard-nosed monetary and fiscal policy that would under no circumstances accommodate a speedup in the inflation rate no matter what the consequences for output and employment, or so the argument goes, then wage settlements would quickly fall into line. Even on its own terms, the proposition is dubious. Moreover, the prescription is virtually impossible to implement. A government cannot credibly commit itself to persist in a course of action no matter what disastrous consequences ensue.

8. In principle, it would be possible to design a package of expansionary fiscal and monetary actions that would both offset the decline in aggregate demand and reduce the once-for-all rise in the price level, thereby complementing direct wage-price policies aimed at containing a wage reaction. A sufficiently large reduction in indirect taxes (payroll, excise, and sales taxes), combined with monetary ease and reduction in interest rates, would help counter both demand contraction and the direct price level increase. Unfortunately, this type of flexible use of fiscal policy, taking advantage of the double-edge character of indirect taxes, would—even if the government knew exactly how to do it—face severe political and institutional obstacles, such as the resistance against using income tax revenues to finance social security, because of the actuarial fiction. Moreover, unless one is prepared to contemplate more elaborate tax-based wage-price policies, negative indirect taxes, and the like, we could not often rely on excise tax cuts. The federal government collects only about $25 billion in excise taxes. For repeated doses of expansionary medicine that would also help contain cost-push inflation, one would have to rely on payroll and state and local sales tax reductions, with compensating federal subsidies.

ing, and the consequent increase in excess supply in goods markets and labor markets, elicits a cutback in wage settlements and in prices rather than in production and employment. (The only direct effects on the price level are actually in the "wrong" direction. Tightening money, increasing interest rates, and increasing indirect tax rates will show up as an increase in the official CPI.)

Unfortunately, whenever the U.S. economy suffers reductions in aggregate nominal demand in orders and sales and a pile-up of unwanted inventory, most mainline industries respond by cutting back output and laying off people; they shave prices only a little and very gradually. The key to this downward stickiness of the price level and, indeed, to the persistence of an "inherited" inflation rate is wages. Speaking broadly and loosely, industrial prices pretty much follow variable costs, normalized for operating rates. Variable costs are dominated by wages. And wage rates tend to keep rising at least at their recent, inherited rate—chasing prices or each other or both, and pushing up prices still further. Wages will respond to unemployment —but only gradually. The evidence does not contradict the qualitative hypothesis that a reduction in nominal demand and increase in unemployment will slow and eventually stop inflation. The evidence suggests, however, that it will do so only very slowly and at large cost in output and income.[9]

MANAGING DEMAND FOR ENERGY
DURING SHORT-TERM CRISES

Whatever we do about energy in the long run, we shall be continually susceptible to "oil crises" for the next fifteen or twenty years. By

9. In a perfectly competitive, stably self-adjusting economy where prices and wage rates in every market respond reliably and quickly to excess supply—for example, to a drop in demand—there would be no problem of secondary burden, at least no macroproblem. Unfortunately, in a modern industrial economy, wages and prices do not so respond.

Because an incipient wage reaction is the villain of the story—that is, an attempt to evade the relative reduction in real income imposed by a deterioration in the terms of trade by securing larger nominal wage settlements that will succeed only in pushing up prices even further—it may be worth noting here a paradoxical possibility regarding the relative undesirability of a large, once-for-all increase in the oil price, as against a persistent upward creep of equivalent discounted primary cost. If one could rely on a full and quick fiscal-monetary response to compensate for the large drop in aggregate demand; if one could count on increases in the oil price being passed through into final product prices quickly, in a few months; and if a temporary wage restraint policy in response to a dramatic one shot oil price increase were politically more acceptable than living with one permanently, then it may be that a large, one shot increment in the oil price would in fact impose a smaller secondary burden due to an inappropriate wage reaction than would an equivalent gradual increase. But that is likely to be many too many "ifs."

"oil crises" we mean sudden interruptions, interferences, or curtailments in the supply of oil, expected to last months rather than years.
A crisis could arise through internal events that interrupt supply,
such as the turmoil in Iran in early 1979. It could arise through attempted embargo by one or more supplying countries, as after the
October War of 1973. The problem would be "short term" in two
respects: there would not be a long warning time, and policymakers
would not believe the situation likely to last for several years. The
problem would be "long term" in two respects: it might last indefinitely, without being recognized as of long duration; and the continuing threat of such crises will be with us through the next twenty
years.

For at least two reasons, it is likely that any such crisis would
bring about some U.S. regulation of oil imports and some U.S. control over refinery products. One reason is that the obligations undertaken in the framework of the International Energy Agency (IEA)
might make U.S. government intervention necessary, not only to
ensure that the United States shared in the reduction of imports but
to demonstrate the readiness of the U.S. government to take steps to
make its IEA obligations a reality. The second reason is that imports
might be coming in at widely different prices, as spot prices jumped
around above long-term contract prices, creating a disorderly import
market, worse than what occurred in the spring of 1979.

Any crisis will entail uncertainties. There will be uncertainty as to
duration, as to severity, and as to whether and how IEA obligations
may be invoked or adhered to by other countries. In any prior contingency planning, there will be uncertainty about inventory levels
and the status of a strategic stockpile, about the season of the year,
and about the state of the economy when the crisis occurs.

Except for efforts to use strategic reserves—a subject discussed in
Chapter 1—or to affect the inventory policy of fuel companies, the
microeconomic problem will be one of matching demand to limited
supply. That is, an oil crisis will pose an unexpected short-term problem in which changes in aggregate supply will not be part of the
solution. Some substitution of one fuel for another can be part of
the solution, but mainly the problem will be to reduce demand.

The ideal system for reducing demand for energy will do the
following:

1. Maintain efficiency in the distribution of oil products and fuels;
2. Distribute hardships, by region, by income level, and so on;
3. Absorb the inflationary impact;
4. Fulfill international obligations; and
5. Avoid adverse effects on longer term energy problems.

The policy variables to be considered are prices, taxes, consumer rationing, allocations to intermediate suppliers and industry, and various demand-reducing regulatory activities. One crucial policy question is what to do about gasoline. A second is what to do about home heating oil.

Gasoline

Many combinations of policies to conserve gasoline are possible. There can be price control with or without rationing. There can be regulations to cut demand, such as restrictions on when people can drive, where they can drive, what vehicles they can drive, who can drive, when or where gasoline can be sold, and how fast people may drive. And these regulations may occur with or without price control, with or without rationing, with or without special gasoline taxes. There can be special gasoline taxes, per gallon or levied on price increments; and these taxes could be imposed with or without specified rebate provisions. There could be gasoline stamps or gas tax stamps for the poor. And there can be specific allocations for privileged and emergency vehicles or drivers.

We might identify at least three different levels of severity or permanence of an oil crisis. The least serious would be moderate curtailment of supply expected to be of short duration, inviting temporary regulatory interventions by the government to moderate demand but not requiring comprehensive rationing or allocation, price controls, or special taxes. The second would be a crisis of sufficient severity and anticipated duration to make desirable some comprehensive system of intervention by the federal government. And a third might be a change in the world supply situation that creates a sudden, drastic, and apparently permanent change in the cost of oil.

The third of those situations we believe should not be treated as a "crisis" but as part of the long-run problem of adjusting to rising fuel costs. Throughout this report we recommend that rising costs of fuel be recognized and adapted to and not suppressed through efforts at permanent regulation. Some of the cost increases may come suddenly rather than gradually, and this may justify some efforts to moderate the inflationary impact, but that fact should not be considered a sufficient reason for abandoning primary reliance on the price system. A system of energy demand management should be designed for coping with short-run crises. It should be temporary, arbitrary but workable, and have a minimum of administrative infrastructure, on the grounds that inequities can be tolerated for a short period and that the crisis should be passed before any detailed perfectionist system could be erected.

Many people appear to think of gasoline rationing as an easy solution to a severe oil crisis. We are impressed with the enormous complexity of any system that would attempt to allocate motor fuel equitably and efficiently among the more than 100 million vehicles that burn gasoline or diesel fuel. The complex and inherently arbitrary system of gasoline rationing that this country experienced within living memory enjoyed a number of advantages that are unlikely to be present during a contemporary fuel crisis. First, and most important, there was a war on. Nobody questioned whether the shortage of fuel was genuine. Gasoline was but one of many vital commodities in short supply; comprehensive price and wage controls and allocation of labor and materials were accepted, and inequality in the distribution of hardships was known to be inevitable. Second, there were shortages of tires, repair parts, and vehicles themselves, making gasoline just one of several limitations on driving. And third, rationing was a comprehensive administrative task of long duration, warranting complex and time-consuming procedures to hear and adjudicate appeals, to elaborate gradations of severity, and to construct systems of enforcement.

Any system of gasoline rationing will be crude, arbitrary, awkward, and imperfect. Any rationing formula must discriminate among urban and rural drivers whose average mileage differs by several hundred percent; people with old cars and those with new cars with gas mileages that differ by 50 percent or more; families with several drivers and several cars and those with only one; and people who drive different distances to work, school, or shopping centers. This is not an argument against rationing in an emergency; it is an argument against believing that any kind of rationing can be other than an arbitrary and crude approximation to an equitable distribution of hardship in an emergency.

It is important to distinguish between mechanisms of demand management that are intended to allow a maximum of flexibility and individual choice, working mainly through a price mechanism, and mechanisms that are primarily regulatory, prohibitive, and allocatory, without a significant role for market prices.

Our recommendation, for any but the most temporary of crises, is to make maximum use of the price system. One way to do this would be to levy a federal tax on price increments in excess of some base date prices, coupled with a rebate system to dispense the revenues. The base price should be uniform by locale and not be the prices of individual vendors, to avoid discriminating against lower base prices. Another approach would be a consumer rationing system with transferable rations, under which the federal government would encourage a well-informed and smoothly working market in rations.

If the distribution of rebates corresponds to the distribution of rations, the two systems may have about the same results.

Marketable rations are far superior to nontransferable rations in three respects. First, they encourage conservation by raising the marginal cost of driving but not the average cost within the rationed amount, permitting those who prefer money rather than gasoline to relinquish gas to those who would rather drive than save money. Second, by minimizing the worst of the hardships that might be associated with an inefficient distribution of rations, transferable rations will make the system politically and administratively more tolerable. And third, though they may possibly make the counterfeit coupon problem somewhat more difficult, transferable rations will eliminate awkward and inefficient black markets in gasoline and reduce gasoline theft.

A ration of ten gallons per week with pump prices at 80 cents and ration coupons going for 30 cents a gallon leaves drivers in the same position as if they paid $1.10 a gallon at the pump and received a weekly rebate equivalent to $3. Assuming that all rations are used, so that the size of the weekly ration is not affected by the choice of system, the least satisfactory of the three systems is the nontransferable ration of ten gallons per week. Those who would prefer to drive less and save $1.10 per gallon and those who would rather spend the $1.10 a gallon to drive more are both frustrated by the absence of a market that allows both sides the gains from trade. Either transferable rations or the flexible tax rebate system would practically eliminate any need for additional arbitrary measures like holiday closings or mandatory car pooling.

The only proper role of these arbitrary interventions is to avoid a major system of taxation or rationing, with all the administrative infrastructure that may be required. The piecemeal approach is therefore acceptable for the smaller crises.

Home Heating Fuel

Heating fuel differs from gasoline in that virtually all of it, at least in urban and suburban areas, is by subscription instead of cash purchase. The suppliers have regular customers and keep excellent records of deliveries in past years. As a result, the government could ration heating fuel to single family or multiple unit residences by allocating the fuel to supplying companies. If the announced purpose is to ration fuel to users in proportion to consumption in prior years, with regional and local allocations partly determined by degree day and wind chill factors, there will remain mainly the problem of new customers. And it seems unlikely that people will deliberately pay the cost of extra fuel consumption to establish a "base year" record that would raise their fuel allowance.

A system of this sort would entail price controls without the problems and inefficiency that price control of gasoline would entail. The price controls would presumably be of about the same duration as any rationing or special tax that might be applied to gasoline.

INTERNATIONAL ECONOMIC AND FINANCIAL CONSIDERATIONS

Rising oil prices are a source of recurring concern about the future of the world economy and the durability of the international economic system. The reasons follow from the special importance of oil trade. First, oil trade is very large, amounting to about 15 percent of the non-Communist world's trade; hence, changes in oil prices can have a substantial effect on balance of payment positions and ultimately on trade policy. Second, the Western (OECD) countries are heavily dependent on oil imports, which constitute about three-fourths of their oil consumption and almost 40 percent of their energy consumption. Third, at present prices, oil amounts to about one-fifth of the value of all imports for the developing countries with oil deficits. These nations contain the predominant portion of the developing world's people, and their economic prospects are usually constrained by the availability of foreign exchange. Fourth, control over world oil exports is inherently oligopolistic, and export supplies are potentially volatile and uncertain. Two-thirds of world exports originate in six countries that have a combined population of less than sixty million and are situated in a region of chronic internal tension and long-standing great power confrontation.

All this came into sharp focus when the Arab oil exporters interrupted supplies and OPEC quadrupled oil prices in 1973. By the end of 1974 the oil price increase had transferred $75 billion to the oil-exporting countries, or about 2 percent of the total output of the oil-importing nations, and added measurably to cost-push pressures in a world already suffering from inflation. Furthermore, since most of this sudden, huge income transfer initially took the form of financial assets (that is, claims on future purchases, or IOUs) rather than current imports of goods and services, it had a depressing effect on demand, output, and employment in the OECD countries and brought about unprecedentedly large deficits in their international payments.

The seeming weakness of corrective forces made doubts about the economic future even more severe. Because of the unique structure of the oil trade, it was widely believed that annual OPEC financial surpluses would continue indefinitely at a high level. If so, the reasoning went, accumulating deficits elsewhere in the world, which were the necessary counterpart of OPEC surpluses, would bring the international economic system close to collapse. Specifically, some industrial countries and many developing countries would become financially insolvent, threatening a banking panic; international institutions and arrangements to finance deficits would come under insupportable strain; the industrial countries would resort to trade restrictions in a self-defeating effort to correct their payment deficits; the OPEC countries would soon own a large and steadily growing proportion of the world's productive assets; and to complete this doomsday scenario, Saudi Arabia and the other sparsely populated oil exporters, having all the financial assets they could contemplate spending, would keep more of their oil in the ground, curtail exports, and thereby further disrupt economic output.

This grim forecast proved to be a gross miscalculation, principally because corrective forces were much stronger and the international economic system more flexible than had been supposed. Nonetheless, the economic cost of the 1973-1974 oil shock has been substantial— certainly large enough to require assessment of possible dangers for the future and consideration of the need for special action, either to reduce oil imports so as to make the system less vulnerable to OPEC actions or to establish additional backstop facilities to cope with future pressure points.

This section addresses the consequences of still higher oil prices on the course of world trade and on markets for foreign exchange. But first, it is instructive to review the adjustments that followed the events of 1973. Then we discuss (1) the connection between oil prices and the market for dollars and other currencies and (2) possible future courses of international adjustment to changes in world oil supplies and prices.

Adjustments to the 1973-1974 Price Increases

How did the international economy adjust in the medium term to the quantum jump in oil prices? There is no simple answer to this question because other important factors were at work. The data presented in Table 4-1, showing the changing composition of the world current accounts in the first five years of higher oil prices, provide a starting point to trace what happened. Several points stand out.

First, contrary to what the pessimists expected, the OPEC current surplus dropped dramatically from $68 billion in 1974 to $9 billion in 1978. The drop would be seen as even more pronounced if allow-

Table 4-1. World Current Account Balances, 1973-1978[a] (In billion U.S. dollars).

	1973	1974	1975	1976	1977	1978
OECD countries[b]	19	-15	16	-4	-8	22
United States	9	7	21	8	-13	-13
Japan	—	-5	-1	4	11	17
Germany	7	13	8	8	8	13
Switzerland	—	3	4	4	6	6
Canada	—	-2	-5	-4	-4	-4
France	—	-5	1	-5	-2	4
Italy	-1	-8	-3	-1	3	4
United Kingdom	-1	-8	-3	-1	3	4
Other	5	-11	-10	-16	-21	-7
OPEC countries	7	68	35	40	33	9
Non-OPEC Developing countries	-11	-35	-45	-31	-28	-33
Oil exporters[c]	-2	-5	-7	-5	-3	-4
Oil importers[d]	-9	-30	-38	-26	-25	-29
Adjustments to balance[e]	-15	-18	-6	-5	-3	-2

[a]Goods, services, and private transfers.

[b]Excluding Greece and Turkey, which are included in nonoil devloping countries.

[c]Egypt, Syria, Mexico, Trinidad and Tobago, Malaysia, Columbia, Bahrein, and Angola.

[d]Including Greece, Turkey, Romania, and Yugoslavia.

[e]Statistical errors, asymmetries in reporting balance of payment statistics, and balances with Communist countries, excluding Romania and Yugoslavia.

ance were made for inflation. Furthermore, adjusting for military grants to Egypt, Syria, and Jordan and for concessional economic aid to developing countries generally would reduce the surplus from 1974 onwards by $5 to $7 billion a year. Thus, by 1978 the net OPEC current surplus that would ordinarily go into reserves or be available for other investments had been virtually absorbed. Continuing but declining surpluses for Saudi Arabia, Kuwait, Libya, Iraq, and the United Arab Emirates in that year were mostly offset by the deficits of Algeria, Venezuela, and Nigeria; the other major oil exporters (Iran and Indonesia) were about in balance.

This sharp decline in the OPEC surplus resulted from three factors—a rapid increase in imports of goods and services, a stagnant market for OPEC oil exports, and a deterioration in OPEC's terms of trade. The surge in OPEC imports of goods and services was the most important factor—and the most remarkable, in view of early doubts about whether these countries as a group could spend a reasonable portion of the huge increase in their oil revenues. Between 1974 and 1978, the increase in these imports averaged 25 percent a year in real terms for all OPEC countries and 35 percent for Saudi Arabia alone, where doubts about absorptive capacity were particularly strong. Purchases of services (notably fees for foreign technical assistance of foreign workers, as well as freight and insurance costs) were an important element, amounting to about two-fifths of the import goods. During 1974–1975, when OPEC countries seemed to be trying to do everything at once and the spending frenzy was at its peak, the volume of imports grew by 40 percent a year. After that, the growth in spending slowed markedly, as the monumental wastage due to bottlenecks came under control, policy shifted toward fiscal restraint to curb inflation, and more countries began to experience balance of payments constraints.

OPEC's oil market prospects also changed sharply in the wake of the price rise. On the basis of pre–October 1973 projections (that is, on the assumption of low oil prices and comparatively high trend rates of economic growth), OPEC's oil exports would have been expected to increase from 30 million barrels a day in 1973 to 40 mbd in 1978. Instead they declined to 28 mbd, principally because higher oil prices and lower world economic expansion curtailed the growth of world energy consumption and increased the supply of substitutes for oil. At the same time, new oil supplies from Alaska, the North Sea, Mexico, and Egypt cut the demand for OPEC oil by 4 million barrels a day.

This generally soft market probably constrained OPEC price policies after 1973. The official OPEC price, expressed in dollars, was increased by 10 percent in October 1975 and again in January 1977 and then left unchanged until the end of 1978. (It was increased by an average of 10 percent, phased in over the year, at the beginning of 1979 and then by much larger amounts when the world oil market became chaotic in the wake of reduced exports from Iran.) In comparison, the price of OPEC imports between 1974 and 1978 rose by 35 percent in dollar terms. It is arguable whether Saudi Arabia's conservative oil price policy or the soft market was the greater restraint on prices, but certainly the two factors reinforced each other. In any event, the moderate decline in the real price of oil, following the quadrupling of prices in 1973, contributed to the rapid rundown of the OPEC surplus from the very high level reached in 1974.

As the current surplus declined, the specter of an overwhelming OPEC accumulation of financial assets receded. Even though smaller than expected, these holdings constitute a large concentration of liquid funds, which, it was once feared, could "slosh around," for political or economic reasons, and greatly intensify exchange market instabilities. This has not happened. The proportion of new OPEC placements invested in the United States declined after about 1975. This decline in dollar investments was neither precipitate nor unusual. More generally, there has been a pronounced shift since 1974 toward longer term, less liquid, and more diversified investments and away from Eurocurrency and other bank deposits, which at least suggests greater stability in these long-term holdings.

Growing Imbalances Among OECD Countries

The massive increase in the OPEC surplus drastically altered the global payments picture in 1974, bringing about major changes in the payments position of most oil-importing countries (see Table 4-1). Somewhat more than half of the $60 billion increase in the OPEC current surplus in 1974 had its counterpart in an extreme swing from a large surplus to a large deficit in the combined account of the OECD countries and the rest in an increase in the combined deficit of the non-OPEC developing countries. After 1974, the sharp rise in OECD exports of goods and services to the OPEC group, particularly exports from the major industrial countries, rapidly reduced the net impact of higher oil prices on their balance of payments. By 1978, the normal OECD position as a net exporter of capital to the developing countries had been reestablished.

Over the same period, however, wide changes in current balances within the OECD group became increasingly important determinants of exchange rate instabilities. The most critical element was the sharp deterioration of the U.S. current balance and growing pessimism about its future. By 1978, these imbalances, rather than the OPEC surplus, were dominant.

This raises the question of how heavily higher oil prices and rising oil imports contributed to the weakening of the U.S. payments position. The answer is not self-evident. It is of course true that the value of U.S. oil imports (F.A.S. basis) jumped from $6 billion in 1973 to $40 billion in 1978, an increase roughly equal to the total U.S. trade deficit and more than twice the U.S. current account deficit in 1978.

These figures, however, do not take into account two first round offsetting effects: (1) the increase in U.S. exports of goods and services to OPEC countries that higher OPEC oil revenues made possible and (2) the reduction in U.S. oil imports from levels that might have been expected at pre–October 1973 prices. Adjusting for these factors, the net cost of the changed situation in oil on the U.S. current balance is very roughly estimated at $6 billion, or almost half the U.S.

current account deficit (see Table 4-2). What this figure suggests is that if the United States had cut its oil imports in 1978 by 1 million barrels per day, the net impact of higher oil prices on the balance of payments would have just about disappeared. Alternatively, the same result could have occurred through an increase in the U.S. share of the OPEC market. This market share (including U.S. exports of military goods and services) seems to have been about the same in 1978 as it was in 1973.

It is clear that the divergent U.S. economic performance during this period, rather than the oil account, became the major factor behind the wide adverse swing in its current account. In 1974 and 1975, the U.S. recession was deeper and its inflation rate lower than in the other major industrial countries. As a result, the U.S. current account showed a large surplus in both years despite the sharp increase in oil import costs. Over the next three years, the U.S. economic recovery was strong while the Western European economy remained soggy and Japan recovered at less than full potential. At the same time, the rate of the U.S. inflation began to turn up while that of the other major industrial countries turned down. Thus, in 1977 and 1978, the U.S. current balance moved into substantial deficit and the U.S. dollar

Table 4-2. Estimated Effect of Higher Oil Prices on U.S. Balance of Payments, 1978.

		In Billion U.S. Dollars
Value of oil imports, 1978		
8.5 mbd @$13 a barrel		−40
Less projected oil imports pre-		
October 1973 price of $2.75		
a barrel ($4.00 in 1978 prices)		16
Incremental oil import costs		−24
Less estimated increase in U.S.		
exports to OPEC between 1973		
and 1978 in:		
Nonmilitary goods	12.8	
Nonmilitary services	3.4	
Military goods and services	4.7	
	20.9	
Adjustment for normal increase		
in exports to OPEC, at pre-		
October 1973 rate of 8		
percent a year	3.0	18
Incremental effect of higher		
oil prices on U.S. balance		
of payments in 1978		−6

weakened, despite a narrowing of the incremental oil deficit as U.S. exports to OPEC continued to expand.

Similar conclusions can be drawn about the current account positions of the other major industrial countries. Soon after the October 1973 crisis, conventional wisdom suggested that Japan and those countries in Western Europe heavily dependent on imported oil for their energy requirements would suffer severe structural damage to their balance of payments position, while the United States and other countries in a more fortunate energy resource position would come out much better.

Again, the actual outcome proved to be different. Each of the major countries incurred substantially higher oil import costs; and initially, the impact of these costs on their economic and balance of payments position did vary with the importance of imported oil in their total energy consumption. Each country, however, fared differently in the subsequent adjustment process, depending on the economic and foreign policies that they pursued. As adjustment to higher oil prices proceeded, the balance of payments position of each of these countries came to be affected much more by their employment and domestic price performance than by oil. Thus, to take the extreme cases, Japan and Italy, the countries most heavily dependent on imported oil moved from substantial current deficit in 1974 to substantial current surplus in 1978. Canada, on the other hand, which is virtually self-sufficient in oil and a net exporter of primary energy fuels, has been in persistent current deficit throughout the period 1974–1978.

Financing the Developing Countries

Higher oil prices added $10 billion directly to the import costs of the developing countries that import oil. This amount alone was equal to their net capital inflow in 1973. In addition, the rise in oil prices resulted in higher prices for their imports of capital equipment and other goods from the industrial countries, adding to the adverse shift in their terms of trade. Finally, the recession in the OECD countries in 1974–1975, partly caused by oil prices, depressed the exports of the developing nations.

Economic growth in the non-OPEC developing countries between 1974 and 1978 ranged between 4 and 5 percent a year. Although this was a costly reduction from the trend rate of more than 6 percent in the previous decade, it still represented a much more satisfactory outcome than was experienced in the industrial countries. Adjustment to the oil shock included a reduction in the growth of energy consumption and a rise in domestic production of primary energy fuels, as in the industrial countries, but such opportunities were limited in the short term and even in the medium term. Exports to OPEC countries increased, although at a much slower rate than those of the OECD countries. On the other hand, remittance from the large number of workers migrating to the Persian Gulf countries became steadily more significant, probably reaching $5 billion in 1978. The major beneficiaries were neighboring Middle East countries, India, Pakistan, and Bangladesh. Initially, however, a substantial increase in capital inflows was the most important adjustment factor, permitting these countries to maintain needed imports of capital equipment and producer goods until other adjustment factors gained force.

Much of this increased capital inflow took the form of borrowing by governments on Eurocurrency and other international private capital markets. The middle income developing countries were the participants in this process, since they alone were sufficiently creditworthy. The poorest developing countries had to continue to rely on receipts of concessional assistance from both bilateral and multilateral sources. This flow was significantly augmented by concessional aid and credits from the OPEC countries.

Contrary to earlier fears, the aggregate amount of borrowing did not get out of hand, although a few countries did get into financial difficulties. As can be seen from Table 4-1, the total capital inflow had declined by 1978 in nominal terms from the peak reached in 1975 and had dropped even faster in real terms. Furthermore, some of these inflows were used to increase financial reserves rather than for current purchases. When we make allowance for inflation and for the higher level of current output, we could argue that the current level of borrowing is not out of line with traditional levels. Thus, although the total level of developing countries' indebtedness continues to rise, the financial position of the developing countries that import their oil generally has not become a reason for alarm.

The Initial Adjustment in Retrospect

In general, therefore, international financial adjustment to higher oil prices took place fairly smoothly and rapidly. When we look at changes in real resource transfers among the major groups in the world system, we can say that the cumulative savings of the OPEC countries over the period 1974-1978 had their counterpart in the capital imports of the developing countries—partly reflected in the growing debt total of the latter group of countries. This flow of resources was made possible by the financial intermediation of private banks and governments in the OECD countries—a part of the process of recycling the OPEC surpluses. In effect, OPEC countries provided the savings, and OECD governments and private financial institutions took the ultimate credit risk.

Through this intermediation process, OPEC savings could be used to maintain imports of developing countries, which in turn both helped these countries to maintain economic growth and added to aggregate demand in the OECD countries, at a time when the latter were in recession. Furthermore, these OPEC savings substituted for capital exports that would normally have flowed from OECD to developing countries and thus made it possible, potentially, for OECD nations to invest more of their domestic savings in the energy sector and in the capacity to produce increased exports for the OPEC countries. Both developments contributed to long-run adjustment to higher oil prices.

Nor did the oil shortage bring about a giant struggle among the major industrial countries to compete for oil supplies or gain control over them. In 1973–1974, oil shortages arising from the oil embargo were shared more or less equitably as a result of the informal allocation of supplies by the international oil companies. After the embargo, the industrial countries adopted formal measures, through the International Energy Agency, to share the burden of future supply interruption and to reduce their economic impact. The cooperative approach is now being put to the test in reaction to the curtailment of supply from Iran. So far at least, multilateral cooperation both in managing physical shortages and in adjusting financially to higher oil prices has helped all nations adjust to the new international situation in oil. As a result, differences in domestic resource positions have been largely neutralized as a factor in adjustment.

To say that the international economic system adjusted to the oil shock more smoothly than had been expected is not to understate the large cost that was incurred in higher inflation and a massive amount of lost economic output. Furthermore, as far as economic policies are concerned, the fact is that governments of most major industrial countries feel uncomfortable with large current deficits, even though they may recognize, as they did in 1974, that such deficits are the logical and necessary consequence of a unique action that was essentially exogenous to the system and could not be corrected in the short run. In any circumstances, therefore, large deficits increase the danger that governments will act to reduce imports unilaterally or to restrict demand excessively with damage to all countries. In varying degrees, this danger materialized on both counts.

Furthermore, a large accumulation of deficits exercises an influence of its own on expectations and markets. The influence can have a disproportionately adverse effect on confidence, investment, and employment. In this respect it was fortunate that in 1974, the year of the extraordinary OPEC surplus, the U.S. current account was in strong surplus—albeit for the wrong reasons. This strong payments position for the world's reserve currency country facilitated the management of the large financial imbalances in that year. This was in marked contrast to the situation in 1977 and 1978, when the weakness of the dollar, for reasons mainly unrelated to oil, added to the difficulties of an international system plagued by large imbalances among the OECD countries, with the OPEC surplus a comparatively minor complication.

Thus, despite the resilience of the international economic system and despite the general success of OECD governments in avoiding widespread resort to restrictionism, the world economy is still far from full recovery. There is a substantial amount of economic slack as well as inflation in the industrial world, and growth in the developing countries is at least one-fifth below trend. The volume of world trade from 1973 onward has grown on average by little more than 4 percent a year, less than half the rate for the previous decade. Even by 1978, world trade grew by only 5 percent. This sluggishness can hardly be attributed solely to oil, because the onset of the present cycle had wider causes. Resilient though it has proven to be, however, the international system cannot be expected to adjust to such shocks without sustaining consideral damage.

Rising Oil Prices and Currency Exchange Rates

A large increase in the world price of oil will, as noted above, cause a large increase in the oil-consuming countries' spending on imports and, in the near term, an increase in their deficits on current account. The consequences for the oil importers' balance of international payments and for the international value of their currencies will also depend on what happens on capital account—that is, on what OPEC does with the money it does not use to purchase imports of goods and services. And here, an arithmetic truism comes into play. Any foreign exchange earnings that the oil exporters do not spend on imports of goods and services from the oil-importing countries, they have to use to make direct investments in businesses or land in those countries or buy stocks and bonds from them, or simply keep in bank deposits or currency. In the aggregate, there is no leakage. It follows that the *aggregate* goods and services deficit of the oil importers, taken as a whole, will finance itself—it will be exactly matched by a surplus of capital inflow on capital account.

For any particular importing country, however, that need not be so; country by country, capital inflows will not exactly offset goods and services deficits—OPEC treasurers may wish to invest their excess earnings mainly in Deutschmarks, or Swiss francs, or Japanese yen and not in dollars or Philippine pesos. As a result, some oil-importing countries' deficits are likely to be overfinanced, and some under-financed, and the exchange rates of the underfinanced currencies will fall relative to the rest.

Until recently, the dollar has attracted more than its share of OPEC money; the outflow of American oil payments has been more than offset by the inflow of OPEC investment in dollar assets and by OPEC purchases of U.S. goods. Consequently, the direct effect of past oil price increases on the dollar exchange rate has not been nega-tive. For the future, it is not at all certain that the effect on the dollar of a new oil shock would be persistently negative. Even if it were, rising oil prices would have less to do with it than the effect of our high relative inflation rate and the special role of the dollar in the world monetary system.

As long as the rate of inflation in the United States remains some two to six percentage points faster than in Japan, Germany, and Switzerland, there will be a downward bias in the value of the dollar measured in terms of those currencies. Currently, that bias is more or less offset by differences in interest rates. Holding dollars earning 9 percent in New York is just as attractive as holding Swiss francs earn-ing 3 percent in Zurich, even though, in terms of goods, the dollar is depreciating at a rate six points faster than the Swiss franc. But if it were thought that the U.S. inflation rate will speed up relative to the Swiss, there would be a further shift into francs. An oil shock becomes relevant, therefore, if there is a general belief that it would worsen U.S. inflation more than inflation in Switzerland—or in Ger-many or Japan.

The widespread concern with the relatively fast U.S. inflation rate is compounded by the enormous holdings of liquid and near liquid dollar assets by foreigners, by international companies, and by the world's central banks. Some of those holdings represent working balances—the dollar is still the principal transactions currency in international trade. But some dollars were accumulated when the dollar was thought to be a more stable form of reserves than it has turned out to be. And some were acquired by central banks in the course of their attempts to prevent "undue" appreciation of their own currencies (that is, operations that involved the purchase of dollars and the sale of their own currencies in the exchange markets).

In a world of fluctuating exchange rates, in which the dollar is not convertible at a *fixed* rate either into gold or into any other asset, a prudent money manager, private or public, will want to spread working and reserve balances among many currencies. The mere fact of fluctuations, combined with uncertainty, is sufficient reason for diversification—probably more diversification than characterizes the current composition of the reserve and working balances of many international businesses and central banks. If that is so, the dollar may continue to be, for a while longer, a no win currency, like sterling in the 1960s. Any strengthening in the dollar will trigger selling by "diversifiers" who want to take advantage of the higher price. Weakness will trigger selling by those inclined to cut their losses. With large "excess" dollar balances abroad, a substantial nonoil current account deficit, large Japanese and German current account surpluses, and a faster U.S. inflation rate than the Japanese, German, and Swiss—a lot of people will sell dollars on any news, whether it is good or bad.

Against that negative "bias" one must weigh the fact that, at current exchange rates and evaluated in terms of its real purchasing power over goods and services and U.S. businesses and common stock, the dollar seems remarkably cheap. Potential returns for the foreign equity investor appear very large. And, not least important, politically the United States is still probably the safest haven for private money in the world. So on balance, the net effect on the dollar exchange rate of another oil shock would be anyone's guess.

What if the effect were negative and the dollar declined? A moderate decline in the dollar, short of a panic, would have two effects. First, it would amplify the direct effect of the higher oil price on the price level. All imports would go up in price, and they now constitute some 10 percent of GNP. A wage reaction would become more likely. Second, after several quarters' delay, a less expensive dollar would tend to alleviate the negative impact of the oil price rise on total spending on U.S. goods. The relative cheapening of American goods in world markets, if it persists (that is, if it is not nullified by faster wage inflation), would discourage imports and stimulate U.S. sales abroad. Nevertheless, the previous qualitative conclusions would still hold. The wage reaction, and the government's fiscal and monetary response, would remain critical.

Future Oil Prices and the International Economic System

What comfort, if any, can we draw from the fact that the heavy flows of funds to OPEC nations, while exacting a substantial resource cost on oil-importing countries, have not severely punished the world trading and monetary system? Should we expect that developments in the world oil market over the next twenty years will have much the same international economic consequences?

The key variable is the prospective strength of demand for OPEC oil, since that would determine (1) the course of oil prices; (2) the distribution of the oil market as between countries that would spend their additional oil revenues rapidly and those such as Saudi Arabia that would tend to add to their financial assets; and hence, (3) the extent to which trade in oil would contribute to pressures on the international financial system.

The size of the OPEC market will depend on the rate of economic growth in the consuming countries, the efficiency with which energy is used, the rate at which alternative sources of energy are developed, and the growth in non-OPEC oil production. Internally consistent, but widely different, scenarios can be postulated.

For example, high demand for OPEC oil could lead to recurring tight markets that are brought into balance by erratic jumps in oil prices, leading to inflationary pressures, sudden deterioration of international payment positions, and reduced economic growth in the importing countries. In this situation, Saudi Arabia and the few other capital surplus countries would have a larger share of the market, which would mean that a larger proportion of OPEC revenues would add to financial surpluses rather than be spent on current imports of goods and services.

Alternatively, if demand for OPEC oil is comparatively restrained because of a rapid and sustained response of the demand for energy and the supply of primary energy fuels to higher prices and to legislatively mandated oil conservation measures, oil prices are likely to rise smoothly over the remainder of this century. In this situation, OPEC countries tending to spend oil receipts rapidly would account for a larger proportion of total OPEC revenues, and net real OPEC financial surpluses would accumulate only moderately, if at all.

The first scenario would put continuing stress on the international economic system; the second would not. Toward which end of the range the world will gravitate over the next ten or twenty years we cannot predict; but we know from the experience of the past five years that adjustments that will ease movement away from OPEC oil will not come rapidly and that the short-term ability of OPEC to tighten the market and boost prices will persist for some time. Such a trend would substantially raise the risk that oil prices would increase by discontinuous jumps rather than smoothly. For one thing, spare capacity among the lower income OPEC countries would be the first to be used up, thus putting the OPEC price hawks in a stronger position to press their case. Second, in a tight market, curtailment of supply from any major exporter—such as has happened in the Iranian case—would make the market vulnerable to a sharp run-up in prices.

The sudden recurrence of large OPEC surpluses would again test the efficiency of international financing arrangements. Until the underlying adjustments to new jumps in energy prices run their course, the investments of OPEC surplus countries will again have to be reshuffled or recycled among industrial countries, and to the importing developing countries, so as to avoid serious disruption of the international economic system.

In the past five years, this recycling for the most part was accomplished through the private international capital markets, with great efficiency. The process, however, depended on an effective backstopping role by the international financial institutions—predominately the International Monetary Fund and, to a lesser but significant extent, the World Bank and the regional development banks. These institutions perform three critical functions in the recycling process: (1) to provide supplementary credit, (2) to use the leverage provided by this credit to encourage the oil deficit countries to adopt adequate internal adjustment measures, and (3) to improve the credit status of these countries so that they will have continuing access to international private capital markets.

During the period 1973-1978, the financial capabilities of these institutions were substantially increased, principally through the addition of roughly $60 billion to the potential lending capacity of the IMF and $35 billion in capital subscriptions and contributions to the multilateral development banks (see Table 4-3). In light of the possible recurrence of large OPEC surpluses, the need for additional international financial backstopping facilities will have to be expeditiously reexamined. These facilities have proven to be cost-effective insurance; when large amounts are put into place, the less likely will it be necessary to use them.

These considerations underline once again the importance of the energy policies of the United States and the other major industrial countries in containing the vulnerability of the international economic system to future oil shocks. A continued increase in U.S. oil imports would insure a growing demand for OPEC oil, erratic oil prices, and sudden reappearances of large OPEC financial surpluses. It would also make the U.S. problem of regaining payments equilibrium more difficult to solve, especially since it is no longer a matter of a disequilibrium between the United States and OPEC, but equally a disequilibrium among the industrialized nations, some of which

Table 4–3.   New International Credit Facilities, 1974–1978 (In billion U.S. dollars).

Date	Facility	Nominal Amount	Estimated Potential Addition to Lending Capacity of International Financial Institutions	Possible Recipients
	IMF			
1974	Oil Facility	8.9	8.9	All countries
1974	Extended Fund Facility	8.3	8.3	LDCs
1975	Compensatory Financing Facility	15.6	4.5	LDCs
1976	Trust Fund	1.1	1.1	LDCs
1978	Sixth Quota Increase	12.6	6.3	All countries
1978	Seventh Quota Increase	25.8	12.9	All countries
1979	SDR Allocation	15.0	7.5	All countries
1979	Supplemental Financing Facility	10.0	10.0	All countries
	Subtotal IMF	97.3	59.5	
1974	Fourth IDA Replenishment	4.5	4.5	Low income LDCs
1976	Selective World Bank Capital Increase	8.4	8.4	LDCs
1976	IFC Capital Increase	0.5	0.5	LDCs
1977	Fifth IDA Replenishment	7.6	7.6	Low income LDCs
1974–1978	Regional Development Banks	13.7	13.7	LDCs
	Subtotal, Multilateral Development Banks	34.7	34.7	
	Total	132.0	94.2	

have been running very large surpluses and others with equally large deficits. Thus, on all counts, additional oil import demand by the United States and other OECD countries would put the international financial system under strain.

The performance of the system over the period 1974-1978 indicates that higher oil prices need not be a forecast of doom. Indeed, in this study we have assumed that real oil prices will not much more than double over the remainder of this century, and we believe that if this occurs smoothly, the strain on the international economic system will continue to be manageable. On the other hand, the potential economic costs that would be involved in recurring oil shocks, much like the potential political damage, puts a large premium on containing the demand for oil imports and hence on national and international actions to supplement what the market alone will do. The fact that the international economic system weathered the oil shock of 1973-1974 is no automatic guarantee that it can do so as smoothly in the future, especially in light of five years of faltering attempts to cope with the conflicting demands of higher oil prices, chronic inflation, and disappointing economic growth.

Regulating Crude Oil Prices: Gainers and Losers

5

This chapter analyzes the effects of crude oil price regulation on both efficiency and equity. The efficiency test is whether the benefits of any regulatory policy, irrespective of whom they accrue to, exceed or fall short of the costs. And because the benefits and costs accrue to different people, questions of equity arise.

We conclude that regulation of crude oil is causing significant inefficiency in our use of energy, which is a real cost for the economy. Costs arise from underproduction of domestic petroleum, overconsumption of imported petroleum, impairment of the economy's ability to adjust to world energy prices, and regulatory administration and enforcement. Avoiding these costs by decontrolling crude oil would represent a gain to the nation as a whole, but it would also bring some change to peoples' income—depending on their wealth, occupation, region, consumption habits—and would also make some of the poor worse off relative to those better off.

Furthermore, price controls and the entitlements scheme have created benefits for some special interests that yield no benefit to the consumer. There is a special subsidy to small refiners that creates incentives for operating small, inefficient plants; there is a subsidy for importers of residual fuel oil that increases residual imports; very small wells, "strippers" that produce less than ten barrels per day, are exempted from controls, so the owners of slightly larger wells have an incentive to cut their production to qualify for the higher prices.

Controls and entitlements lower the price of oil to refiners, encouraging the consumption of oil. In 1978 the subsidy was over $1.60 per barrel. The OPEC price increases of 1979 may have raised it to somewhere around $3. Demands for oil were being satisfied that were worth less than the cost of the oil imported to satisfy them. Domestic production was reduced, and imports correspondingly increased.

The *current* pure efficiency costs, that is, the costs arising from consuming more and producing (domestically) less than would be the case without price controls, are in the range of $1–$2 billion per year. The uncertainty is caused by the fact that the outcome depends crucially on what is assumed with regard to the short-term responsiveness of both demand and supply to a given difference in the price; on this score opinions vary strongly. However, these measurable efficiency costs are only part of the total cost of a system of price control. To the extent that they can be quantified, the public and private costs of operating the program approach three-quarters of a billion dollars. This excludes the cost of efforts by those affected to get around it and bend it to their own ends. Even adding these costs understates the drain on the economy because it assumes that current controls affect only current production and consumption. Since "new" and "future" oil are not immediately subject to controls, the current controls might seem to have no supply disincentives; but even "old" oil production is affected by the price it receives. Moreover, faced with the history of price control over the past several years, investors might reasonably expect that the oil that is "new" today or tomorrow will be "old" next year or the year after. The same political logic by which this year's "windfall gains" are controlled away or taxed away may be just as appealing when oil prices rise another 20 or 50 percent. The investor may conclude that the risk of failure will be borne alone, while the benefits of success will be shared through the reclassification of the oil discovery as "new old oil" or some similarly inventive regulatory designation. Though large, it is difficult to estimate the magnitude of this chilling effect on incentives, and it is not the only indirect effect of regulation.

Reliance on price regulations in normal times creates a false expectation for the benefits of regulation increases. Many of the oil shortages that occurred during 1973–1974 were probably created by the regulations hastily implemented to alleviate the supply curtailment; end of month shortages were more likely caused by the incentives to wait for the first of the month changes in the price ceilings than by any physical shortage. The economic cost of waiting in line to fill one's gas tank is no less real than the price of gasoline, and it is created by the price and distribution rigidities imposed by regulations.

PETROLEUM PRICE AND
ALLOCATION REGULATIONS

Petroleum price regulations did not begin in the 1970s. They started in the 1930s with state conservation regulations that, in effect, organized oil producers into a state-run cartel. The control system fixed the market price by varying the amount of production. After World War II, world prices dropped below U.S. prices, and imports began to encroach on domestic markets. An import quota was implemented in 1959 that limited imports to 12.2 percent of domestic production; a continuing battle over its application resulted in a complex of regulations and exceptions. Many became embodied in the control programs that succeeded the quota program. For example, quota licenses were not allocated in proportion to refinery capacity, but on a sliding scale that favored smaller refineries; this small refiner bias was carried over into the entitlements program. Residual fuel oil imports were exempted from the quota limitations for use as a boiler fuel in the northeastern states. Special import licenses were granted to refiners to locate in Puerto Rico and the Virgin Islands, ostensibly to promote economic developments and to process Venezuelan oil for U.S. markets. A preference was established for imports from Canada that led to the location of refineries in border states; these refineries faced extinction with the end of the quota and received special treatment in the entitlements program. In short, the entitlements and price control system contains features that are descendants of earlier regulations.

The embargo of 1973 prompted additional regulations. The two tier pricing system for domestic oil that followed defined "old" crude oil in any month as the quantity produced from domestic oil fields in the corresponding month of 1972. Output in excess of 1972 levels and from fields not yet producing in 1972 (and some other oil) was defined to be "new" oil and not subject to control. Moreover, for each barrel of "new" oil, a producer was permitted to "release" a barrel of "old" oil from its "old" classification. "Stripper" oil was defined as production from properties that produced less than ten barrels per well per day. "Old" oil from each property was allowed to sell for its price on May 15, 1973, plus $1.35, leading to an estimated average price for the country as a whole of $5.03 per barrel. Imported oil, as well as "new," "released," and "stripper" oil, were uncontrolled. Suppliers were required to continue to supply domestic oil to their precontrol buyers.

New rules were set in January 1976 by the Energy Policy and Conservation Act of 1975 (EPCA), which brought all but stripper oil under control and, based largely on time of discovery, set up a new category of oil for pricing purposes. EPCA established three tier pricing. Production not in excess of a field's 1975 output of old oil or of its 1972 total output is "lower tier"; output in excess of lower tier, and from fields not producing prior to 1976, is "upper tier." Imported oil constituted the third and highest tier. Oil owned by the federal government is not subject to controls, and Alaskan crude oil sells as "upper tier" (although transportation costs keep the wellhead price below upper tier ceilings).

Lower tier oil sells at a maximum of $1.35 plus its May 15, 1973, price plus monthly incentive and inflation adjustments determined by the Department of Energy. Upper tier sells for $11.28 plus monthly adjustments. The average of lower and upper tier cannot exceed $7.66 plus the cumulative monthly adjustments. Only imported oil, government oil from federal lands, and stripper oil sell at market price.

The prices of refined products were subject to controls that limited them to their 1973 level plus dollar-for-dollar adjustments for changes in production costs; during 1976, refined products except gasoline and jet fuel became exempt from price controls, and jet fuel became exempt in early 1979.

Without some program to equalize costs, a multitier system of crude oil prices would result in different costs to refiners. Refiners granted access (through the regulatory freezing of buyer-supplier relationships) to large amounts of controlled oil would have lower costs than refiners whose only source of crude is imported oil. The entitlements program adopted in 1974 was designed to equalize crude oil costs among refiners. The Department of Energy makes monthly issues of "entitlements" for low-priced crude oil to each refiner. The amount of entitlement granted to each refiner is equal to the controlled crude oil that the refiner would have used at national average proportions of controlled and uncontrolled crude oil. If a refinery uses more controlled oil than that, it must purchase entitlements. And if a refiner processes less controlled oil than the entitlements it is issued, its costs are greater than the national average, and it is permitted to sell entitlements. The purchase price of an entitlement is approximately the price difference between controlled oil and the imported price. The result is as if low cost oil were allocated proportionately among refiners.

Entitlements do not exactly equalize costs. Smaller refiners (less than 175,000 barrels per day) are given extra entitlements, importers of heavy fuel oil receive extra entitlements, and numerous "exceptions and appeals" are granted. Finally, the federal government allocates itself saleable entitlements to lower the cost of its strategic petroleum plan for storing crude oil.

THE COSTS OF CONTROLS AND ENTITLEMENTS

Inefficiency on the Demand Side

Price controls reduce the proceeds to domestic crude oil producers. On the assumption that the controls have had no impact on the world price, the reduction has averaged $13-$14 billion per year. With the narrowing gap between world and domestic prices, the difference was approximately $11 billion in 1978. It will be larger in 1979 with the sharp OPEC price increases.

These sums are a saving to processors and users of crude oil over what they would spend if all oil were priced at the world price. Division of these savings among large refiners, small refiners, and ultimate consumers is determined by the entitlements program and by market forces.

Under the program, all refiners pay less than the market price. The entitlements program has the same impact as if the Department of Energy taxed away the proceeds and used the tax proceeds to subsidize imported oil.

The magnitude of the subsidy varies with the difference between the world price and the national average of imported and domestic oil. That difference was $3.55 in 1975, $2.59 in 1976, $2.57 in 1977, and $2.11 in 1978. Actual subsidies averaged $2.84 in 1975, $2.51 in 1976, $2.27 in 1977, and slightly more than $1.60 in 1978. The discrepancy was due to the small refiner bias, the special subsidy to industrial fuel importers, the "exceptions and appeals" process, the subsidy to the strategic petroleum reserve, and so on. Subsidies have been paying for between a tenth and a fifth of imported crude oil.

The entitlements subsidy induces refiners to produce products that consumers value less than their cost. In 1978, incremental barrels of crude oil were acquired from foreign sources at a cost of about $14.50 but produced added value of only about $12.90 (the difference being the subsidy). The magnitude of the inefficiency depends on how responsive consumers are to price changes. A 1 percent reduction in the price would induce an increase in use of crude oil of about 0.5 percent within the year of the price reduction.

The price difference ($14.50 minus $12.90) is about 12 percent. If it induced a 6 percent increase in consumption, the difference would be about 6 percent of 5 billion plus barrels during the year, or something over 300 mill. bbl. of imports in 1978. Assuming that these 300 mill. bbl. were valued between $12.90 and $14.50, they were worth, on average, about $1 less than they cost, for a net loss—the "demand side inefficiency"—of about a third of a billion dollars.

These costs are underestimated if there are additional costs of overdependence on imported oil—problems of national security and supply interruptions. If there is an "overdependence" cost of $1 or $2 per barrel of imported oil, for example, the net cost of the entitlements subsidy rises to $600 or $900 million per year. And if the enlarged imports have any effect on OPEC prices, now or in the future, the added cost of all imports must be attributed to these incremental barrels.

Decontrol would raise the paid cost to the true cost (although, as just mentioned, it would possibly lower somewhat the true cost). The distributional impacts would be felt rapidly; the gains to the economy would appear over some adjustment period.

Inefficiency on the Supply Side

Current regulation induces not only overconsumption but domestic underproduction, another cost to the economy. If the incremental cost of domestic is less than that of imported oil, resources could be saved by reducing imports and increasing domestic production. It is costly to hand over $15 to foreign producers if a barrel can be acquired domestically for $12 or $13. This criterion is violated by controls.

The multitier pricing scheme reduces efficiency insofar as it prevents the incremental costs of all domestic producing fields from equating to the price of imports. Nevertheless, the multitier system produces less distortion than a system that controls *all* domestic prices at the current domestic average price. The discouragement to production is reduced by allowing higher prices on upper tier oil. Further incentives are provided by allowing lower tier wells that produce above base period levels to sell incremental output at upper tier prices. The lack of controls on stripper wells offers mixed incentives: it permits continued production from properties that might otherwise be shut in, but induces some producers to reduce production to qualify for stripper prices! Except for this stripper effect, the multitier system discriminates most against the least responsive production, as it should.

Regulations discourage production not only from existing fields but also from fields that go unexplored or undeveloped because of the inability of such fields to capture a price above the upper tier price. Over the longer run, the discouragement of exploration and development probably exacts greater costs on the economy than does the reduction in current production.

The effects of controls depend on the responsiveness of production to decontrol and on the period of time over which costs are measured. In the nearer term, the costs depend on the responsiveness of output from existing fields. A conservative estimate is that the near-term cost of discouraging supply from both new and existing sources would be close to $1 billion annually. But controls discourage exploration and development. If investors view the average price or even the upper tier price as the price at which they can expect to sell their output, the annual cost of controls in unrealized new supplies might be several billion. And when the effects on investment of the uncertainties generated by controls are added in, the costs rise substantially again. Finally, controls distort the decisions of investors and producers, causing them to direct their efforts where they promise the greatest revenues, not the greatest amount of oil. Unnecessary expenditures are made, while some that would be worthwhile to society (but not to the investor) are not. These costs, while not readily quantifiable, can be large.

OTHER COSTS OF PRICE REGULATION

Other costs of the control program include the impairment of the economy's ability to adjust to world energy prices and costs of administration, enforcement, and compliance.

Administrative Costs

For the regulated industry, whose costs will be borne largely by consumers, the burdens of regulation arise from (1) maintaining records and making reports, (2) disruption of normal business by the need for federal approval, and (3) shifting competition from the market to the political and bureaucratic realms. During 1977 the federal compliance program imposed requirements on over 300,000 firms. These firms had to file over a million reports, at a cost estimated at $160 million. When the costs of maintaining records, monitoring compliance, and attorneys' fees are included, the administrative costs are estimated at nearly $500 million.

Numerous aspects of refinery operations, including capacity changes, shutdown for maintenance, and sale of a refinery, require regulatory approval, since refinery operations have pricing, entitlement, and buyer-supplier obligations associated with them. Transactions involving crude oil producers require extensive certification as to quantity and applicable price ceilings; and changes in buyer-supplier relationships with pipelines and refiners are subject to lengthy review and appeal processes. These raise costs.

The viability of a firm depends on its ability to "produce" favorable regulation and rulings—as opposed to its ability to produce crude oil, refined products, or marketing services. A premium can be expected on management who know more about getting along with and influencing regulators than about producing for their customers. In general, regulatory mechanisms lead to lobbying, campaigning, and appealing for favorable regulatory decisions.

The administrative burden is on the taxpayer, too. The budget allocation of the Economic Regulatory Administration for 1979 is $133 million. If the ERA's share in the administrative expenses of DOE's energy information, policy, and regulation programs is equal to its share in direct expenses, the tax burden of the ERA would total over $200 million. Although not all of that would be avoided by deregulating petroleum, a large portion of it could be.

Adjusting to Price Changes

Regulatory policies impede the ability to adjust to sudden changes. Prices and profits in unregulated markets are the signals and inducements that allow producers and consumers to adjust when shocks to their environment occur. The appropriate response to a sudden increase in oil prices is to increase domestic production and reduce consumption, thus reducing imports. So long as the incremental cost of domestic output is below the world price, allowing domestic production to push out imports results in a saving for the economy. Allowing rising prices to reduce consumption avoids the costs associated with having petroleum put to uses less valuable than the cost on the world market.

SUMMARY: EFFICIENCY EFFECTS OF OIL PRICE CONTROLS

Proposals for deregulation are met with claims that decontrol will "cost" billions of dollars. If such claims refer to the whole U.S. economy, they are incorrect. Regulations impose a net loss on the economy. The value of goods and services available to the U.S. public are reduced. Deregulation would result in a net economic gain.

The gains arise from several sources. On the demand side, the annual waste from controls is estimated at $300 million. On the supply side, price regulations impose an efficiency loss of up to $2 billion annually. Over the longer run, the discouragement of exploration and development impose larger costs; and both supply and demand side costs are increased by dependence on foreign suppliers. In addition, the private costs of administrative obligations may be $500 million annually plus the distortions in business transactions and competitive behavior. The federal administrative burden may be $200 million. Finally, the inflexibility and inefficiency of current policies magnify the problems in adjusting to sudden shocks in world energy prices.

These are real costs. Removal of controls would eliminate them, but would entail distributional consequences.

EQUITY CONSIDERATIONS IN ENERGY-PRICING POLICY

In broad terms, three groups have a distributional stake in oil-pricing decisions. These are crude oil producers, intermediate users of crude oil (notably refiners), and final consumers. The interests of these groups conflict. Even within these groups interests are not identical: some oil companies have large domestic reserves; others depend on imported oil. The distributional stakes within and across these groups have been crucial in energy policy and will continue to be, apart from how equitable and efficient proposed policies may be.

In the absence of price controls, rising oil prices would have resulted in a sizable transfer of wealth to domestic producers of crude oil. Oil from wells producing prior to the price increases would have had its price bid up to world levels. The increased value of this output would have been not due to the expense of its production, but the result (a "windfall") of the actions of OPEC, which increased the value of the oil in the ground. And "windfalls" to crude oil producers are a loss to the users.

Petroleum is consumed directly and indirectly. While automobile drivers and homeowners are direct consumers of gasoline and heating oil, industrial, commercial, and transportation purchasers of energy also pay higher prices, and these prices raise production costs. The effects are on stockholders to the extent that the market prevents the costs from being passed on in the prices of goods and services and on consumers to the extent that they are passed on.

The effects of oil price increases are thus wider than might be inferred from direct fuel consumption. The result is to make the burden of higher prices more nearly equal across income classes than it would be from fuel purchases alone. Evidence on the direct and indirect consumption of energy indicates that total expenditures on energy make up a slowly decreasing proportion of income as income increases, although the proportionate expenditure on fuel alone decreases more rapidly as income rises. Thus the subtraction from real income is absolutely greater on rich than poor but proportionately greater—somewhat—on the poor than on the rich.

There is great variation in the burden of energy price increases within income classes. People in cold climates or in nonurban areas spend more, in relation to their incomes, for heating homes or driving automobiles. Urban dwellers spend relatively less on fuel but consume public transportation that has large indirect energy components. Some low income people are dependent on automobiles for their living or for what recreation they get; others are too poor, or too old, to drive.

Heating, driving, and industrial heating use tend to impart a regional pattern to the impacts of rising prices. The geographical basis of the Congress gives these impacts political expression. Interregional differences may influence policy more than income differences.

Rising prices induce compensating adjustment in production processes and consumption patterns. Firms, industries, and sectors of the economy most able to make adjustments find a competitive advantage in doing so; consumers most able to make adjustments will find their incomes less vulnerable to erosion. The burden of rising prices is most severe for users whose behavior is least responsive to price changes.

The burden of rising oil prices is particularly severe for one industry —crude oil refining. The depressing effects on consumer demand tend to leave refining capacity underutilized and to discourage expansion; competitive pressures prevent the industry from passing on the full price increases. The result is some decline in earnings. Thus the burden of price increases tends to rest partly on the owners (stockholders) of refineries as well as on the final consumers of gasoline, heating oil, and other refined products.

Neither consumers nor refiners welcome increases in oil prices, and it is to be expected that they exercise influence to prevent them. But the struggle over the distributional impacts is not just between consumers, refiners, and crude producers. As federal policy prevented crude oil producers from capturing a large prospective windfall from OPEC's pricing actions, the beneficiaries have included all the users of crude oil.

Without any increase in output, producers would have captured approximately $11 billion in additional proceeds, were it not for controls. (And by increasing output at higher prices they might have realized another gain of nearly $1 billion; this gain would not have been at the expense of consumers or refiners, but would have arisen from added production gainful to sellers and buyers.) This $11 billion "financed" the entitlements program. Some $2–2.5 billion was absorbed by the special programs already noted, (e.g., the small refiners' advantage and others). So about $8.5–9 billion was transferred to crude oil users (of which some $0.3 billion was wasted through the demand side inefficiency of the subsidy.)

Most of the benefits of those special programs went to the refining industry. The industrial fuel import subsidy, however, directly reduces the cost of imported fuel and was equivalent to a transfer of $0.4 billion to users of industrial fuel, of which $0.3 billion was directly financed by the entitlements program and about $0.1 billion by reducing prices of domestic refiners. The incidence of the $0.1 billion subsidy to the strategic petroleum reserve is problematic. Were it not financed out of the windfalls associated with crude oil, the subsidy would be raised from taxes proportional to income and paid by consumers.

The difference between the approximately $11 billion total and the special programs component, less the demand side inefficiency previously estimated at $0.3 billion, was the value of the subsidy to the users of crude oil. In a rounded figure it totaled about $8.5 billion. The users, however, include both refiners and consumers, and the division of that sum between them is not self-evident. The entitlements subsidy lowers incremental costs and encourages expansion of refining; this reduces the prices that consumers pay for refined products. Estimates vary widely as to the proportion of the subsidy passed on to consumers. Most evidence suggests slightly more than half. That would mean that $4.5 to $5 billion of the 1978 subsidy was captured by consumers, while $4 billion was retained by refiners.

The immediate beneficiaries of current policy then are refiners and consumers. The gain of these groups, however, is less than the cost to crude oil producers. (This is a restatement of the conclusion that current policies are inefficient.) If oil prices were allowed to rise to world levels, the size of the nation's economic "pie" would increase by more than $1 billion—the difference between the gain to crude oil producers and the loss to users. This sum does not include the gain from added import security, greater investment certainty and enhanced future production of oil, and greater use of alternative fuels. Neither does it include the administrative cost of controls borne both by taxpayers on the part of government and by stockholders and consumers on the part of business.

Policy faces a trade-off between the equity implications of benefiting some groups at the expense of others and the anticipated gains in efficiency. It is necessary to examine the equity implications.

EQUITY AND DECONTROL

There are several principles of equity, fairness, or welfare involved in price control, and there is no consensus on the weights they deserve. There is the rather extreme view that the distribution of income is not the government's business but that promoting efficiency is; there is the view that the government should avoid policy changes that have severe adverse effects on some part of the population; there is the view that freedom of contract and access to the free market should be an inviolable right; there is the view that particular elements of the population are especially unable to defend themselves from adversity and should be preferred beneficiaries of any policy change; there are views that certain regions of the country or age groups or ethnic groups or urban groups or rural groups should be affected equally by any policy changes; and there is a widespread though not universal view that policies are to be preferred that benefit the poor rather than the rich, even those that benefit the poor at the expense of the rich or at least that do not benefit the rich at the expense of the poor.

Despite the substantial differences in the impact of energy price changes on different regions of the country according to climate or population density, on urban and rural populations, on large families and small, on the employed and the retired, on different industries and occupations, and on different tastes in living style or recreation, energy so permeates the economy and is so broadly consumed directly and indirectly that there is no important division within the population between those who consume energy and those who do not. As indicated earlier, the rich consume more energy than the poor, both directly as fuel or electricity and indirectly in the goods and services they purchase, because they consume more in total than the poor, while the poor consume proportionately more, but only modestly, so that rising energy prices tend to be of somewhat more concern to the poor than to the rich.

On the supply side of the market, while increased energy prices will sometimes be associated with higher wages to mine workers or with greater demand for rail transport or shipping or pipeline construction, crude oil deregulation transfers income from consumers in general to the owners of oil-producing properties or to the companies that have long-term leases and contracts for crude oil. In other words, the transfer of income that results from decontrol is from consumers generally, but proportionately slightly more from poorer consumers, toward the stockholders of industries that own or have contractual rights to crude oil. (To the extent that crude production and refining companies are not identical and that decontrol will transfer some income from stockholders in refining companies to stockholders in crude-oil-producing companies, there may be some difference in income levels but not one that can be documented and probably not an important one.)

The key empirical fact is that the distribution of stockholdings in the United States is very different from the distribution of income. Data on stockholding in crude-oil-producing and refining companies are scarce, but it is not likely to be an egregious error to assume that the distribution of such stockholdings across income classes is similar to the distribution of stockholding in general. General stockholding is very disproportionate to income.

Of all the equity and welfare effects of decontrol, the one that appears to be of most universal interest, and properly we believe, is the distribution of gains and losses by income level. Some people are particularly concerned about the very poor; some are concerned about income equality in general; some are particularly exercised about the impact on the very wealthy; but there is a widespread if not universal interest, which we share, that the general direction of policy should be toward reducing poverty, reducing extremes of wealth, and achieving a less unequal distribution of income and wealth. Since decontrol works in the opposite direction, its distributional impact is undesirable while the improvement in economic efficiency is desirable, and the two must be weighed and compared to reach a judgment.

Decontrol would transfer about $5 billion from consumers generally to stockholders. (The transfer from refiners to crude oil producers would be at most from one group of stockholders to another and perhaps neutral from the point of view of income distribution.) Of the $5 billion transferred from consumers, corporate and severance taxes would capture somewhat more than half. Assuming that tax revenues produce benefits or allow reductions in other taxes whose incidence is roughly in proportion to income, there would be a net redistribution of $2 or $3 billion (before personal taxes) from consumers to stockholders.

It should be emphasized that while the general direction of this income transfer is from poorer to richer, it is not a transfer from the poor to the rich. It is a transfer from consumers generally, some of whom are poor, toward stockholders, not all of whom are rich—some stockholders being pension funds, university endowments, and non-wealthy individuals, including retired persons living on fixed incomes. Since the nonpoor consumers spend more on petroleum products than poor consumers, the larger burden of this transfer in dollar terms is from the nonpoor to the comparatively well to do.

Weighing a shift of this general direction and magnitude against the efficiency gains of decontrol is not an operation for which there exists any universally acceptable formula. But perhaps there is no need for one. The most striking feature of the comparison is that both terms, though opposite in value—one positive and one negative —are small in absolute magnitude. The net shift in income after corporate taxes, in the direction from poorer to richer, is $2 or $3 billion, a tenth of 1 percent of GNP, and personal income taxes reduce the impact even further. The short-run efficiency gain, calculated without regard to import vulnerability, national security, or the effect on the OPEC price itself, was estimated at a couple of billion dollars. However one strikes the balance, the magnitude is not great. Whatever formula one might use for comparing the positive gain in efficiency with the welfare loss due to the distributive effect, any "net gain" or "net loss" would apparently be equivalent to no more than a billion dollars a year in either direction.

The modest size of this net figure might lead to either of two conclusions. Since it doesn't matter much, why not go for the efficiency of decontrol. Or, since it doesn't matter much, why transfer income in the wrong direction? But there is another way to look at the issue.

If we ask how far wrong we may be about the income transfer and what we may have left out of account, it is hard to see that we could be wrong by more than 100 percent. That is, the adverse effect of the income transfer may be twice as great as in our rough calculations, but it cannot be ten times as great. When we look at the efficiency gain, what we have left out of the reckoning may be much greater than what we have included. To reiterate four things left out of the arithmetical calculation, there is, first, the significant likelihood that a reduction of U.S. oil imports—or a slower rate of increase, at worst—can result in lower future OPEC prices—or a less rapid rise. The above arithmetic took no credit for savings due to any softening of the world oil market. Second, there is the likely reduced vulnerability to interruptions in supply if imports can be reduced as a result of removal of that subsidy. Third, there is the foreign policy and national security interest in helping, through lower U.S. imports, to reduce world demand for oil imports and to minimize the OPEC price impact on other oil-importing countries and on international financial institutions. Finally, if exploration for new fuel sources and development of new fuel technologies, even consumer adaptations in housing and automobiles and other consumer goods technologies, depend on realistic anticipation of future oil prices, decontrol may have the exceedingly salutary effect of reassuring suppliers about the market value of the new supplies that they may bring on the market, of inducing consumers to anticipate the genuine higher future costs of liquid fuels, and of stimulating technology in response to the genuine higher cost of domestic or imported liquid fuels.

Together, these effects, which were not explicitly taken into account in the short-range nondynamic estimates of what we called "efficiency" above, could easily be ten times, or more than ten times, as important as the narrowly defined efficiency gains estimated above at $1 to $2 billion.

We unhesitatingly conclude that the advantages of prompt deregulation wholly eclipse its short-range adverse distributive effects (the worst of which can, and should, in any event be ameliorated by other policies dealing generally with income supplements). The poor, and consumers generally, have far more to gain ten to twenty years from now from the benefits of decontrol than they have to lose immediately in beginning to pay the true cost of petroleum. Whatever may be done about a windfall tax or an excess profits tax on old oil as decontrol goes into effect, the potential benefits of decontrol dominate the choice.

We would caution, however, that an important part of these benefits depends on decontrol with some finality. Taxing future oil discoveries so that they cannot compete in the market with OPEC oil would miss much of the purpose of decontrol. A firm cutoff date, so that any windfall tax applies to "old oil" with a fixed historical definition, and not to any future oil that may be declared old shortly after it is discovered, would be an essential part of genuine decontrol.

Oil Imports

Oil is special among the fuels. Restrictions on oil provoked the "energy crisis" of 1973-1974. Reducing oil imports was the focus of President Carter's National Energy Plan in 1977. Oil is the fuel that the International Energy Agency is about. Oil, being naturally liquid, has been uniquely relevant to modern air and surface transport. Oil is by far the most important fuel in international trade. Oil is, for most countries including the United States, the fuel most vulnerable to interruption in supply. Oil reserves are distributed with remarkable unevenness over the earth, nearly half being in the politically unstable and, until recently, poor and undeveloped countries surrounding the Persian Gulf, which contain 1 percent of the world's population. Imported oil is half of U.S. oil consumption, up from near zero in the last two decades and up by a third from the winter of 1973-1974. Both the politics and the economics of oil trade are uncertain. Oil accounts for more than $40 billion a year of U.S. payments to foreigners, and the payments are rising; and oil proceeds during the past half decade accounted for the most spectacular acquisitions of liquid assets that have occurred in modern times.

Like other countries, the United States has developed a technology and lifestyle very much dependent on petroleum. There was a steady shift toward oil, and in the last couple of decades imports have burgeoned. Although there are important substitutions among fuels, modern transport, both surface and air, is technologically dependent on liquids. It is not easy to shift from liquid to other fuels in the short run for many reasons, including the tendency for the capital to be fuel-specific and slow to turn over. Cars, buses, trucks, and agricultural vehicles are not going to burn anything but liquid fuels within the next couple of decades. Thus oil has a central role in the national and world energy problems; oil is not merely one of several fuels, but one that is specific in important uses in which substantial changes can occur only over periods about as long as a generation.

Because imports are expensive and uncertain and create economic and political vulnerabilities and because the trend is toward greater imports despite the elevated prices of recent years, a central question of U.S. energy policy is whether oil imports should be singled out as a target of direct government action.

Oil imports are a usefully flexible source of energy. The world market in petroleum is a reservoir from which the United States can draw more or less according to sudden changes in domestic supply and demand—a coal strike, a harsh winter, delayed construction or shutdown of nuclear reactors, an upsurge of automobile tourism. There is no comparable market for coal or gas that is available for discretionary use. Imported oil, besides meeting essential needs for which gas or coal or hydroelectric or nuclear power could not quickly substitute, is a balancing residual source of energy.

Oil imports have an ambivalent effect on domestic reserves. In a geological if not an economic sense, reserves are a strictly determined physical supply subject to depletion. (Economically, the quantity of crude oil available for exploitation depends on the prices people are willing to pay; crude that is of no interest at today's prices may be worth recovering at prices prevailing a generation from now.) Imported crude oil is an almost perfect substitute for domestic crude. In meeting a given demand, the more we import, the less we produce from domestic reserves. Thus, imports conserve future domestic supply by inducing lower rates of withdrawal. But the domestic oil does not sit in reservoirs that can be tapped at an arbitrary rate. It has to be "produced." Oil has to be found; wells have to be sunk; pipelines have to be constructed. Modest increases in production can be effected in a matter of months; large increases entail lead times measured in years. Thus, while imports enhance potential future domestic supply by displacing domestic oil, they simultaneously reduce the rate of current production, even the rate of exploration, reducing short-run domestic availability while enhancing long-run supply.

Of all the characteristics of fossil fuel that concern us, by far the most worrisome is that about half the world's petroleum reserves (and more than half of the reserves outside the Soviet bloc) are in the

Middle East, mostly in the countries on the Arabian Persian Gulf. Twenty million barrels per day pass through the narrow Straits of Hurmuz! For decades, the poverty, political instability, and lack of cohesion in that area, together with the competing interests of major oil firms (many of them representing different national governments), appeared to give assurance against any organized economic or political action that would jeopardize the supply of oil or permit the region to behave like a monopoly. The area is still unstable and unpredictable. But instability and poverty no longer provide assurance against the organized manipulation of supply. Oil exports can be vulnerable to the internal affairs of Iran, to Middle East politics against Israel, and to Soviet-American rivalry. The Middle East is the only area of the world in which economic interest—oil in this case— could lead to dangerous military confrontation. And while it is Middle East oil that endows such remote places as Afghanistan and Ethiopia with strategic significance, the risks are not only the risks of interrupted oil but of escalation in the event of violence or confrontation.

A WORLD OIL MARKET

It is important to keep in mind that the market for oil is indeed a world market. Great attention is paid to the fraction of U.S. imports from different areas and countries—Iran or Nigeria, Venezuela or Canada or Mexico. And there are political interests that fear the diversion of Alaskan oil to Japan. There are threats that particular countries, like Libya or Algeria, will refuse their oil to particular countries or favor others with their exports. And it is recalled that the Arab countries, after the October War of 1973, earmarked particular countries—the United States and the Netherlands—for a boycott.

But the shipping and marketing of crude oil and refinery products are decentralized. End use controls on refinery products are nearly impossible. Oil exports can be shifted in destination; transshipment of oil and refinery products is easy. Keeping Venezuelan or Nigerian or Iranian oil from particular countries such as Israel or the United States cannot make much difference if the remaining supply is adequate to the needs of those countries and available to them. The tightness or looseness of the world oil market is something that the oil-importing countries share. Boycotts are not easily directed against particular countries. A country that acquires most or all of its imported petroleum from "safe" parts of the world is a little less vulnerable—but only a little less—than other importing countries.

When contracts run out, there is no reason why a supplying nation need be loyal to old customers when new customers, recently rebuffed by their own suppliers, are willing to pay the same or, if necessary, more.

Indeed, the International Energy Agency, established in 1974, is an institutional reflection of the global character of the market for petroleum. To prevent competitive scrambling for scarce oil in the event of supply disruption, a scrambling that would merely raise the price at which consuming countries would acquire their oil, the agency has established a formula for sharing any sizeable shortfalls, together with commitments for accumulating and, when necessary, drawing down strategic reserves of petroleum.

OIL AND THE BALANCE OF PAYMENTS

Oil imports, recently at the rate of 7, 8, or 9 million barrels per day, could easily be on the order of 10 or 12 million barrels per day over the coming decade, at prices upwards of $20 per barrel, with a gross annual outflow of $75 billion or higher on current account attributable to oil. Holding imports to 8 million barrels per day, even with no effect on prices, could reduce that gross figure by $20 billion and could reduce it by another $5 billion if the reduced imports caused prices to be lower by a dollar or two.

The net difference in the current account balance of payments would depend on the growth of current spending by the oil-exporting countries, some of which may not promptly spend any further increment in the oil proceeds they would acquire as the U.S. import total rises beyond the first $50 or $60 billion per year.

Just how to assess this effect on the balance of payments is a matter of judgment. To the extent that the proceeds of U.S. oil imports are spent on currently produced U.S. goods and services, there is no effect additional to what is already been counted as the "cost" of imported oil. To the extent that the proceeds are held in liquid balances, there can be (1) some reduced stability of international short-term capital markets; (2) the equivalent of short-term lending to the U.S. economy, permitting higher levels of consumption and investment here; (3) some income depression in this country if fiscal and monetary policies fail to offset the subtraction from current domestic spending; and (4) pressure on the dollar's exchange value if oil-exporting countries elect to hold nondollar currencies and to invest mainly in other countries.

The overall net effect is more likely to be adverse, the faster the growth of U.S. spending abroad; so a continued growth of imports, coupled with continued increases in oil prices, should be judged adverse on balance.

THE VULNERABILITY OF IMPORTS TO INTERRUPTION

The embargo of 1973-1974 and the Iranian shutdown of 1979 demonstrated that exports can be reduced unexpectedly in a few months by as much as a fifth or a quarter. Inventories, including oil in transit, provide a few months' cushion; spot prices can jump suddenly in anticipation of expected shortfalls in delivery. In addition to the inevitable economic losses arising from the unexpectedness of such disruption, there is a management problem, especially in a regime of energy regulation, in coping with sudden price increases, in misallocation or unfairness in fuel distribution, in regional differences in hardship, and so on.

There are several mechanisms for dealing with such interruption. Through the International Energy Agency, nations agree to share the shortfalls in oil deliveries by reducing consumption and drawing on oil inventories. There is also agreement on holding strategic reserves. A U.S. government program, barely under way, is intended to accumulate a strategic reserve of a billion barrels of crude oil by the mid-1980s—a quantity that could offset three-fifths of a 25 percent import reduction for more than a year and a half at least, if it could be drawn down with perfect foresight of the severity and duration of the crisis and could be distributed smoothly to refineries. (Arithmetical calculations of what can be accomplished with strategic stockpiles tend to suggest an ideal drawdown of reserve stocks, as though the severity and duration of the shortfall, including the behavior of other countries, as well as of suppliers, could be reliably converted into a schedule of optimal depletion.)

If it were thought that the worst that the United States might face in the future would be the sudden loss of, say, 3 million barrels per day of imports over a protracted period and that the effect could be anticipated and mostly averted by reducing prospective imports from, say, 12 million barrels per day to 9 million, thus "absorbing in advance" any unexpected disruption and being immune thereafter, the case for doing so would be powerful. But immunity of that kind is unattainable. The United States would find itself participating in a world shortfall below anticipated deliveries whether our "take" or "dependence" on world exports were 12 million barrels per day or 9 million. The lesser U.S. imports, by making the world market somewhat less tight, could make such supply interruptions less likely and less severe if they did occur, especially a crisis caused by deliberate withholding on the part of exporters; and the United States might be better able to negotiate shares in reduced consumption with other importing nations if U.S. imports were at the more modest level. And a given percentage reduction in U.S. imports would be a lesser percentage of total U.S. consumption, if U.S. imports were at the lower level and, especially, if domestic production were at a correspondingly higher level. So the argument for reducing vulnerability by reducing in advance the level of imports is valid and of some strength. But it is not as though—in getting gradually accustomed to import levels several million barrels lower than otherwise—we could let the entire interruption be absorbed by that nonexistent margin of imports forgone, and thus excuse ourselves from the crisis.

OIL IMPORTS AND INTERNATIONAL COOPERATION

There are at least three reasons why other oil-importing countries focus attention on U.S. imports. One is that U.S. imports are large enough to have an impact on the world price. If U.S. imports grow another several million barrels per day over the next five years, and if the effect is to cause the price of oil to be a dollar or two more than if U.S. imports stabilized at present levels, Japan and Western Europe and developing countries will be paying a dollar or two per barrel more for the oil they import, for a difference in their foreign exchange expenditures of some 20 billion dollars per year attributable to the increase in U.S. imports. They have, of course, the same interest in each other's imports, and we in theirs, but none of them individually matters as much as the United States.

The second reason is their concern for financial stability. While the calamitous predictions for world money markets that arose in 1974 and 1975 with the great upheaval in world oil prices did not come to pass, and probably will not, concern with international financial stability and with the international transmission of inflation among countries has not abated. The value of the dollar is of concern to many countries, and U.S. oil imports are seen as a major component in the U.S. balance of payments.

And third, for coping with unexpected interruptions in oil supply, it will be important to have standby controls that could be put into effect promptly. During a short-run supply crisis, international commitments may not permit a laissez-faire approach to a shortage even if U.S. domestic politics allow it, and it is not evident that domestic politics will. Workable controls demonstrably in place could enhance the U.S. ability to commit itself credibly in advance to prompt action, in accordance with international agreements.

THE PRICE OF OIL

There is dispute and uncertainty about the intentions and capabilities of the OPEC countries, individually and together, to affect or to determine the price of oil. There is dispute and uncertainty about the extent to which, in the long run, market forces determine the outcome, so that OPEC reflects opportunities and constraints rather than deciding historical trends. But there should be no dispute that unforeseen events have played an important role in prices during recent decades; and future prices will probably display the impetuous quality of the past.

The price of petroleum rose following the Suez War of 1956; prices were then steady or gradually declining until the early 1970s. In 1971, three "unforeseen" or unappreciated events enabled OPEC nations to raise prices. The Suez Canal was closed, as a result of the 1967 war. The major oil companies underestimated demand in Europe. And a Syrian bulldozer broke the Trans-Arabian Pipeline so that Libyan crude, close to Europe, became more valuable. This event enabled Libya to obtain higher prices. This move was possible because many independents were heavily dependent on Libyan oil. In contrast, the major companies had widely diversified sources. Further, King Idris, the ruler before Qaddafi, had been thrifty and had put several years' oil revenue in the bank, enabling Qaddafi to be aggressive when the time came. (And of course a fourth event, the replacement of Idris by Qaddafi, was part of the situation.)

Between 1971 and 1973, prices rose gradually with the gradual decline of spare capacity in the world. Effective market prices actually rose above "posted prices," encouraging OPEC nations to raise prices to recoup profits rather than let them go to the oil companies.

In 1973, the October War caused the Arab OPEC members to cut production by 5 million barrels daily, triggering the mammoth price rise of January 1, 1974. Just what happened to prices is not easily described; markets were disorderly, and different prices were charged to different customers. But the overall effect after several months was the greatest boost in any commodity price of recent times.

From then until the end of 1978, as world inflation raised the nominal prices of the goods and services that OPEC countries purchased with their oil proceeds, there was a gradual erosion of oil prices in real purchasing power. World consumption declined during 1975 and grew relatively little in 1976, 1977, and 1978 because of the twin effects of the price rise and economic slowdowns.

In January 1979 another unforeseen event upset the world oil market—the turmoil in Iran and the temporary shutdown of its oil industry. The loss of production in Iran and the possibility of continuing turmoil had encouraged OPEC at its December 1978 meeting to announce price increases larger than expected. On January 1, prices were to be increased 5 percent, with further increases through 1979 equal to 14.5 percent. But with Iranian oil absent in early 1979, the price for crude not covered by long-term contracts jumped several dollars above the OPEC price. At an OPEC meeting in March, another price rise was announced—9 percent, plus various surcharges. In mid-1979, the market was still in a confusing state, erratically but rapidly moving upward, with one clear conclusion: an unforeseen event had again caused a rise in the price of oil.

There is no theory for predicting the decisions of OPEC. The members agree on a price for the crude oil used as a basic reference, called the "marker crude" (Saudi Arabian Light), with other crude oils worth more or less according to distance from markets and their sulfur content, gravity, and composition. Since OPEC does not have, and never has had, a system of prorating production, the members voluntarily restrict their output to maintain the price.

Saudi Arabia is the key. As long as Saudi Arabia holds the price of its marker crude at the agreed level, the OPEC price is said to be steady. Because of changes in transportation rates and demand for different products (reflecting changing weather, for example), the various crude oils continually change in their value to different refineries. To obtain a target volume, each OPEC nation adjusts its own price (directly or by changing terms of payment). Saudi Arabia, because it cannot readily adjust the price of its marker crude, is in a less flexible position. It can adjust the price of its heavy crude oil and adjust the ratio of heavy to light that it requires companies to take.

Although the other nations allow their sales volume to vary somewhat with demand for OPEC oil, Saudi Arabia bears the brunt of adjustment. There is general agreement on this point and that Saudi Arabia is the principal influence on OPEC's price; but there is no general agreement on what the objectives of the Saudis are. Royal family politics aside, any government would have to balance (1) future oil proceeds against present proceeds, (2) the costs of rapid modernization and development against the benefits, (3) the value of current oil proceeds against the value of investments abroad in considering the effect of oil prices on world inflation and exchange rates, and (4) the advantages against the disadvantages of varying oil production to maintain discipline in the world market and to exercise price leadership.

THE RELATION OF QUANTITY TO PRICE

If a country imports half a million barrels of oil per day, or less, it does not pay attention to the effect that a 50 percent expansion or a 50 percent reduction of its imports would have over the course of several years on the price it pays for oil. The difference would be 1 percent of internationally traded oil; and even if the price of oil changed proportionately, the 1 percent change in price would not have a noticeable effect. The case is different if the country imports 10 million barrels per day.

U.S. imports have expanded ever since the OPEC price increases of 1974. In 1979 they were around 8 million barrels, and in the absence of measures to inhibit imports or a further drastic increase in oil prices, imports could continue increasing. They could reach 10, 12, or 14 million barrels a day by the late 1980s. Deliberate policies to conserve liquid fuels, to enhance the domestic supply of petroleum and synthetic liquid fuels, and otherwise to discourage imports could easily make a difference in the late 1980s of 3 or 4 million barrels per day of imports, possibly more. A difference of 4 million barrels— the difference between, say, 9 million and 13 million barrels per day by the middle or late 1980s—would be equivalent to about 10 percent of the petroleum in international trade. Such a difference would be bound to make a difference in the conditions of sale and specifically in the export price of crude oil. And a substantial difference in the price of oil would be of major consequence.

Only to illustrate the principle, and not to suggest magnitudes, the following arithmetic can be done. Consider two alternative paths for imports over the next dozen years. One is continued steady growth by a million barrels per day each year for half a dozen years, with imports leveling off at around 13 during the second half of the 1980s; the other is a slow rise to 9 million barrels, where imports are stabilized through the 1980s and continue through the decade and beyond.

Suppose that at the lower level of imports, the price of oil might reach the neighborhood of $24 per barrel by the middle 1980s and at the higher level of imports—higher by some 10 percent of internationally traded oil—the corresponding price would be $26.50 (proportionally increased by 10 percent—or, alternatively, $30). The cost of the incremental 4 million barrels at the $26.50 price would be the difference between $26.50 times 13 million and $24 times 9 million, or $345 million minus $215 million, or $130 million per day. Thus the 4 million barrels add $33 apiece to the total cost of imported oil. At the $30 price it would be $30 times 13 million compared with $24 times 9 million, or the difference between $390 million and $215 million—a difference of $175 million or $44 per barrel.

Most commodities are expected to be higher priced, the greater the demand for them. For most commodities, estimating the effect on price of a 10 percent difference in quantity demanded over the course of a decade would be difficult. With oil, the problem is more difficult. First, the governments of importing countries will not be passive to changes in the price. Second, the policies of the exporting countries, especially those in the Middle East, are uncertain. Both their individual interests and their collective interests are hard to perceive, even in purely economic terms. And their interests are not purely economic.

Most oil in the Middle East has the character of a depletable stock-pile that can be pumped at almost negligible cost. And although the stockpile increases with new discoveries, a barrel sold today is a barrel less to sell next year or twenty years from now. If oil prices are expected to rise, a government has a choice between converting oil today into foreign assets, which will earn some return, and holding oil in the ground, which will also yield a return with rising prices. The poorer governments will impute a high rate of return to current earnings for economic development; a government that has already accumulated foreign assets may impute to additional financial assets a return lower than the expected appreciation of oil itself. Rising oil prices increase revenues relative to current imports; and rising prices may or may not reduce a government's estimate of the future rate of price increase, depending on what appears to be the cause of the increase. In some countries the regime of government may face the further uncertainty about whether, if the price of oil should rise substantially in the future, the benefits will accrue to the existing regime or to some alternative regime.

Middle Eastern governments are aware that their oil reserves are likely to last decades rather than centuries, that without oil they have meager resources, that too rapid domestic development can waste resources and generate untenable income expectations, and that how much oil to put on the market is an investment decision in which the future must be balanced against the present.

Furthermore—thinking of U.S. policies to inhibit imports—the responses of both exporting countries and importing countries would be influenced by the particular manner in which American imports were curtailed. So the question cannot be posed in purely numerical terms.

The greatest uncertainty is our inability to predict the production, price, and revenue strategies of the key oil-exporting countries. There is, to begin with, a lack of intelligence about the way policy has been made by Saudi Arabia, Iran, Iraq, Kuwait, and other important countries. There is also the dramatic effect that changes in regimes and politics can have on pricing and production strategies. Iran is an example. Policies will continually change because within most of the exporting countries there are different views and contending factions. There are controversies about the pace of economic development, the political orientation of the government, the importance of an Arab bloc, and the relation between foreign investments and the financial stability of importing countries, which in turn depends on the price of imported oil. Now one and then the other view may gain predominance, and the world price may be affected.

Nevertheless, we expect that oil prices will, as a trend, vary positively with demand. But it has to be admitted that where Middle East oil is concerned, even that is not certain.

Any conclusion about the effect of U.S. imports on the price of oil has therefore to be a matter of judgment. Better intelligence into the way that policies and plans are made within the individual OPEC governments or the way that collective plans are negotiated in OPEC itself could not dispel the fact that we are talking about the plans and policies of governments over the next dozen years or more in a part of the world that is unpredictable. Even the political leaders in those countries have a poor basis for estimating the political, economic, diplomatic, and strategic situations in which their countries will find themselves a decade hence. And in some cases even the regimes that will be responsible for the decisions will change.

But a judgment has to be made. Responsible opinions on the "correct" estimate of the marginal cost of imported oil might be anywhere in the range between the price of oil and perhaps twice that price (and could in principle be even below the price of oil). Inasmuch as the future price is itself uncertain, whether at 9 or at 13 million barrels, the incremental cost of the difference between 9 and 13 million is doubly uncertain. In our judgment the incremental cost that should be in mind when estimating the value of measures to reduce imports would be above the price, but nothing like 100 percent above it. Any estimate, besides being difficult to support except as a matter of judgment, is subject to change. So we abstain from proposing a number, which might convey exaggerated definiteness and permanence.

What are the implications of such a number when we have an estimate? The answer depends on the number itself—if the price effect is small, the incremental cost is little different from the price, and the distinction is not important. But if the effect on price is the substantial difference between, say, $24 and $30, the incremental cost would be $44 or so—greater than the price by nearly half—and there would be important implications.

POLICIES FOR REDUCING OIL IMPORTS

The discussion to this point strongly suggests that a high and rising level of imports is a serious liability. One can wish that U.S. imports showed a downward, not an upward, trend, but wishing will not make it happen, and any policies intended to reverse that trend will have domestic consequences. Indeed, there are reasons for the trend toward larger oil imports—some have been mentioned—and reversing the trend will mean significant changes in the way energy is marketed and consumed in the United States. What to do about oil imports, therefore, depends at least partly on how the level of imports might be affected, on the alternative means by which oil imports could be discouraged or reduced, on the problems of domestic economic management that they would entail, and on the economic costs that would go along with the various economic benefits.

The first set of questions is by what means imports might be reduced, if over the coming decade there were a decision that imports should be significantly below the level that would occur in the absence of programs to reduce imports.

Almost any policy that reduces the demand for fuel, that enhances conservation, that improves fuel efficiency, or that elicits new supplies will have an effect on the demand for imported oil. Subsidized home insulation, fleet gasoline mileage standards for automobiles, accelerated development of solar home heating, and restricted thermostat settings in public buildings are bound to have some effect. But these measures work slowly and unpredictably. Their precise effect will never be known. They will not appear to be oriented specifically toward oil imports. And in some cases, their main effects will be on other fuels than oil.

While oil can often substitute for gas and coal and even hydroelectric power, there are limits, especially in the short run, to the substitution of those other energies for liquid fuel. Some measures will more directly affect oil imports than others. Two of particular interest are policies that enhance the gasoline mileage of automobiles and that accelerate the availability of synthetic liquid fuels. But even those measures are indirect and uncertain as to timing and effect. There are other measures—fees or tariffs on imported oil, or quantitative restrictions on imports—that work directly, immediately, and more predictably.

A further distinction is between the direct measures that work through the price system and those that impose quantitative restrictions, allocations, price controls, and prohibitions. We recommend that the price system be utilized even in direct measures aimed at oil imports. A conventional way is to put a duty on imported oil. A duty involves determination of a price, not a quantity; but the duty can be selected for its anticipated effects and adjusted up or down. A less conventional measure, still using the price system, is to auction import licenses for crude oil, determining a quantity to be auctioned and letting refineries bid the import fee to where it clears the market. Either way the price to the importer is higher than the world price. (If the effect on imports softens the market and lowers the price below what it would have been, the actual price rise to importers will be less than the duty.)

A crucial difference between measures that work through price and measures that work through quantity, is flexibility. It was mentioned earlier that a characteristic of imports is their cushioning effect on changes in supply and demand within the country—or shortfalls or excesses of energy policies. Oil imports have been a swing variable. Programs that freeze imports deny that capability. Programs that work through the price, allowing swings in demand and supply to be met by fluctuations in imports, will preserve that flexibility.

A duty raises the price that consumers pay and thus affects the price index. If no steps are taken to control the price of domestic oil, that price will rise to equal the import price (including the duty or license fee). Similar in its effect on imported oil would be a tax on all crude oil. While nondiscriminatory with respect to imports, it would discriminate by type of fuel. The tax on domestic oil could, of course, be less than the import duty; it could apply to some domestic crude but not all—oil from old wells but not new, oil equivalent to original production but not to additional production, and so on. If a domestic tax reduces the supply response to the higher price of imports, imports will be reduced less.

"Nonprice" techniques for inhibiting imports would all directly limit the quantities that could be imported. If the price is held at the previous level and no duty is imposed, but imports are restricted in quantity, there will be excess demand. That has to be accommodated somehow. Imports could be allocated to traditional importers; but if there are no further price regulations on their product, they will enjoy the higher prices required to bring the demand for refinery

214 Energy and the Economy

products into balance with supply. If refinery prices are controlled, something will have to suppress the excess consumer demand.

To recapitulate, if the price system is used, the price on all oil will rise, discouraging consumption, encouraging domestic supply, discouraging imports. For a given duty, the effect on imports will be less if domestic supply is prevented (through the tax system) from enjoying the entire price increase. If prices are not allowed to rise, price controls with one or another technique of rationing will have to be instituted, and price equalization techniques may be needed to take care of differentials between domestic and imported oil. Those are the main alternatives.

Domestic Options

If imports are restricted by direct means, price or nonprice, the result is bound to increase demand for domestic fuels—especially for domestic petroleum but, because of interfuel competition, also for gas and coal.

With respect to domestic crude, there are three choices: (1) the price of domestic oil can be allowed to rise, the proceeds going to the producers; (2) the domestic price can be allowed to rise, with a tax diverting some or all of the increased revenue to the federal treasury; or (3) the price of some or all domestic oil can be regulated so that it does not rise to the competitive market level. By far the least complicated is to allow the price paid by refineries to be uniform and equal to the price of imported oil (with appropriate allowance for transport and quality), clearing the market without control. This policy, simplest with respect to fuel, complicates macroeconomic policy by raising the price index. And it is politically divisive because it transfers income from consumers of oil to the owners of oil reserves.

The redistribution from consumers to owners of crude oil can be reduced (but not the effect on the price index) by a wellhead tax that captures some of the additional proceeds attributable to the price rise. The ordinary tax system will capture a substantial portion, probably something more than half, through the corporate profits tax and the personal income tax on dividends and capital gains. And of course, to the extent that oil companies' stocks are held by pension funds, university endowments, or individuals with middle-sized incomes and assets, another modest fraction can be thought of as returning to the consumers. But there will remain an amount, probably on the order of $3 or $4 billion per year for every $2 per barrel rise in the price of domestic oil, that could be (and, of course, has been in the president's tax proposals) a "windfall" target for taxation.

Price controls can avoid most of the impact of domestic price increases on the price level and avoid divisive questions about what to do with tax proceeds. But if they attempt to allow higher prices selectively on the more supply-responsive sources of oil, they entail severe difficulties of enforcement and equity. And there has to be a system for evening out the cost differentials between imported oil and domestic oil, and among different kinds of domestic oil, if refineries are not to enjoy or suffer arbitrary advantages and disadvantages in the prices of the crude that they purchase. Something like the present system of "entitlements" would presumably be continued if prices were controlled below what would be their market levels.

Under the present system of entitlements, with imports unrestricted, the market for petroleum is cleared by the price system without any need for the rationing of refinery products. Oil imports rise or fall with changes in demand and supply serving as the market stabilizer. The price system moderates demand because the domestic price of all oil, being the average price of domestic and imported oil together, goes up as the imported share increases.

But with imports directly restricted and domestic crude prices controlled below market levels, there would be no source of crude to play the market-stabilizing role that imports currently play. Some form of rationing would have to be imposed on refinery products.

Targets and Programs

If measures are put in place that work directly on imports, targets need to be set. The target variable could be the quantity of imports, it could be the net domestic price (inclusive of duties), it could be the dollar volume of imports, or there could be some combination of price and quantity targets. Any price or quantity goals would be trajectories over time rather than fixed values and based on estimated developments over many years to come.

The simplest measure to administer would be a specific duty per barrel—or ad valorem duty—representing an increment over the world price, with the domestic price depending on the world price and the quantity of imports a dependent variable. A duty that was varied continuously or periodically could be adapted to the ensuing level of imports to approximate quantitative targets on some time schedule measured either in months or in years. Quantitative goals, if achieved by inflexible licensing methods, would have to treat the price as a variable (e.g., by auctioning off current and future import permits) or would have to treat the regulatory system as the place where something had to give. Generally, there will be a policy choice between import restrictions that are administered flexibly and adaptively to changing circumstances and changing goals and restrictions operated to meet announced commitments in a determined fashion.

Diplomacy will be involved in setting goals. Other oil-importing countries will observe what the United States does, and even more important, oil-exporting countries will react to an announced U.S. program. Any import targets announced by the United States will become part of the world energy discussion. Price and revenue targets may look like signals or announcements of dollar proceeds withheld from foreign suppliers through the techniques of import duties. Success or failure in reaching targets will also enter into the ongoing petroleum diplomacy.

The timing of import restrictions has two aspects. One is that a program target would reach into an uncertain future and be a time path rather than a number. The second is when such a program might go into effect. There will be some impact from letting the domestic price go to the level of the world price, abolishing entitlements, and letting consumers respond to the higher fuel prices. Just phasing out the current controls as proposed by the president could take years rather than a few months, and the market's reaction to world price levels will be slow to appear. (The "base case" not yet having been achieved—domestic prices at world market levels—the "further" restriction of imports is not yet at the top of the agenda.) The timing of any restrictions that go beyond the dismantlement of the present subsidy to imports involves the question of when to begin. A related question is when to erect the institutional structure of the forthcoming import restrictions—the fee or duty system, the licensing system, the tax system, or whatever would be involved—to give notice, to be prepared to initiate on short notice, or to take advantage of such latent measures to meet an unexpected crisis.

Targets without Controls

An alternative policy, implicit in the president's National Energy Plan of 1977, is to reduce imports by means that do not bear directly on imports themselves.

There are a multitude of measures that would reduce the demand for imports. They include conservation of liquid fuel, stimuli to the production of crude oil or synthetic liquids, and measures that encourage the switch to alternative fuels and discourage the switch to liquid fuels. Income tax credits for home insulation, right turn on red traffic lights, federal outlays for the development of coal-based liquids, fleet mileage standards for automobile manufacturers, and prohibition on the uses of petroleum in new electric power plants are

all measures that would have some effect, prompt or slow, predictable or indeterminate, on the demand for oil.

None of the measures either currently in effect or proposed will
have a prompt effect that is large and predictable. Some could have
large effects, some could have prompt effects, some could have predictable effects. But no mix of them would add up to a program confidently geared to import targets below the levels expected during
the next five years.

Two areas of domestic policy that work indirectly on imports and
that can make a sizable difference illustrate the uncertainties and the
lead times and also illustrate the way market principles can be used.
One is synthetic liquid fuels, and the other is improved motor vehicle
fuel efficiency.

There is no doubt that liquid fuel suitable for motor vehicle
engines can be obtained from shale and from coal. There is uncertainty about the price at which synthetic liquid fuel would become
competitive. Especially, little is known about the environmental
consequences of mining the shale and the techniques of disposing of
the residue. For a number of reasons (discussed in Chapter 15), the
private sector may be slow to develop or to experiment with the
technology of shale oil on a commercial scale. Some stimulus from
the federal government can be worthwhile, but heavy governmental
investment in the production of a standard consumer commodity
should be avoided.

We believe that the government should offer to contract for some
significant volume of synthetic liquid fuel, of specified quality,
preferably from diverse sources, for delivery beginning probably in
the late 1980s and should let contracts by competitive bidding. The
federal government could do this within a ceiling price derived along
the lines discussed earlier in this chapter, a price that should properly
be indexed for inflation of the general price level. We believe that
this could be a reasonable bargain at a price greater perhaps by one-
fifth than the expected price of natural petroleum fuels in the late
1980s and early 1990s. Quantities up to a million barrels per day
could prove to be a sound investment, good insurance, and a major
beginning for a potential large supply of liquid fuel for the last
decade of this century. But it should be done through competitive
bids on firm contracts, with the fuel so obtained being distributed
through normal private channels. How speedily such a program could
be enacted would depend on the smooth functioning of procedures
relating to environmental impact. Since such plants require very large
capital outlays, it might also be necessary to set up some loan guarantee facility, especially, but not only, if broad bidding is desired.

Average fleet mileage standards for the major American auto producers appear to be working successfully and may well make substantial improvements in fuel efficiency during the coming half dozen years. This method of inducing improved gas mileage may look like direct regulation that bypasses the market system, but in reality it is not. It is a remarkable characteristic of the American automobile industry that the major manufacturers produce full lines of vehicles, from light to heavy, from economy to luxury, from sport cars to working vehicles. Anyone who buys an automobile buys it voluntarily, comparing the prices of alternative cars, both alternative cars of the same manufacturer and the cars of other manufacturers. The only way that the prescribed average can be met by a manufacturer is by pricing the cars so that the customers take off the market a mix that meets the average. If the mandated average is effective in raising gas mileage, it means that consumers are induced to buy, on average, cars with better gas mileage than they would have bought. They can only be induced to do that by the manufacturers pricing the cars so that the high mileage cars are more attractive to consumers, more of them are purchased, and the average is raised. Thus, the auto companies follow a pricing policy—not altogether unlike a mixture of "taxes" on low mileage cars and "subsidies" on high mileage cars—in a way that allows consumers individually a market choice of vehicles and collectively a way to achieve the required average.

A Caveat on Import Management

One objection to direct import controls is that the U.S. government has demonstrated a great capacity to mismanage oil imports and, for that matter, fuel policy in general. There are protectionist interests in import restrictions; there are regional interests that attach to different fuels and different sources of fuels; there are large stakes in any system that involves duties and taxes; there are ideological interests involving oil and trade, nuclear power, and relations with Arab countries; there are national security implications; and generally there is a history of oil import management that does not stimulate confidence that a soundly conceived import restriction program would be administered with competence, wisdom, and statesmanship. Even people who would strongly favor a simple system of import duties on foreign oil, with a freely working price system at home coupled with limited wellhead taxation, can have no confidence that, were such a program to be enacted, it would not soon become overburdened with a superstructure of special provisions, exceptions, and controls. At least some credence has to be given to the notion that the propensity to mismanage oil imports is sufficient to give warning against such a program.

RECOMMENDATIONS

Taking together all of the considerations raised in this chapter, we conclude that there is a compelling case for proceeding promptly to eliminate the subsidy to imported oil that is embodied in the present regulation of oil prices. The present system subsidizes imported oil out of the forgone proceeds on domestic oil. Consumers not only pay less than the incremental cost of imported oil, they pay even less than its nominal price. The market price to refineries should no longer be held below the world price of oil. The inflationary effect on the general price level is small and worth incurring in the interest of economizing oil. Even the true effect on the price level will be less than the nominal effect, to the extent that the higher prices paid by domestic consumers lead to lower imports and less upward pressure on future world oil prices.

We do not advocate taxing away the proceeds on domestic oil in any fashion that would reduce the incentive to bring in new oil supplies. New supplies should be free of any such tax, and even old oil should be taxed, if at all, in a way that does not threaten taxing the new oil that comes on the market as soon as it becomes "old."

We recommend that new oil, upper tier oil, and stripper wells be uncontrolled—that controls be removed and taxes not be imposed on any except the genuinely "old" oil. We propose that controls on old oil be phased out within two years (preferably by lowering the reference point by which old oil is defined) in order that price regulation in the oil industry not acquire a permanent character.

We would go further. The billion barrel strategic reserve of petroleum that the Congress has authorized and that we consider a justified investment can be financed in different ways. One would be a fee on imported oil. A tax of 10 percent on the value of imported oil would finance an accumulation for strategic storage of a billion barrels in a little over three years. Such a fee would be closely related to an already announced policy (on building a strategic reserve); it would not appear as an aggressive effort to capture OPEC's oil proceeds; it would raise the price of oil in the general direction indicated by all the considerations raised in this chapter; and it contains a rationale that helps to protect it from political maneuver. Other ways of handling the strategic reserve are discussed in Chapter 1.

Finally, it is of the highest importance that the legal and administrative basis for the IEA sharing arrangements be kept intact, that supporting legislation be not allowed to lapse, and that additional authority be put in place as necessary. These measures, of course, are in addition to measures that will work indirectly on oil imports, like the development of synthetic liquid fuel, automobile fleet mileage standards, conversion of electric power facilities from oil to coal, and the many measures discussed in the chapter on conservation.

We believe that these recommendations constitute minimal steps. Rather than discuss in detail what further steps to reduce imports directly might be taken after deregulation has been accomplished and after the proposed fee for financing the strategic storage had a chance to take effect, we would emphasize the urgency of deregulating promptly and decisively and of eliminating the subsidy to imports—a subsidy that becomes more costly with every increase in the OPEC price of oil.

APPENDIX

A few benchmarks can help to identify the uncertainties about OPEC price and production strategies. One possibility is that the OPEC nations determine, in the form of a schedule over time, the annual quantities to be produced, letting the prices then be determined by demand. A second is that they determine, as a schedule over time, the annual prices that they will charge, managing the quantities made available as the means of bringing about the target prices. An intermediate possibility is that they plan prices and quantities and, if demand at the planned prices exceeds the projected quantities, that they increase both prices and exports. And quite a different hypothesis is that, having planned prices and quantities, they act to keep their planned revenues (prices times quantities) on target.

Consider the first case, an OPEC quantity projection with the price to be determined by demand in the market. This could either be agreed quantities for all OPEC countries or an agreed total with particular countries responsible for quantity adjustments to keep the supply on target. It would have to be a trajectory over time, rather than a single number. And as a benchmark we assume that the quantity program is adhered to and not revised with unanticipated price developments. In this case, the determinant of what a 4 million barrel difference in American imports would do to the price is the net elasticity of demand in the rest of the world. The only way that the United States could acquire 4 million barrels per day more than otherwise would be by bidding up the price to the point where other countries were taking 4 million less off the market. Alternatively, if the United States were to import 4 million barrels less than otherwise, the price would drop until the 4 million barrel difference had been absorbed by other countries. Those net demand elasticities would depend not only on consumer demand at different prices and on the non-OPEC supply response, but on government policies, including policies toward the balance of payments, because so much foreign currency would be involved.

Without regard to foreign exchange policies, net elasticities (of demand over supply) of about -0.7 is a reasonable estimate; this would mean that over the long run, the quantities imported would differ by 70 percent of the proportionate price difference—that is, that if the price were higher by 10 percent, the quantity imported would be 7 percent smaller. A quantity difference of 4 million barrels would entail a price change of about 20 percent. So this particular benchmark case gives a rise in price per barrel from the postulated $24 to something like $30, yielding about $44 per barrel as the incremental cost of U.S. imports associated with the incremental 4 million barrels. This benchmark case entails additional foreign payments of 10-15 percent for countries with no production of their own and is not implausible.

If instead, as in the second benchmark case, the OPEC countries resolved on a price projection over time that was independent of the quantity demanded, the price in any year would be independent of the quantity imported by the United States. The difference then between our illustrative 9 and 13 million barrels would make no difference to the price, and the added cost would be identical to the price per barrel.

The intermediate case, in which OPEC adjusts production by, say, 2 million barrels for a 4 million barrel difference in U.S. imports, would entail a price difference more like 10 percent, perhaps corresponding to the illustrative $26.50 used above.

Less plausible, but not to be ruled out altogether, is that at higher levels of demand OPEC follows a more moderate price trend, holding to a revenue target. The incremental cost per barrel would then be even less than the price.

Each of those benchmark hypotheses assumes that the oil-exporting countries determine their price and output policies without regard to the perceived intent or stated policy of the U.S. government and without regard to the manner in which U.S. imports might be inhibited if there were a U.S. policy to keep imports at the lower level. If to achieve that lower level of imports—9 million barrels compared with 13 in our illustration—the United States had to impose strong and conspicuous restrictions on imports, publicly justifying them on grounds that it would force OPEC to lower the export price of oil, OPEC would be likely to make some policy response rather than merely proceed with price and production plans that had earlier been determined without regard to U.S. policy.

If consumers are free to consume as much petroleum products as they wish, they will be employing petroleum in uses in which it is worth the $30 per barrel that the importer pays for it. But collectively, consumers will be consuming products that, over the range between 9 and 13 million barrels, are worth $30 but costing them $44. If a tax of, say, $6 per barrel could induce a lower level of imports by a million barrels per day and the price fell from $30 to $28.50 (one-quarter of the way from $30 to $24), the price paid by consumers, including tax, would be $28.50 + $6, or $34.50. If the tax proceeds could then be redistributed without affecting the level of imports, consumers would be forgoing oil worth $29 or $30 per barrel but saving $48 per barrel. ($30 $\times$ 13 − $28.50 $\times$ 12 = $390 − $342 = $48. The fourteenth million costs a little more than the $44 average over the 4 million.)

To take another example, if some substitute for imported oil were available at $35 per barrel, it would be unattractive at an oil price of $30; but if consumers bought a million barrels of it for $35, the cost of liquid fuels would be less by $13 million, the $8 savings per barrel of imports exceeding the $35 per barrel of the substitute fuel.

Energy in an International Setting

Introduction

Part II explicitly deals with some international questions that were only incidentally touched on in Part I. The first, treated in Chapter 7, sets forth what we believe it is important to know about energy resources—their location, their estimated amounts, and, especially with regard to nonconventional sources, when they will become available and roughly how much of the supply requirements they will be able to provide. The time frame of the study makes these considerations less significant, since individually and much more so in the aggregate events in the next twenty years will put little stress on resources as such. Rather, resource-related problems are likely to emerge from the clash with objectives such as environmental goals, political motivations, and other factors that influence supply, rather than from physical resource limitations. The chapter is brief, not because the topic is unimportant, but because it is overshadowed by others in the twenty year perspective.

Chapter 8 is international in a different sense—that is, it deals with energy positions, developments, and policies outside the United States. As we have stressed throughout this volume, energy is an international problem. What we do, or fail to do, affects Europe, Japan, Africa, and so on, and by the same token, their activities affect us. All oil importers feel the consequences of OPEC decisions. The planned economies are not immune to our fate nor we to theirs.

Embracing all this in one chapter of reasonable length has been a difficult problem. Thus, we have written just enough to illustrate whether and how the situations differ from those prevailing in this country. The reader will see the much greater dependence abroad on

OPEC oil, the ambitious plans to reduce it, and efforts to move to coal and to nuclear power. Success has been greater in discouraging oil consumption than in developing alternative energy sources. Coal and nuclear power are struggling in most of Europe and in Japan just as they have been here, despite the greater vulnerability of these areas to oil supply interruptions and other shocks.

Chapter 8 deals next with the Soviet Union and China and the continuing controversy on whether Russia will cease to be a net oil exporter—we tend to think not, or not soon in any event—and how quickly China may become a significant contributor to the world market. Data are poor, and political motivations strong and subject to rapid change. Beyond the near future, there is little that transcends intelligent speculation.

As for the OPEC countries, the chapter describes possible scenarios and judges that even economic rationale does not provide a satisfactory key to future supply and price decisions. Attempts to probe the Saudi mind or the deliberations of OPEC are bound to produce intriguing, but not very definite, findings, and so one ends up with the weak conclusion that reduced demand for oil is more likely to act as a price damper than is increased demand, though even that conclusion has been challenged.

Finally, the chapter tries to bring some structure to the discussion of how energy stringency is likely to affect the developing countries. Three thoughts deserve mention. First, energy is only one input into these economies, and its future role depends largely upon their development strategies, which must be assessed on a case-by-case basis. Second, no energy sources are especially appropriate for developing countries. And third, because noncommercial sources (for example, firewood, dung, and so on) play a larger role in many of these countries, the potential pressure on commercial sources, when the transition from traditional sources occurs, can be very large.

There are, of course, many chapters in this report besides these two that have an international flavor. One cannot deal with nuclear or solar energy, or with the role of oil imports or certain environmental issues (such as increased carbon dioxide in the atmosphere), to name just the most obvious, without transcending national boundaries. But Chapters 7 and 8 have an overwhelmingly global aspect and are thus offered as a separate part.

Global Energy Resources

7

The 1973 oil embargo and the events that it set in motion caused those concerned with matters of energy supply to reevaluate the state of knowledge of energy resources both in the United States and in the rest of the world. The major question to be answered was whether there was a sufficient resource base available to allow an orderly transition from present patterns of energy resource use to some new pattern that would be required in the future.

No attempt has been made in this study to develop a new resource assessment, nor has it been necessary, in the context of a twenty year horizon, to even engage in a careful critique of the resource data base. Work in progress at Resources for the Future and estimates from a large number of published sources were relied on for information on specific sources.

The rapidly increasing prices for conventional fuels and the recurring concern about the adequacy of supplies of oil and gas—the two fuels used most widely at present—have shifted attention to the availability of nonconventional fossil fuel resources. As a result, more emphasis than usual has been placed on evaluating the probable size of these resources, because as real prices of the remaining natural gas and oil rise, these resources should become economically recoverable and be extremely useful in making the inevitable transition to renewable resources that will occur sometime in the next century.

There is no universally accepted set of definitions for making resource assessments. The term "reserves" in nearly all classifications includes that part of the resource base (all of the resources that exist in the crust of the earth) that has been identified with what is considered a high degree of accuracy and can be recovered econom-ically using existing technology. The reserve base increases as real prices rise, with the development of new and lower cost extraction technology, and as new information about resources is developed. "Recoverable resources" is that part of the resource base that it is believed will be discovered and eventually recovered using technol-ogy that will be developed and at prices that will prevail in the future. A diagram widely used to classify resource information is shown in Figure 7-1, which illustrates the meaning of many of the commonly used resource terms.

CONVENTIONAL RESOURCES

Coal

World resources are shown in Table 7-1, which gives the most recent estimates of "geological resources" and reserves of bituminous coal and anthracite by continent and by countries with major coal resources. The term "geological resources" as defined by the World Energy Conference includes only that part of the resource base that may be expected to become of economic value in the future. In addi-tion there are estimated to be an additional 2.4 trillion (10^{12}) tons

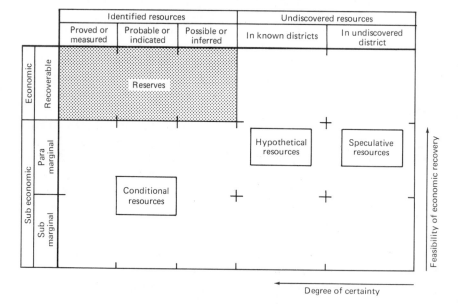

Figure 7-1. Classification of Mineral Resources.

Table 7-1. World Resources of Bituminous Coal and Anthracite (10^6 tons of coal equivalent).[a]

Continent (countries)	Geological Resources	Reserves
Africa	172,714	34,033
Botswana	(100,000)	(3,500)
South Africa	(57,366)	(26,903)
America	1,308,541	126,839
United States	(1,190,000)	(113,230)[b]
Canada	(96,225)	(8,706)
Asia	5,494,025	219,226
China	(1,424,680)	(98,883)
USSR	(3,993,000)	(82,900)
Australia and South Seas	213,890	18,164
Australia	(213,760)	(18,128)
Europe	535,664	94,210
Federal Republic of Germany	(230,300)	(23,919)
Poland	(121,000)	(20,000)
United Kingdom	(163,576)	(45,000)
Total	7,724,834	492,472

Source: World Energy Conference, *World Energy Resources 1985-2020* (New York: IPC Science and Technology Press, 1978), pp. 66-67.

Note: Numbers shown in parentheses reflect only the resources of the country with the largest reserves within that continent and are not addable to the totals.

[a]"Equivalent coal" has a heating value of 25.2 million (10^6) Btus per short ton.

[b]These values are lower than those in Table 7-2 as they do not include subbituminous coal and lignite and represent recoverable coal, not coal in place, as does Table 7-2.

of coal equivalent of subbituminous coal and lignite, for a world coal resource total of 10.1 trillion tons of coal equivalent.

From Table 7-1 it is clear that three countries dominate the bituminous coal and anthracite geological resources and reserves of the world. The USSR has 51.7 percent of all the geological resources, China has 18.4 percent, and the United States has 15.4 percent; these countries together have more than 85 percent.

The world resources of all ranks of coal are so large that the known reserves could supply the total energy demand of the world for 180 years at current rates of consumption, while the geological resources would last for more than a thousand years.

The data on U.S. resources and reserves of coal are more complete than for most other countries. The coal resources of the United States are generally divided into three regional groupings. The eastern region includes the deposits found in the Applachian area; the interior region includes the deposits from Indiana to west Oklahoma and Texas; and the western region includes all the coal west of the interior region deposits.[1]

The most useful resource classification for understanding the potential role for coal in the United States over the next fifty years is the "demonstrated reserve base." This is the quantity of coal in place under certain specified depth and seam thickness criteria and for which there is a high degree of geologic information and engineering evaluation available. All the coal in the demonstrated reserve base can be recovered economically and legally. The proportion that will actually be recovered depends on the method of mining and other factors and can range from 40 to 90 percent of coal in place.

Table 7–2 shows the demonstrated reserve base by region, method of mining, and sulfur content. Approximately 68 percent of all reserves are suitable for underground mining. The strippable reserves are concentrated in the West, with 60 percent of the total in the states of Montana, Wyoming, and North Dakota. These reserves consist entirely of subbituminous coal and lignite. Low-sulfur coals occur mainly in Montana and Wyoming in the West and in West Virginia in the East. The low-sulfur West Virginia coals are very high quality bituminous coking coals—that is, coals that produce a coke when heated in the absence of air.

Total U.S. resources of coal are very large. Even if just the demonstrated reserves are considered, and using an average recovery rate of only 50 percent, the resource base would be able to supply coal at current U.S. coal consumption rates for about 350 years, while it would last 6,600 years if all the coal resources were included.

Petroleum

Table 7–3 gives the most recent estimates of the proven world petroleum reserves by geographic region. Nearly 67 percent of the non-Communist proved reserves are in the Middle East and about 14 percent are in the Western Hemisphere. The largest reserves are in Saudi Arabia (27 percent), but Kuwait (12.4 percent), Iran (11.3 percent), Iraq (6.2 percent), and Abu Dhabi (5.7 percent) all have reserves as large as those of the United States.

Many estimates of the world's total ultimately recoverable reserves

1. The states included in each region are: *the East*—Alabama, Georgia, eastern Kentucky, Maryland, North Carolina, Ohio, Pennsylvania, Tennessee, Virginia, and West Virginia; *the Interior*—Arkansas, Illinois, Indiana, Iowa, Kansas, western Kentucky, Michigan, Missouri, Oklahoma, and Texas; and *the West*—Alaska, Arizona, Colorado, Montana, New Mexico, North Dakota, Oregon, South Dakota, Utah, Washington, and Wyoming.

Table 7-2. Demonstrated Coal Reserve Base of the United States as of January 1, 1974 (10⁶ tons).

Mining Method and Percentage of Sulfur	Region			Total[a]
	East	Interior	West	
Underground				
Up to 1.0 percent	25,988.4	1,682.1	99,258.3	126,928.8
1.1–3.0 percent	39,576.4	9,976.2	9,847.6	59,400.2
More than 3.0 percent	20,742.5	51,177.8	1,800.0	73,720.3
Unknown	11,120.6	18,608.1	10,033.0	39,761.7
Subtotal[a]	97,456.4	81,448.9	120,934.6	299,839.9
Surface				
Up to 1.0 percent	5,219.8	984.1	67,048.7	73,252.6
1.1–3.0 percent	4,593.2	4,713.0	24,291.1	33,597.3
More than 3.0 percent	3,188.7	14,917.6	844.9	18,951.2
Unknown	2,811.3	5,956.8	2,308.2	11,076.3
Subtotal[a]	15,826.7	26,572.7	94,486.4	136.885.8
All Methods				
Up to 1.0 percent	31,208.2	2,666.2	166,307.0	200,181.4
1.1–3.0 percent	44,169.6	14,689.2	34,138.7	92,997.5
More than 3.0 percent	23,931.2	66,095.4	2,644.9	92,671.5
Unknown	13,931.9	24,564.9	12,341.2	50,838.0
Total[a]	113,283.1	108,021.6	215,421.0	436,725.7

Source: U.S. Department of the Interior, Bureau of Mines, *Coal—Bituminous and Lignite Annual 1974* (Washington, D.C.: Government Printing Office, January 27, 1976), pp. 5–6.

[a]Data may not add due to rounding. "Equivalent coal" has a heating value of 25.2 million (10⁶) Btus per short ton.

Table 7-3. World Proved Reserves of Petroleum as of January 1, 1978.

Geographic Region	10^9 Barrels
Asia-Pacific	20
Western Europe	24
Middle East	370
Africa	58
Western Hemisphere	76
Subtotal	548
Communist areas	94
Total	642

Source: Oil and Gas Journal 76, no. 52 (December 25, 1978): 102–103.

of crude oil have been made. Some of those made between 1942 and 1975 are shown in Table 7-4.

Except for the four earliest estimates—made between 1942 and 1948, before the discovery of the large Middle East oil deposits—the estimates fall in the range of 1,000 to 2,500 billion (10^9) barrels.

Table 7-4. Estimates of Total World Ultimately Recoverable Reserves of Crude Oil for Conventional Sources.

Year	Source	10^9 Barrels
1942	Pratt, Weeks, and Stebinger	600
1946	Duca	400
1946	Pogue	555
1948	Weeks	610
1949	Levorsen	1500
1949	Weeks	1010
1953	MacNaughton	1000
1956	Hubbert	1250
1958	Weeks	1500
1959	Weeks	2000
1965	Hendricks (USGS)	2480
1967	Ryman (Esso)	2090
1968	Shell	1800
1968	Weeks	2200
1969	Hubbert	1350–2100
1970	Moody (Mobil)	1800
1971	Warman (BP)	1200–2000
1971	Weeks	2290
1975	Moody and Geiger	2000

Source: Workshop on Alternative Energy Strategies, *Energy: Global Prospects 1985-2000* (New York: McGraw-Hill, 1977), p. 115. Reproduced with permission.

Estimates made since 1960 tend to be at the upper range, with most of them at approximately 2,000 billion barrels. These include reserves in Communist countries. If one assumes that ultimate reserves in the Communist countries are about 15 percent of the total (the same percentage as proved reserves), then the non-Communist ultimate recoverable reserves will be about 1,700 billion barrels.

Starting in the mid-1960s the rate of addition to world oil reserves slowed down appreciably, and most geologists expect this trend to continue. The continuing increase in world oil demand combined with this decline in reserve additions has been projected to result in a peak in world oil production sometime in the period 1990 to 2010.

Figure 7-2 shows the estimated proved reserves of crude oil in the United States during the period 1945 to 1976. Between 1955 and 1969, proved reserves remained relatively constant at about 30 billion barrels. The Alaskan discovery at Prudhoe Bay increased reserves to nearly 40 billion barrels in 1970, but a sharp and continuous decline in the reserve values has occurred since then.

A number of estimates of U.S. undiscovered renewable petroleum resources starting in 1959 are shown in Table 7-5. Most of the earlier estimates were much higher than those that have been published

Billions of barrels

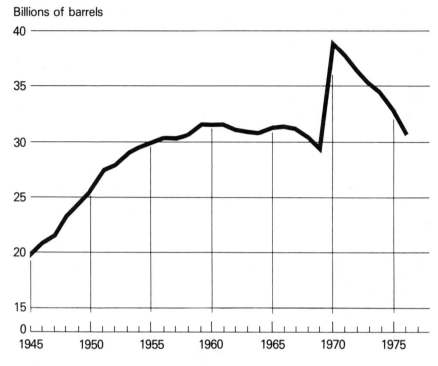

Figure 7-2. Proved Reserves of Crude Oil in the United States, 1945-1976.
Source: American Petroleum Institute, *Reserves of Crude Oil, Natural Gas Liquids, and Natural Gas in the United States and Canada as of December 31, 1976* (Washington, D.C.: API, May 1977), p. 84.

since 1974. The five most recent ones range between 50 and 127 billion barrels (or from seven to nineteen years of supply at current U.S. rates of consumption), where the two extremes reflect differences in the probability of the stated amount being in fact discoverable and recoverable.

Natural Gas

Historical data on world natural gas resources are very limited. The large resources of natural gas are, except for the United States and, more recently, the United Kingdom, concentrated in regions where energy markets are small. Unless natural gas supplies could be delivered to large markets that could be reached by natural gas pipelines, they had limited commercial value, and as a result, they were poorly developed.

234 Energy in an International Setting

Table 7-5. Estimates of U.S. Undiscovered Recoverable Petroleum Resources (adjusted to 1974).

Date	Source	10^9 Barrels
1959	USGS—Bulletin 1136[a]	387
1960	Weeks[b]	74
1962	National Academy of Sciences[c]	91
1962	National Fuels and Energy Study[d]	289–399
1962	Energy Policy Staff—DOI[e]	648
1965	USGS Circular 522[f]	264
1972	USGS Circular 650[g]	450
1974	Mobil[h]	88
1974	Hubbert[i]	67
1975	USGS—Resource Appraisal Group[j]	98
1975	National Academy of Sciences[k]	113
1975	USGS Circular 725[l]	50–127

[a]Paul Averitt, *Coal Reserves of the United States—A Progress Report*, January 1, 1960, U.S. Geological Survey Bulletin No. 1136 (Washington, D.C., 1969).

[b]L.G. Weeks, "The Next One Hundred Years Energy Demand and Sources of Supply," *Geotimes* 5, no. 1 (July-August 1960): 18.

[c]Marion King Hubbert, *Energy Resources* a report to the Commission on Natural Resources, National Academy of Sciences--National Research Council (Washington, D.C., 1962).

[d]Committee on Interior and Insular Affairs, *Report of the National Fuel and Energy Study Group on an Assessment of Available Information on Energy in the United States*, 83 Cong., 1 sess., September 31, 1962.

[e]Department of the Interior Energy Policy, *Supplies, Costs, and Uses of the Fossil Fuels*, prepared for the Atomic Energy Commission (Washington, D.C., June 1962).

[f]T.A. Hendricks, *Resources of Oil, Gas, and Natural Gas Liquids in the United States and the World*, U.S. Geological Survey Circular 522 (Reston, Va., 1965).

[g]Paul K. Theobald and others, *Energy Resources of the United States*, U.S. Geological Survey Circular 650 (Reston, Va., 1972).

[h]Robert Gillette, "Oil and Gas Resources: Did USGS Gush Too High?" *Science* 185, no. 4146 (July 12, 1974): 127.

[i]Marion King Hubbert, *U.S. Energy Resources: A Review as of 1972*, Committee on Interior and Insular Affairs, 93 Cong., 2 sess. (Washington, D.C., 1974).

[j]The Oil and Gas Branch Resource Appraisal Group, *USGS Mean* (Reston, Va.: U.S. Geological Survey, 1975).

[k]National Academy of Sciences–National Research Council, *Mineral Resources and the Environment* (Washington, D.C., 1975).

[l]Betty M. Miller, *Geological Estimates of Undiscovered Recoverable Oil and Gas Reserves in the United States*, U.S. Geological Survey Circular 725 (Reston, Va., 1975).

Proved reserves of natural gas for the world have only been published since 1967. In 1967, proved world reserves were 1,042 trillion cubic feet (about 30 percent of the size of the world oil reserves in that year), and this value increased steadily to 2,546 trillion cubic feet by 1975. In 1976 and 1977, estimated proved reserves declined, and by 1977 they were reported as 2,115 trillion cubic feet.

Table 7-6. Estimates of World Ultimately Recoverable Reserves of Natural Gas.

Year Made	Source	Reserves in 10^{12} cu. ft.	Reserves in 10^9 Barrels Oil Equivalent
1956	U.S. Department of the Interior	5,000	850
1958	Weeks	5,000-6,000	860-1,035
1959	Weeks	6,000	1,035
1965	Weeks	7,200	1,240
1965	Hendricks (USGS)	15,300	2,640
1967	Ryman (ESSO)	12,000	2,070
1967	Shell	10,200	1,760
1968	Weeks	6,900	1,200
1969	Hubbert	8,000-12,000	1,380-2,070
1971	Weeks	7,200	1,240
1973	Coppack	7,500	1,300
1973	Hubbert	12,000	2,070
1973	Linden	10,400	1,800
1975	Kirby and Adams	6,000	1,030
1975	Moody and Geiger	8,150	1,400

Source: Workshop on Alternative Energy Strategies, *Energy: Global Prospects 1985-2000* (New York: McGraw-Hill, 1977), p. 150. Reproduced with permission.

Estimates of the world's ultimately recoverable natural gas reserves made between 1956 and 1975 are shown in Table 7-6. The more recent estimates range from 6,000 to 12,000 trillion cubic feet (including reserves in the Communist area). Reserves of undiscovered recoverable natural gas in the Communist areas were estimated in 1975 to be 30 percent of the world total, compared to 15 percent for petroleum. The major large sources of potential supply are the United States and the Middle East (approximately 20 percent each) with the balance of the non-Communist world having about 30 percent, collectively.

Table 7-7 shows estimates of U.S. undiscovered recoverable resources made between 1956 and 1975. As with petroleum, the most recent estimates are much lower than the earlier ones. The two extreme values for the USGS estimates reflect differences in the probability that these resources will actually be found. The lower value would supply the current U.S. demand level for seventeen years; the upper value, for thirty-four years.

Uranium

Data for uranium resources outside the United States are reported in two classes: "reasonably assured," which is similar to the U.S. "reserves" classification; and "estimated additional," which corresponds to the U.S. "probable" classification. Table 7-8 summarizes

Table 7-7.   Estimates of U.S. Undiscovered Recoverable Natural Gas Resources.

Date	Source	10^{12} cu. ft.
1959	USGS—Bulletin 1136[a]	1,804
1962	National Fuels and Energy Study[b]	1,250
1962	Energy Resources Report[c]	800
1962	Department of Interior to AEC[d]	2,318
1965	USGS[e]	1,080
1972	USGS[f]	2,100
1973	Potential Gas Committee[g]	880
1974	USGS[h]	1,000–2,000
1974	Mobil[i]	443
1974	Hubbert[j]	361
1975	NAS[k]	530
1975	USGS[l]	332–655

[a] Paul Averitt, *Coal Reserves of the United States—A Progress Report*, January 1, 1960, U.S. Geological Survey Bulletin No. 1136 (Washington, D.C., 1961).

[b] Committee on Interior and Insular Affairs, *Report of the National Fuel and Energy Study Group on an Assessment of Available Information on Energy in the United States*, 83 Cong., 1 sess., September 21, 1962.

[c] Marion King Hubbert, *Energy Resources*, a report to the Commission on Natural Resources, National Academy of Sciences–National Research Council (Washington, D.C., 1962).

[d] Department of the Interior, *Energy Policy, Supplies, Costs, and Uses of the Fossil Fuels*, prepared for the Atomic Energy Commission (Washington, D.C., June 1962).

[e] T.A. Hendricks, *Resources of Oil, Gas, and Natural Gas Liquids in the United States and the World*, U.S. Geological Circular 522 (Reston, Va., 1965) adjusted to 1970.

[f] Paul K. Theobald and others, *Energy Resources of the United States*, U.S. Geological Survey Circular 650 (Reston, Va., 1972).

[g] Potential Gas Committee, "Potential Supply of Natural Gas in the United States" (1973).

[h] USGS news release, March 26, 1974.

[i] Mobil Oil Corporation, "Expected Value," *Science*, July 12, 1974.

[j] Marion King Hubbert, *U.S. Energy Resources: A Review as of 1972*, Committee on Interior and Insular Affairs, 93 Cong., 2 sess. (Washington, D.C., 1974).

[k] National Academy of Sciences–National Research Council, *Mineral Resources and the Environment* (Washington, D.C., 1975).

[l] The Oil and Gas Branch Resource Appraisal Group, *USGS Mean*, U.S. Geological Survey (Reston, Va., 1975).

the world uranium resource data. For "reasonably assured" resources at forward costs up to \$30/lb. of uranium oxide (U_3O_8), the major deposits are in the United States (32 percent), South Africa (18 percent), and Australia (17 percent). For "estimated additional" resources at that same cost cutoff, the major deposits are in the United States (53 percent) and Canada (27 percent). At costs of up

Table 7-8. World Uranium Resources (not including Communist countries) (thousand tons U_3O_8).

Country	Reasonably Assured	Estimated Additional
	Up to $30/lb. U_3O_8	
South Africa	398	44
Australia	376	57
Canada	217	510
Niger	208	69
France	48	31
India	39	31
Algeria	36	65
Gabon	26	6
Brazil	24	11
Argentina	23	—
Other	67	50
Foreign Total	1,462	874
United States	690	1,005
World Total	2,152	1,879
	Up to $50/lb. U_3O_8	
South Africa	452	94
Sweden	391	4
Australia	385	64
Canada	237	853
Niger	208	69
France	67	57
Argentina	54	—
India	39	31
Algeria	36	65
Other	142	119
Foreign Total	2,012	1,355
United States	920	1,505
World Total	2,932	2,860

Sources: DOE News, No. 79-47, April 25, 1979; and Organization for Economic Cooperation and Development, *Uranium: Resources, Production and Demand*, A Joint Report by the OECD Nuclear Energy Agency and the International Atomic Energy Agency (Paris, December 1977). The OECD-IAEA statistics employ a cost concept that is closer to full costs than that of "forward cost," used by DOE, except for Canadian resources, which are based on prices. The resulting lack of comparability, according to OECD-IAEA, is "not significant" within the general range of uncertainty in these estimates.

to $50/lb. of U_3O_8, the total amount increases from 2.2 to 2.9 million (10^6) tons in the "reasonably assured" category and from 1.9 to 2.9 million tons in the "estimated additional" category.

The widespread concern about the adequacy of uranium reserves and resources that was prevalent only two years ago has now abated.

Projections of installed nuclear generating capacity, both for the United States and worldwide, have been revised downward sharply, so that even known uranium resources will be adequate for twenty-five years or more.

Table 7-9 gives additional information on U.S. uranium resources and reserves and indicates that if potential resources are included, up to 4.1 million tons of $U_3 O_8$ might be available at costs up to $50/lb. Additional very low grade uranium resources are known to exist in very large deposits in the Chattanooga shales and in New Hampshire granite.

UNCONVENTIONAL OIL AND GAS RESOURCES

Petroleum

Enhanced Oil Recovery (EOR). The amount of unrecovered oil in place in the United States in reservoirs in which primary and secondary production has already taken place is estimated to be as much as 300 billion barrels, or nearly ten times the proved petroleum reserves of the United States. The location of these resources is known, but what is unknown is how much of this oil can be recovered economically.

Estimates of the amount of this oil that could be recovered at $15 per barrel generally range from 15 to 30 billion barrels. Table 7-10 shows the estimates by the Office of Technology Assessment of the amount of oil that could be recovered by EOR at various prices and for two different types of performance from new tech-

Table 7-9. U.S. Uranium Resources as of January 1, 1979 (10^3 tons $U_3 O_8$).

Cutoff Cost[a] ($/lb.)	Reserves	Potential			Total
		Probable	Possible	Speculative	
15	290	415	210	75	990
30	690	1,005	675	300	2,670
50	920	1,505	1,170	550	4,145

Source: DOE News, No. 79-47, April 25, 1979.

[a]Each cost category includes material in the lower category or categories. Not shown are an estimated 140,000 tons of $U_3 O_8$ that could be produced as a by-product of phosphate and copper production. The cost concept underlying these estimates is that of "forward cost," which, roughly speaking, refers to capital and operating costs not yet incurred at the time an estimate is made and specifically excludes past expenditures for acquisition of land, exploration, and mine development. Thus forward costs are lower than production costs.

Table 7-10. Estimates of Ultimate Recoverable Oil and Daily Production Rates from Enhanced Oil Recovery (EOR): Advancing Technology Case with 10 Percent Minimum Acceptable Rate of Return.

	Price per Barrel	Ultimate Recovery[a] (10^9 barrels)	Production Rates 10^6 barrels/day		
			1985	1990	2000
High Process Performance					
Upper tier	$11.62	21.2	0.4	1.1	2.9
World oil	$13.75[b]	29.4	1.0	1.7	5.2
Alternate fuels	$22.00[c]	41.6	1.3[d]	2.8	8.2
More than	$30.00	49.2	—	—	—
	$30.00	51.1	—	—	—
Low Process Performance					
Upper tier	$11.62	8.0	0.4	0.5	1.1
World oil	$13.75	11.1	0.5	0.7	1.7
Alternate fuels	$22.00	25.3	0.9	1.8	5.1

Source: U.S. Congress, Office of Technology Assessment, Enhanced Oil Recovery Potential in the United States (Washington, D.C.: OTA, 1978), p. 7.

[a] These figures include 2.7 billion barrels from enhanced recovery processes that are included in the API estimates of proved and indicated reserves.

[b] 13.75 is the January 1977 average price ($14.32 per barrel) of foreign oil delivered to the east coast, deflated to July 1, 1976.

[c] 22.00 per barrel is the price at which the Synfuels Interagency Task Force estimated that petroleum liquids could become available from coal.

[d] Production rates were not calculated for oil at prices of $30 per barrel or higher.

nologies under development. For oil at about $13.75 per barrel (1977 dollars), the ultimate recovery from EOR is estimated to range from 11 to 29 billion barrels (depending on process performance) or from approximately one-third to about the same amount as the proved U.S. petroleum reserves.

Recovering this known oil in the ground has been an intriguing problem for a long time, but achievements have been disappointing, partly because there is no technology uniformly applicable— apparently each field requires special treatment or nearly so. Yet the amounts involved are very large, and the absence of chance, as occurs in drilling new fields, continues to make this source highly attractive.

Tar Sands and Heavy Oils. Bituminous tar sands are now being produced commercially in Canada; production in 1976 was reported as 14.9 million (10^6) barrels per year, with new and larger capacity under construction. Tar sand deposits have also been reported in Venezuela, the United States, Zaire, and Trinidad and Tobago, but no commercial operations exist in those countries. Total world reserves of tar sands have not been estimated systematically, but in Canada alone the potential recoverable reserves are estimated to be 163 billion (10^9) barrels. Tar sand deposits in the United States are concentrated in Utah, and total tar sand reserves are estimated to be approximately 30 billion barrels—about equivalent in size to current estimates of petroleum reserves.

World deposits of heavy oils are believed to be very large, but very few countries have attempted to estimate the quantity, since under current conditions most of them cannot (except in a few special instances) compete with conventional sources of liquid fuels. Very large deposits of heavy oils are known to exist world-wide; those in Venezuela alone are reported at 890 billion barrels.

Some heavy oils are currently being produced in the United States, particularly in California and Texas, by thermal methods; these reservoirs are included in conventional oil reserve estimates. Heavy oils will have to be produced by some form of enhanced oil recovery technique, perhaps by extensions of currently successful techniques of in situ combustion or steam injection. However, reservoirs where these techniques have been successful have a combination of oil viscosity and reservoir permeability that make this feasible. Many other reservoirs are not porous enough, and thermal reduction of viscosity does not lead to sufficient mobility of the liquid in place to permit pumping. Perhaps some combination of hydrofracturing and thermal recovery will be required.

Oil Shale. The world oil shale resources of all grades are extremely large, as shown in Table 7–11. Approximately one-third of the high grade deposits are found in Asia and about one-fourth in Africa, with the others approximately evenly distributed among the remaining continents. The total amount of energy in even the high-grade world oil shale resources is also very large—about 350 times that of the estimated ultimately recoverable resources of petroleum.

Although shale oil has been produced commercially for more than one hundred years and is still being produced in a few countries, costs have generally been higher than for conventional petroleum production. As petroleum prices rise, shale oil production will become economic. In 1978, best estimates are that production might become possible only at a large premium over conventional oil, and some severe environmental problems need to be solved. Shale oil has, of course, been "in the wings" for many decades, but has yet to appear on the stage with any impact. Whether its time has come should become apparent in the balance of the century. If so, the large resource base could provide liquid fuel for decades, at higher prices.

Natural Gas

Very little information exists with respect to world unconventional natural gas resources or reserves. In the United States, large resources of methane are known to exist in Devonian shales, in coal seams, in tight gas formations in various parts of the country (particularly in the West), in geopressured brines along the Gulf of Mexico Coast, and perhaps in other places. There is currently some amount of commercial production of gas from Devonian shales and tight sand formations. Recently, considerable attention has been paid to potentially large gas reserves in deep sedimentary basins.

There is a wide range of uncertainty, both with respect to the size of the resource in place and with respect to what the supply-price curve might be for methane from each type of deposit. However, it is now believed that considerable unconventional methane could be obtained at prices below that required to produce methane from coal or to import liquefied natural gas.

Devonian Shales. The Devonian shales in the United States are concentrated in the eastern portion of the country. The estimated amount is 285 trillion cubic feet in the Appalachian Basin that extends from Michigan and lower New York State to the northern part of Alabama. This reserve is about equal in size to the proved reserves of conventional natural gas.

Table 7-11. Order of Magnitude of Total Stored Energy in Organic Rich Shale of the United States and Principal Land Areas of the World.

Continent or Country	Approximate Area Underlain by Sedimentary Rocks (10^6 square miles)	Shale Containing 10-65 Percent Organic Matter			Shale Containing 5-10 Percent Organic Matter		
		Shale in Deposits (10^{12} short tons)	Minimum Organic Content (10^{12} short tons)	Combustion Energy Content Q (10^{18} Btu)	Shale in Deposits (10^{12} short tons)	Minimum Organic Content (10^{12} short tons)	Combustion Energy Content Q (10^{18} Btu)
United States	1.6	120	12	310	1,200	60	1,600
Africa	5.0	370	37	960	3,700	190	4,900
Asia	7.0	500	50	1,300	5,000	250	6,500
Australia	1.2	90	9	230	900	45	1,200
Europe	1.6	120	12	310	1,200	60	1,600
North America (including United States)	3.0	220	22	570	2,200	110	2,900
South America	2.4	180	18	470	1,800	90	2,300
World total	20	1,500	150	4,000±	15,000	750	20,000±

Source: D.C. Duncan and V.E. Swanson, "Organic-Rich Shale of the United States and World Land Areas," U.S. Geological Survey Circular 523 (1965), table 1.

Note: Estimates and totals are rounded.

Coal Seams. Many coal seams contain significant amounts of natural gas that are released during mining. This gas constitutes a major safety problem, so that its removal and use prior to mining would increase safety and increase the size of the energy resource base. Some commercial use of the methane contained in coal has occurred in Europe, and several projects are now under test in the United States.

The total resources of methane in coal beds in the United States, varying with stipulated price and degree of optimism regarding this resource, are estimated to be in the range of 300 to 700 trillion cubic feet or from about the same size to more than twice as large as estimated resources of conventional natural gas.

Tight Gas Formations. Most of the tight gas formations in the United States are in the western part of the country. Table 7-12 shows the estimated quantities of gas in each of the major tight gas formations. Total resources in these formations are about twice those of conventional natural gas.

Geopressured Brines. Most of the extensive geopressured zone formations that have been identified with any degree of reliability in the United States have been in the Gulf Coast Basin along the coast of Texas and Louisiana, both onshore and offshore. One of the largest deposits is in a zone 200 to 300 miles long in sedimentary deposits up to 50,000 feet thick. The gas is now believed to be dissolved in the hot water that is found in these deposits and in the shales within the geopressured zones.

A variety of estimates exist for the size of the natural gas in geopressured brines. The estimates range from 3,000 to 49,000 trillion cubic feet. There is great uncertainty in the estimates of the size of the resource base. Even greater uncertainties are associated

Table 7-12. Estimates of Gas Resources in Tight Formations.

Area	10^{12} *cu. ft.*
Green River Basin, Wyoming	240
Piceance Basin, Colorado	210
Uinta Basin, Utah	150
San Juan Basin, New Mexico	63
Northern Great Plains, Montana	130
	793

Source: U.S. Department of Energy, *Non-Conventional Natural Gas Resources*, National Gas Survey, U.S. DOE/FERC-0010, June 1978.

with the estimates of the amount of the resource base that can be economically recovered. One estimate puts the recoverable portion at only 5 percent of the total resource in place, which would give a range of recoverable resources of from 150 to 2,450 trillion cubic feet. In another study, recoverable reserves were estimated at 260 trillion cubic feet, but the upper limit was given as 1,150.

Geopressured zones are also found in South America (Colombia, Venezuela, Brazil, and others), Europe (France, Holland, Germany, United Kingdom, Norway, and others), Africa (Nigeria, Egypt, Algeria), Asia (Pakistan, India, Russia, Iran, Iraq, and others), Australia and Oceania, and in Taiwan and Japan. For these countries even less is known than for the United States about the size of the gas resources in place and the quantities recoverable at different costs, but with the large amount of gas potential, some production from these types of reservoirs should be developed over the next twenty to thirty years, at least from some of these deposits.

RENEWABLE RESOURCES

A wide variety of renewable resources have been used extensively in the past, and many continue to be used in various parts of the world. These include noncommercial fuels—animal dung, agricultural wastes, and wood—as well as water power, wind power, geothermal, and solar energy. Interest in the use of renewable resources has increased sharply in step with growing concern about the depletion of world oil and gas resources and about the environmental, health, and other risks associated with the use of coal and nuclear fuels. The size of the renewable resource base and the prospects for greater reliance on renewable resources are discussed in Chapters 13 and 15.

THE ADEQUACY OF WORLD AND
U.S. ENERGY RESOURCES

From the preceding review of energy resources in the United States and the world the following conclusions can be reached:

1. Resource estimates are affected by a large margin of uncertainty, as the history of such estimates indicates. Because liquid and gaseous resources are hidden to a much greater degree than is true for coal, the uncertainties are correspondingly greater in the case of oil and gas. The discovery of Alaskan oil and gas in the 1960s and the Mexican discoveries of the recent past make the point forcefully. It is doubtful that surprises of this magnitude are ahead for coal. Given the large areas of the world that have not so far been

adequately explored, one cannot say with certainty that we are approaching the end of the discovery of major oil and gas resources— except, in all likelihood, in the more thoroughly explored oil and gas regions of the world such as the United States, the countries bordering on the Persian Gulf, and parts of the Soviet Union.

2. Geographic concentration is especially strong in oil, less so in gas and coal. Table 7–13 shows world reserves and resources of conventional nonrenewable energy sources. In oil, the Middle East accounts for more than half of the reserves and between one-third and one-half of the estimated remaining recoverable resources. The North American continent and the Soviet Union run a relatively poor second and third, though the recent discoveries of oil in Mexico may push this particular resource there substantially closer to the Middle East. In coal, the United States, the Soviet Union, and China account for about two-thirds of reserves and for practically all the remaining recoverable resources. However, because these quantities are extremely large, even in countries with only a small share of reserves, the amount of energy is very large. For example, the remaining recoverable resources in Australia are estimated at over 100 billion (10^9) tons, which is almost as large as world petroleum reserves. For gas, the distribution is substantially more even across the globe. In the reserves category, the Soviety Union has by far the largest amount, about 35 percent of the world total, with the Middle East close behind with another 30 percent. The two together account for about 65 percent of the world total. North America and Western Europe are next largest, though the Mexican situation is likely to change these relationships somewhat. Distribution of remaining recoverable resources is even less concentrated.

3. Reduced to the common denominator of Btus (see Table 7–13), the world's fossil energy sources turn out to consist three-fourths of coal in the case of reserves and about 90 percent in the case of remaining recoverable resources. The apparent degree of accuracy is somewhat deceptive. There is much disagreement, especially when estimates are made on a world scale, and the "true" ranges are probably much wider. Nonetheless, the dominance of coal is an inescapable fact, both for reserves and, even more so, for the less certain categories of remaining recoverable resources. It changes only, and radically at that, when uranium use in breeders is included.

CONCLUSIONS

It is in the light of these facts that one must evaluate the warning that "the world is about to run out of" any specific fuel. For one

Table 7-13. World Reserves and Resources of Coal, Oil, Gas, and Uranium (10^{15} Btu).

	Coal		Crude Oil		Natural Gas		Uranium [b]	
	Technically and Economically Recoverable Reserves	Remaining Recoverable Resources,[a] Excluding Reserves	Technically and Economically Recoverable Reserves	Remaining Recoverable Resources, Excluding Reserves	Technically Economically Recoverable Reserves	Remaining Recoverable Resources Excluding Reserves	Reasonably Assured Resources, at $50/lb. or Less Cost	Estimated Additional Resources at $50/lb. or Less Cost
United States	4,945	30,780	168	1,856–2,230	210	1,672	276	452
Canada	250	1,333	35		60		71	256
Mexico	28	56	93		33		2	1
Other Western Hemisphere	278	83	145	302–534	83	788	24	5
Western Europe	2,528	3,445	139	145–261	147	358	153	38
Middle East	—	—	2,146	2,030–3,654	749	1,038		—
Africa	945	1,445	336	261–545	191	833	223	78
Asia-Pacific	1,861	2,917	116	313–603	123		131	28
USSR	3,056	64,450	412[c]	365–713	933[c]	2,332	N.A.	N.A.
China	2,750	17,224	116		26		N.A.	N.A.
Other Communist	1,028	1,195	17		10		N.A.	N.A.
Total	17,668	122,927	3,723	5,272–8,630	2,565	7,021	880	858

Sources: Coal reserves and resources: World Energy Conference, Conservation Commission, *Coal Resources* (Guildford, England: IPC Science and Technology Press, 1978). Oil and gas reserves: *Oil and Gas Journal* 76, no. 52 (December 25, 1978). Oil resources: Richard Nehring, *Giant Oil Fields and World Oil Resources*, Report R-2284-CIA (Santa Monica, Calif.: Rand Corporation, 1978). The estimate shown for North America includes revision made by the author since the publication of this report. Gas resources: Joseph D. Parent and Henry R. Linden, *A Survey of U.S. and Total World Production, Proved Reserves and Remaining Recoverable Resources of Fossil Fuels and Uranium as of December 30, 1975* (Chicago: Institute of Gas Technology, 1977). Uranium: *DOE News*, No. 79-47, April 25, 1979; and Organization for Economic Cooperation and Development, *Uranium: Resources, Production, and Demand*, A Joint Report by the OECD Nuclear Energy Agency and the International Atomic Energy Agency (Paris, December 1977).

Note: The values shown here differ somewhat from some of those exhibited elsewhere in this chapter, owing to reliance on different sources of information. These differences reflect the persistent uncertainty of energy resource estimates, but are not large enough to alter the conclusions drawn in the chapter. In all categories of the table, past production is excluded, as are the so-called unconventional sources of petroleum and natural gas. N.A. = information not available.

[a] Assumes 50 percent recoverability. As surface mining increases, the estimates become conservative.

[b] The values shown are quads (quadrillion Btus) of thermal energy generated in a light water reactor, assuming a burnup of 30,000 MWd/metric ton or uranium, 0.2 percent tails, 3 percent enrichment of U-235, 15 percent losses during conversion and fabrication. If converted into electricity, the available energy would be only about one-third as much. If used in a fast breeder reactor, the same amount of uranium would yield between sixty and one hundred times as much heat or electricity as it would if fissioned in a light water reactor.

[c] USSR figures are "explored reserves," which include proved, probable, and some possible.

thing, the warning is hardly novel; it was first heard at least two centuries ago. The fact that it has so far proved wrong makes it legitimate to suspect it, but not to discount it entirely.

In the context of the twenty-year horizon of this study, we surely need not be concerned over physical inadequacy of coal, be this on the domestic U.S. scale or for the world as a whole. We cannot make the same assertion for petroleum or natural gas. Although surprises cannot be excluded and may even be expected, the present scale of consumption is such that even if we successfully throttle demand for these two fuels, the world will consume them at such a rate that conventional deposits will be depleted at least sometime in the next generation, if not sooner, and rising cost will reflect this prospect. Thus, the unconventional resources of petroleum and natural gas will loom increasingly large as an important part of the resource base.

For uranium, if breeder reactors are developed, the known uranium resource base would be extended by a factor of sixty to one hundred, so that the resource base would be comparable to remaining recoverable coal resources. Fuel other than uranium (i.e., thorium) would further extend the time horizon.

It is difficult to comprehend the size of the energy reserves that the world already has in hand. If only the reserves of fossil fuels were used and the growth rate in energy demand were 2 percent per year, then ten years of supply would still remain in the second decade of this century; if the remaining recoverable resources of fossil fuels are included, ten years of supply would still be available near the end of the next century. These calculations do not include the nuclear fuels; the large quantities of unconventional oil and gas resources, peat, or shale oil; or the large amount of renewable resources.

The United States is well enough endowed with fossil fuel resources and uranium to enable it to move without panic to a different pattern of energy sources (which is not to say that it will do so). What might that pattern be like? Barring absolute limits to the burning of fossil fuels due to the carbon dioxide problem, coal is likely to persist longest, among conventional fuels; fission-based electricity would endure if acceptable breeders can be designed; nonconventional sources of oil and gas would gradually take the place and extend the lifetime of conventional ones; renewable sources, especially solar energy, would find widening application; and possibly at a date substantially into the next century, fusion might offer a source of energy with a long horizon. A time profile of this kind would also hold for a number of other countries, but cannot be generalized for the world as a whole.

Barring misguided public policies, nations will presumably proceed from conventional to unconventional resources along a path that minimizes the rate of increase of cost. This means, for example, that nearly conventional oil and gas are likely to be exploited before there is large-scale production of synthetic fuels from coal, while timing of the entry of shale remains as uncertain as it has been for a long time. The picture is complicated not only because we need the wisdom to recognize the optimal path and the time to proceed on it, but also because so much is yet to be learned about the production costs of unconventional supplies and even more about the social, environmental, and health costs associated with their exploitation. Yet it is our judgment that we must obtain a much better hold on the nature, costs, and problems of this promising array of sources if we hope to make orderly transitions. Government leadership is an important ingredient in this endeavor.

Trends and Policies Abroad

United States energy policy cannot be made without reference to the rest of the world. The interconnected world market for oil illustrates the point. In the summer of 1978, when world oil production was exceeding consumption, supplies at U.S. gasoline pumps were plentiful, prices had stabilized, and for Americans the "energy crisis" was but an unpleasant memory. Events in the Middle East swiftly dispelled that illusion. A revolution toppled the shah of Iran and shut down that country's oil production for sixty-nine days, removing 5 million barrels a day from the world supply stream.[1]

Consuming nations made up 2 million barrels by drawing down reserves. Saudi Arabia, Kuwait, and a few other exporting countries increased their output. The net result of the Iranian shutdown was

---

1. There are abundant statistics and a vast literature on energy trends and developments. In addition, the study group commissioned a number of background papers that will be published in a companion volume entitled *Selected Studies on Energy: Background Papers for Energy: The Next Twenty Years* (Cambridge, Mass.: Ballinger, forthcoming). The relevant papers are: Dennis W. Bakke, "Energy in Developing Countries"; Marshall I. Goldman, "Energy Policy in the Soviet Union and China"; Gerald Manners, "Prospects and Problems for an Increasing Resort to Coal in Western Europe, 1980–2000"; Gerald Manners, "Prospects and Problems for an Increasing Resort to Nuclear Power in Western Europe, 1980–2000"; Horst Mendershausen, "Energy Prospects in Western Europe and Japan"; and Dankwart F. Rustow, "OPEC and the Political Future of the World Oil Market." Especially useful among published sources (not referenced further in the text) were J.C. Sawhill, K. Oshima, and H.W. Maull, *Energy: Managing the Transition* (New York: Trilateral Commission, 1978); International Energy Agency, *Energy Policies and Programs of IEA Countries: 1977 Review* (Paris: OECD, 1978); and International Energy Agency, *Steam Coal: Prospects to 2000* (Paris: OECD, 1978).

a world shortage of 1 to 2 million barrels a day compared with 1978.

Some nations began paying inflated prices on the spot market. At its March 1979 meeting, OPEC boosted the price of oil from $13.34 to $14.55 per barrel. The thirteen OPEC members were allowed to impose surcharges if they wished, thereby institution-alizing the free for all pricing of the post-Iran spot market. Several exporting countries have availed themselves of that opportunity. Perhaps of even greater significance, three OPEC members seriously considered limiting production.

The variability among nations' indigenous energy supplies is reflected in their differing ability and willingness to pay increased oil prices. Moreover, there exists a complex of relationships among energy, economic, and foreign policy objectives. For example, the Egyptian-Israeli peace treaty aroused the ire of several OPEC nations on whom the United States relies to maintain the international sup-ply of oil. Some OPEC members wish to maintain their huge earnings by increasing prices, yet not damage the Western world's financial system, in which they have invested their profits. Some OPEC nations rely on the United States to protect their security, while others are ideologically hostile and have close links to the Soviet Union.

Finally, domestic politics affects individual nations' domestic and international energy policies. Indigenous energy supplies may not be used for any of a variety of economic or social reasons. Environmental constraints are a major element. Coal is expensive and dirty, and the nuclear power upon which some nations' energy plans have been relying is mired in controversy. The accident at the Three Mile Island nuclear plant increased antinuclear sentiment and opposition to ambitious nuclear building plans.

The Iranian revolution provided an example of how domestic decisions in one country can seriously affect the rest of the world. The world's second largest oil exporter decided to slash production, largely because the new rulers felt that oil was the nation's patrimony and should not be depleted at headlong speed to finance a moderniza-tion with which they did not concur. Other oil exporters may follow suit. Mexico's long-range development plan calls for a moderate increase in oil production, far below full capacity levels. Saudi Arabia may decide not to expand its production in the 1980s as much as originally anticipated. Even western producers like the United Kingdom, Norway, and the Netherlands (in natural gas) are choosing to lower export deliveries to guarantee their ability to fulfill domestic demands far into the future or to exert better control over domestic economic performance.

THE ENERGY POLICIES OF WESTERN EUROPE AND JAPAN

In 1976, the European Economic Community relied on imported oil to meet 55 percent of its total energy demand, and in Japan oil imports represented nearly three-quarters of all energy consumed. By contrast, U.S. oil imports, although large in absolute terms, were a substantially smaller fraction of total primary energy consumption—only 19 percent. Data on the role of oil imports in individual nations are exhibited in Table 8-1.

The magnitude of energy consumption, and in particular the large and growing dependence on energy imports, dates only from the 1950s and 1960s. Although depletion and rising costs of conventional resources would eventually have forced a transition to alternative fuel forms, it was the oil embargo and price rise of 1973-1974

Table 8-1. Net Oil Imports as Percentage of Total Primary Energy (TPE) Consumption.

Country	Net Oil Imports (percentage of TPE)		
	1960	*1973*	*1976*
Austria	5	41	42
Belgium	28	60	51
Denmark	56	91	82
Finland	26	61	59
France	30	71	68
West Germany	19	54	51
Greece	76	77	71
Iceland	67	58	45
Ireland	29	77	70
Italy	40	73	68
Luxembourg	7	35	31
Netherlands	42	44	39
Norway	39	34	-28[a]
Portugal	59	82	88
Spain	25	64	69
Sweden	48	59	57
Switzerland	36	64	62
Turkey	10	36	44
United Kingdom	26	49	38
Japan	31	75	72
United States	5	16	19

[a]Note that Norway is a net exporter of oil.

Sources: International Energy Agency, *Energy Policies and Programmes of IEA Countries, 1977 Review* (Paris: OECD, 1978). For Iceland, Finland, France, Turkey, Portugal: International Energy Agency, *Energy Balances of the OECD Countries, 1974/1976* (Paris: OECD, 1978), pp. 26 and 27.

that suddenly and decisively persuaded the nations of Western Europe and Japan that the energy structures evolved over the preceding two decades were becoming untenable.

The peril was immediate. As a result of the actions of the Organization of Petroleum Exporting Countries (OPEC), in the near term Western European and Japanese oil supplies suddenly became considerably more expensive and at the mercy of political troubles that these nations could not control. Moreover, a decade or two into the future, continued worldwide economic growth seemed likely to trigger additional energy price increases, balance of payments strain, and perhaps conflicts over the allocation of oil.

In the search for a way out of their predicament Western Europe and Japan looked toward the following options:

- Oil import assurance from existing suppliers;
- Development of alternative oil sources (indigenous or new external);
- Development on nonoil energy sources (indigenous or new external); and
- Reduction of the energy inputs, particularly oil, required per unit of economic output—that is, energy conservation.

Bringing about major changes of this sort in a short time was fraught with uncertainty and confusion over suitable means of management. Each nation's choice of a strategy depended largely on the extent of its reliance on energy imports, the prospects for increased indigenous supplies, and a host of factors not directly related to energy policy. These factors included expectations for economic growth, balances of trade and payments, and domestic political institutions and actors. Many Western Europeans felt that the United States should take the lead in striking an overall bargain with OPEC to ensure continued supplies, a strategy that the United States rejected. Individually or through multinational institutions, several nations tried unsuccessfully to assure themselves preferential access to OPEC oil by offering attractive economic and political services in return. There was also some hope that cooperative multinational initiatives, such as through the Commission of the European Economic Community (EEC) or the newly created International Energy Agency, could help to ease the transition.[2] Meanwhile, sub-

2. Members of the European Economic Community are Belgium, Denmark, France, West Germany, Ireland, Italy, Luxembourg, Netherlands, and United Kingdom. The International Energy Agency's member countries are Austria, Belgium, Canada, Denmark, West Germany, Greece, Ireland, Italy, Japan, Luxembourg, Netherlands, New Zealand, Norway, Spain, Sweden, Switzerland, Turkey, United Kingdom, and the United States.

stantial oil finds in the North Sea were about to come into production. Moreover, Alaskan, Mexican, and offshore oil in various places were beginning to promise supply relief that could benefit Europe and Japan either directly or by filling some of the demands of others, most notably the United States.

Among alternatives to oil, the best hopes for expanding supplies seemed to be coal, gas, and nuclear power. Britain and West Germany possessed ample coal reserves and mature coal industries. Western European mining costs were high because the cheapest reserves had already been mined, labor was expensive, and most mines were deep rather than surface. But that region could always import coal from North America, Australia, Poland, the USSR, and some other countries.

In natural gas, Dutch wells had just begun to produce from substantial reserves, a pipeline system was nearing completion, and North Sea wells could be expected to add more gas soon. Soviet, Iranian, and perhaps Algerian gas could be tapped with the help of other pipelines leading to Western Europe, and additional gas from various overseas sources could be transported in liquified form. France, Sweden, and Spain had significant uranium reserves, with additional supplies likely to be available from the United States, Australia, Canada, and Africa; and nuclear power seemed increasingly ready to displace the petroleum being burned to produce electricity.

Under the pressure of uncertain and expensive energy supplies, most European countries and Japan saw opportunities for significant savings of energy in industry, transportation, and space heating. During the winter of 1973–1974, most countries adopted drastic energy-saving measures, though the measures were relaxed later.

The Organization for Economic Cooperation and Development (OECD) estimated in 1974 that as a result of the new oil price and energy policies, the share of oil imports in OECD-Europe's total primary energy consumption (TPE) would decline significantly. But implementation of energy programs proved difficult. Almost as soon as governmental programs appeared, swiftly moving events put into question the assumptions about future demand and feasibility on which those programs had been based. The need for flexibility and cautious experimentation did not permit adherence to fixed time schedules.

DEMAND REDUCTION AND CONSERVATION

The post-1974 economic recession tempered the growth of energy demand and lowered expectations for future economic and energy demand growth. Although no one rejoiced over slow growth and unemployment, Western Europe and Japan had plenty of oil, gas, coal, and electricity and idle capacities in their oil refineries. These developments took the edge off the urgency that had pervaded energy planning in 1974.

Progress in energy conservation can be recorded—albeit only grossly and to be interpreted with caution—as a drop in the ratio of a nation's total primary energy use to its gross domestic product. A review by the International Energy Agency in 1977 showed that from 1973 to 1976, about a third of the nations of Western Europe (and also Japan) recorded essentially no change in their TPE-GDP ratios. Another one-third recorded declines of 8 percent or more, and the remainder showed smaller declines.

The extent to which conservation plans will be realized remains uncertain. Conservation in Western Europe and Japan is growing, but at a slower rate than in the period immediately following the Arab oil embargo. The reescalation of the cost of energy imports subsequent to the March 1979 OPEC price increase decision should raise interest and investment in energy conservation. An early, though inconclusive, test will be provided by the OECD countries' ability to achieve the jointly agreed target of a 5 percent reduction in oil imports by the end of 1980.

Coal

In 1977, coal supplied 24 percent of Western Europe's total primary energy consumption. For Japan the corresponding figure in 1976 was 17 percent. Coal has held roughly constant since 1973, notwithstanding expectations that higher oil prices would revive ailing domestic coal industries. There has been some growth in the number of coal-burning electric plants, but there has been an offsetting decline in residential and industrial coal consumption. Decline in production has been even greater, and imports have grown.

Through the 1980s the demand for coal should increase, provided that Western Europe and Japan will have moderate economic growth, that coal prices will be low relative to oil and gas, and that delays in nuclear power will benefit coal. A 10 percent increase through the 1980s is possible. But coal output prospects are not good. In the United Kingdom, productivity, cost, environmental, and market problems suggest that production is more likely to remain constant than to expand, and it may slip back. West German industry expects to stabilize its hard coal output, and lignite production is not expected to grow; it will be up to imports to sustain any increase in consumption. The outlook for Japan, with its small and shrinking coal base, is similar.

Japan appears willing to encourage coal imports, although demand may be limited by logistical and environmental constraints. Western European nations—especially those with abundant coal reserves—are ambivalent in their attitude toward imports. Domestic coal industries and concerns over balance of trade tend to support protectionism and subsidy of domestic production.

After the 1980s, the decisive questions are whether demand for a bulky, dirty, and inconvenient fuel can be sustained and, if so, whether and at what price coal imports will be available. In any event, international coal trade will stagnate unless potential importers commit themselves to a policy of steady acceptance of substantial tonnage. Without such commitments, potential exporters may not risk the large investments needed to expand output.

Nuclear Power

The jump in energy prices after 1973 appeared at first to remove any doubts among electric utilities about the advantages of nuclear energy. The OECD in 1974 foresaw vigorous growth in Western European and Japanese installed nuclear capacity to the year 1990. The ambitious programs have since been severely curtailed by reduced estimates for future electricity demand and by public opposition to nuclear power. At one extreme, little or no nuclear power is planned in Denmark, Norway, Portugal, Ireland, Greece, or Turkey, and in a closely contested national referendum, the Austrians decided not to use the nation's first completed nuclear plant. There has been little opposition to nuclear power in the United Kingdom, but that country's North Sea oil and gas reserves are slowing growth in nuclear capacity. A militant West German antinuclear movement, adept at using a decentralized governmental structure to delay power plant approvals through court action, is significantly reducing nuclear plans in that part of the world. Among Western European countries, France remains the most enthusiastic about nuclear power; nearly half of the region's nuclear power now projected for 1985 is French. Japan's nuclear power program has been progressing, but there is public resistance, and the accident in Pennsylvania may mean that a slowdown is in the offing. Chapter 12 contains some related discussion of nuclear energy trends and issues outside the United States.

Consequences for Oil Imports

Whether national governments will be willing or able to take the steps necessary to hold down oil import demand is far from certain. Imported oil has a significantly adverse impact on some countries' balance of trade, as detailed in Chapter 4. On the other hand, from 1974 to 1977 the price of imported oil dropped in real terms, helped for several countries by the depreciation of the dollar, which is the currency in which OPEC oil is quoted. Until the Iranian cutoff, oil flowed smoothly from producer to consumer. Most Western European countries and Japan have had a long experience with importing large quantities of oil; they have so far been able to offset the import costs through a competitive export trade and have discovered that their economies can survive, even flourish, while remaining dependent on large imports of energy supplies. In this context, the difficulty of implementing national energy plans has not been severe. Indeed, energy policy in these countries has come to be submerged within a broad spectrum of economic issues and in all likelihood will continue in this condition unless the wave of OPEC price rises that have followed the Iranian events prove sufficiently damaging or ominous to impel the governments to pursue conservation measures more aggressively.

Indigenous oil and gas resources in Western Europe can temporarily alleviate the pressure of demand for imported oil, but significantly so only for the United Kingdom, Norway, and the Netherlands and, to a minor degree, for such other countries as may obtain imports from these countries. Total contribution to energy consumption may reach 10 percent for all OECD countries by 1990, but resources are limited, and output is expected to begin declining then if not sooner. Clearly, this is not a long-term solution, and even in the short run the alleviation will be only slight.

Table 8-2 provides a sketch for a few selected countries of energy trends as they have developed between 1960 and 1976. The figures illustrate both the original swing to imported oil and the difficulty in reducing oil dependence.

Summary

Since the oil crisis of 1973-1974, surprise developments have changed perceptions of the future state of energy affairs, and more surprises can be expected.

Shortly after the crisis, analysts predicted that Western Europe and Japan would significantly substitute other energy sources for OPEC oil to satisfy continued growth of energy demand. So far there is little evidence of such a trend, and more recent forecasts anticipate considerably less substitution. Now, analysts foresee that energy conservation and slower economic growth will take the place of expanded supplies of such alternatives as nuclear power, coal, and non-OPEC oil and gas. This has happened not because slow growth is wanted nor because Europeans and Japanese have come to agree that they are wasting energy conspicuously. It happened principally

Table 8-2.   Energy Trends, 1960-1976, Selected Industrial Countries (million tons of oil equivalent).

	West Germany			United Kingdom			France			Italy			Norway			Japan		
	1960	1973	1976	1960	1973	1976	1960	1973	1976	1960	1973	1976	1960	1973	1976	1960	1973	1976
Total Primary Energy (TPE)	146	267	260	170	224	207	90	184	178	49	133	136	9	20	21	95	338	366
Domestic Production	129	120	118	126	112	128	58	42	39	23	26	26	5	13	27	58	39	46
Oil	6	7	6	—	—	13	2	2	2	2	1	1	—	2	14	1	1	1
Gas	1	15	15	—	25	33	3	6	6	5	13	13	—	—	—	1	3	3
Coal	118	92	89	124	79	72	40	19	17	3	2	2	—	—	—	43	16	13
Nuclear	—	3	6	1	7	9	—	3	4	—	1	1	—	—	—	—	2	8
Other (hydro, etc.)	4	4	3	1	1	1	14	11	11	12	9	10	5	11	12	14	18	22
Net Oil Imports	27	144	132	44	111	79	27	130	120	20	97	92	4	7	-6	29	255	266
Net Gas Imports	—	12	22	—	1	1	—	8	12	—	2	10	—	—	—	—	3	8
Net Coal Imports	—	—	—	-5	-1	1	10	10	14	8	8	10	—	—	—	7	42	47
Net Oil Imports as Percentage of TPE	19	54	51	26	49	38	30	71	68	40	73	68	39	34	-28	31	75	72

Source: International Energy Agency, Energy Policies and Programmes of IEA Countries, 1977 Review (Paris: OECD, 1978).

because all of the contemplated substitutions for OPEC oil have difficulties.

Although less oil undoubtedly will be imported in the 1980s and 1990s than would have been the case had oil prices remained at their 1972 level, it is also clear that Western Europe and Japan will remain more dependent on oil imports than will the United States. Nonetheless, to the Europeans and Japanese the most important single step that the United States can take to help them cope with their energy problems would be to curb its own appetite for imported oil. Over the long run, reduced demand for oil imports can help restrain price increases regardless of the part of the world in which that demand reduction occurs, but other nations feel that their own efforts to limit oil imports will prove futile unless the United States—the world's largest oil importer—reduces its demands for petroleum. That at least is the clear message arriving in the United States from both Western Europe and Japan.

THE SOVIET UNION AND CHINA IN INTERNATIONAL ENERGY MARKETS

The USSR and China each hold one-third or more of the world's coal and natural gas reserves and a significant although smaller proportion of world natural gas and petroleum reserves. Both are net oil exporters, and petroleum sales are by far their largest source of hard currency.[3] At what level these nations will export energy in the future is a debated issue.

The USSR

The Soviet Union is the world's largest producer of petroleum and the second largest producer of natural gas. Its petroleum exports are second only to those of Saudi Arabia (unless Iran resumes exports at its pre-1979 level), and its natural gas exports are surpassed only by the Netherlands. Most Soviet energy exports go to Eastern Europe, Mongolia, Cuba, and Vietnam, but about 1 million barrels per day are sold to the hard currency world, and gas deliveries to Western Europe are growing rapidly.[4]

Total Soviet exports to hard currency countries amounted to about $10.4 billion ($10^9$) in 1977, half of which consisted of petroleum sales. (Natural gas exports earned about $570 million ($10^6$)

---

3. Energy reserve, resource, production, and trade figures are very uncertain, because the Soviet and Chinese governments classify much of this information.
4. Goldman.

in hard currency and, although due to increase in the near future, are unlikely to exceed $1.5 billion.)[5] If the Soviets were to lose their hard currency petroleum revenue, they would be unable to pay for their nonenergy imports from the hard currency nations; these amounted to roughly $12 billion in 1977. Moreover, if they were to begin importing petroleum, this added demand would boost international oil prices, perhaps sharply, and the Soviet annual hard currency import bill could jump to $30 billion or more.[6]

Should their trade balance move from deficit to surplus, as it did in 1974, then the Soviets might conceivably curb their petroleum exports. But as long as they have a large foreign trade deficit, they must at least try to sustain their oil exports.

The Soviets have several strategies that are conceptually no different from energy programs in force in other industrialized nations—energy conservation, expanded reliance on alternative indigenous resources, and increased efforts to attain petroleum production targets. Although the Soviet Union is relatively efficient in its use of energy in some sectors—notably in its use of district heating in residences and in its small per capita automobile use—observers agree that other sectors are energy squanderers. For example, buildings are poorly insulated, sloppy work habits create energy waste in many Soviet factories, and drivers in Soviet motor pools make many unnecessary trips. Evidence that the Soviets are trying to conserve petroleum include their decision early in 1978 to double the price of gasoline and reports of fuel shortages at pumps and depots. Both of these events occurred at the same time that petroleum exports increased, suggesting that the Soviets are not about to let these exports slide.

The Soviets are trying to convert oil-fired boilers to coal wherever it is available, to substitute gas for oil in the far Northeast where coal is unavailable and in West Siberia where gas is abundant, and to use nuclear energy for electricity and (eventually) district household heating in the more densely populated areas of the western Soviet Union.

The Soviets are also moving vigorously to increase petroleum output. Vast expanses of territory are as yet unexplored. Although Soviet technology is second to none in some industries, their fossil

---

5. Ibid.

6. Ibid. For example, at $14 per barrel, oil imports of 3.5 million barrels a day (mbd) would cost the Soviets $18 billion annually. At $20 per barrel, oil imports of 4.5 mbd would cost $33 billion. This is in addition to Soviet nonpetroleum imports and must be compared to Soviet nonpetroleum hard currency earnings of only $5.1 billion in 1977.

fuel exploration and production technology lags behind.[7] But the Soviets have not hesitated to bring in foreign experts and technology whenever it will help. They turn first to the United States; if unacceptable political or economic obstacles arise, they look elsewhere. Purchases of advanced petroleum technology have in the past and can in the future contribute significantly to increasing Soviet petroleum output. (Reform of some severely inefficient aspects of the Soviet industrial pricing and incentive system would help as well, but politically sensitive, sweeping economic reforms do not seem imminent.)

Granting that the Soviets have the incentive and at least potentially the means to maintain their petroleum exports, some in the United States have argued that advanced petroleum technology should be withheld from the USSR to limit its economic, political, and military strength or else should be used as a bargaining chip in human rights or other noneconomic negotiations.[8] Balancing these arguments is beyond the scope of an energy policy study, but it must be noted that if the Soviets were forced to reduce their oil exports or even to become oil importers, theirs would not be the only economy to suffer. The loss of Soviet oil supply—or worse, the addition of their demand to the world petroleum market—could trigger a sharp increase in the price of oil that would hurt all importing nations, including the United States. Thus, if the United States wants to help create a larger and more diversified future petroleum supply, it should facilitate, not hinder, cooperation in exploration and the export of petroleum technology to the USSR.

In summary, petroleum production increases will not come easily for the USSR, but it is likely that the various far-ranging measures it is adopting will provide exportable surpluses through the early 1980s and possibly beyond. Information is not, unfortunately, good enough to permit a firmer and longer run judgment to be made.

China

The Chinese have become dependent on petroleum exports to finance their foreign trade. Oil exports may soon constitute 90 percent of Chinese foreign trade earnings.[9] Increased industrial efficiency has recently become a priority target; this means increased reliance on foreign technology, which in turn means a need for

7. John W. Kiser, "What Gap? Which Gap?" *Foreign Policy* 32 (Fall 1978): 90-94.

8. See, for example, Samuel P. Huntington, "Trade Technology and Leverage: Economic Diplomacy," *Foreign Policy* 32 (Fall 1978): 63-80.

9. Goldman.

exports, although nothing is certain where Chinese foreign trade policy is concerned.

Since 1960, Chinese petroleum production has increased from almost nothing to 2 million barrels per day (mbd). The Chinese manage on a very small energy base per capita. As their domestic production has increased, they have set aside a large share for export. In recent years petroleum output has grown at a rate of more than 10 percent per year, with exports keeping pace. The Chinese sell oil to Japan, the Philippines, Hong Kong, North Korea, and Brazil. Assuming that they make no political shifts, the Chinese will be petroleum exporters of growing importance in the future. (Some Japanese project Chinese output at about 9 mbd by 1985, with annual exports equaling 4 mbd.)[10]

China's export to Japan promises to be particularly important. Japan's imports from China were 130,000 barrels a day in 1977. This figure is expected to grow to at least 300,000 barrels a day by 1982 and possibly to twice that.[11] Moreover, the Chinese have begun to negotiate actively with Japanese and American companies for joint undertakings offshore. Like the Soviet Union, China is seeking to increase its production of other energy sources, in part to enable the Chinese to set aside petroleum for export. They are building a huge hydroelectric project on the Yangtze Kiang River with total eventual capacity of 25,000 MWs—equivalent to the capacity of twenty-five standard size nuclear power stations. The Chinese have also signed a $4 billion contract with West Germany to open up two strip mines and six underground mines that are part of a general plan to boost coal production to 1 billion (10^9) tons by 1988.[12] If these projects are built as planned, the Chinese could meet their more ambitious oil export targets.

THE OPEC COUNTRIES

All thirteen members of OPEC are developing countries. The non-petroleum sectors of their economies are preindustrial. The members also share a colonial heritage that largely accounts for their self-image as the leaders of a worldwide movement to redress past injustices inflicted by Western imperialism. Beneath this common world view, however, there is scant political cohesion within OPEC.

Aside from the politically motivated Arab embargo of 1973–1974, the divisions among OPEC producers on price and availability

10. Ibid.
11. Ibid.
12. Ibid.

of petroleum have been primarily economic. These divisions have pitted "high absorbers" against "low absorbers"—that is, countries that also possess nonpetroleum assets and have relatively low per capita incomes from petroleum against countries with small populations, few nonpetroleum assets, and large foreign exchange reserves. The former are eager for additional income for development, but the latter are primarily interested in preserving the value of their existing foreign assets rather than in raising prices to add to their monetary accumulations. The principal "high absorbers"—Indonesia, Algeria, Venezuela, Nigeria, and, until the 1979 revolution, Iran—have been producing near capacity and have generally advocated large price rises. The "low absorbers"—chiefly Saudi Arabia—have had considerable spare capacity and hence have favored modest price increases or even periodic price declines in real terms. Although Libya and Kuwait are "low absorbers," they have for ideological and domestic political reasons, respectively, sided with the "high absorbers" in advocating large price increases.

The Iranian oil cutoff in 1978-1979 reinforced the economic power of OPEC. Initially Saudi Arabia, Kuwait, Nigeria, and Venezuela increased their production to help make up for the nearly 5 million barrels a day no longer being exported by Iran. Soon, however, they decided to cut back somewhat, thereby guaranteeing a continued shortage. At its March 27, 1979, meeting in Geneva, OPEC announced a 9.05 percent price rise, from $13.34 to $14.55 a barrel, and granted individual members the right to levy surcharges at their discretion.

Meeting privately after the meeting officially ended, Saudi Arabia, Kuwait, and Iraq agreed to cut production as the new government in Iran increased it—Saudi Arabia by 1 million barrels a day, Iraq by 500,000 barrels, and Kuwait by 400,000 barrels. Previous attempts to limit production as a method of firming up increased prices had failed because parties to the agreements had broken ranks. But political developments in the Middle East—the revolution in Iran and the signing of the Egyptian-Israeli peace treaty—at least temporarily impelled the members to greater solidarity.

A nagging worry to the oil-importing nations has always been the internal political instability of many of the OPEC member countries. Until the Iranian revolution, the relatively frequent political upheavals had generally had only marginal effects on the production of oil or on the oil policies adopted by successive governments. Oil production and exports appeared the one permanent feature of the changing scene, lending weight to the contention that no matter the regime, oil would always flow to willing customers. Iran's about-face under the Khomeini regime provided an example of possible new directions that political instability might take. Unalterably opposed to the shah's rapid modernization program and particularly offended by its social side effects, the Islamic revolutionaries scaled down the massive development schemes underway. With a greatly diminished interest in modernization, the Khomeini government allowed oil production to revive but decided to restrict it. In addition, it announced a new policy of selectivity in its oil exports—henceforth Israel and South Africa would be denied supplies. Whereas the shah had maintained oil exports in defiance of the Arab embargo in 1973–1974, the Islamic republic of Iran could be expected to act differently in any serious future embargo situation.

Aside from domestic instability, there is potential for regional conflict in the Middle East and North Africa. The Shatt-al-Arab, Oman, South Yemen, and the former Spanish Sahara constantly simmer and occasionally boil over in crises. Attempted subversion by Libya in Egypt, the Sudan, and Tunisia, territorial claims of Iraq against Kuwait and smaller neighboring states and somewhat defused claims of Iran against Iraq and Bahrain, indicate possible problems in the future.

The two most serious threats to the Western world's oil supply in the short term would be an overthrow of the royal house in Saudi Arabia and a Communist takeover in Iran. The United States has recognized the former danger and has attempted to reassure the Saudis of its commitment to their survival by such moves as selling advanced aircraft to them and demonstrating American power in neighboring Yemen. Tensions within the royal family were heightened by the events that toppled the shah, and the aftermath of those events themselves appear not to have run their course.

In summary, OPEC seemed in 1979, as securely in control of the world petroleum market as ever before. Although economic factors seemed likely to remain the crucial determinants of the price and availability of oil for the foreseeable future, the revolution in Iran and chronic instability elsewhere showed that political and even religious events could exert a profound influence on the market.

Moreover, even in the matter of economics, there is ample opportunity for speculation as to motivations and reasoning on the part of OPEC members. Each move permits different interpretations. Thus, Saudi Arabia might want to curtail output to raise the price and maximize revenues or merely to depress activity and thereby limit revenues to rein in internal development and change. Conversely, it might wish to raise output to keep price increases down and thereby protect its foreign investments and world stability generally or in order to raise revenues. Each scenario is realistic and probably has its protagonists in the country. The mere exposition of this diversity highlights the difficulties of suiting U.S. policy and that of oil importers generally to the putative objectives of oil exporters. Chapter 6 deals in detail with this subject.

ENERGY AND NON-OPEC
DEVELOPING COUNTRIES

The non-OPEC developing countries (NODCs) currently are estimated to consume about 10 percent of the world's commercial primary energy and 15 percent of global petroleum output. Despite rising prices since 1973, energy demand is growing swiftly in these nations, outpacing their economic growth rates. Consequently, the NODCs will play an increasingly important role in energy markets. This role poses challenges to developing and developed countries alike.

The challenge to the developing countries is to obtain the energy they need for development at the lowest possible economic and social cost. Because they are short of capital and dependent on extensive nonfuel imports to achieve economic growth, the NODCs can ill afford wasteful energy expenditures. There is not, however, any one energy "solution" for all of the developing countries. Rather, because their immediate needs, long-term development objectives, and natural resource endowments are diverse, these countries must, and can be expected to, pursue a wide array of energy strategies.

Some NODCs are emphasizing expanded exploration and development of indigenous fossil fuel resources. More than a dozen non-OPEC developing countries are currently net oil or gas exporters, and a few, such as Mexico, will contribute significantly to future world petroleum supply. Others that are less well endowed can nonetheless use domestic production to reduce the cost of imports. Many NODCs are relatively unexplored for fossil fuel resources. A five year, $4 billion program recently initiated by the World Bank should provide some needed stimulus by granting poorer developing countries low interest loans for exploration and development of petroleum, natural gas, and coal.

To NODCs not blessed with fossil fuel adequate to their needs, the principal alternatives are imported fossil fuels, nuclear fission, or renewable resources. Choices will usually be determined by the structure of end uses; a nation with growing automobile and truck use must look to imported petroleum or perhaps, as Brazil is attempting, to liquid fuels from crops. Each alternative, however, has drawbacks and limitations.

Nations considering increased imports must weigh the likelihood of higher prices in the future, the effect of higher prices on their ability to import other goods, and the prospect of sudden price jumps or shortages should there be disruption in world energy markets. Today's nuclear power reactors are too large to be accommodated in most NODC electrical grids. Developing countries currently obtain over 40 percent of their electricity from hydroelectricity. Their unexploited hydroelectric potential is large. Small-scale renewable resource technologies (e.g., minihydroelectric, windmills, firewood, and biogas) may be well suited to rural locations where the infrastructure for centralized energy production and distribution is lacking, unaffordable, or undesired. But renewables are no panacea. Many renewable energy technologies require further development to be economical, and biomass plantations if improperly planned can cause deforestation and severe erosion as, in fact, has been happening in parts of Africa and Asia.

Strong forces drive the NODCs to increase their consumption of commercial energy. These forces include industrialization, population growth, rising income levels, and growing scarcity of traditional fuels such as firewood and agricultural wastes. In both modern and traditional sectors of developing countries, opportunities abound for improving energy productivity—that is, for making each barrel of oil (or its nonoil equivalent) perform more services. In urban areas, where the number of new appliances, buildings, vehicles, and factories is growing rapidly, close attention to energy efficiency can yield an evolving capital stock that is increasingly better adapted to the post-1973 world of high and steadily rising energy costs and prices. Traditional fuel use, which comprises between 30 and 90 percent of all energy consumed in developing countries, can also be made more efficient. For example, it takes three to six times more energy to cook a meal with wood on an open stove than to cook the same meal on a gas range in the United States—although one should note that the open wood fire may at the same time perform non-cooking services such as heating and lighting.

Conservation does not appear prominently in NODC energy plans, and at a minimum, research, development, and demonstration are needed to identify practical and economical means of achieving theoretically available improvements in NODC energy productivity. But as energy costs rise, it is essential to compare alternative energy supply and demand strategies so that development needs are satisfied in the most cost-effective manner.

In short, NODC planning and management—particularly in nations that cannot become substantial fossil fuel producers or exporters—will require:

- Improved data on energy consumption patterns, resources, supply, and demand elasticities;
- Energy analysts and research centers;

- New energy technologies suited to NODC needs, resources, climates, and fuel costs; and
- Financial and technical assistance from developed countries and international institutions.

Implications for the United States

There are at least three reasons for the United States and other developed nations to take an active interest in the non-OPEC developing countries' energy situation.

1. Rising NODC petroleum demand exerts the same upward pressure on international oil prices as do the growing oil imports of Italy, Japan, the United States or any other nation, and NODC energy demand has the potential for rising steeply. As oil prices rise, importing developing countries have an increasing interest in restraining their oil imports to conserve foreign exchange so that they can finance other imports. Some already impose severe restrictions on the use of petroleum for electricity generation or for industrial process heat. Because reduced oil demand anywhere will restrain future price increases, the industrialized world can benefit from helping the NODCs to develop alternative energy sources and energy-economizing consumption patterns.

2. Some non-OPEC developing countries are, and others can become, net energy exporters. Energy exporting NODCs not only are spared the rising expense of energy imports, but they also earn foreign exchange that can finance further economic development. But NODCs often need help in finding, developing, and producing fossil fuel resources. Oil-importing nations have an economic interest in providing this assistance.

3. Sound energy planning and management have become necessary conditions for successful economic development. Because the United States is committed to assisting global economic development, its concern extends to the energy problems of other NODCs as well.

The energy research, planning, and management assistance needed by the NODCs can in general best be provided through international institutions such as the United Nations Development Program and the World Bank. Not only do they have the experience and expertise, but support channeled through international institutions is more acceptable both in the United States and abroad than in bilateral aid. The United States especially should urge the World Bank to fund a broader range of the activities suggested in the preceding paragraphs. In special situations, necessary financial and technical support can best be provided bilaterally. For example, higher income NODCs have difficulty qualifying for assistance from international organizations. The proposed U.S. Institute for Technological Cooperation could become a useful vehicle for funding energy research and development in and for developing countries.

CONCLUSIONS

The United States will continue to be dependent on fossil fuels. The market for these fuels—and the point cannot be made too strongly— is a world market. A shift in supply or demand anywhere in the world can have an impact on the United States and vice versa. A Western European's desire to see the United States institute effective measures to reduce oil imports is as legitimate and proper as an American's wish for increased oil production in Saudi Arabia or Mexico.

The United States is in the curious position of being by far the largest importer of OPEC oil and yet ultimately less vulnerable to OPEC pressure than nearly all other OECD countries. Wealthy nations like Japan, West Germany, and France all depend on OPEC oil for a much higher percentage of their energy than does the United States. This fact accounts for the frequently surfacing desire of some Western Europeans to strike an overall bargain with OPEC under American leadership, a bargain that would presumably set supply targets, allocations, prices, and so on. We see many problems in such a scheme—among them its potential divisiveness among oil consumers; the complete abandonment of market forces in favor of political negotiations in a situation in which oil consuming countries appear to be holding few chips if any; and the likelihood that the bargain, if struck, would be superfluous in times of slackness and would prove inoperative in times of strain.

Energy is related to a variety of other issues of concern to the United States. Oil, for example, can be a divisive force between nations, a ready and effective weapon, but also potentially an impetus to cooperation as in the matter of Chinese and Soviet petroleum production. What happens regarding energy to Western Europe, Japan, the Soviet Union, China, and the less developed countries will continue to be matters of extreme importance to the United States—quite apart from energy per se—by virtue of its influence on international politics.

Coal: An Abundant Resource— with Problems

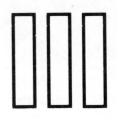

Introduction

Energy may not be responsible for the bulk of today's environmental problems, but it surely is a major source. The use of fossil fuels, hydropower, and nuclear energy adversely affects the land surface, water, the quality of air, and human health. Oil-carrying tankers pollute the ocean both routinely and in accidental spills; transmission lines and cooling towers offend visual beauty; waste piles accumulate. At the end of the line, there are the spectacular possible hazards of nuclear accidents and climatic change. In sum, it is a large and varied assortment of adversities and calamities and one that by now has been researched and written about in a large and growing body of literature.

Consequently, our group had to make some decisions: What did we wish to address? Obviously only a few, select topics. In the twenty year perspective, when competing, conventional oil and gas are under pressure of depletion as well as objects of potential denial and when nuclear energy's public acceptance may well undergo further and perhaps fatal erosion, we have concentrated our attention on coal-connected environmental issues, in the belief that if we fail to find broadly tolerable ways of living with coal we are in serious trouble. However, both Chapter 12 on nuclear and Chapter 13 on solar energy touch on selected environmental and health aspects of these sources.

While coal has been with us far longer than the other sources that now furnish our energy, we have found it useful to begin with a description of the industry's salient features: its past performance, its structure, its markets, its capacity for substantial expansion, and most important, the obstacles that may stand in the way of its becoming once again a major energy source. Chapter 9 thus provides the background not only for what follows in this part, but also for some of the material that appears elsewhere, such as in Chapter 12 dealing with nuclear energy, in Chapter 14 dealing with decision-making, and in Chapter 8 where non-U.S. trends and policies are discussed. Conversely, at least one item, the coal resource picture, is placed not in this part, but in Chapter 7, where resources are discussed across the board.

This part's opening chapter thus establishes that there are no physical obstacles to a large expansion in output; that the solution of a number of important institutional and environmental problems depend less on especially novel information or methodology than on political wisdom; and that progress is needed all along the front if coal consumption is to expand. It then zeroes in on what we discern as the critical dimension in coal use—air pollution from combustion.

Both Chapters 10 and 11 deal almost exclusively with that environmental problem, in the belief that we must learn to manage efficiently emissions from coal-burning power plants and industrial facilities; otherwise, more heat may be generated from an unceasing debate than from boilers and furnaces.

Chapter 10 draws together several related issues. It shows why it makes sense to focus on air pollution; it describes the process in which the undesirable pollutants are generated, transformed, and transported and how these emissions cause illness and death; it stresses the difficulties and large areas of ignorance in information and analysis along the entire chain of events that link combusion to air quality to health impairment to costs; it singles out the impact of carbon dioxide on global climate as perhaps the one future environmental hazard that might conceivably limit, if not eliminate, the burning of fuels; and it illustrates the difficulties encountered in relating estimated costs of coping with these hazards to estimated benefits from easing or eliminating them.

Chapter 11 develops further the issue of air pollution resulting from coal combustion. Drawing on the general theme of uncertainty, especially as it relates to the emission-transport-damage-control chain, it critically examines the approach we have taken in this country to achieve cleaner air and to reduce adverse impacts on health and the environment.

With a critique of past philosphy as a starting point, the chapter develops a set of guidelines for an improved pollution control policy, both within the existing legislative framework and beyond it. Its principal earmarks are the establishment of a cost-effective process that moves in the right direction and uses to the greatest extent feasible market or marketlike mechanisms in place of targeted standards and dates of achievement, rigid regulations, and inadequate attempts to articulate and be guided by cost and benefit calculations.

Coal Production: Potential and Constraints

THE OUTLOOK: A SUMMARY

Although a large number of possible constraints, such as environmental effects or lack of manpower, could limit U.S. coal production in the next twenty years, none of these should prove so restrictive as to prevent expected demand from being met. However, careful planning will be required to overcome these constraints and will increase the cost of coal production. At the same time, if potential government actions were more certain, it could significantly enhance the demand outlook for coal.

A serious potential supply constraint on meeting future coal demand is the availability of western federal coal leases. There has been a moratorium on federal leasing since 1971 and, unless leases are offered shortly, the anticipated long-term growth in western strip mining will be delayed because of the extensive preplanning required and the standards and regulations that must be met.

For the short term, up to 1985, coal use will be constrained by demand because of the physical limitations of converting many existing oil- or gas-fired units and the unattractive economics of converting others. Coal will be able to penetrate the market for new major fuel-burning facilities at a rate that will be determined largely by growth in economic activity; the overall cost of using coal compared with oil, gas, and nuclear reactors; and the anticipated relative increases in the prices of these fuels over time.

In the near term, coal will compete in the utility and industrial markets where it can be used as a raw material to generate steam and, for some industries, to provide process heat. In the future, if converting coal to synthetic oil and gas becomes economic, this would also provide large markets for coal. However, unless major changes in energy policy occur, it is unlikely that there will be as much as even 1 or 2 quads of energy from coal conversion used by the year 2000.

The major fuel sources for electric utilities in the future will be nuclear energy and coal, since the Energy Act of 1978 provides for oil and gas to be largely phased out of this market. Coal use by the utility sector in 2000 could be from 22.5 to 25.9 quads, or 46 percent to 53 percent of energy used by the utilities. This compares with 47 percent of the energy input in the form of coal in 1977. If the average Btu content of a ton of coal were 20 million (10^6) Btus (increased quantities of western coal with low heating value will lower the average Btu slightly by 2000), then 1.1 to 1.3 billion (10^9) tons of coal would be used in this market.

Energy use by industry is projected to be about 60 quads in 2000, of which electricity would supply 20 quads. This would leave 40 quads to be supplied by other fuels, of which coal could supply 16.5 quads, or 41 percent of the nonelectric demand. This compares with only 16 percent of nonelectric demand supplied by coal in 1975 in the industrial sector.

Total coal consumption in 2000 could be in the range of 1.9 to 2.1 billion tons (not including exports of perhaps 150 million tons), and coal could provide about 35 percent of total U.S. energy demand. There could be sufficient production capability to supply this amount of coal by 2000. Even if nuclear energy fails to supply the total 13 to 15 quads it is projected to do, coal should be able to make up the deficiency if enough lead time is provided. Since the development of major facilities utilizing coal requires longer lead times than the development of coal production and transportation facilities, this is not an unreasonable qualification.

The resource base for domestic coal is so large that there need be no concern over the physical availability of supplies for much longer than the period covered in this study. However, there are a number of constraints on the greatly increased use of coal. One of these is the potential for creating adverse health and environmental effects at all stages of the coal cycle, from production of coal to its ultimate consumption. Another is the need to convert coal to other forms before it can be used in some of the major energy-consuming sectors, such as transportation and chemical feedstocks. Other potential problems that we believe are less serious and susceptible to policy solutions of a more conventional kind are the regional, social, and economic problems that could be created (for example, boom-towns); the occupational health and safety impacts on coal miners;

and the demands for complementary resources, such as water, that may be in limited supply in some regions.

The prospect of increased coal use has created adverse reactions among some policymakers and a part of the general public. It is viewed as difficult to handle and transport without posing many serious environmental problems. In addition, the history of the industry evokes images of economic depression, poverty, labor violence, and the proliferation of social ills that characterized the coal regions and coal-mining towns for many years. Many of the problems that have beset the coal industry in the past have received attention from local, state, and federal agencies, and solutions for them are being found and put into action. Others still need to be addressed, but none of them are a necessary characteristic of the coal industry as such. It should be possible to find means of producing, transporting, and using coal in a socially acceptable way.

Specifically it is believed that existing standards and regulations will keep the adverse effects to acceptable levels, although more stringent provisions could be adopted if required. Except for the longer term concern over the possible effects of carbon dioxide (resulting from combustion of fossil fuels) on the global climate (see Chapter 10), none of the environmental concerns appear to be limiting.

In order to make coal more attractive to the industrial sector, methods of using it at low cost and with reduced emissions of air pollutants must be demonstrated. This could be done by developing some combination of small, packaged fluidized bed boilers for direct burning of coal, by small low or medium Btu gasifiers that could produce a clean fuel gas for use in the industrial sector, by producing a synthesis gas from coal that could be used as a chemical feedstock in place of oil and gas, and by developing techniques of supplying very clean coal to the user (e.g., solvent-refined coal). Improved pollution control technology suitable for small installations would also help increase coal utilization.

The two main markets for future coal use are highly competitive. New developments that would reduce the cost of mining, transporting, and utilizing coal would permit coal to compete more widely with nuclear energy in the electric utility market. Similarly, lower cost coal and improved methods of burning it directly while minimizing air pollutants would enlarge the industrial market in which coal competes with the cleaner burning fuels—oil and gas.

Coal is not viewed as an important energy source because of love for this particular resource. Rather, it seems the logical choice. Oil and gas supplies, on which both this country and most of the rest of the world have come to rely heavily in the past two decades, seem now to be much closer to depletion than was thought even a decade or so ago by most analysts. They also are insecure and expensive. Supplementary unconventional sources that could substitute for them are much talked about but are at best in their infancy. Renewable energy resources are receiving greatly increased R&D support and may well be the wave of the future, but they are not yet on the scene. Finally, the outlook for nuclear energy has continued to dim as public acceptance has diminished. All of this is described in some detail in Chapters 7, 12, 13, and 15. The end result is that coal appears as an abundant, existing, and quickly available fallback source of supply.

Until now, the United States and other industrialized nations have met a large part of their growing energy demand and supplemented their declining domestic fuel production by imports. Natural gas has not as yet been a major commodity in world trade because of the large capital investment required to transport it, including processing at both the shipping and receiving points, and some concern over safety, though a liquefied natural gas (LNG) industry is making itself felt. Exceptions to this are situations in which gas can be exported by pipelines—for example, from Canada and Mexico to the United States and from the USSR to Western Europe. Moreover, increasing imports of gas from Mexico and Canada may not furnish a completely satisfactory alternative for imported oil, even if it were low cost, since this too would continue the country's reliance on foreign supplies and contribute to the U.S. trade imbalance. Also, gas is not a suitable substitute for oil in transportation.

For these reasons, a number of countries have begun to investigate other means to meet their energy requirements and to reduce total energy consumption, with the greatest emphasis on reducing petroleum use. Most of the industrial countries that together consume nearly 60 percent of all energy have been large producers and consumers of coal in the past. Many hoped that they could again turn to coal. The opportunities are particularly good for countries such as the United States, USSR, and China, which have large reserves of coal and the potential for producing it at low cost (see Chapter 7).

THE SETTING: PRODUCTION, INDUSTRY STRUCTURE, MARKETS, AND TRANSPORTATION

Coal's Past Role in Energy Supply

In 1925, coal supplied nearly 83 percent of all world energy requirements and about 75 percent of those of the United States. Most of this consumption (86 percent) was in the industrialized nations of

North America and Western Europe. After 1925, two major trends developed in world coal consumption: (1) the percentage of total energy consumed in the form of coal declined sharply, until by 1976 it had reached only 32.4 percent of total energy supply; and (2) the USSR, Communist East Europe, and Communist Asia increased their coal consumption from 7.2 percent of the world total in 1925 to 51.3 percent by 1976.

The sharp growth in world energy consumption of all fuel forms after 1925 was largely due to a greatly increased use of petroleum and in the United States, at a somewhat later time, to the consumption of increasing amounts of natural gas. These increases were a result of discoveries of large petroleum resources that could be produced and transported at low costs and greatly expanded demand for liquid fuels, which are currently indispensable for transportation. Toward the end of the 1960s, the greater convenience that liquid and gaseous fuels offered the consumer was supplemented by a new factor—their greater cleanliness compared with coal. Table 9-1 shows the proportion coal contributed to the nation's 1978 energy consumption.

U.S. Coal Policies

Current U.S. energy policy calls for greatly increased use of coal in the period to 1985 and beyond. The National Energy Plan of 1977 projected coal production to reach 1.2 billion (10^9) tons by 1985, about double the 1978 level of 660 million tons. This amount of coal might be producible by 1985, but only if the large number of new laws and regulations affecting coal production and utilization that have been enacted over the past ten years are adequately interpreted and clarified, if they are not changed frequently, and if

Table 9-1. U.S. Primary Energy Consumption, by Source, 1973, 1978 (in 10^{15} Btu).

	10^{15} Btu		*Percent of Total*	
	1973	*1978*	*1973*	*1978*
Coal	13.3	14.1	17.8	18.1
Natural Gas (dry)	22.5	19.8	30.2	25.4
Petroleum	34.8	37.8	46.6	48.5
Hydro	3.0	3.1	4.0	4.0
Nuclear	0.9	3.0	1.2	3.8
Other	0.05	0.2	0.1	0.3
Total	74.6	78.0	100.0	100.0

Source: Department of Energy, Energy Information Administration, *Annual Report to Congress, 1978*, Vol. 2 (Washington, D.C.: GPO, 1979).

Note: Due to rounding, sum of items may not add to total.

construction of new productive capacity starts immediately and proceeds on schedule.

Finding markets for this amount of coal by 1985 is another matter. The National Energy Plan anticipated a large growth in coal use by the electric utility industry, and we expect this to take place. However, a much larger growth rate for coal use was predicted in the industrial sector. The latter's share of total coal consumption is projected to grow from 28 percent in 1976 to 38 percent by 1985, or an increase of over 10 percent per year. This rapid growth will not occur unless a number of difficult problems can be resolved. These include the limitations on demand that result from the uncertainty introduced by changing air pollution standards (which are especially difficult for the smaller industrial users to meet), the increased costs of meeting environmental regulations in producing and using coal compared with other fuels, the lack of a coal distribution infrastructure, the high cost of coal transportation, and the high cost or physical inability to convert some existing oil- and gas-burning equipment to coal. Taken together, these factors have led us to judge that coal in the next ten to fifteen years is "demand limited," particularly in the industrial sector. Increases caused by new environmental regulations could also affect the amount of coal used by electric utilities by making coal less competitive with nuclear energy, though the cost of the latter is also moving upward.

While the National Energy Plan did not project energy use beyond 1985, its implications appear to be clear: coal will be used in even greater quantities after 1985. The constraints on this use need not be as great as in the shorter term. There should be adequate time to install the larger productive capacity in an environmentally acceptable way, to find ways to burn coal cleanly, and to convert coal more economically to more convenient and cleaner fuel forms— oil and gas.

Our judgment of coal's large potential is based on the fact that the current delivered cost of coal per million Btu in most of the United States is less than that for oil and gas. We expect this differential to increase. Increases in the cost of coal production and use will result mainly from the control technology that will be required to assure adequate health, safety, and environmental protection to the miners and to the general public, although the costs of meeting all current or future requirements have not yet been completely determined. In contrast, finding and producing domestic oil and gas is certain to become more difficult and expensive. Unless there are unexpected discoveries of large reservoirs worldwide, oil and gas prices in the United States can be expected to increase substantially.

As a result, compared with oil and gas, even with new regulations, standards, and legislation, costs of coal should be more favorable in the future.

Coal Production

Worldwide. Table 9-2 gives coal production in millions of tons of coal equivalent[1] for the major producing countries in 1977. Twelve countries produced 92 percent of all the coal extracted. The USSR, the United States, and China, the countries with the largest coal reserves, were also the largest producers. Together they produced 59 percent of the world total.

As Table 9-2 indicates, there are about as many large coal-producing nations as there are major oil-producing countries. However, except for the United States and the USSR, the major foreign coal producers are not generally the same countries as those that produce the largest quantities of oil. This could help lead to vigorous competition for the world energy markets in the future both between oil- and coal-producing countries and among the coal-producing countries.

Table 9-2. Estimated Coal Production of Major Producing Countries in 1977 (10^6 tons coal equivalent).

	Metric Tons Coal Equivalent	Short Tons Coal Equivalent
United States	623	687
USSR	516	569
China	500	551
Poland	197	217
United Kingdom	123	136
Federal Republic of Germany	121	133
India	100	110
Czechoslovakia	84	93
Australia	82	90
Republic of South Africa	81	89
German Democratic Republic	76	84
North Korea	50	55
Total above	2,553	2,814
World total	2,775	3,060

Source: United Nations, *World Energy Supplies 1972-1976*, Series J, no. 21 (New York, 1978), table 2.

1. Tons of coal equivalent is the amount of coal that has an average heating value of 25.2 million Btu per short ton.

Table 9-3.   U.S. Bituminous Coal and Lignite Production, 1920-1978.

Year	Production (thousands of short tons)
1920	568,667
1930	467,526
1940	460,772
1950	516,311
1955	464,633
1960	415,512
1965	512,088
1970	602,932
1971	552,192
1972	595,386
1973	591,000
1974	603,406
1975	648,438
1976	678,685
1977	688,575
1978	653,800

Sources: 1920-1970: Bureau of the Census, *Historical Statistics of the U.S., Colonial Times to 1970* (Washington, D.C.: Government Printing Office, 1975), Series M93. 1971-1975: Bureau of Mines, "Coal—Bituminous and Lignite in 1975," *Mineral Industry Surveys*, February 1977, table 1. 1976-1978: Department of Energy, *Monthly Energy Review*, February 1979, p. 52.

U.S. Coal Production. Table 9-3 shows changes in U.S. bituminous coal and lignite production from 1920 to 1978.[2] Between 1920 and 1930, coal production ranged from 400 to about 550 million tons per year; from 1931 to 1939, the years of the Depression, it ranged from 300 to about 450 million tons; from 1940 to 1949, the years of World War II, it ranged from 440 to 619 million tons (until 1975, the highest annual production ever recorded); from 1950 to 1966, a period in which there were major shifts in the degree of mechanization of coal mines and changing federal policies toward imported oil, production ranged from 390 to 530 million tons; and from 1967 to 1977, coal production increased steadily as a result of the rapid growth in coal used by the electric utility industry. In 1978, coal production declined to 654 million tons per year, largely as a result of the extended strike at the beginning of the year.

Coal production by states and by type of mining for 1977 is

2. During the period from 1920 to 1954, anthracite production (not included in Table 9-3) varied from 30 to 90 million tons per year and made significant contributions to solid fuel supply. From 1954 to the present, production declined steadily, reaching a level of less than 10 million tons per year.

shown in Table 9-4. More than half (52 percent) of the coal produced in the United States came from the East (Appalachia)—West Virginia, eastern Kentucky, Pennsylvania, Ohio, and Virginia. In the interior region, the major coal supply is from western Kentucky, Illinois, and Indiana, which together supplied 19 percent of total U.S. production. Wyoming and Montana were the largest producing states in the West but together accounted for only 11 percent of total production. Installation of new mines in the West is proceeding rapidly. Wyoming and Montana are projected to have about 160 million tons of new capacity by the end of 1980 or approximately 2.2 times their 1977 production, and total western capacity is pro-

Table 9-4. Production of Bituminous Coal and Lignite in the United States in 1977, by State and Type of Mining (thousand short tons).

State	Underground	Surface	Total
Alabama	6,580	14,640	21,220
Alaska	—	665	665
Arizona	—	11,475	11,475
Arkansas	20	550	570
Colorado	4,205	7,715	11,920
Georgia	—	185	185
Illinois	29,590	24,290	53,880
Indiana	525	27,470	27,995
Iowa	—	525	525
Kansas	—	630	630
Kentucky:			
Eastern	41,005	51,145	92,150
Western	23,010	27,785	50,795
Total Kentucky	64,015	78,930	142,945
Maryland	260	3,030	3,290
Missouri	—	6,625	6,625
Montana	—	29,320	29,320
New Mexico	—	11,255	11,255
North Dakota	—	12,165	12,165
Ohio	13,925	32,280	46,205
Oklahoma	—	5,345	5,345
Pennsylvania	38,365	44,860	83,225
Tennessee	4,675	5,645	10,320
Texas	—	16,765	16,765
Utah	9,240	—	9,240
Virginia	26,200	11,650	37,850
Washington	—	5,055	5,055
West Virginia	74,030	21,375	95,405
Wyoming	—	44,500	44,500
Total	271,630	416,945	688,575

Source: Department of Energy, Energy Information Administration, *Energy Data Reports*, "Weekly Coal Report No. 30," April 28, 1978, table 4.

jected to reach between 400 and 500 million tons by 1985. Growth in productive capacity in other regions will be at a much slower rate so that by about 1990 as much as 50 percent of total coal production is expected to come from western fields.

A related major shift is the rapid rate at which surface mine production has grown compared with underground mining. From 1950 to 1969, the share of total production obtained from surface mines increased slowly but steadily from 23.9 to 36.2 percent or a growth of 2.2 percent per year. Following the passage of the Coal Mine Health and Safety Act of 1969, which increased the cost of mining coal underground, the amount of coal produced by surface mining increased to 60.6 percent in 1977, or 7 percent per year. These trends are expected to continue as new western strip mines come into production. As much as 65 to 75 percent of coal may come from surface mines by 1990.

Much of the western coal that will be mined over the next twenty years can be extracted at low cost by surface mining methods because the seams are thick, with a thin overburden (material that lies over the deposit). It is for this reason, as well as the low sulfur content of these coals and the increased demand for coal by western electric utilities, that the large expansion in production capacity in the West is expected to occur.

The U.S. Coal Industry

Throughout most of its history, it has been easy to enter the coal industry. Coal reserves are very large and widely distributed, and before mechanization of the mines took hold in the mid-1950s, the capital required to open a mine was small. Small operators were able to enter the market when demand was high and leave without major losses when coal requirements declined. As a result, the industry consisted of a large number of small operators and a few companies that had a large share of the total production.

By increasing capital requirements, mechanization of the mines made entry into coal mining more difficult. While the number of operating mines had normally ranged from 5,500 to 9,000, starting in about 1955 the number declined steadily until it was down to about 4,700 in 1973. The effects of the OPEC oil embargo and the impending wage negotiations between coal industry management and the United Mine Workers created a sudden demand for coal and a brief reversal. Prices escalated, and a number of new small mines entered the market, so that by 1976 there were 6,200 mines operating. Small mines, however, will be unable to compete under normal market conditions. Moreover, the impact of the Surface Mining Act of 1977 on both strip and underground mines will adversely affect costs at small, older mines much more than at some of the new large ones.

Another change relates to ownership patterns. For many years the ownership of coal mines was largely of two types—companies engaged only in coal mining and the captive coal mines owned by the steel companies that supplied the coking coal needed for their steelmaking operations. Ownership patterns have changed dramatically since the early 1960s when Consolidation Coal Company was acquired by Continental Oil Company. Captive coal production continues among the major steel producers, but the electric utilities, the largest coal consumers, are now increasing their share of captive coal, and this trend is expected to accelerate.

The trend toward ownership of coal companies by conglomerates that are basically noncoal corporations has significantly altered the structure of the coal industry since the mid-1960s. In 1960, twenty of the top twenty-five coal-producing companies were independents. By 1975, only four were independent, six were captive operations of steel or utility companies, six were owned by conglomerates, and nine were owned by companies engaged in other energy activities—mostly oil companies. In the past fifteen years, in addition to Continental Oil Company, other major companies including Gulf, Exxon, Occidental, Sohio, and Ashland Oil have acquired or created new coal company interests. By 1976, coal companies owned by oil and gas interests produced about 25 percent of all coal mined.

The effect of the change in patterns of ownership is difficult to assess. Although the acquisitions of coal companies by other interests has virtually stopped since 1975, this may or may not be a temporary lull. Congress has expressed concern over how the creation of "energy companies" affects competition among the various fuels. So far, the evidence has not been conclusive either way.

On the other hand, the acquisition of coal companies has led to an infusion of capital into the coal-producing companies that should strengthen the coal industry and enable it to compete with the much larger and more lucrative oil and gas industries. As mine sizes become larger to achieve economies of scale and large dedicated tonnages[3] are needed for electric utilities and for future plants to convert coal to oil or gas, the financial resources of the oil industry and other large firms that have acquired or created new coal companies should help coal compete more effectively.

3. Dedicated tonnage is that amount known to be required by specific plants and hence assigned to them.

The U.S. Coal Market

The coal industry has been faced with major market changes in the past thirty years (see Figure 9-1). Two of its main markets were virtually eliminated—the residential-commercial and railroad markets, each of which once used over 100 million tons per year. The residential-commercial market was lost because of the low price, cleanliness, and convenience of natural gas once it became available to the large consumer markets in the East. The railroad market was lost to diesel fuel because diesel locomotives were four times more efficient than coal-fired steam locomotives.

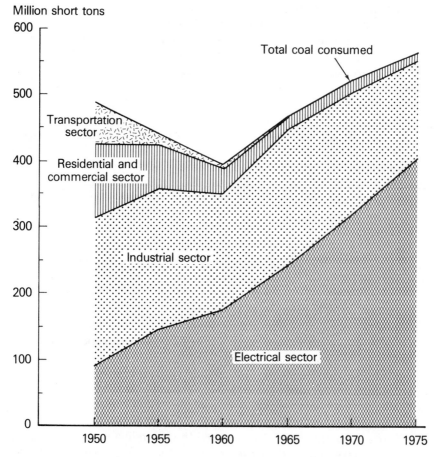

Figure 9-1.   Coal Consumption by Sector, 1950–1975.

Source: Energy Perspectives 2 (Washington, D.C.: U.S. Department of the Interior, 1976), pp. 184–189.

On the other hand, the utility market has grown rapidly, while the size of the metallurgical market, which is entirely dependent on steel demand, has begun to decrease. The industrial market for coal has declined slowly, but this trend could be reversed, as noted earlier.

The size of the export market for U.S. coal has varied between 30 and 80 million tons per year between 1950 and 1976 and now consists almost entirely of high quality metallurgical coal. The size of this market depends mainly on world steel demand, and there is little the U.S. coal industry can do to affect world requirements for U.S. coking coal. Most of the coal is shipped from the Central Appalachian area to Canada, Western Europe, and Japan. In recent years Australian coal has captured a much larger part of the Japanese market because of the lower mining costs and shorter transport distance.

Coal Transportation

Of the coal consumed in the United States, 12.6 percent is used at the mine or at mine site facilities for operating electricity and local consumption. The usual means of transport under these conditions is by conveyor belt or truck. Most coal, however, must be transported much longer distances to where it is consumed. Railroads shipped 63.9 percent of the tonnage in 1976, with the average ton moving about 295 miles. Barges and other vessels accounted for 10.3 percent and trucks for 13.2 percent. A small quantity was transported by slurry pipeline.

On the average, rail transportation costs represent about 20 percent of delivered costs, but this percentage can vary widely. Barge costs are about 0.4 to 0.5 cents, rail costs about 1 cent, and trucks about 6 to 7 cents per ton mile. For a 1,000 mile haul, coal transport uses 2.5 percent of the energy in the coal if transported by barge and rail, but about 12.5 percent if delivered by trucks.

One slurry pipeline is currently in operation in the United States, transporting nearly 5 million tons of coal per year 273 miles from northeastern Arizona to southern Nevada. Where large tonnages of coal need to be transported from a concentrated producing area to consumers 300 or 400 miles away, slurry pipelines can be the most economical way to move the coal. Two major problems exist—water availability and rights of way needed to construct the pipeline. If the coal originates in a water-short area, objections have been raised to utilizing the limited water supplies for this purpose. Returning the water to the source by pipe is technically possible, but it increases transport costs by 30 or 40 percent.

In order to obtain the rights of way necessary to construct most long distance pipelines, condemnation rights will be needed. A pipeline will almost always cross a railroad at some point, and railroads have resisted granting rights of way to competitive transportation. Attempts to give coal slurry pipelines the right of eminent domain so as to obtain these rights of way have failed in the Congress, but have succeeded in a number of states. Although they have limitations, in some instances slurry pipelines can transport coal at the lowest cost and with the least risk of environmental damage.

Special attention will have to be given to those railroads and routes on which new coal freight traffic will develop as western coal reaches utility and industrial markets that have not traditionally used coal, such as Texas and possibly the West Coast. The railroads serving these markets will need new equipment and coal-handling facilities. In addition, when coal traffic becomes large, it may be necessary to double track sections of the main line to provide reliable delivery service. All this will require careful planning to avoid delays in coal deliveries, but there is no need for new technology or institutions. A heavy increase in rail traffic can result in delays of grade crossings and cause noise and air pollution, but these problems can be alleviated (see the section on Transport Issues).

Waterway traffic is less flexible than the railroads. First, there are geographic limits as to where the coal can be moved by water but, more important, saturation of the waterways with coal and other freight could set an upper limit on the amount that can be moved this way even if the capacity of locks is not exceeded, a concern already being voiced about certain locks that are now in operation. There is no similar constraint on truck traffic, although the condition of coal mine roads in Appalachia has deteriorated significantly. Funds must be raised on an equitable basis to keep the roads in repair.

FACTORS SHAPING FUTURE COAL PRODUCTION

Although vast quantities of coal reserves are distributed widely under rather favorable geologic conditions, several constraints could limit the production of coal to less than that needed. They include restrictions on leasing of western federal coal lands; the limited availability of capital, equipment, and manpower for both producing and transporting coal because of government-induced uncertainties; labor-management relations that could lead to interruptions in production; issues of health, safety, and productivity; and environmental impacts.

Examination of these problems suggests that only a few are likely to be of major importance (although all will have to be resolved to some degree if desired production levels are to be achieved). Availability of capital, equipment, and manpower should present no problems, and methods exist for removing each of the other constraints. However, because of the large number of possible barriers, careful planning and execution of programs will be needed to prevent any one of them from limiting production. The next sections discuss a number of important issues.

Federal Leasing Policies

As much as 70 percent of the large coal reserves west of the Mississippi are in federal ownership and the federal government can influence the mining of much of the other coal reserves through the permits that are required to economically develop privately held resources. Much of the western coal is low in sulfur and can be surface mined at low cost. Federal leases awarded in the past contain 17 billion (10^9) tons of recoverable coal reserves. Preference right[4] lease applications were filed for an additional 10 billion tons, with 180 million tons granted as of January 1979. In addition, leases have been granted on Indian lands that contain another 3.5 billion tons of recoverable coal.

In May 1971, the Department of the Interior imposed a moratorium on further federal coal leasing while it considered reforms in the management of federally owned coal lands. In September 1977, a U.S. District Court ruled that the Department of the Interior had not complied with requirements of the National Enviironmental Policy Act in its 1975 leasing program and ordered the department to prepare a new environmental impact statement (EIS). In the meantime only short-term leases were issued and only to aid ongoing operations or to permit production where land had been developed and was fully ready for operation.

The Department of the Interior was scheduled to issue a new impact statement in May 1979, with the first lease sale now scheduled for 1981. Past leasing policies of the department have been reexamined, including such issues as how to schedule new lease sales and how to treat old leases and preference right leases.

4. In areas where the existence or workability of coal is unknown and mineral rights therefore are unlikely to enter competitive bidding, the federal government issued prospecting permits to encourage exploration. These permits expired after two years and could be renewed only once. If commercial quantities of coal were discovered via this process, a firm could then apply for a "preference right" to lease, given its prior interest and investment. These "rights" are now not being issued.

Responsibility for setting leasing and production goals for federal coal lands was assigned to the Department of Energy when it was created. A number of public interest groups have argued that there is no need to lease additional federal lands at this time because of the large reserves already in private hands and because even at a high projected rate of western coal development, the existing leased reserves would be sufficient for many years of mining.

This view is oversimplified. There is some uncertainty about the total amount of coal under lease and even more about how much of it could be recovered economically. Even on the best tracts, recovering 80 percent of the coal in place is considered good recovery efficiency. Much of the coal is unquestionably in units too small to justify opening the type of large mines needed to achieve the economies of scale necessary for western coal to be competitive. Other deposits are almost certainly found under geologic conditions that would make exploitation uneconomic at this time, and some leases may have been issued on lands that are now disqualified for mining (alluvial valley floors, prime agricultural land) under the terms of the Surface Mining and Control Act of 1977. Some leases are undoubtedly held by lessees who acquired the reserves under favorable economic terms and are unwilling or financially unable to start mining at this time. Finally, while 17 billion tons under lease appears to be a large amount of coal, a mine with a capacity of 10 million tons per year requires at least 200 million tons of recoverable reserves or up to 300 million tons of coal in place in a single block or in a "logical mining unit." Even if most of the 17 billion tons were available for use and could be mined economcally, only about 500 million tons could be produced in a year—about the level expected in the West by 1985. For all of these reasons, the amount of coal that can be produced yearly from reserves now under lease is much smaller than it first appears, and leasing should be resumed by the Department of Interior. The rate at which leases would be offered should be determined by the anticipated growth rate in demand for western coal and a judgment as to what leasing rate would maximize the return to the government consistent with protecting other values of the land.

Labor-Management Relations

Labor unrest, prior to a twenty year period of labor peace from 1950 to 1970, has been characteristic of the industry. Because of intense industry competition, operators have been extremely cost conscious and have tried to limit wages and fringe benefits, since on the average, labor costs have been between one-third to one-half of production costs. The social pressures in the closed company towns have reinforced conflicts between miners and their supervisors that start in the work place.

The changing nature of the work force also created new problems for union leadership. The long period of shrinking employment in the coal industry ended in 1969, but most of the miners who had been displaced in underground mines by mechanization were too old to be rehired when demand for coal increased. The industry recruited younger workers, creating a generation gap between the remaining older miners in their later forties and fifties and the new younger men in their twenties. The new employees have different objectives than the older men in negotiating with the operators. They have been much less interested in pension and health benefits than in high wages. They also have been impatient even with the new leadership in the United Mine Workers union, which had instituted local autonomy and permitted contested local elections, but was administratively weak at operating the union and failed to keep some promises to the rank and file.

The strained labor-management relations are serious because they affect customers' perceptions of the reliability of coal supplies. President Carter has recognized this and appointed a commission to assess the basic causes of these labor-management disputes and to recommend methods to overcome them. The friction between miners and operators has a long history that is now an inherent part of the fabric of labor-management relations, so that it is unlikely that a single simple quick solution will be found. Until relations are improved, some consumers will be reluctant to rely on coal, even though the effects of the longest strikes have been cushioned by large stockpiles and many of the new and growing number of western mines are nonunion.

The above discussion of labor-management relations has dealt exclusively with those between the operators and the largest labor union, the United Mine Workers (UMW). Fifteen years ago about 80 percent of all miners were members of the UMW, but the percentage of UMW miners has declined steadily. Now many coal miners either belong to other much smaller unions or are not organized at all. This downward trend will probably continue, partly because of the internal dissensions in the union and partly because the large western mines are mainly non-UMW operations. As western coal mine production becomes a larger share of the total coal produced, unless changes occur in past labor patterns, union-management relations and contract negotiations will assume decreasing importance on a broad national basis.

Health and Safety

Since the passage of the Coal Mine Health and Safety Act of 1969 (amended in 1977), coal mine fatalites have dropped markedly in both underground and surface mines. The largest improvement has been in the more hazardous underground mines. The underground fatality rate is now about half of what it was in 1969—0.35 deaths per million man hours compared to 0.84 in 1969. While the occupational safety record of the coal industry has improved substantially, its record is still among the poorest of any major industry, and further improvement should be possible.

The second major health problem of coal workers, not publicly recognized until the 1950s in the United States, is a chronic disabling lung disease known as coal workers' pneumoconiosis or "black lung." The 1969 legislation has led to improvements in dust conditions, and the greatest improvement in occupational health and safety should result from a reduction in lung-related disorders among underground miners. Both coal dust and silica standards were a part of the 1969 Health and Safety Act, and it was believed that the standards were set at a level to prevent future cases of coal worker pneumoconiosis. However, the extent of the impact on health is still uncertain because of the long induction time between exposure and the onset of the disease. Thus it is too early for statistical evidence on whether even more stringent dust standards are needed.

Other factors also contribute to the uncertainty as to how much improvement in health conditions will occur. The composition of coal and hence coal dust and its toxicity vary substantially with different locations; the current standards of 2 mg/m^3 may not be sufficiently stringent. Also, the possible synergistic effects of cigarette smoking and inhaled coal dust are not fully known. The current estimate for the industry is that between 10 and 15 percent of miners suffer from black lung disease.

Another source of uncertainty is the effect of new technology, including greater use of diesel engines and increase of long wall mining. There is particular concern about the presence in diesel exhaust of nitrogen dioxide and the polynuclear aromatic hydrocarbons, which may be carcinogenic or mutagenic. Although diesel engines have been used extensively in other types of mines, there has been insufficient study to quantify the potential hazard from diesel exhaust in coal mines.

Deaths from accidents and disability and premature death from respiratory disease are the major health problems associated with deep mining, but there are other concerns as well: mortality studies in both the United Kingdom and the United States have shown excesses of total mortality on the order of 35 to 70 percent (observed over expected deaths), with large excesses for nonmalignant respiratory diseases.

It should be remembered that the 1969 Act was the first legislation to control dust exposure of miners in this country. Although implementation may be imperfect and incomplete and the benefits not yet fully quantifiable, there is no doubt that marked improvement has taken place in the prevention of serious disease. Obviously, continued efforts are necessary, both to ensure compliance and to demonstrate that the benefits have in fact been achieved. New technologies must be examined from the same perspective.

Fewer studies have been done of the health aspects of surface mining compared with deep mining, but there is evidence that accidental deaths and coal workers' dust diseases of the lungs are much less frequent under surface mining conditions. Heat stress and cold stress are more common with surface mining, but these problems are relatively easily controlled. The expected increase in surface mining should thus be accompanied by an improved health situation.

Boomtowns

Boomtowns are those communities that undergo rapid demographic, economic, and sociological change as a result of a relatively sudden and/or extreme increase in the development of resources. The eastern coal-mining regions have severe social problems of varying degrees, many of them arising from the activities of the coal industry over the years; but since coal production in the East is expected to grow much more slowly than in the West it is unlikely that they will face the boomtown phenomenon. The major problems will be related to correcting past damages and ensuring that the new influx of labor will not aggravate any existing problems.

Although the total number of people that will be needed to mine coal and provide services in the West will not be very large (between 300 to 400 men for a 10 million ton per year modern mine), it represents a large percentage increase in existing populations near production sites. Local governments will need to greatly expand schools, roads, police, libraries, and so forth to serve a population that may easily double or triple in a few years. Those who are currently living in the area may resist the increased public expenditures, if they are even able to meet them, and the changes that would result from the different lifestyles of the new residents.

These problems will require careful planning and a method of providing funds for the new facilities early on, until the tax base provided by the new mine or conversion facility is large enough to generate income. The absolute numbers of people that will have to be accommodated are small. Even in Rock Springs, Wyoming, a community that has been cited as an example of the adverse social impacts that can occur when a small town grows very rapidly, the population only doubled with the addition of 13,000 people between 1970 and 1977. Providing new and adequate facilities for these relatively few people should pose the kind of problem that advanced, careful planning can satisfactorily cope with.

Environmental Impacts of Coal Production

While the extraction of coal can have major environmental impacts on land and water, its effect on air quality is only minor. At a surface mine, noise, dust from blasting and from haul roads, and air pollutants from machinery and trucks are the main sources of environmental impacts other than those on land and water. Underground mines create even less air pollution. On the other hand, coal mining, unless controlled, is a major source of land disturbance and of water quality degradation.

A justifiable concern exists that a resurgence in coal mining will unleash the same types of destructive effects that characterized it in the past. While these fears cannot be dismissed, we do not believe that expanded coal mining over the next twenty to thirty years, or beyond, will again devastate the landscape.

The principal potential environmental impacts of coal mining are disturbance of the land surface and pollution of rivers by runoff of mine drainage containing toxic soluble metal salts and sediment. Environmental problems of surface mine reclamation and water quality control are both regional and site-specific, varying with climate, topography, vegetation, and the chemical quality of the coal itself. In the drier areas of the western United States, it is difficult to reclaim and revegetate land surface after mining because there is sparse rainfall and poor soil. In very arid regions, little or no revegetation occurs without irrigation. To date the evidence suggests that areas mined for coal in eastern Montana, Wyoming, and Colorado can be successfully reclaimed if surface soils are handled carefully and if the vegetation and land surface are properly managed. As a result, we believe that increased surface mining in the semiarid West can be carried out without damaging the environment, assuming careful site planning and regulatory control and monitoring during and after mining.

The potential impacts of coal mining on aquatic ecosystems are shown in Table 9–5. The disruption of water systems in western regions is likely to pose primarily local problems of depletion of limited quantities of water. Federal legislation already limits mining on

Table 9-5. Potential Impacts on Aquatic Ecosystems from Coal Mining.

| | Direct Site-specific or Regional Effects | | | Indirect Effects | | |
| | *Mining* | | | *Transportation* | | |
	Surface	*Subsurface*	*Beneficiation*	*Rail*	*Water*	*Combustion*
Water Quality Impacts						
Surface water degradation	++	+	++	P	++	++
Ground water degradation	++S	++	++	P	+	+
Toxic discharges	++S	++S	+		+	+
Temperature alterations	+	+S	+		++	
Elevated turbidity (TSS)	++		++	C	++	
Habitat Alteration						
Hydrograph alteration	++	+				P
Stream bed siltation	++		++	C	++	P
Channel alteration	++/++S			+	++C	
Migration barriers	++				++	
Water balance alterations (watershed)	++	+				
Water consumption	+	+			P	P

Source: Edwin E. Herricks, "A Report on Potential Impacts of Increased Surface Mining in the Eastern United States," August 15, 1978 (unpublished).

+—Minor impact.
++—Major impact.
S—Impact site-specific or geographically related.
C—Construction phase only (may include maintenance activity).
P—Potential (may be unrelated to mining).

alluvial bottomlands. In addition, potential disturbance of local ground water systems can be mitigated by proper advance planning and by control of mine location: permits to mine could be denied for sites where disruption of water systems is unavoidable.

Opportunities for surface mining in the East are also great. While the wetter climate makes it easier to reclaim and to revegetate surface mines, in the past, severe water quality problems have been generated by runoff from surface mines and from disposal of waste materials from underground mines at some surface sites. While a wide range of potential impacts on aquatic ecosystems from coal mining is shown in Table 9-5, it is important to stress that many of these problems can be controlled and thus are potential only in a qualified sense.

The wide variety of environmental conditions make it impossible to consider specific coal mine sites here; instead, we discuss two regional eastern types of conditions. These are the relatively flat-lying areas of the Midwest (such as Illinois and western Kentucky) and the steeper terrain of the Appalachian Plateau in western Pennsylvania and the valley and ridge provinces of central Pennsylvania.

Southern Illinois is a good example of a region that has had acids and sulfates dumped into streams from surface as well as underground mines. Surface mining is likely to expand in this area. A recent water pollution planning report by Boris and Krutilla, to which we return later in this chapter, noted that "even with present mining controls some pollutant discharges from active mining areas must be expected . . .," but it also noted that "in some cases new mines could actually improve water quality by bringing an abandoned site up to the more stringent existing water quality standards." The planning study concluded that mining controls, coupled with the presence of limestone in the overburden (which would neutralize the acid produced), should result in little increase in acidity, although increased sulfate loading in streams could be expected.

Viewed as a whole, experience over the past five to ten years indicates that surface mining in the Midwest can be carried on with little or no increase in acid runoff from mined areas where known mining and reclamation techniques are employed. We believe that current experience supports the view that the impacts of surface mining of coal in the Midwest in the next several decades are likely to be controllable and that the damages resulting from mining alone will be minimal.

Similarly, experience in several parts of the Appalachian region has shown that the restrictions set by the Surface Mining Control and Reclamation Act of 1977 have begun to reduce the environmental impact of surface mining. In the relatively steep terrain of western

Maryland and much of Pennsylvania, for example, state laws have been thoughtfully administered and enforced to permit modern mining of coal with limited impact on streams and slopes. Likewise, problems of erosion and sediment in northern and northeastern Alabama have been reduced by federal and state regulations that require preliminary evaluation of prospective hazards before mining and continuous monitoring and surveillance during mining.

In view of estimates by the Appalachian Regional Commission that perhaps 5,000 miles of rivers and streams in Appalachia have been made acid by drainage from mines, it may seem paradoxical to suggest that increased coal mining over the next several decades need not further degrade streams in the area. We believe that this is possible, however, given both the techniques of mining available and the known potential effectiveness of methods of controlling runoff and erosion. In addition, many of the coal reserves that will be exploited in the near future in many regions of the East and Midwest are in basins that have already been exploited. For that reason, damage is likely to be concentrated in areas and in river systems previously degraded by unregulated coal mining. Legislation now requires operators to reclaim abandoned mines before they resume mining on or near them, so there is some prospect that, although renewed mining might retard the recovery rate of streams in some areas, others may benefit from reclamation and control extended to previously degraded terrain. Many, if not most, severely degraded stream ecosystems do recover over time from the damaging effects of excessive sediment and chemical loads once the pollution ceases.

Reclamation Issues

Although experience in the eastern United States has provided information on environmental damage and its control, it is still not clear to what degree achievable controls of modern technology will actually mitigate environmental damage. Some observers continue to feel that recent state legislation, often matching current federal legislation in objectives and controls, has frequently been less effective in action than its potential on paper suggests. They point out that controls have not worked everywhere, citing continuing erosion and sedimentation from surface mines in Kentucky despite relatively strong state legislation passed in the 1970s.

Proponents responsible for the passage of the Federal Surface Mining and Reclamation Act of 1977 clearly believe that federal oversight will not only provide broader coverage for the country as a whole, but will also result in stronger enforcement of regulations and hence better environmental results.

The principal disagreements over the new regulations for the 1977

Act concern the cost of implementation compared to the environmental benefits to be gained. To date, environmental controls on surface mining have added approximately 20 percent to the price of a ton of coal in the West and 15 percent to the price of coal mined in the East, while the cost added by reclamation in underground mines appears to be closer to 20 percent. These added costs are significant for all mines except some in the West with the thickest coal seams and least overburden. The increment also slightly increases the price of electricity, its principal outlet.

The debate over costs and benefits has raised several issues of pollution control. First, regulations should specify an objective and be designed to achieve it. To the extent possible, the means of meeting this objective should be in the hands of the mine operators, who will try to minimize the costs of meeting the requirements. Second, competent and experienced inspectors in sufficient numbers must be given authority to close down operations that violate the rules or use technically unsound means to conform to the regulations. Such an approach should eliminate the need for overly detailed regulations that cannot take into account wide regional differences in conditions.

Successful experience at the state level in reclaiming surface mines indicates that the task requires a larger force of competent inspectors in the field than many previous efforts have provided. Poor surface mine reclamation continues in many locations today because local citizens and officials are not interested or because they believe that their livelihoods will disappear if mining operators do what the law requires. Since both industry and environmentalists agree that environmentally sound coal mining can be achieved, there appears to be no reason why agreed-upon goals cannot be met at reasonable cost.

In summary, with proper enforcement of existing state and federal legislation, it is likely that increased coal production over the next several decades will not reproduce the adverse impacts that still remain on thousands of surface-mined acres. Nevertheless, realistic appraisal of the direct impacts of coal mining indicates that some areas will be denuded or left pockmarked and that some streams will be polluted with acids and toxic materials, but that such damage will not be extensive. A large surface mine in the West with a life of twenty-five to thirty-five years disturbs roughly 200 acres per year, and the maximum area disturbed at any one time is about 600 acres (roughly the area of a small farm in the United States). We believe further that, with effective reclamation and environmental controls, such degradation will be modest in relative proportion to the energy provided and that the costs will be small compared with the benefits achieved from the use of this energy resource over the next several decades.

Transport Issues

There is some concern that increased rail and barge transportation of coal may cause significant environmental damages. Many towns in the western United States on, or dissected by, railroads are concerned that large numbers of unit trains of coal cars may pass through each day and disrupt social and economic life. It might be possible to avoid this by either rerouting rail lines, depressing them, soundproofing rights of way, or bridging some crossings. These solutions, of course, beg the questions of money and politics. In many instances the problems may prove to be "insurmountable," but it should be possible to devise mechanisms to deal with these local and regional problems. Such problems are not unique to energy, but are part of the U.S. railroad transportation problem.

If increased coal traffic requires the construction of locks and the improvement or channelization of the waterways, these projects will be sources of potential environmental impacts on aquatic ecosystems. Between 45 and 98 percent of the barge traffic on major Ohio River waterways consists of coal. Additional facilities may well be needed to accommodate increased traffic on some rivers. Studies on the possible impacts on aquatic systems have just begun; clearly there will be some effects, although one might assume that new damages to large avenues of navigation from barge traffic alone will be limited. If dams are required that significantly raise water levels, however, there will be controversy as to their location and even their desirability.

FACTORS SHAPING FUTURE COAL DEMAND

Competitive Fuels

The growth in total coal demand will be determined by (1) how rapidly total energy demand increases; (2) the relative costs of producing, delivering, and utilizing competitive fuels (including social costs that have been internalized in the fuel price); and (3) national policies that influence the market share that each form will capture.

Other fuels do not compete with coal in the metallurgical or export markets, and coal no longer is a significant factor in the residential-commercial or transport sectors. Thus, coal's competition is with oil, gas, and nuclear fuel in the electric utility market and with oil and gas in the industrial market.

Electric Utilities Market. The average delivered price of the three fossil fuels to electric generating plants for April 1977 and May 1978 by region is shown in Table 9-6. Adjustments must be made in order

Table 9-6. Delivered Fuel Prices to Electric Utilities, by Region, April 1977 and May 1978 (cents per 10^6 Btu).

Region	April 1977			May 1978		
	Coal	Oil	Gas	Coal	Oil	Gas
New England	127.9	218.5	200.1	146.8	195.3	184.3
Middle Atlantic	102.5	230.8	155.4	118.7	207.8	162.5
East North Central	93.9	256.3	184.7	116.6	262.0	191.7
West North Central	72.5	298.7	96.0	86.6	189.3	118.5
South Atlantic	108.4	217.8	85.7	129.1	198.4	112.3
East South Central	96.5	180.5	154.7	116.2	182.8	155.2
West South Central	60.2	200.3	113.7	69.0	182.0	135.8
Mountain	42.4	220.6	134.9	51.3	226.1	150.2
Pacific	70.8	235.8	204.5	78.3	250.3	220.4
National average	90.1	226.2	125.6	110.6	209.6	143.5

Source: Department of Energy, *Monthly Energy Review*, October 1978, p. 82.

to compare the delivered cost of different fuels on a competitive basis. For coal, we must add the extra capital and operating costs of handling and storage, crushing and pulverizing, the larger size boiler needed for coal, ash collection and disposal, and meeting air pollution regulations for particulates and sulfur oxides. For oil, we must add the costs of oil storage and whatever air pollution control equipment is needed for the particular residual oil being used.

As a result, for coal to be competitive in a large utility plant using coal and flue gas desulfurization (FGD), its price must be approximately 80 cents per million Btus less than the price of low-sulfur oil and about 90 cents per million Btus less than the price of gas. As Table 9-6 indicates, in May 1978, high-sulfur coal with FGD was competitive with oil for the United States as a whole, but the trade-off varied among the regions. The controlled price of gas in May 1978 rendered coal with FGD not competitive with gas.

The Energy Tax Act of 1978 grants an additional 10 percent investment tax credit through 1982 for business expenditures for certain types of alternative fuel equipment. It also provides for rapid depreciation for early retirement of oil- or gas-fired boilers if they are replaced by alternative fuels, which should encourage early replacement with coal equipment. The investment tax credit for coal is an estimated 10 to 20 cents per million Btus credit. The quantitative impact of the rapid depreciation provision on coal demand will vary with regional energy availability and costs for the different fuel forms, the size of the stock of equipment eligible, environmental contraints, and the capacity of the equipment to be retired.

The relative delivered prices of fossil fuels can be expected to change over the next twenty years. In what follows we offer some reasoned speculation from the vantage point of early spring 1979. Unless countermanded by Congress, the president's phased removal of price controls on domestic oil proposed in April 1979 will increase average U.S. crude oil prices to the refiner by about $2 per barrel or about 35 cents per million Btus. By the end of 1979, world oil prices will be increased by OPEC by 14.5 percent above their 1978 levels (about $1.80 per barrel or 30 cents per million Btus), and probably beyond that level in the wake of the Iranian shortfall. Under decontrol gas would also rise by approximately the same amount. Residual oil prices to the end user could then be expected to increase 65 cents per million Btus for high-sulfur residual and 75 cents per million Btus for low-sulfur residual.

What about increases in world crude oil prices beyond those already announced and possibly still to come in 1979? Based on recent experience with world oil prices and the general expectation and consensus that world oil production will peak sometime in the next twenty years, the price of oil will surely rise more rapidly than inflation. U.S. prices, once controls are removed, can be expected to follow world prices. A gradual doubling of real oil prices over twenty years, making crude oil prices in 2000 about $5 per million Btus (1979 dollars), has been the upper boundary general assumption made by our study group.

Domestic natural gas prices will rise gradually as a result of the provision of the Natural Gas Policy Act of 1978, and most domestic natural gas supplies will be decontrolled by January 1, 1985. At that time, natural gas prices will approximate or possibly exceed the world price of the refined oil products with which gas competes unless large volumes of low cost gas from conventional or unconventional resources should become available or be a near certain prospect. Even in the unlikely event that real world oil prices do not increase between the end of 1979 and 1985, natural gas prices paid by the pipelines would increase from about $1 to about $2.50 per million Btus or more. If real world oil prices increase to $5 per million Btus by 2000, natural gas prices would approximate that same price, unless unexpected large, low cost supplies are developed. Imported natural gas can be delivered to the east coast of Canada for $3.97 per million Btus or $3 higher than the national average purchase price of gas by major interstate pipeline companies in 1978.

Coal prices (f.o.b. the mine) can also be expected to increase because of the new environmental and health and safety regulations that will have to be met by both surface and underground mines. The costs of these new requirements have not yet been calculated, but most of the impacts of the health and safety standards were reflected in the 1978 prices of coal through reduced productivity as well as higher capital investments. In fact, productivity may start to increase slowly, but we assume that this will be balanced by the retirement of older mines with their much lower capital investment.

Reclamation costs for strip mines imposed by the 1977 reclamation act will vary from region to region. If one assumes a cost as high as $5 per ton of coal, (there should be ample strippable reserves nearly everywhere at a cost as high as this), then the average price per ton of utility coal (f.o.b. the mine) would increase from less than 90 cents[5] per million Btus in 1978 to about $1.10 per million Btus by 1985, when all the new surface mine reclamation regulations have been implemented. If the real price of coal then increases by 1 percent per year, its price (f.o.b the mine) in 2000 would be $1.28 per million Btus.

Transport costs for coal, already a larger share of delivered costs than for oil and gas, can be expected to rise more rapidly than transport costs of these competing fuels. Railroads have already sought and been granted increases in coal freight rates, while existing oil and gas pipeline rates should increase more slowly. That, together with an expected increase in the average distance that coal will be transported as a larger share of the coal is produced in the West, could increase average rail rates by as much as $2.50 to $5.00 per ton or 10 to 20 cents per million Btus.

Under these circumstances and assumptions, and after adjustments have been made for taxes, differences in costs of use, and other factors discussed above, the national average prices for the three fuels as they would be delivered to utilities in the year 2000 could be:

	Price per Million Btus (1979 dollars)
Oil	5.20
Natural gas	5.20 +
Coal	2.60

Thus, it is apparent that on a price basis alone, coal may continue to be by far the least expensive fossil fuel for utilities in most regions. Indeed, coal prices could increase even faster and still remain competitive. In addition, the Power Plant and Industrial Fuel Use Act of 1978 requires high volume users of fuel, which include almost all

5. The 90 cents per million is the average price for all coal and includes about 20 percent of coking and export coal, which sells for much higher prices.

electric utility plants, to use coal in place of petroleum and natural gas. New power plants (with certain permanent or temporary exceptions) over 10 MW in size cannot use oil or natural gas as a primary energy source and must be capable of using coal. Existing power plants cannot use natural gas as a primary energy source after 1990, and there are restrictions on the use of natural gas in existing power plants between now and then.

Thus for both economic and regulatory reasons, utility plants will be moving rapidly to the use of either coal or nuclear energy. Chapter 12 discusses the relative costs of generating electricity using light water nuclear reactors compared with coal and concludes that for many parts of the country, light water reactors produce electricity at lower cost. However, for reasons given in the same chapter, projections of nuclear power capacity have dropped dramatically.

Chapter 2 reviews current thinking on future energy demand. If we take total energy demand for 2000 to center at 120 quads (a possibly high value that now appears in many recent studies), then electricity demand in 2000 might range from about 43 to 49 quads. If we assume that obsolete fossil fuel plants are replaced largely by new plants using coal or coal-derived fuels (80 percent of the replacements are coal plants), then total coal usage by the utilities in 2000 could range from 22.5 to 25.9 quads.

The Industrial Market. Estimating coal use in the industrial market is more difficult than for the electric power market. New large industrial boilers should be able to use coal as easily as small electric utility plants. The Power Plant and Industrial Fuel Use Act of 1978 has requirements for oil and gas use by major fuel-burning installations similar but less restrictive than those for electric generating plants. However, as the size of the industrial installation decreases, the difficulties and cost of burning coal compared with other fossil fuels increase. For small installations, the cost of coal stockpiling, and handling, ash disposal, and especially air pollution control will become an important and possibly dominant part of the total fuel cost. Capital and operating costs per million Btus of energy consumed increase rapidly as unit size decreases. If a major industrial market for small users is to be developed, new transportation and distribution networks for coal will have to be created in many places. For most small industrial users the fuel form must be either a clean gas, oil, or electricity. If coal is to be substituted as the primary energy source, it will have to be gasified or liquefied or, for some uses, converted to electricity either on site or at a central location, and the clean products distributed.

In 1974 it was estimated that about 7 percent of the oil and gas used for nonboiler applications in industry could be replaced by coal using proven technology and that another 21 percent could be replaced by coal with only a low conversion risk.[6] Thus coal in nonboiler uses could substitute for 28 percent (2.2 quads) of the 8 quads of energy supplied by oil and gas to industry in 1974 for these purposes. In practice, however, the picture may be different. A survey of eight major energy-consuming industries found that converting existing plants to coal would save only about 0.9 quads of oil and gas.

Total process energy use by the industrial sector in 2000, after adjusting for the expected reduced consumption per unit of output, would be 32.2 quads. If we subtract the energy consumed by the electricity purchased (since it was included in total electricity demand above), the consumption of energy for process uses (from all fuels) in new plants would increase by 6.2 quads by 2000. Based on previous studies, the amount of coal that could be used as a fuel for process use by industry in 2000 would be about 2.5 quads[7] out of the total of 6.2 quads. To this must be added the 2.0 quads of coal already used directly by industry for metallurgical purposes.

Industrial boilers that produce steam now use about 1 quad of coal per year. Unless the boiler was designed for coal, the conversion costs will be prohibitive and there will be few conversions. Other economic and environmental factors also make conversions unlikely, and at most 1 quad of oil and gas would be saved. However, if older, larger oil and gas boilers are retired faster because of the investment tax credit in the Energy Tax Act of 1978, then coal could provide 3 and possibly 4 additional quads of energy in the industrial boiler market.

If all the new boilers that produce more than 100,000 pounds of steam per hour were to use coal and if half of the new boilers of a smaller capacity also used it (another optimistic assumption), then the industrial sector would consume an additonal 5 quads of coal by 2000. Total industrial coal consumption (excluding the coal used indirectly in the form of purchased electricity) in 2000 could then be as high as 14.5 quads, but probably would be less because of the optimistic assumptions used. On the other hand, our estimates have not included the coal that may go into the production of medium or low Btu gas for industrial use or the manufacture of synthesis gas for producing chemicals.

6. No insurmountable technical obstacles are foreseen, but coal-burning equipment must be built and demonstrated in the United States.
7. This it the total of the 0.5 quads of coal used by industry for proven uses in 1974, the 0.9 quads convertible in the eight major consuming industries, and the 1.1 quads of coal used in their plants.

The National Energy Plan of 1977 calls for use of coal in the industrial sector to grow by 11 percent per year until 1985. However, the 14.5 quads in 2000 estimated above would result from a steady 6 percent growth rate through 2000.

When the total 14.5 quads for industrial use of coal and the previously estimated range of coal requirements for electricity generation of 22.5 to 25.9 quads are added, total coal use in 2000 could be 37 to 40.4 quads. Production could be somewhat higher because of increased demand for exported coal for metallurgical purposes, which could rise from the current 54 million tons per year to 150 million tons by 2000. Total coal production could then be 40.5 to 43.9 quads per year. This range falls near the high estimate of demands projected by various models, shown in Table 9-7. Note that our projection does not include any coal that may be converted to gas (low, medium, or high Btu) or to liquids. The prospects for coal conversion are discussed next.

Conversion to Gases and Liquids

Using coal directly can create difficult environmental problems in all markets, and it cannot be used in some markets, such as in the transportation sector and by some industrial consumers where product quality demands a clean fuel. Converting coal to gas is an old art that was used commercially for more than one hundred years but was displaced wherever low cost natural gas became available. Conversion of coal to liquids was done on an industrial scale in Germany

Table 9-7. Projections of Coal Production Made in Recent Major Studies (coal production in 10^{15} Btus per year).

Study	Scenario	1985	1990	2000
1. National Energy Plan (NEP)	Reference	23	27	—
	High	31	—	—
2. Workshop on Alternative	Low	20	—	28
Energy Strategies (WAES)	High	26	—	51
3. Dartmouth COAL 2 Study	Reference	24	—	34
	High	29	—	50
4. National Energy Outlook	Low	22	—	—
(NEO)	Reference	23	—	—
	High	28	—	—
5. CONAES Modeling Re-	Low	16	19	23
source Group (MRG)	Reference	22	24	34
	High	27	30	55

Source: Adapted from Energy Modeling Forum, *Coal in Transition: 1980-2000,* vol. 1 (Palo Alto, Calif.: Stanford University, September 1978), p. 2.

in the 1930s and 1940s, and one plant is in operation today in South Africa. Another is nearing completion, and plans for a third have been announced.

It would not now be possible to use in the United States the commercial coal gasification processes that were used in the past. These processes were suitable only for much smaller plants than would be required by most users. They would produce unacceptable levels of pollutants, and the costs of the liquid and gaseous products would be too high. As a result, for the past twenty years, a considerable research and development effort has been directed toward devising a lower cost and more environmentally acceptable processes for making gases and liquids from coal.

The lowest cost gas that can be made from coal has a low heating value and must be used on site because transportation costs would make its use uneconomic. Low Btu gas from coal can be produced to make a fuel free of sulfur compounds and of particulates. However, at this time producing gas from coal just to meet environmental standards for these pollutants costs more, in most instances, than burning the coal directly and using emission control technology to meet the air pollution standards. This type of gas made from coal could be used at commercial installations that require a clean fuel for process reasons or at utility plants using combined cycles for generating electricity.

Medium Btu gas from coal is somewhat more expensive to produce than low Btu gas, but can be transported economically for distances of fifty miles or more. (The Democratic Republic of Germany has an extensive distribution system in operation.) This would permit a number of industrial establishments to receive a clean gaseous fuel at a centrally located gasification plant. No plants have as yet been constructed in the United States since today's cost of medium Btu gas from coal would be in the range of $4 to $5 per million Btus, much higher than current prices of natural gas and oil. However, such plants could become economic if oil and gas prices rise sufficiently high or if regulations forbid certain types of industrial installations to use oil and gas.

High Btu gas, a substitute for natural gas, is the most useful and expensive of the three types of gas that can be produced from coal. This gas could be used to supplement and extend natural gas supplies and thus help to maintain a high load factor in the extensive existing natural gas transmission and distribution systems. High Btu gas from coal today would cost $5 or $6 or more per billion Btus and is not competitive with oil or natural gas at current prices. However, the gas industry is urging that the federally regulated gas transmission companies and locally controlled distribution companies be permitted to "roll in" this high-priced gas with their lower priced supplies and be given a guaranteed rate of return on the gasification plant investment. If this is done, high Btu gas plants could be constructed before the price of the products would normally justify their use (see also Chapter 15).

Coal can be liquefied by turning it into a gas, then synthesizing the carbon monoxide and hydrogen to a liquid. The technology is known, but the cost of liquids produced using existing processes are not competitive with world oil prices. No plants have been constructed (except for a new plant in South Africa) for more than thirty years, so that all cost estimates are speculative. Liquids produced by existing gasification and synthesis technology would probably cost about $35 per barrel or more. Coal can also be liquified directly by using hydrogen (with or without a donor solvent) and estimates as low as $25 per barrel or less have been published for some of the newer processes still under development.

Direct conversion of coal to a very heavy liquid or to a solid that is low in sulfur and ash and that might be burned directly is also being investigated. Since the primary purpose of producing a fuel of this type for boiler use is to meet environmental standards for sulfur and ash, the cost must be less than that for burning coal directly and controlling pollutant emissions by other means. A secondary purpose of such processes is to produce a fuel that can be used in an environmentally acceptable manner in existing oil-burning installations without major boiler modifications. Estimates of costs of producing this fuel range from $3.50 to $4.50 per million Btus, so that wherever control technology can be used, it should be more economic than using clean fuels derived from coal. In small plants or in existing large plants that were designed for either oil or gas, this type of fuel produced from coal could be an acceptable method of meeting environmental standards while using a coal-based fuel.

Liquid fuels can also be produced from oil shale, as discussed in Chapter 7 and 15. At present, suppliers of liquids produced from coal or oil shale cannot "roll in" the high cost of products, nor do they have the guaranteed rate of return on investment that high Btu gas from coal might have. This situation could be changed, however, by new federal policies that could require all refiners to use a prescribed percentage of coal- or shale-derived liquids in their refinery feedstocks or products. In general, "rolling in" (see Chapter 3) puts any technology that cannot take that route at a competitive disadvantage. Solar energy is a conspicuous example.

There appears to be little in the current R&D programs of both government and private industry that assures any major reduction in the high cost of producing either gases or liquids from coal. In view of the long lead times for design and construction of plants of this type, the high capital costs required to achieve economies of scale, the great uncertainties about costs, the environmental problems, and the significant technological risks if the newer untested technologies are used, it is unlikely that major quantities of synthetic gas or oil will be produced from coal by 1990. Beyond 1990 the situation is less clear. If oil and gas prices have risen sharply by then and are likely to continue to rise (for the reasons described previously) or if the cost of the last million barrels of oil imported is extremely high, then construction of coal conversion plants will become economic. In that event, the amount of synthetics produced in 2000 will depend on when the incentives are offered, and on their size and nature.

Since the rate of commercialization of coal conversion plants depends on technological developments and government policies that are as yet undetermined, a range of estimates of the size of a coal conversion industry by the year 2000 is the best we can offer. In a "business as usual" situation, and if conversion becomes economically attractive by about 1985 due to a rapid increase in oil and gas prices, about 2 or 3 quads worth of coal-derived synthetic fuel energy (equal to about 1–1.5 million barrels per day) might be produced by the year 2000. This would require about 4 to 5 quads (approximately 200 million tons) of additional coal production. Second, if we assume that no truly competitive process is available, that no incentives are offered, or that oil and gas prices do not rise sharply, then only an insignificant amount of synthetic fuels from coal would be available in 2000. Finally, if we assume a "forced draft" to conversion and construction of the first plants is started shortly, under the most optimistic assumptions, a maximum of about 10 quads of synthetic fuels might be produced in 2000. This would require about 15 quads (approximately 700 million tons) of additional coal production in 2000. Based on current policies and on current technological and economic considerations, it seems optimistic to expect even as much as 2 or 3 quads of additional coal production in 2000 to meet demands for substitute synthetic fuels from coal. Moreover, management of pollutants specific to conversion processes, which are discussed below, could cause significant delay.

Constraints on Coal Utilization

Direct Use of Coal. The least expensive way to use coal in large installations is to burn it directly. When coal is used in this way in boilers to produce heat, steam, and power, it can create several en-

vironmental and health problems, of which by far the most important is the effect of air pollutants on the general public. The health
effects of these pollutants are discussed fully in Chapter 10, as are
the possible long-term effects of carbon dioxide on climate.

Water pollution problems, in the form of thermal pollution and
reduced water quality, are similar for most fuels used directly. In
fact, thermal pollution is greater for nuclear plants generating electricity than it is for plants burning coal or other fossil fuels. Solid
waste disposal problems are much more serious for coal than for oil
and are of little consequence when gas is used.

The difficulty and cost of providing satisfactory solid waste disposal for coal ash is now under active debate in connection with
establishing regulations under the provisions of the Resource Conservation and Recovery Act. A major concern is control of trace metals
known to exist in coal and ash. Under certain circumstances, these
can be leached from the waste disposal areas and enter into aquatic
environments. Many of the metals can then be concentrated by
aquatic biota and enter the food chain of animals and man. Little is
known about the potential magnitude of this problem or its possible
significance. The extent of the problem depends on the chemical
composition of the coal ash (which varies widely from seam to seam)
and on how the ash wastes are produced and disposed of. Under
some possible regulations that might be issued under the Resources
Conservation and Recovery Act, the cost of coal ash disposal might
be increased greatly.

Until the passage of the Clean Air Act of 1977, sulfur oxide
standards could be met by methods other than scrubbing.[8] These included the use of naturally occurring low-sulfur coals, more intense
coal preparation, (i.e., cleaning to reduce the sulfur content of
medium-sulfur coals to acceptable levels), and blending several coals
to make a fuel that would meet sulfur oxide emission standards. The
final regulations to implement the 1977 Clean Air Act amendments
have not yet been adopted, but more stringent regulations are certain
to be issued. Continuous removal of sulfur oxides is required for
large plants by the Clean Air Act amendments of 1977. The provisions for small plants using coal are still uncertain and will be decided
at a later date than those for large plants.

Control of sulfur oxide emission by using throwaway flue gas
scrubbing processes will greatly increase the amount of solid waste
to be disposed of. Sludge production rates (tons/kWh) are a function
of the heating value and the ash and sulfur contents of the coal being
burned, the size of the coal plant, scrubber reagent type, particulate

8. For a detailed discussion of the Act and subsequent developments, see
Chapter 11.

and sulfur flue gas cleaning efficiency, and whether or not mechanical sludge dewatering devices are used. A ton of coal with no sulfur to be removed and a 10 percent ash content produces 0.1 ton of solid wastes. If the sulfur content of the coal increases to 2.5 percent, the amount of dry solids to be disposed of will increase by nearly 100 percent. If most of the ash is removed separately, the scrubber solid waste will contain little of the trace heavy metals. Even then, it may be difficult to dispose of the sludge from scrubbers because of the large quantities involved, the physical properties of the sludge, and the need to prevent leaching of soluble substances contained in the sludge. Until now, most scrubber sludge has been put into ponds; this does not represent a final solution because the sludge does not become a solid nonleachable material, and in urban areas the acquisition of land for disposal is difficult and expensive.

The cost of flue gas scrubbing is high; installed costs of equipment are in the range of $75 to $125 or more per kilowatt, the exact cost depending mainly on whether the installation is new or old. In addition, there are significant operational costs associated with scrubbers. These include the costs of limestone, water, energy, and manpower. Total costs for flue gas desulfurization are on the order of 6 to 8 mills/kWh, depending on the assumptions used in making the estimates as well as on the sulfur content of the coal used and the load factor and location of the plant.

Many processes other than lime/limestone scrubbing have been proposed to control sulfur oxide emissions. Few have been tested on a large scale and none has been tested thoroughly. They do not appear to have major advantages over the lime/limestone process, although some produce a useful product rather than a sludge that is difficult to dispose of.

An alternative method of controlling sulfur oxide emissions is to burn coal in a fluidized combuster. This allows the coal to be burned at lower temperatures in a bed to which limestone has been added so as to capture the sulfur oxides as they are formed. The intimate contact between the sulfur oxides and limestone reduces the size of the control equipment by eliminating the need for scrubbing dilute sulfur oxides from large volumes of flue gases. The lower combustion temperatures also reduce the amount of nitrogen oxides that are emitted.

Fluidized bed combustion of coal at atmospheric pressure is particularly well suited for smaller coal-fired installations, such as those used by the industrial sector. In these facilities, operating problems and costs of controlling sulfur oxides with scrubbers would be extremely high and thus would impede large increases in coal use. A number of fluidized bed pilot plants have been tested, but the optimum design of such a system and its cost remain to be determined. How efficiently the sulfur oxides can be removed and how much limestone will be required to meet emission standards are among the major unresolved problems.

Atmospheric fluidized combustion for large utility boilers requires additional development. Even if such large fluidized bed boilers are developed, the cost advantages over conventional combustion systems (pulverized fuel boilers) that use scrubber control technology are still uncertain.

Pressurized fluidized bed combustors are also under development for electric utility use. Not only can they control sulfur oxide and reduce nitrogen oxide emissions, but the pressurized operation would permit electricity to be generated at higher efficiencies than are now possible through the use of a combined cycle that uses both gas and steam turbines.

Coal Conversion. Conversion of coal to liquids and gases produces combustible fuels that avoid most of the air pollution problems associated with the direct combustion of coal. However, the conversion plants themselves create a new set of environmental, general health and safety, and socioeconomic impacts. In the absence of any commercial installations, the magnitude of these impacts can only be estimated. Indeed, one of the justifications for constructing a demonstration or full-scale plant is to establish the exact nature and magnitude of the environmental problems that may be created.

At present the products of coal conversion plants are too expensive to justify construction on economic grounds. Even if this barrier did not exist, there are other constraints to the rate at which plants could physically be constructed. These include the problems of locating suitable sites for a large number of plants (each of which would be enormous in order to achieve economies of scale), obtaining the required water, meeting the socioeconomic problems that the plants could create, and preventing unacceptable environmental effects.

The difficulty of overcoming some of these problems will depend on whether the plant is located in the West with its limited water supply and sparse population or in the East where these problems should be more manageable. The trade-offs involved are complex. Western locations would normally be preferred, since it is less costly to move the product from the conversion plant than it is to move coal to the plant and western coal is generally cheaper. On the other hand, the western locations will require plant designs that minimize water use, and this will increase capital costs. Also, the conversion plants require much more manpower to operate than is needed to mine the coal. Thus, the socioeconomic problems will be greater if conversion plants are located in the sparsely settled West.

Temporary shortages of particular types of equipment and of certain resources—such as some kinds of construction skills—could develop locally. Their severity would depend on the rate of construction as well as on the amount of supporting facilities and infrastructure. However, no sustained constraints should arise nationwide from scarcity of manpower, water, coal, other raw materials, or equipment.

It may be difficult to locate suitable sites for a large number of coal conversion plants of the size that are considered standard—50,000 to 100,000 barrels of liquids per day and 125 to 250 million cubic feet per day of a substitute natural gas. It will be at least as difficult to find a site for such a plant as for a 1,000 MW power plant, which is no longer easy. For the few plants that have been seriously studied to date, the siting problem has not presented difficulties, since most of the proposed locations were selected some years ago when competition for suitable sites was less intense and criteria less demanding, though this does not necessarily preclude the possibility of contest and delay, once the site is activated under current conditions and procedures. If as many as one hundred plants were to be constructed (these would supply about 10-15 quads of clean fuels from coal), finding sites that could meet all the zoning and air and and water pollution requirements and that had coal and water available would become extremely difficult. While this is a longer term problem that may not develop for twenty years or so, attention needs to be called to it now.

As with the direct combustion of coal, the major constraint on the construction and operation of coal conversion plants will be their possible adverse impacts on air, water, and land; the disposal of the solid wastes that are created; and the environmental, health, and safety problems that they create for both the general public and workers.

From 10 to 20 percent of the coal used in a conversion plant is burned directly to provide the steam, heat, and power required for the process. The environmental problem of coal are discussed above and in Chapter 10, and the same control technology can be used to reduce emissions; the amount that must be controlled, however, is much smaller.

In addition, water, solid waste, and air pollutants of various types are created in the conversion process itself. Based on pilot data and a general knowledge of the chemical reactions involved, some estimates of the probable levels can be made. The solid wastes from coal burned directly are identical to those produced at electric generating plants, and solid wastes can be disposed by the means described previously. The balance of the solid waste from conversion plants is created under different conditions, and the potential pollutants that could be generated through leaching of the waste piles could be different than those that normally occur at power plants. Depending on the conversion process used, the wastes will contain heavy metals in different chemical states of oxidation or reduction and might contain leachable hydrocarbons not found in the coal ash from combustion systems. Unfortunately, precise quantitative data are not available on the potential pollutants from individual processes.

The amount of solid waste that must be disposed of per million Btus of useful energy produced is greater for a plant that converts coal to gas or oil than when coal is burned directly to raise steam or supply process heat because the efficiency of conversion is usually in the range of 55 to 70 percent compared to 95 percent or more when coal is burned directly.[9] Because of the greater uncertainty as to what compounds will be leached, waste piles from conversion plants should be disposed of in a manner that will ensure that no water that has passed through the waste reaches either streams or aquifers. It may be learned later that such precautions are unnecessary, but the present state of knowledge does not permit such a conclusion.

Some of the water pollutants that are generated by conversion processes are more complex chemically than those obtained from the direct combustion of coal. However, the kinds of contaminants and their quantities depend in large part on the process used. For example, one high temperature coal gasification process produces virtually no liquid streams that must be treated, while a fixed bed low temperature process produces streams that are contaminated with tars, oils, soluble organic materials (such as phenol), polynuclear aromatic hydrocarbons, dissolved inorganic salts, and suspended solids.

The sources of water pollutants at conversion plants are the quench water used to cool hot products, condensates from various steps in the process, and "blowdown" water from the boiler and cooling streams. Some of these are common to all conversion processes, but the exact composition of the composite waste water stream is unknown, and water quality standards for many of the compounds contained in it have not yet been established.

To reduce the water pollution problem, one possible approach now being studied is to route the water in the plant in such a way

9. When coal is burned to generate electricity, however, more coal is required and more solid waste is created for each useful unit of energy, since conversion efficiencies are about 35 percent.

that as it becomes more degraded, it is used for plant processes that require lower quality water. After its last use, the water could be evaporated and the remaining solids from the evaporation discharged with the balance of the solid wastes.

Potential emissions to the air will consist of the sulfur and nitrogen oxides and particulates that result from the combustion of the coal that is burned, as well as hydrogen sulfide, organic sulfur compounds, carbon monoxide, and trace elements resulting from the gasification and gas purification steps. The amount and nature of these air pollutants depends on the coal conversion process used. Most of the air pollutants produced during the coal conversion step must be reduced to low levels for process reasons or to meet product quality requirements, so the amount finally discharged or contained in the product should be small.

However, the conversion plants are very large so that even low concentrations of impurities in the plant discharge streams can amount to large volumes of potential contaminants. The exact nature and quantity of some of the materials that will be released is still uncertain and will vary for the different processes, so that care will be needed in operating the first plants to determine if any unexpected hazardous materials need special control.

COAL AND ENERGY DEVELOPMENT ISSUES IN THE WESTERN UNITED STATES

Water Resources

Introduction. Persistent questions have been raised about the adequacy of water resources for energy development in the western United States since coal mining and, more important, coal used for the generation of power or synthetic fuels requires a reliable water supply, and this resource is limited in the west.

There is no simple definition of "adequate" water. To some, adequacy means a quantity of water sufficient to satisfy the real or presumed needs of existing and future users without conflict or economic adjustments. To others, economists, for instance, the term is misleading; they assert that supplies are always adequate, assuming that users compete for the scarce water by paying a price commensurate with the value of the water in a competitive market. Still a third viewpoint suggests that because society sees water as a scarce or potentially scarce commodity of vital importance to different kinds of competing users, special legal institutions are needed to protect and assure rights to limited quantities of water, and these institutions, more than the market, should determine the availability of water for new users.

It is our view that the question of adequate water for development of energy resources in the West involves a comparison of the demand for water at different prices, the quantity available at different prices, and the appropriate institutions for evaluating and allocating this scarce resource, including markets and legal and political institutions. These three facets are used as a basis for the following discussion.

A Simple Hydrologic View. Much of the Rocky Mountains, Great Plains, and southwestern United States is an area in which the total amount of precipitation is less than the amount of evaporation and transpiration, and hence, stream flow is often low and sometimes zero. Water is primarily derived from high mountain areas from which rivers flow into regions of increasing aridity.

In addition to limited availability of water from rainfall, claims upon available supplies, including augmented flows from reservoir storage, appear to exceed the available water. The classic example is the Colorado River, whose waters have been allocated to the upper and lower basin states and to Mexico through two Colorado River basin compacts and associated international agreements. The amount of water in the river is somewhat less, based on long-term averages, than existing commitments to Mexico and to the lower and upper basin states.

A common approach to the determination of the "adequacy" of water for projected future uses in a river basin is to balance presumed supplies against expected consumptive uses, where consumptive use is the difference between the amount of water withdrawn and the amount returned to stream flow. "Supply" is not a single number, but rather a choice of a particular statistic involving a particular flow and the frequency or duration of its occurrence. A commonly used, but wholly arbitrary, flow frequency is the consecutive seven day low flow recurring, on average, no more than once in every ten years. Additional flow derived from regulation of reservoir storage may or may not be considered in calculating an available supply.

When the balance between estimates of consumptive uses and expected available supply is applied to major river basins throughout the United States, clearly defined "shortages" appear likely in such areas as the Colorado River basin, other parts of the southwestern United States, some areas of the Great Plains (Arkansas River basin), and parts of Texas and California. Assuming reasonable economic growth, some projections suggest that even in the northeast, southeast, and in the Ohio valley, consumptive uses will exceed available supplies by the year 2000. Because energy conversion (particularly wet cooling towers) and synthetic fuel industries can consume large quantities of water, some projections of the demand for energy throughout the United States suggest that virtually every area will experience "shortages." Most projections emphasize shortages in the plains areas and particularly in the southwestern United States and even in the East where cooling towers on smaller streams may consume large quantities of water.

A more detailed inquiry of water needs or expectations in the upper Colorado River basin estimates that future potential uses, including expanding current uses and energy needs, will exceed available supplies in a number of component drainage areas by the year 2000. The study does not speculate about the availability of additional water during low flow periods from new reservoir construction or from ground water or about the use of water-saving technologies such as dry cooling towers, but it points to the basic uncertainty of the statistical estimates of available supplies throughout the basin and the importance of analyzing water availability in stream segments for in-stream uses. At the level of individual power plants, utilization of storage and cooling water ponds, rather than cooling towers, may increase the availability of flows downstream for uses during critical late summer low flow periods.

Most analyses of water availability do not include the possibility of transfer of water rights from high consumptive uses such as agriculture to others, such as municipal, industrial, and recreational activities that require less water. In addition, opportunities for water storage and regulation, for conservation and more efficient utilization, or for importation of water from other sources have not been explored. Thus, estimates of uses projected against the supply of water in streams, in the absence of economic factors, are likely to err in the direction of "shortages" and scarcity. To make a realistic assessment of the adequacy of water resources in the West, assumptions about transferability and the mechanisms by which changes in water use and demands may come about through economic, legal, and political institutions require more extended analysis than that provided by a simple budget of needs and hydrologic supply.

Enhancing the Availability of Water by Market Mechanisms. Economists have noted that the economic return to a variety of activities, particularly agriculture, in many parts of the western United States is low compared with economic returns if the water is used in other activities. Estimates have been made that show that in one part or subbasin of the upper Colorado River basin, the marginal value of water for irrigation ranges from about $0.50 to $22 per acre foot, while the value for use in cooling towers ranges from $870 (60 percent dry cooling) to $536 (100 percent dry cooling) per acre foot. These are amounts that users would pay for the use of one more unit of water in these activities (the marginal value). The huge difference between the marginal value of water in irrigation and in electric power generation attests to the pressure that the market could exert to transfer water from irrigated agriculture to energy conversion. The figures do not prove that water will be made available to higher valued uses, nor do they suggest a time scale for such transfers.

With these facts in mind, because of the very large differential in the economic value of water in energy compared with other activities in some areas of the West, we believe that water is likely to become available in time at prices sufficiently high to reduce the implied shortage of supplies. Support for this view comes from the change under way in parts of the West and from some evidence that the allocation of beneficial uses of water in the West does eventually respond to the market. From this standpoint, then, a projected hydrologic shortage based on equating available supply and consumptive uses is faulty in that projected consumptive uses are often not true demands in the economic sense, but closer to "wants" at low prices. The counter to this presumption, that water could be made available (at higher and higher prices) if markets were allowed to operate, lies in the complex institutions embodied in law and history, which while they do not negate market forces, markedly constrain the allocative process and its timing. Some of these institutional issues and the way in which they may influence energy development in the western United States are discussed in the next section.

Institutional Effects on the Availability of Water. Rights to the beneficial use of water in the West are based on priorities established in time. Water may be appropriated for beneficial uses, including drinking water and water for livestock, irrigation, mining, cities, and other activities. Historically, as settlement proceeded, the largest quantities of water have been appropriated for irrigation to develop agriculture where land was plentiful but water scarce. Withdrawal of water for agricultural use results in a diminution of the quantity, and often the quality, of the water not consumed and returned to streams. For example, in irrigation agriculture, about 50 percent of the water withdrawn is consumed by evaporation and transpiration, and the return flow often contains salts leached from the soil. In addition, the return flow may reappear some distance from the point of withdrawal. In arid regions throughout the world, then, the appropriation doctrine grew up to allocate a scarce water resource and to assure protection of the rights of downstream users not directly involved in water transactions at an upstream location.

"First in time, first in right" defines the hierarchy of rights to available water, but in many instances, the quantity of water to which each user is entitled during periods of limited stream flow has not been firmly established. At present, a wide variety of claims is being adjudicated in the courts. A number of rivers in the West, such as parts of the Colorado and the Yellowstone, flow through areas rich in coal, oil shale, or both, where there is a large potential for energy development. The availability of water for these new activities, however, depends, as noted earlier, not only on a highly variable supply and on the economic value of the water, but also on the determination of existing rights and their priorities.

Determination of existing water rights has been made more complex by claims of the federal government to have "reserved" rights to water for use on various lands, including Indian reservations.[10] Because a number of Indian reservations were established in the early 1850s, they are among the very "first in time," and should the claimed rights to water be confirmed, their rights would take precedence. Beyond the question of the priority of the reserved rights themselves, there is a further question requiring legal clarification—whether the water can be used only to irrigate land on the reservation (a restrictive interpretation) or whether such water can be used for any economic activity that the Indians select. The latter, more expansive, interpretation of the law by the courts could make water available for energy development should the Indians so choose.

Competing uses of scarce water include not only agricultural and energy interests, but also existing towns and industries, as well as fish and wildlife, recreation, and water quality interests. These last seek to assure adequate flow in streams and rivers to assure maintenance of viable habitats and survival of fish and animals dependent on flow and wetlands. Because of the immense variety of conditions influencing potential uses, a wide variety of plausible scenarios can be constructed to simulate the availability of water for energy development. A number of such simulations have been constructed for the Yellowstone basin, an area that contains some of the most readily mined coal deposits in the West, deposits with coal seams 100 feet thick

10. This discussion is drawn from Constance M. Boris and John V. Krutilla, *Water Rights and Energy Development in the Northern Great Plains: An Integrated Analysis* (Baltimore: Johns Hopkins University Press for Resources for the Future, forthcoming).

beneath a thin overburden at the surface. The results of this analysis, which by no means exhausts all the alternatives, suggest a wide variety of opportunities, as well as a number of doubtful prospects, for energy development in major tributary areas and regions of the Yellowstone system. For example, on the Bighorn River, a tributary of the Yellowstone River on which there is already a major reservoir, sufficient water apparently exists to support significant energy development, including major electric power production, after meeting all legal claims by current prospective users. However, because the Bighorn basin itself does not contain coal beds, water would have to be diverted to serve possible energy development where coal is available, but the economic and legal aspects of such a diversion have not yet been studied. In contrast, on other tributaries of the Yellowstone within the coal-bearing regions, flow is insufficient to meet irrigation and in-stream demands, along with high projections of energy development, without a large amount of additional storage. However, with some storage and reduction in in-stream requirements, a lower level of energy development would appear possible. On the mainstream of the Yellowstone itself, considerable water could be made available for future energy development (it is the present water source for several major mines and power plants) provided, again, that the lower of two conflicting estimates of required in-stream uses is accepted.

These brief descriptions are merely illustrative. They cannot do justice to the variety of alternatives, the complexity of the problems, or the uncertainty of the outcomes of the adjudication of water rights and the ultimate allocation of water to competing uses, including the energy industry. The uncertainties are very great in determining whether any given level of coal development will be constrained by water availability because of differences of opinion about the rights under the laws among the legal fraternity and because there are no clear indications that the courts lean heavily in one direction.[11]

After evaluating a number of alternatives in light of existing conditions, attitudes, and the pace of decisionmaking, some conclusions can be drawn regarding the prospects for specific energy types and timing in the Yellowstone basin. Reflecting upon the fact that the high intensity energy development, including coal conversion, within the Yellowstone basin postulated in one study involves, among other things, (1) major storage in some areas, (2) a suit currently in the appellate courts contesting use of water from federally owned reservoirs for energy purposes, (3) unresolved issues of the amounts of

11. Ibid.

water that might be made available for use by the Indians and uncertainties about their choices should water be available, (4) environmental constraints on eventual (let alone rapid) construction of dams and reservoirs, and (5) a five to seven year period to fill the reservoirs after construction, it was concluded that "It would seem that for the near term, i.e., 1985–1990, what energy development we should consider realistic in eastern Montana would be primarily the mining of coal for export."[12]

Over a longer time span, such as the twenty year period considered in this study, the constraint on expansion of coal conversion plants in the Yellowstone basin appears less severe. Water can be made available in some parts of the basin without sharply restricting prior uses, and new reservoirs as well as resolution of Indian water rights may make additional supplies available. In the much shorter term, because coal mining requires much less water than conversion to electricity or synthetic fuels, significant expansion of coal production is possible. The extent to which such expansion may take place, however, depends in part on the stringency of surface mine regulations and on what air pollution control regulations will require with respect to removal of sulfur oxides from low-sulfur western coal.

In the Colorado River basin not only is water scarcer than it is in the Yellowstone, but the degree of commitment to various uses is even greater. The total projected needs in the upper basin will exceed the surface water supply in little more than a decade.

Availability will differ from state to state, and the total for the basin will obviously fluctuate with the natural river flow. The Colorado River example, along with the Yellowstone, suggest that a simple computation of the availability of water or even of the transferability of water from one use to another does not provide an answer to the question of the adequacy of water for energy development in the West. With minimal basis for firm conclusions, it appears that over time water can and will be made available to supply demands for energy development. Potential Indian water rights coupled with new storage of stream flows will begin to make some additional water available for energy development. Over time, water is likely to shift from older to newer uses as the very much higher value of water for new energy uses is reflected in higher prices that agriculture will not wish to pay.

Limits on water availability are unlikely to be approached in mining, and in all likelihood, real or presumed restrictions will limit the technologies that can be used in coal conversion so as to minimize

12. Ibid.

consumptive use of water for cooling and in the production of synthetic fuels. In some regions, energy conversion and synthetic fuel developments may be significantly inhibited, but limitations on electric power generation as a result of air quality regulations appear to be even more imminent. However, the uncertainty regarding both the hydrology itself and the outcome of complex legal proceedings regarding the use of water are likely to inhibit rapid expansion of major coal development beyond mining over the next decade. Whether such inhibitions or delays will be significant in meeting the nation's energy needs is entirely dependent upon the extent to which western coal fulfills a significant proportion of the nation's needs and the way in which it does so.

A preliminary estimate based on the assumption of an increase of about 40 percent in coal development by 1985, with perhaps 30-50 percent of that increase coming from western coal, would suggest that water will not be limiting in the next ten years. If the courts adopt the more expansive view of Indian water rights permitting water to be used for development beyond irrigation, strictly speaking, the "availability" of water then becomes less an issue of hydrology and law and more an issue of social choices: what Indian tribes may choose to do with the waters under their rights, whether reservoir construction will be permitted by society, and whether other holders of land and water rights will consent to sell or transfer their rights either to develop coal and shale or for other energy purposes.

The Clean Air Act

When plans for the greatly increased coal production in the West were originally made in the early 1970s, it was clear that part of the coal would be used in the West near where it was mined, but that much of it would have to be transported and used in the population centers in the Southwest, the Midwest, and even in some eastern states. A large part of the original rationale for producing this coal and transporting it to distant markets was that it could be used in many existing coal-burning facilities and even in new power plants without the need for stack scrubbers and still meet the air pollution standards for sulfur oxide emissions.

The Clean Air Act Amendments of 1977, however, changed the emissions standards for sulfur oxide for new plants and required the use of the best available control technology for continuous emission reduction. This means that sulfur oxide must be controlled by either reducing the sulfur content of the original fuel before combustion or by scrubbing to remove the sulfur oxides from the flue gases or by some combination of both precombustion and postcombustion control technologies, rather than by substituting a naturally occurring low-sulfur coal or oil. This change could reduce the anticipated demand for low-sulfur western coal.

By mid-1979, the final regulations had not been issued, so it is not possible to estimate what combination of control methods for new plants would be the lowest cost in different consuming areas. There could be a combination that uses western low-sulfur coal and a smaller scrubber than would be required when using a higher sulfur coal. Alternatively, a nearby high-sulfur coal, from which part of the sulfur has been removed by coal preparation, and a small scrubber could be used.

Other provisions of the Amendments could also lead to a reduction in the use of western coals. One section prohibits the use of coal "which is not locally or regionally available" if the use of such coal is causing "local or regional economic disruption or unemployment." How this provision is to be administered is very unclear. There are no definitions of what a "local or regional" coal is, nor are there any criteria established about what will be considered "economic disruption."

Another section of the amended Act that will affect the East-West mix of coal production is the requirement that states must redo their State Implementation Plans (SIPs) to be approved by the EPA before July 1, 1979. Any reduction in the emission limits of sulfur dioxide at older plants covered by the SIPs could affect the economics of using western coals to meet the standards in many plants.

These factors as well as others discussed earlier (particularly federal leasing policies) make quantitative determinations of how much western coal will be produced uncertain, but the impact of most of these new considerations can only reduce the amount of western coal that can be used economically outside of the western regions.

WORLD COAL DEVELOPMENTS AND TRADE

As noted in the introduction, coal's contribution to the worldwide commercial energy supply has been declining steadily—from 83 percent in 1925 to about 29 percent in 1973. The decline was especially rapid in Western Europe, where coal's contribution declined from 62 percent in 1960 to about 20 percent in 1977, and in the United States, where coal's share of energy consumed declined to only about 17 percent in 1972 from over 50 percent thirty years earlier. The oil embargo of 1973-1974 revived interest in using domestic resources, and coal's share worldwide increased by about 2 percentage points to nearly 31 percent by 1975.

World coal trade for the past twenty years has ranged from 160 to 220 million short tons per year with 210 million short tons being traded in 1977. Most of the coal exports were from the United States, Australia, South Africa, Federal Republic of Germany, and Poland. In 1977, the United States was the largest exporter (24.8 percent) and Poland was second (19.7 percent). The major importers in 1977 were Japan, France, Canada, and Italy, which together received over one-half of total world imports. Seven other countries, of the twenty-six countries importing coal, imported more than 5 million tons in 1977.

A large part of the coal moving in international trade has been metallurgical coal used to produce the coke for steel making. Some steam coal has been traded, mainly when it was the lowest cost coal that could be obtained from a neighboring country by users near an international border.

The large increase in the real price of oil has made coal competitive in many areas where it had not been before. This, together with a desire by most of the oil-importing countries to reduce their dependence on oil imports, has led to expectations that the declining importance of coal worldwide would be reversed. How these expectations have fared is described in detail in Chapter 8. In summary, immediately following the oil embargo of 1973–1974, it was believed that coal demand and coal production would increase steadily in Western Europe, but this has not occurred. Despite a United Kingdom decision to increase domestic production and a more modest plan by the Federal Republic of Germany to at least maintain current production levels, hard coal production declined between 1973 and 1976 in the European Economic Community countries from 300 to 274 million tons, and imports increased from 33 to 48.5 million tons. Thus for the Western European countries, an increased role for coal has so far failed to materialize.

Coal:
Climate and Health Hazards

INTRODUCTION

No option for providing additional energy supplies for the next twenty years and beyond is without drawbacks. Oil and gas, nuclear power, coal, and even various solar possibilities all entail adverse social, economic, and environmental effects of differing types and magnitudes, on different time scales, and affecting different regions in different ways. Even the "best" choice for the nation will have drawbacks to some people, in some places. The policy problem is to find ways to weigh the various effects, to develop a mix of supply and conservation options that roughly maximizes the excess of benefits over costs overall, and then to distribute the benefits and costs equitably.

Some of the most important and difficult parts of this problem are the relationships between energy and the environment. A complete treatment of them would require consideration of the environmental consequences of all energy resources and of all parts of the resource system from extraction through transport, cleaning and conversion and ultimately to disposal of wastes. However, we have not attempted such an encyclopedic treatment in this book. Instead, we have emphasized three themes—the value of establishing, pursuing, and maintaining environmental goals in a flexible, economical, and effective manner (see especially Chapter 11); the importance of improving processes for making decisions on local or regional issues, such as land use and the impacts of energy facilities (Chapter 14); and finally, and perhaps most important, the necessity of understanding and dealing with some physical relationships that may threaten the global environment or human health. This chapter deals with the last of these three subjects, focusing primarily on coal. Environmental issues related to nuclear and solar energy are discussed in Chapters 12 and 13, which deal with those sources individually, and other environmental aspects associated with the production and use of coal are dealt with in Chapter 9.

The preceding chapter demonstrates that coal is a plentiful resource, that the technology is available for its extraction in environmentally acceptable ways at reasonable costs, and that it can be adapted to the immediate needs of a variety of users, particularly in the production of electric power. The bad news is that using the coal may be harmful to the environment. Adverse effects may vary widely in geographic scale. At one extreme is the possibility that carbon dioxide (CO_2) from the burning of any fossil fuel (but especially coal, since it releases 40–80 percent more CO_2 per unit of heat than does oil or natural gas) could accumulate in the atmosphere and produce an increase in the mean temperature of the earth. Near the other extreme are the site-specific and local problems of disposal of ash and sludge from coal-using facilities. And there are regional and hemispheric effects between these extremes. Table 10-1 lists important sources of coal-related environmental effects and the geographic scope of these effects. Solving any of the problems listed in Table 10-1 will not be easy, cannot be assured, and certainly will not happen without serious attention and hard work. Coping with such problems demands the use of sophisticated science and technology and taxes the capacity of society to agree upon and to implement enlightened decisions. Nevertheless, as noted in the preceding chapter, most of these problems can be managed reasonably well without additional or major changes in national policy—although there is no guarantee that they will be well managed. However, we believe that air pollution from the increased burning of coal, because of its threat to the global climate and to human health, demands special attention. These problems are potentially the most threatening. They are the least understood, and they are poorly addressed by current policy.

In the discussion of two central relationships—between carbon dioxide and global climate and between air pollution and health—this chapter (and the next, dealing explicitly with air pollution policy) attempts to illustrate the characteristics of a range of environmental issues. We believe that our analysis suggests ways in which pollution control and environmental management might be made more effective and efficient. Thus, the discussion of selected

Table 10-1. Geographic Scales of Some Enironmental Problems Associated with Coal Use.

Environmental Problem	Geographic Scale			
	Local	*Regional*	*Hemispheric*	*Global*
Carbon Dioxide			x	x
Sulfur Dioxide Emissions from Power Plants		x	x	
Particulate Emissions from Power Plants	x	x		
Emissions of Nitrogen Oxides	x	x		
Surface Mining and Drainage	x	x		
Waste Heat from Power Plants	x	x		
Cooling Water Discharges	x			
Sludge Disposal	x			
Coal Trains	x			
Boomtowns	x			

environmental issues here is directly related to the development of energy and environmental policies appropriate to the next twenty years—policies that will not foreclose options beyond twenty years and that assure the effective use of the twenty year period to establish new and evaluate old options for the longer term future.

CARBON DIOXIDE AND GLOBAL CLIMATE

Within the past ten years, small fluctuations in climate have produced major disturbances in world food production and food markets. Grain harvests fell in Russia and droughts caused crop losses in the Sahel and in parts of Latin America. Similar climatic events have happened before, of course. But happening together in the 1970s and widely reported, they dramatically illustrated the sensitivity of world and regional food production to climatic variations.

Although nobody is certain, there has been a convergence of opinion among many scientists studying global temperature changes and climatic processes that the burning of fossil fuels could conceivably produce a permanent warming of the atmosphere, accompanied by an even larger change in the temperature differential between the poles and equator, leading to marked changes in global weather patterns and climate. Because such changes could have far-reaching effects on mankind, the carbon dioxide problem has been placed at the center of the debate over the appropriate choice of energy resources to meet rising world energy needs. Thus, it is useful to review the evidence on the earth's climatic history, the effects of carbon dioxide, and the possible implications for global energy options.

History of the Earth's Climate

The climate of the globe throughout much of geologic time was considerably warmer than it is today. More extensive oceans, including large shallow seas and smaller continental areas, with the mean temperature of the globe perhaps 8 to 10°C warmer than it is now, characterized much of geologic time. At other times in the earth's history, average temperatures were considerably cooler than they are now. During the period beginning roughly two million years ago and coinciding with the period during which man evolved, the midlatitudes were roughly 5°C warmer in interglacial periods and 5° cooler during maximum glaciation than they are now.

The last major deglaciation took place about 10,000 years ago. There is abundant scientific evidence that, since then, average temperatures have fluctuated several degrees centigrade over periods of about 1,000 years: oxygen isotope ratios in ice of the Greenland ice cap, sediment cores from the ocean bottom containing remains of ocean organisms adapted to cool and warm water in alternating layers, pollen accumulation in lakes and bogs indicating land vegetation adapted to different climates, tree rings recording growth rates and atmospheric carbon isotope ratios, and geological evidence on the expansion and retreat of glaciers. Historical and archaeological records add evidence of more recent variations in weather and climate in different parts of the world extending back hundreds and occasionally thousands of years. The period from about 1430 to 1850 A.D. in Europe has been called the Little Ice Age in recognition of the cold, wet climate; records of the flowering of cherry blossoms in Japan show similar fluctuations in temperature and in growing seasons; and early records in Greenland provide evidence that both the growing season and the extent of land available for agriculture have been considerably greater in the recent past than they were some hundreds of years ago.

The range of historical fluctuations in regional temperatures has averaged on the order of 0.5 to 1°C with occasional larger excursions. Bryson has pointed out that, within the past forty to fifty years, climatic conditions in the midlatitude regions of the Northern Hemisphere have, in general, been relatively "mild" and conducive to high agricultural production in the corn belt of the United States and in parts of the Soviet Union.[1] This is one immediate reason for the concern that a turn for the "worse" is likely, given the marked variability of the past record.

The Role of Carbon Dioxide in Climate Fluctuations

Since the scientific world first accepted the idea that the earth's climate is not constant, many hypotheses have been put forward to explain the variations. These include small changes in the earth's orbit relative to the position of other planets or in the energy output of the sun, changes in atmospheric particulate levels due to volcanoes or dust from the land surface, combinations of random fluctuations in meteorological phenomena resulting in cumulative changes (positive feedback), and alterations in the chemical composition of the atmosphere due to changes in chemical and biochemical processes. Some scientists, particularly those supporting the astronomical and solar insolation hypotheses, believe that major long- and short-term oscillations are both cyclical and periodic.

During the nineteenth century, John Tyndall first suggested that small changes in atmospheric composition could bring about climatic variations. Other investigators subsequently suggested that increases in the concentration of carbon dioxide in the atmosphere could increase the temperature of the earth's surface, inasmuch as carbon dioxide gas absorbs and then reradiates heat in the long wavelengths radiated from the earth's surface. Some scientists suggest that unexplained and hypothetical fluctuations in atmospheric carbon dioxide levels might account for part of the variations in climate that have been experienced in the past and that the general warming (most pronounced in the Northern Hemisphere) observed from 1880 to the 1940s might be due to increasing atmospheric carbon dioxide from the burning of fossil fuel. But other scientists, while conceding that some variations in climate could be due to fluctuations in carbon dioxide in the atmosphere, argue that neither recent changes in temperature in the Northern Hemisphere nor the cyclical character of climatic variations can be explained by the carbon dioxide hypothesis:[2] for example, between the 1940s and mid-1960s, there was a

1. R.A. Bryson, "Some Lessons of Climatic History," in R.J. Kopec, ed., *Atmospheric Quality and Climatic Change* (Chapel Hill: University of North Carolina, Department of Geography, Studies in Geography, No. 9), pp. 36-51.
2. See, for example, H.C. Willet, "Do Recent Climatic Trends Portend an Imminent Ice Age?" in Kopec, pp. 4-36.

small decline in average global temperatures, while fossil fuel use was increasing rapidly worldwide.

It has been estimated that from the end of the nineteenth century to 1975, the CO_2 content of the atmosphere increased approximately 18 percent (from about 280 to 330 ppm). Estimates also indicate that the amount of fossil fuel burned during the same period paralleled the increase in CO_2, but that the increased amount of CO_2 in the atmosphere is only about one-half the amount of CO_2 emitted by fossil fuel burning over that period; direct atmospheric CO_2 measurements since the late 1950s also indicate that annual increases in atmospheric CO_2 are only about half the estimated annual input from fossil fuel burning. So about half the emitted CO_2 is going somewhere besides into the atmosphere. The CO_2 not going into the atmosphere must be going into one or both of the two other major accessible carbon reservoirs of the earth—the upper zones of the ocean and living biomass on land. But the scientific evidence suggests that not much of the missing CO_2 can be going into the oceans, because of the slow rates of mixing in the ocean. And it is hard to see how much of the CO_2 could be going into net additions to biomass—indeed, some scientists suggest that clearing and burning of forests in developing areas of the world over the past century has made terrestrial biomass a net source of, not a sink for, atmospheric carbon dioxide. Thus, it is unknown just how much CO_2 has actually been put into the atmosphere by human activities or where it is going. Until such basic scientific uncertainties are reduced, it will be very difficult to determine even how fossil fuel use might increase atmospheric CO_2 levels, let alone what the effects might be.

Despite these uncertainties, concern about the future effects of continued burning of fossil fuels is warranted because of the possibility that projected increases in the input of CO_2 could result in global warming. For example, a committee of the National Academy of Sciences concluded that a fivefold increase over the next one hundred years in the annual amount of fossil carbon fuels burned could roughly double the amount of CO_2 in the atmosphere and could produce an increase of about 2 to 3°C in mean surface temperatures and perhaps 5 to 10°C at higher latitudes.[3] Similar projections to a time roughly 200 years from now suggest that "it is not implausible" that, with continued growth in fossil fuel use, the concentration of CO_2 in the atmosphere could be four to eight times preindustrial levels, resulting in a mean temperature increase

3. National Academy of Sciences, Panel on Energy and Climate, *Energy and Climate* (Washington, D.C., 1977).

of 6°C, which is comparable to temperatures not experienced on the globe in the past 70 to 100 million years.

The effects of such a warming are impossible to predict. Clearly, if the warming went far enough, it could have disastrous effects, melting ice caps, raising sea levels, and changing ecological systems at rates too rapid to allow smooth evolutionary adaptation. It might be possible to adjust to less extreme warming—a degree or two on average over a century, for example—and such warming could conceivably even be beneficial in some ways, such as increasing rates of plant growth. Some projections to the mid-twenty-first century suggest that such a warming might be accompanied by warmer but drier conditions in the agricultural areas of the U.S. Midwest, perhaps with improvements in agricultural potentials farther north and elsewhere on the globe (e.g., China, Argentina, Africa). On the other hand, warming increases the possibility of major expansion of deserts elsewhere.

Clearly, these projections are not predictions. The tentative conclusions regarding the potential impacts of the burning of fossil fuels are based upon limited knowledge. In its report the NAS committee emphasized that there are many unknowns in this complex chain. The uncertainties include the possible mitigating effects on global temperatures of dust and moisture in the atmosphere. And there is the problem, referred to above, of simply determining where the CO_2 is coming from and where it is going. In addition, in the absence of a true understanding of the many factors controlling climate change, it cannot be ruled out that any effect due to CO_2 might be overwhelmed by other driving forces such as those associated with solar insolation or astronomical motions—a possibility denied by others who believe that the natural fluctuations are on a longer time scale and are not likely to mitigate the more immediate heating associated with a buildup of CO_2 in the atmosphere. The problem is potentially serious, and the uncertainties are huge.

Policy Implications of Carbon Dioxide

The obvious and safest policy implication of this discussion is that more research needs to be done. It is essential that comprehensive inquiries be carried out to reduce the uncertainties in present knowledge and hence to enhance the ability to determine the future consequences of action. Such inquiries require the international cooperation and participation that they are receiving. Of course, new knowledge will not necessarily demonstrate that present estimates of potential warming and its effects have been too alarmist—

it is also possible that they have understated the problem. But, whatever the reality may be, the sooner it is understood the better.

An equally important, but less obvious, implication is that the CO_2 problem does not provide a warrant for turning away from the use of fossil fuels at this time. Scientists generally agree that whatever the precise relationships and ultimate limits may be, the effects of CO_2 will be small and gradual over the next several decades and appear to be reversible if they do not go too far. Projected increases in worldwide use of coal over the rest of this century are not likely to take the world to a point where it is physically or economically impossible to control atmospheric CO_2 buildup.

For this reason, and since coal is plentiful and economically advantageous in many applications, its use should not now be discouraged because of concern about atmosphere CO_2 buildup. The possibility of adverse environmental and economic consequences in the future is no different in principle for fossil fuels than for other energy forms. As the NAS committee said in the foreword to its report:

> The consequences of using fossil fuels as a principal source of energy over the next few centuries comprise part of a family of assessments that should be made to consider the environmental impact of the most attractive alternative sources of energy. Potential difficulties per se in the use of fossil fuels as a principal source of energy should not be used as an argument for turning to any specific alternative source. Policy decisions with respect to the most desirable source of energy should be based on satisfactory information on the long-term environmental impact of each source and the impact of several combinations of energy sources.
>
> The results of the present [NAS] study should lead neither to panic nor to complacency. They should, however, engender a lively sense of urgency in getting on with the work of illuminating the issues that have been identified and resolving the scientific uncertainties that remain. Because the time horizon for both consequences and action extends well beyond usual boundaries, it is timely that attention be directed to research needs and to the anticipation of possible societal decision making now. The principal conclusion of this study is that the primary limiting factor on energy production from fossil fuels over the next few centuries may turn out to be the climatic effects of the release of carbon dioxide.[4]

We concur with this judgment. The potential danger and the uncertainty must both be recognized in making policy decisions within the next decade.

4. National Academy of Sciences, pp. vii–viii.

A third important implication of the CO_2 problem is that the world should not count on being able to use all the fossil fuel resources that are physically and economically available. There is a high probability that increasing global fossil fuel use, say, 5 percent per year for a century or more—which is quite feasible, given the resource base—will have clearly unacceptable climate consequences eventually. While there are conceivable technological ways to burn fossil fuels without releasing CO_2 (e.g., capturing CO_2 in stack gases and injecting it into the deep ocean or into oil and natural gas reservoirs), these are now hypothetical, and their use would surely raise the cost of using fossil fuels. Thus, the world must develop and maintain nonfossil energy options for the long run of a century or so and must soon begin thinking about the international political and economic issues involved in trying to control CO_2 release on a global basis.

We conclude, then, that despite the importance of the potential CO_2 problem, coal can be an important element in reducing reliance on oil and gas in the United States and the world over the next twenty years. The world can well afford to begin expanding the use of coal to the levels generally projected for the year 2000 without severely affecting the ability to reverse the expansion in coal use or the buildup of atmospheric CO_2 should either of these become necessary. Most observers expect that the uncertainties surrounding the atmospheric accumulation of CO_2 and the average rate of warming likely to be associated with the CO_2 buildup (but probably not the details of the resulting climate changes) will be reduced significantly over the next several decades. Thus, any decision about limiting fossil fuel use for CO_2 reasons can safely be put off—but only if the intervening years are spent improving scientific knowledge about CO_2 and developing options and contingency plans for dealing with energy and climate problems on a cooperative international basis when and if it becomes necessary to do so.

COAL-RELATED AIR POLLUTANTS IN THE ENVIRONMENT

A variety of pollutants is emitted to the air when coal and other fossil fuels are burned or processed. These include compounds of sulfur and nitrogen; particulates; trace elements such as cadmium, arsenic, lead, and nickel; and a variety of hydrocarbons. The quantities emitted, their chemistry and concentrations, and the effects on the environment and human health depend on the composition of the fuel, the conditions under which it is burned, the control technology utilized, and such factors as temperature, moisture, sunlight, other pollutants, and wind. The interrelationships of these factors are exceedingly complex, making it very difficult to identify precisely the effects of any particular pollutant emission; this fact suggests some important things about the appropriate form of air pollution control policy.

In this section we discuss several of the most important coal-related air pollutants. The emphasis is on the combination of sulfur and particulates, which illustrates the complexity of the problem, both in the behavior of the pollutants themselves and in the variety and uncertainty of effects on the environment. The next section discusses the problems of determining the effects of these pollutants on human health. The discussions are illustrative rather than exhaustive; our intent is to provide some insight into both the quality of information and the uncertainty involved in making decisions about environmental management, not to provide detailed information. Some implications of these scientific realities for the formulation of policy are set forth in a concluding section of this chapter preparatory to the analysis of air pollution policy in Chapter 11.

Sulfur, Particulates, and Sulfates

Most fossil fuels contain sulfur. High-sulfur ("sour") natural gas is routinely desulfurized at the wellhead, primarily to prevent corrosion of pipeline equipment. Some high-sulfur crude oils and coals are partially desulfurized before burning, and oil and coal naturally low in sulfur are used where available and economic. Flue gas desulfurization (FGD) equipment or advanced combustion technologies (such as fluidized bed combustion or low Btu coal gasification) can control most of the sulfur emission from combustion at a cost. But greater use of fossil fuel, and especially coal, may significantly increase the amount of sulfur discharged to the atmosphere.

During combustion, 95 to 99 percent of the sulfur in a fossil fuel is converted to sulfur dioxide (SO_2) and emitted to the atmosphere unless controlled. Because sulfur has long been considered damaging to human health, vegetation, and materials, emissions of SO_2 have been regulated for several decades. The U.S. Clean Air Act strategy for controlling sulfur pollution is based primarily on accomplishing precise "ambient air quality standards" that limit the concentration of SO_2 in the "ambient" or surrounding air. However, recent studies indicate convincingly that the sulfur air pollution problem involves sulfates and other particulates in addition to SO_2 and that simply trying to limit ambient SO_2 concentrations may be an inappropriate approach.

The Importance of Sulfates. Sulfur dioxide is emitted from exhaust stacks along with particulates, such as acid mist, smoke, ash and dust, and other substances. After being emitted, some of the SO_2 is oxidized to sulfuric acid by complex reactions involving moisture, oxygen, and metallic ions in particulates from the same or other pollution sources. Some of the sulfuric acid in turn reacts with metallic ions or other substances in the air to form sulfate compounds, which remain in the air as small (submicron) particulates. There is growing evidence that it is the aerosol or "mist" containing sulfuric acid and metallic sulfates, and not the SO_2 gas itself, that is the major culprit in sulfur pollution. It impairs visibility, damages vegetation and materials, and harms human health. Thus, the sulfur-particulate-sulfate complex must be treated as a whole in analyzing both its development and its effects.

Although there is clearly some relationship between sulfur dioxide emissions and atmospheric sulfate levels, it is a poorly understood one. There are many sulfates in the atmosphere, ranging from sodium sulfate derived from evaporation of seawater (which accounts for about 23 percent of atmospheric sulfur) to a variety of others such as calcium, manganese, iron sulfates, and ammonium sulfate derived from ammonia in the atmosphere. It is not known how these various sulfates are formed, which are the most dangerous, or even whether some of them may be less dangerous than the atmospheric sulfuric acid from which they are (presumably) formed. Indeed, it is not even clear that reducing SO_2 emissions or ambient SO_2 concentrations in a region significantly decreases ambient sulfate concentrations in the region.

This last anomaly is suggested by several facts. Hidy et al. report that urban sulfate concentrations are correlated with local ozone, moisture, and aerosol concentrations, but are not significantly correlated with SO_2 concentrations.[5] They also note: "During the 1960s SO_2 emissions and ambient concentrations decreased in cities, while urban and nonurban sulfate levels remained nearly constant over several years, although rural SO_2 emissions increased in the eastern United States." In a belt extending roughly southwest to northeast from Illinois to Albany, New York, the sulfate levels reach a maximum in the sunny summertime; in places as different as New York and Los Angeles, sulfate levels are highest in the summer, when SO_2

5. G.M. Hidy, P.K. Mueller, and E.Y. Tong, "Spatial and Temporal Distributions of Airbourne Sulfate in Parts of the United States," in R.B. Husar, J.P. Lodge, Jr., and D.J. Moore, eds., *Sulfur in the Atmosphere*, Proceedings of the International Symposium held in Dubrovnik, Yugoslavia, 7–14 September 1977 (New York: Pergamon, 1978), pp. 735–52.

levels are lowest. All this suggests a photochemical origin for sulfates, with some factor other than SO_2 (e.g., ultraviolet light) being the limiting factor in sulfate creation. Thus, controlling ambient SO_2 concentrations may be an ineffective way to get at the real problem— sulfate levels.

Long Distance Transport and Acid Rains. A series of studies in Sweden in the 1960s demonstrated conclusively that numerous lakes and streams in relatively pristine areas of the country had become highly acid. In many, fish populations had been destroyed or severely damaged.[6] Sweden's case study to the United Nations Conference on the Human Environment held in Stockholm in 1971—"Air Pollution Across National Boundaries: The Impact on the Environment of Sulfur in Air and Precipitation"—estimated that in some parts of Scandinavia the acidity of rainfall had increased as much as 200-fold since 1956. These findings dramatized both the intensity and the large regional scale of atmospheric sulfur problems in the atmosphere, and they led to remarkable cooperative international studies by the OECD. These studies have documented sources of sulfur emissions, transport in the atmosphere, chemical transformations of sulfur occurring during transport, the distribution of sulfate in the atmosphere, deposition of sulfur in various forms throughout the region, and injury to vegetation and to fish in streams and lakes. They clearly confirm that a large part of the acidity of the rainfall is related to sulfates in the atmosphere produced by chemical transformations of sulfur dioxide emitted hundreds of miles away.

High acidity has also been observed in some Canadian lakes downwind from the high sulfur emissions of smelters in the large Sudbury mining district in Ontario. Because concentrated sulfur emissions have long been known to damage vegetation and waters severely in the immediate vicinity of smelters, these observations do not necessarily indicate longer distances of transport. However, comparison of the acidity (inversely related to pH) of a lake in the Adirondacks, far from any sulfur emission sources, reveal that the pH dropped from a range of 6.6-7.2 in 1938 to a range of 3.9-5.8 in 1960.[7] By 1970, observations at a number of sites in the Finger Lakes region of New York showed that the pH was about

6. S. Odén, "The Acidification of Air and Precipitation and its Consequences on the Natural Environment," (in Swedish), *Ecol. Comm. Bull.* Vol. 1, (1968) trans. by Transla. Consultants, No. TR-1172.

7. G.E. Likens, *The Chemistry of Precipitation in the Central Finger Lakes Region* (Ithaca, N.Y.: Cornell University Water Resources and Marine Sciences Center, Technical Report No. 50, 1972), p. 18.

3.9 on average, with even lower values in the summer. Since about half the increased acidity was due to sulfates (the other half being nitric and hydrochloric acid) and since there were no significant local sources of sulfur, Likens suggested that sulfates were probably transported from quite distant areas, such as the Ohio valley.[8] The presence of nitrates implicated gasoline and natural gas as contributors to acid rain. But determination of the relative contribution of nitrates and sulfates, and of their actual sources, remains to be accomplished, particularly in areas where acid levels are not high.

The existence of high sulfate levels and of acid rain far from known sources has also been confirmed by a variety of regional and global measurements in Scandinavia, New England, and elsewhere. In some instances specific air masses containing high SO_2 and sulfate concentrations have been followed for days as they move hundreds of miles. Over the open Atlantic Ocean in the northern midlatitudes, the ratio of SO_2 to sulfates (mostly natural sodium sulfate from sea salt) has been increasing and is now strikingly higher than the normal background ratio over the oceans. Analyses of yearly snow layers from inland Greenland ice indicate that the sulfate content has increased about threefold, from 0.06 μg/liter before the nineteenth century to 0.2 μg/liter in 1977.[9] Thus, the evidence is persuasive that sulfur dioxide and sulfates are increasing in the atmosphere and travel over distances of many hundreds of miles, where they may impair visibility, adversely affect ecological systems or materials, or damage human health.

American and European observers have also noted that while mean concentrations of sulfur dioxide, sulfate, or acid levels in the air expressed as annual or monthly averages provide useful synoptic pictures, not infrequently the movement of air masses and particular weather conditions produce abrupt increases in sulfates and in the acidity of rain, unrelated to local SO_2 levels. Such changes in meteorological factors may give rise to specious correlations or may mask real correlations between long-term trends in sulfur emissions and sulfate levels over a broad region.[10] And they certainly call into question management methods based on predicting and/or preventing any specific local ambient concentrations.

8. Ibid., p. 14.

9. B. Ottar, "An Assessment of the OECD Study on Long-Range Transport of Air Pollutants," in Husar, Lodge, and Moore, p. 453, citing M. Koide and E.D. Goldberg, "Atmospheric Sulfur and Fossil Fuel Combustion," *Journal of Geophysical Research* 76 (1971): 6589-95.

10. L. Granat, "Sulfate in Precipitation As Observed by the European Atmospheric Chemical Network," in Husar, Lodge and Moore, pp. 418-19.

Implications for Sulfur Control Strategies. To say that the processes and effects involved in sulfur air pollution are complex and poorly understood is not the same as saying that too little is known to make rational policy. It is, after all, important for policy purposes to know that the problem is not simple. And, in fact, quite a bit is known. For example, it is known that there is no simple, quantitative relationship between emissions of sulfur dioxide at one point and ambient concentrations of SO_2 or sulfates or both at another point. It is known that reducing SO_2 emissions in a region may not lead to lower ambient sulfate concentrations in the region, either because sulfates are being transported into the region from elsewhere or because some factor other than SO_2 (e.g., ultraviolet radiation) is limiting sulfate production. It is known that SO_2 emissions at one point may cause damage hundreds of miles away, even though it may be impossible to predict just where or when.

Knowledge of this type has important policy implications beyond the inevitable, but still valid and important, call for more research. For example, the existing information has made it more difficult, if not impossible, to argue convincingly that the impact of atmospheric sulfur could be reduced or eliminated simply by building tall stacks. While tall stacks might moderate local problems by dispersing emissions to higher elevations in the atmosphere, significant quantities of sulfates are transported over very large distances, perhaps exacerbating problems elsewhere. It may be possible to identify certain situations where prevailing winds carry the sulfur out to sea, into less populated areas, or into areas where (as in the western United States) lack of atmospheric humidity results in less sulfate formation; but there is little basis for a general presumption that dispersing the sulfur dioxide results in less overall damage.

More generally, the complexity and poor understanding of the sulfur-particulate-sulfate complex suggests that it is not advisable at this time (and probably for a long time to come) to use sulfur pollution control strategies that require knowledge of any precise relationship between particular SO_2 emissions and ambient SO_2 or sulfate levels at some specific point. There are too many variable and unknown factors in any such relationship for it to be an effective management tool—even if there were some logical way to determine the "best" level of or increment in ambient concentrations. As discussed in the next chapter, it is our view that current U.S. pollution policy puts too much emphasis on quantitative ambient standards and that there are preferable alternative policy approaches.

Nitrogen Oxides, Oxidants, and Hydrocarbons

When fossil fuels are burned, nitrogen compounds in the fuel and nitrogen gas in the air are oxidized by the high flame temperatures—usually to nitric oxide. Nitric oxide is further oxidized, within the combustion process or photochemically in the atmosphere, to a variety of oxides, of which nitrogen dioxide (NO_2) is the compound of greatest concern for environmental and human health reasons.

Natural sources of nitrogen dioxide, from bacterial processes, fires, and volcanic activity, are estimated to be ten- to fifteenfold greater than those from anthropogenic sources on a global basis, but community levels of NO_2 pollution are primarily due to man-made sources. Emissions from stationary heating or power-generating equipment and operation of motor vehicles are typically the primary local sources, although some communities also have specific sources, such as nitric acid and explosive manufacturing facilities. In addition, there are important indoor sources such as tobacco smoke and un-vented gas cooking or heating appliances. Increased electrical power production by fossil fuels could substantially increase nitrogen oxides, compared with minor changes in sulfur oxides and suspended particulates, because of the slow development of emission control technology for NO_2 compared to SO_2 and particulates.

Health Effects of Nitrogen Oxides. More so than sulfur dioxide, nitrogen dioxide is known to be toxic at relatively high concentrations. Acute and chronic injury of the lungs has been observed, for example, in South African gold mine workers who entered areas too soon after detonation of trinitrotoluene (TNT), which releases nitrogen oxides. A few farmers were killed or severely poisoned in the U.S. Midwest in the 1950s when they entered silos in which ensilage was fermenting;[11] it had not previously been recognized that nitrogen dioxide could be formed in large quantities under these conditions of anaerobic bacterial action. The concentrations of nitrogen dioxide in these industrial exposures are extremely high: underground blasting in mines has resulted in concentrations as high as 167 ppm, and concentrations of NO_2 in freshly filled silos have been found as high as 4,000 ppm. These values are not relevent to air pollution effects since ambient concentrations in air are less than 1 ppm; but they can be compared to values in tobacco smoke between 80 and 120 ppm.

11. T. Lowry and L.M. Schuman, "'Silo-Fillers' Disease:' A Syndrome Caused by Nitrogen Dioxide," *Journal of the American Medical Association* 162, no. 3 (1956): 153-60.

Very few studies have been conducted on the toxicity of urban levels of nitrogen dioxide, and as in most epidemiologic studies of air pollution (see below), the results are not conclusive. The possible effects of nitrogen oxides cannot be definitely separated from the possible effects of other pollutants present at the same time, since the areas with the highest NO_2 had relatively high values for sulfates and suspended particulates as well. However, toxicological experiments involving mice and human asthmatics (see below) indicate that NO_2 concentrations near those experienced in the ambient air can have adverse biological effects.

A prospective field study currently being conducted includes nitrogen dioxide as one of several air pollution variables and has recently yielded findings that strongly emphasize the importance of indoor air pollution in any study of community air pollution health effects. Whereas some pollutants have indoor concentrations lower than outdoor ambient levels, nitrogen dioxide is likely to be more highly concentrated indoors since it is commonly generated by tobacco smoke and by use of unvented gas cooking stoves or space heaters. The current findings indicate that the NO_2 level is three to seven times higher in homes with gas (compared to electric) cooking stoves and that children in such homes have more frequent illnesses of the lower respiratory tract and lower scores on pulmonary function tests.[12]

These results are similar to those found by Melia and her associates earlier in Britain.[13] The followup of these children will be very important both in confirming the findings and in determining possible implications to future growth, development, and health of their lungs. The study confirms the need to pay greater attention to what people are actually exposed to (in this case, indoor pollution), instead of relying on standard monitoring data. It also raises questions about the cost effectiveness of trying to improve public health by cleaning up the community air, while simultaneously reducing ventilation in gas-fueled homes as an energy conservation measure.

Nitrogen Oxide Reactions in the Atmosphere. Nitrogen oxides are involved in many complex chemical reactions, particularly in the

12. B.G. Ferris, Jr., F.E. Speizer, Y.M.M. Bishop, and J.D. Spengler, "Effects of Indoor Environments on Pulmonary Function of Children 6–9 Years Old," *American Review of Respiratory Diseases* 119, no. 4 (1979): 214 (abstract).

13. R.J.W. Melia, C. duV. Florey, D.S. Altman, and A.V. Swan, "Association Between Gas Cooking and Respiratory Disease in Children," *British Medical Journal* 2 (1977): 149–52.

presence of sunlight. The reactions are much influenced by the presence or absence of hydrocarbons, linking the chemistry of nitrogen oxides closely to that of ozone and other photochemical oxidants. Although the processes are even less well understood than those involving sulfur compounds and metallic particulates, there is growing evidence that nitrogen oxides and hydrocarbons interact over periods of days and distances of hundreds of miles to produce high ambient levels of oxidants and nitrates and acid rains far from emission sources.

In the atmosphere, much of the nitrogen oxide may be converted to the very toxic intermediate, peroxyacetyl nitrate (PAN); and the most likely final product is nitric acid, a powerful oxidizing agent and strong acid. The nitric acid presumably reacts with other contaminants of the air to form nitrates. Although nitrates in the atmosphere have been relatively little studied, one report from the Los Angeles basin showed ammonium nitrate to comprise 10 to 15 percent of the total suspended particulates.[14] Nitric acid has been identified spectroscopically in the Los Angeles air, along with the related nitrogen compounds formaldehyde and formic acid. Nitric acid and its salts are therefore of concern with respect to health effects of inhaled particulates and as contributors to acid rain.

For these reasons, the nitrogen oxide air pollution problem is about as complex as the sulfur dioxide problem. Nitrogen dioxide ambient levels may or may not be a good indicator of the health and other effects of PAN, nitric acid, oxidants, and nitrates; and reducing nitrogen oxide emissions will have uncertain effects on the ambient concentrations of NO_2 and related compounds. Ambient levels of any of these substances will be influenced by emissions hundreds of miles away. There is no simple way to say that emissions at a certain point will or will not cause any particular effects at any particular point.

THE RELATIONSHIP BETWEEN AIR POLLUTION AND HUMAN HEALTH

The idea that products of coal combustion are harmful to health was well established centuries ago. People spending a lifetime in cities were expected to have an increased risk of lung infections and to have black lungs at autopsy. Nevertheless, there were few attempts

14. R.J. Gordon and R.J. Bryan, "Ammonium Nitrate in Airborne Particles in Los Angeles," *Environmental Science and Technology* 7, no. 7 (1973): 645-57.

to measure health effects of air pollution before the Donora, Pennsylvania, air pollution episode in 1948 and the much more severe episode in London in 1952.

In the three decades after the Donora episode, substantial research was conducted and much has been learned about the health effects of burning coal. Our review of that research will be selective, since numerous comprehensive reviews of air pollution health effects have been published in the past five years.[15] Our intent is to indicate the nature of the information and to demonstrate that much of what has been learned suggests the need for a reassessment of present U.S. air pollution policies along the lines set forth in the next chapter.

Protecting Human Health: General Concepts

Current air pollution decisionmaking, legislated by the Congress in the Clean Air Act (CAA) and implemented by the U.S. Environmental Protection Agency, is based on the concept that there is a threshold concentration level for some air pollutants below which there are no adverse effects. Furthermore, the CAA assumes that ambient air quality standards can be set that, if met, will protect the public health completely, with an appropriate "margin of safety" for even the most susceptible groups in the population. For the "criteria" air pollutants—that is, the pollutants for which the government has set up air quality standards (hydrocarbons, carbon monoxide, photochemical oxidants, sulfur oxides, nitrogen oxides, and particulate matter)—regulated in this way, benefit-cost considerations in attaining the standard are excluded by law; however, in practice it is difficult to conceive of their being ignored. For other pollutants the law requires a certain weighing of risks and benefits. Each of the concepts involved in these procedures is far from clear-cut, however.

Threshold.   The concept of a threshold is well established in classic experiments in physiology or pharmacology: a muscle does not contract until an applied stimulus reaches a certain level of intensity;

---

15. See John McK. Ellison and Robert E. Waller, "A Review of Sulphur Oxides and Particulate Matter As Air Pollutants With Particular Reference to Effects on Health in the United Kingdom," *Environmental Research* 16 (1978): 302-25; "Air Quality and Automobile Emission Control," A Report by the Coordinating Committee on Air Quality Studies, National Academy of Sciences, National Academy of Engineering, Prepared for the Committee on Public Works, U.S. Senate, *Health Effects of Air Pollutants*, vol. 2 (September 1974), Serial No. 93-24; and Benjamin G. Ferris, Jr., "Health Effects of Exposure to Low Levels of Regulated Air Pollutants: A Critical Review," *Air Pollution Control Association Journal* 28 (1978): 482-97.

for a given chemical, a specific physiological response diminishes or becomes unmeasurable when the stimulus (dose or exposure level) reaches some low level. But this threshold concept is of dubious applicability to the health of whole organisms and particularly to populations exposed to air pollution.

As an example, consider the health implications of carbon monoxide (CO). A dose-response curve can be obtained for a specific function of the body (vision, heart function, athletic performance, and so on) for a specific individual. Such curves seldom show sharp discontinuities or thresholds within ranges of exposure common in air pollution. And when one considers a wide variety of effects on a population of individuals at all ages, containing smokers and nonsmokers, and the full range of infirmities to which flesh is heir, the concept of a threshold becomes even more suspect. In any population there are individuals who are severely, even fatally, affected by any amount of carbon monoxide in the air that raises their body carboxyhemoglobin levels above the natural background levels; and low concentrations for long times, or in combination with other stresses, will have different effects on different people. As each pollutant is examined in this way, it becomes doubtful that any pollutant has a meaningful threshold in terms of its overall effects on a population. The implication is that the present clean air legislation comes close to requiring the control of air pollutants down to a natural background level.

There are few environmental stresses for which the low end of the dose-response curve has been examined quantitatively, for the practical reasons that very large numbers must be studied to verify small effects statistically, biological effects may be long delayed, and multiple factors may be involved in any given response. Not unexpectedly, the lack of precise information about the low end of the curve is a source of controversy, as witnessed by recurring concern about the effects of low levels of ionizing radiation exposure and extensive debate about whether there is a threshold for chemical carcinogens.[16] Many scientists, while recognizing the need for research to improve understanding, take the pragmatic view that a linear, no threshold model is a prudent one to adopt for predicting the biologic effects of ionizing radiation and of carcinogenic environmental factors. The linear no threshold model has not been as thoroughly studied for noncarcinogenic substances, but is just as valid theoretically as for carcinogens and mutagens.

16. Research News, "Chemical Carcinogens: How Dangerous Are Low Doses?" *Science* 202, no. 4,363 (October 6, 1978): 37.

As a logical and practical matter, then, it is doubtful that a meaningful public health threshold exists for any pollutants: fewer (more) people exposed to lower (higher) concentrations for shorter (longer) times will, on average, lead to fewer (more) and less (more) severe health effects in the population. Forcing scientists or administrators to pick a number that purports to be a threshold of some kind may or may not (we think not) be a useful administrative procedure for setting a policy goal; but it is certainly not a scientific procedure and should not be disguised as such.

Nevertheless, environmental decisionmakers often choose or are forced by law to incorporate the concept of "threshold concentrations" into the process for setting standards. The "threshold concentration" is typically chosen operationally somewhere between a concentration at which an effect on health or function has been looked for but not observed and a concentration at which such an effect has been demonstrated. Information of this type is usually available from laboratory studies, and it may be available for controlled exposures of healthy human subjects, but it is inherently unavailable for the most susceptible groups—the very old, the very young, and the very ill. And because only certain effects on certain groups are considered, the chosen level will always be somewhat arbitrary. It is almost invariably true that by looking harder for different effects on different groups, evidence can be found that the truly "safe" threshold must be lower.

The "Margin of Safety." A "margin of safety" is, in theory, a compensatory device to protect the public health in the presence of scientific uncertainties. In practice, however, it also permits the exercise of value judgments about the seriousness of the threat, the number of people affected, the potential benefits and costs involved in a regulation, and so on. Thus, it often becomes something of a safety valve or "fudge factor," allowing costs and benefits to be considered even when they are not supposed to be.

Considering only the biological effects component of the margin of safety, many kinds of uncertainty are included—the quantity and quality of the information; gaps in information that inherently cannot be filled, such as effects on most sensitive groups, the range of individual variabilities, and uncertainty about what kind of exposure is important (occasional peaks versus long-lasting low concentrations, intermittent versus continuous exposure and so on); extrapolation of laboratory findings to possible effects in human populations; the possible synergism of multiple environmental stresses; and the possible findings of future research. Given all these scientific factors, in addition to the surmised cost-benefit issues that often enter implicitly, it is not surprising that margins of safety vary widely for different chemicals and that judgments about "safe" levels are often very important and very controversial decisions.

Using judgmental margins of safety to incorporate many nonscientific issues in the setting of a threshold standard is better than ignoring these other issues altogether. The problem is that doing so can disguise important social value judgments as scientific determinations. Although biological and health scientists have as much right, obligation, and qualification as anyone else to take an active part in these social value judgments, they have no special rights, obligations, or qualifications in this regard. Thus, in our view, greater care should be taken to separate scientific from non-scientific issues in setting standards, so that social issues can be weighed openly in the social processes best suited for that purpose, not treated as purely scientific matters.

Estimation of Risks.   It is not easy to estimate the risks to public health or welfare from an activity when scientific knowledge is poor. And it is not easy to weigh such risks, even when they are understood reasonably well, against each other or against other factors such as economic welfare. Thus, the search for scientific definitions of safe levels or regulations is understandable. If they can be found and accomplished at reasonable cost, many decisions can be made less difficult.

The problem is that in a modern world, where there are many new and old dangers and increasingly sensitive ways of identifying these, absolute safety cannot be assured. So risks must be estimated, compared, and evaluated relative to benefits, either openly or not. Even the Delaney amendment, which requires an absolute ban on the use in food of any chemical shown to be carcinogenic in any animal species at any dosage level, has to be reconsidered when it affects a valuable and popular chemical such as saccharin. So risks must often be estimated, evaluated, and compared; they cannot always be eliminated.

In the Clean Air Act, the setting of the safe threshold standards with an adequate margin of safety has involved scientific judgments about the effects on the most susceptible groups of the minimal concentrations of the pollutant found to have a biological effect. However, as the cost of meeting the initially chosen levels has become apparent, EPA has kept an eye on the implications of using one population group or health effect rather than another to define "safe." For example, evidence on sulfates that could have been used to set unattainable sulfur dioxide emission limitations has long been down-played, and a controversial rereading of the scientific data allowed EPA in 1978 to relax the oxidant (ozone) standard to a somewhat less impossible level. EPA stated that the adverse health effect threshold concentration for oxidants is actually unknown and the EPA used its best judgment about a "probable effects level";[17] one wonders how much further the standard could be relaxed without undue concern for increased illness or deaths (ozone is *not* the cause of the eye irritation so troublesome in the Southern California air basin).

For substances designated by the Clean Air Act as hazardous air pollutants, the procedure for incorporating scientific risk estimates and trade-offs into the control program is more explicit. The first step is a thorough review of the known biological effects; the second is an attempt to define exposure levels for various segments of the population. The third step is the estimation of the total health effects (mortality or morbidity) to be expected at different levels of postulated exposure. The resulting information, reflecting the best scientific knowledge but not attempting to pick a fictitious "safe" level or to make social value judgments, is then used by the administrative agency in developing a control strategy.

Procedures of this type need to be developed and applied more widely, not just to hazardous or carcinogenic substances, but to other pollutants, including those that are now regulated on the basis of threshold concepts. Some changes in philosophy and in legislation are required before this can be done; the next chapter presents some of our suggestions along these lines.

Health Effects of Air Pollution in Perspective

Policymakers would like to have the best evidence possible on which to base decisions, though they know decisions often have to be made when evidence is incomplete or faulty. For example, one of the most important and least well-known pieces of information is the number of deaths that would be prevented (or the number of lives prolonged) if air quality were improved by specified amounts. This number may be as high as estimated in one recent publication by Lave and Seskin, who predict that the U.S. total mortality rate (deaths per year per thousand people) would be reduced by 7 percent if sulfur dioxide were reduced by 88 percent and particulates were reduced by 58 percent.[18] Assuming that the number of deaths

17. Environmental Protection Agency, "Threshold Concept," *Federal Register* 44 (February 8, 1979): 8203.
18. Lester Lave and Eugene P. Seskin, *Air Pollution and Human Health* (Baltimore: Johns Hopkins University Press for Resources for the Future, 1977).

per year is approximately 1 percent of the population, a change of 7 percent in the mortality rate would mean approximately 140,000 fewer deaths per year at current levels of the mortality rate and population. Or the number may be, as estimated by others, only a few percent of this Lave and Seskin estimate. But it is important that the range of possibilities and the many uncertainties be put in perspective.

Estimates of deaths postponed or years of life added have been made by other investigators and for factors in addition to air pollution. It should be of concern to public policy to look at the health effects of air pollution compared with other factors, as well as to consider the range and uncertainty of the various predictions. This can be done for total mortality and for specific causes of death, such as cancer. Cancer is a useful surrogate for environmental effects because of two characteristics: it has attracted a great deal of scientific study, and most important, the vital statistics for cancer are more accurate and reliable than for most other diseases.

Figure 10-1 presents the data on the number of cancer deaths per year in the United States graphically, with the increase since 1900 adjusted for the larger size and older average age of the population. (Cancer is expected to be more prevalent in older age groups.) When these demographic adjustments are made, there is a growing residual amount of cancer, presumably due to "environmental" factors. Higginson has observed that perhaps 80 to 90 percent of cancer is caused by (or significantly influenced by) "environmental" factors.[19] Such data and observations have led many people to infer that environmental pollutants in food, water, air, or places of work have caused an epidemic of cancer in this country.

What these graphs and data do not show is that "environmental" factors include much more than pollution. For example, Higginson quite properly included personal behavior or "lifestyle" among the important environmental variables; accordingly, he included tobacco smoking, alcohol ingestion, and diet as important "environmental" causes of cancer. If lung cancer is subtracted from total cancer deaths in Figure 10-1, there is no residual increase in death rates after adjusting for demographic factors; it is known that the great increase in lung cancer in this century is associated with the rise of tobacco smoking and not significantly related to other environmental factors. Thus the data, correctly interpreted, do not suggest that pollution in the sense that we are considering it here is a major factor in the increasing cancer death rate.

19. J. Higginson, "Present Trends in Cancer Epidemiology," *Candian Cancer Conference* 8 (1969): 40-75.

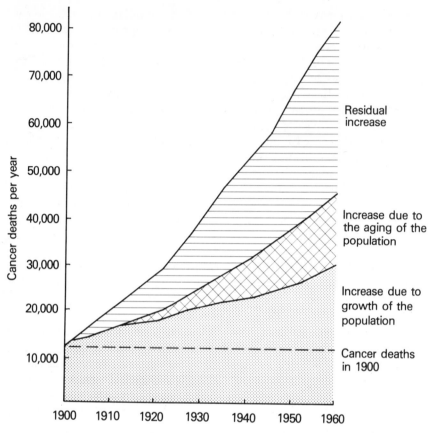

Figure 10-1. Number of Deaths from Cancer. U.S. death registration area of 1900, 1909–1960. U.S. Public Health Service, *Cancer Rates and Risks,* Second Edition, 1974 Washington D.C.

A report published by the National Science Foundation in 1973 attempted to link cancer and other causes of death to many of the suspected environmental factors.[20] Table 10-2 summarizes some of the results. The estimate of deaths linked to air pollution was an approximation based on several considerations. Deaths due to chronic nontuberculous lung disease (e.g., emphysema) have been rising for decades, particularly among males; however, in the 1960s they accounted for only about 1 percent of deaths in males and 0.5 per-

20. National Science Foundation, Science and Technology Policy Office, Report of the Panel on Climate and Health of the President's Science Advisory Committee, *Chemicals and Health* (Washington, D.C.: Government Printing Office, September 1973).

Table 10-2. Numbers and Percentages of Deaths in 1967 Linked to Various Chemical Factors (100 percent = total deaths from all causes = 1,850,000).

Factor	Numbers of Deaths Linked to Factor in 1967	Percentage of all Deaths in 1967 Linked to Factor
Cigarette Smoking	300,000	17
Alcohol Abuse[a]	56,000	3
Adverse Reactions to Medications in Hospitals	75,000	4
Narcotic and Addictive Drugs	10,000	0.6
Community Air Pollution	9,000	0.5
Airborne Particles (occupational)	9,000	0.5

Source: National Science Foundation, Science and Technology Policy Office, Report of the Panel on Climate and Health of the President's Science Advisory Committee, *Chemicals and Health* (Washington, D.C.: Government Printing Office, 1973), p. 35.

[a]Most deaths linked to alcohol are secondary effects of alcohol abuse—largely fatalities from highway accidents in which alcohol was a primary factor.

cent in females (or about 10,000 annual for males and 5,000 for females). Air pollution could have been a factor, but all studies in which cigarette smoking has been included have indicated a much stronger effect for cigarette smoking, with respect to nontuberculous lung disease as well as lung cancer; the estimate of deaths linked to environmental factors has to reflect this. The same is true of occupational dust exposures compared with community air pollution, though to a lesser extent than cigarette smoking. The estimate of 9,000 deaths linked to air pollution also includes 5 percent of the deaths due to lung cancer, although no epidemiologic study in the United States has demonstrated an effect of air pollution on lung cancer.

According to this National Science Foundation study, in 1967 about 9,000 deaths, or 0.5 percent of all deaths, were "linked to" community air pollution, while about 300,000 deaths, or 17 percent of all deaths, were "linked to" cigarette smoking. This result suggests that a 3 percent reduction in cigarette smoking would reduce mortality as much as total elimination of air pollution. There are, of course, uncertainties and differences in interpretation in all such estimates, many of which are discussed below. The contribution of air pollution to public health effects may be, as suggested by Lave and Seskin, ten or more times higher than estimated in this National Science Foundation study, or it may be lower. Nonetheless, numbers and comparisons such as these should be kept in mind when making decisions about how best to use resources to protect public health, and efforts must be continued to improve these estimates.

Problems in Estimating Health Effects

Information about the actions of pollutant chemicals on biologic systems, including health effects, comes essentially from two sources—controlled laboratory experiments and observation of events in the uncontrolled environment. However, the utility of controlled experimentation is limited, and most of the relevant information comes from epidemiological studies—direct observations of acute air pollution episodes or statistical studies. All of these methods have their problems, with the result that uncertainty and disagreement are inevitable.

Controlled Experiments.   Most basic information on the biological effects of pollutant chemicals is derived from controlled toxicologic experiments in laboratory animals, tissue cultures, and other standardized biologic systems. Specific information is obtained about dose- or exposure-response relationships, uptake, retention, distribution in body tissues, metabolism, excretion, and mechanisms of action, with emphasis on adverse effects. Many kinds of toxicity are looked for, including organ system malfunction; neurobehavioral effects; carcinogenicity; effects on reproduction, growth, and development; genetic material; and so forth. Valuable as such studies are in advancing science, they are of limited use to decisionmakers, primarily because dose or exposure levels are usually too high to be relevant to the real world of air pollutant exposure; they also do not reflect the multiplicity and complexity of stresses to which people are exposed in daily life.

Even when effects are observed in a test system with concentrations and combinations of pollutants similar to those that occur in community atmospheres, the interpretation must take into account all the problems of extrapolation—from the test system, which might be a tissue culture, for example, to the response of the whole organism, with all its complicated ways of adjusting to stress; and from findings in one species to the response of another species, including the human being.

A special kind of toxicologic study is the controlled exposure of human beings to pollutants or combinations of pollutants. The subjects may be normal or may have specific impairments, conditions, or disease states that may be thought to influence susceptibility to the action of pollutants. Where ethical considerations permit such studies, the results are invaluable because they reduce some of the problems of extrapolation. However, increasing restrictions on human experimentation in recent years have greatly discouraged such study.

One pollutant for which controlled toxicologic experiments have provided more useful information than usual is nitrogen dioxide (NO_2). The principal toxicologic findings of interest are several studies suggesting increased sensitivity to infections (based on a mouse infectivity model for bacterial or viral infection) and one study of human asthmatic patients.[21] The animal infectivity model demonstrated an effect at 0.5 ppm (compared to a present ambient air quality standard of 0.05 ppm on an annual average), making it the most sensitive parameter observed to date; however, there are serious problems in extrapolating this finding to man, since the level of NO_2 used in the studies is not often exceeded in the real world and the dose of infectious agents was unrealistic in terms of everyday human exposures. The asthma study involved twenty asthmatics in whom the resistance to air flow in the lungs was shown to increase more in response to a certain drug after exposure to air containing 0.1 to 0.2 ppm of NO_2; this is the kind of result that, if confirmed, could have a decisive effect on the setting of an air quality standard. Whether society should devote large resources to reducing air pollution because of such laboratory results is another question.

Although they occasionally yield useful information, for purposes of establishing an information base from which "safe" levels of pollutant exposure can be determined, such animal and human toxicologic studies are typically seriously deficient, owing to the artificiality of experimental conditions. Agents are usually studied one at a time, rarely in combinations of two or three. The duration, fluctuations, and intermittency of exposure have little resemblance to conditions in the real world. Therefore, great dependence has been placed on another kind of study—the attempt to associate disease, impairment, or death in population groups with prior or current exposure to specific environmental variables, using the methods of epidemiology.

Epidemiological Studies. To say that the information base on health effects of air pollution must come primarily from epidemiological studies is to state a necessity, but it is not to imply that such

21. R. Ehrlich, J.C. Findley, J.D. Fenters, and D.E. Gardner, "Health Effects of Short-Term Exposures to Inhalation of NO_2-O_3 Mixtures," *Environmental Research* 14 (1977): 223-31; and J. Orehek, J.P. Massari, P. Gayrard, C. Grimand, and J. Choxpin, "Effect of Short-Term, Low-Level Nitrogen Dioxide Exposure on Bronchial Sensitivity of Asthmatic Patients," *Journal of Clinical Investigation* 57 (1976): 301-307.

information is easily obtained or readily interpretable. The limitations of epidemiologic methods are in fact severe. The difficulties include the identification and measurement of both the pollutant exposure and the health effects and the determination of a valid relationship between exposure and effects.

Estimating Exposure. The problems of estimating environmental exposure arise from poor data and from the complex nature of atmospheric processes discussed in the preceding section. In early British studies of the health effects of burning coal, exposure was inferred from the consumption of coal in a region. When air monitoring was developed in the 1950s, place of residence became an improved but still very crude index of exposure. As methods were developed to measure sulfur dioxide and fine particulates in air, it was found that these pollutants tended to vary together. Their concentrations are highly correlated with each other in any given setting, making it impossible to separate the effects of one from those of the other. Even today, the methods available and systems in operation for monitoring sulfates of various kinds, particulates of different sizes, and the composition of the individual components of the nitrogen oxide-hydrocarbons-oxidant complex are inadequate to provide reliable data on ambient levels of these pollutants individually.

The estimation of exposure is also complicated by the fact that pollution levels vary with time and place. Variations of concentration with time are determined mainly by the weather: when there is a temperature inversion and wind speeds are low, concentrations of most pollutants are high, regardless of their sources. People work, live, and play in different (and changing) locations within any region, and they move from indoors to outdoors and back again. It is not generally known whether repeated, short-term, high exposures are more or less harmful than constant, low level exposures, and no very good way exists to determine what percentage of a population experiences how much of which type of exposure. Many improvements in methods of measuring and modeling the dispersion of pollutants and the movements of people have been made, but even the best methods permit only inferences about what people are exposed to now or have been exposed to in the past several years.

Because of these difficulties, any measure of *the* pollution level in a region is at best a crude proxy for, or index of, the real levels of pollution. For example, the annual average SO_2 concentration, averaged over three or ten monitoring sites in the region, probably, in a statistical sense, moves more or less with the annual average levels of sulfates and particulates and with the peak daily levels to which people, on average, are exposed. But health or environmental effects can generally be related only to the average mixture and the average levels, not to individual components or to particular time patterns or exposure.

Of course, pollutant monitoring, even with all the problems, is essential and should be improved; without it, there is no way to improve epidemiological information or to know whether current programs are having any effect. We emphasize the difficulties in order to shed some light on what existing data do and do not reveal.

Estimating Effects. The definition and determination of adverse health effects of air pollution is at least as difficult as estimating exposures. Opinions and judgments have profound effects on decisions about the levels of pollutants considered "safe"—or where absolute safety is admittedly impossible, the levels to which regulatory efforts are aimed. Definitions of health are very imprecise, ranging from the negative definition—"absence of disease or disability"—to the broad and positive definition adopted by the World Health Organization (WHO), involving "total physical, emotional and social well-being." The WHO definition is not useful operationally; but it is implicitly accepted when the loss of visibility due to air pollution, odors, and other aesthetic aspects of our reaction to environments are used to define "safe" levels.

More objectively measurable reactions to pollutants range from direct irritation of membranes (usually eyes and respiratory tract) through reflex changes in the lungs, impairment of organ system function, impairment of physical or mental performance, increased likelihood of infection, overt disease, and premature death. But how many of these are "adverse health effects"? For example, the eye irritation so annoying to residents of the Southern California air basin is not known to have residual effects, and it has not been demonstrated that Los Angeles smog increases death rates. Are transient functional changes in response to a pollutant exposure, with complete recovery when the stimulus is withdrawn, to be defined "health effects"? There is no consensus of answers to these questions in the context of "protecting the public health."

Two measures of health effects have proven particularly valuable, both in demonstrating the health effects of coal combustion products and in demonstrating improvement in health as air quality is upgraded. Both indexes are particularly relevant to air pollution effects and control, because they depend for their sensitivity on the most susceptible groups in a population, the groups that have their protection specifically mandated by the Clean Air Act. These methods are the use of daily mortality records and the use of a daily diary by a selected group of individuals.

The daily mortality methods have been extensively used in Britain and to a lesser extent in the United States. These methods use data on daily death rates in different groups, at different times and places, and under different conditions to measure "healthfulness"; presumably, if more than the expected number of deaths occur when air pollution is high, the "excess" deaths can be attributed to air pollution. An extreme example—the 4,000 excess deaths during the acute air pollution episode in December 1952 in London—is described below. For less extreme cases, the basic methods have been refined in order to improve the specificity of attributing deaths to many factors that influence daily death rates, such as season, epidemic infections, heat waves and cold waves, and so forth. In the United Kingdom, procedures have been developed so that daily mortality can be examined for any part of the country.

The diary method was invented by Lawther and has the virtue of simplicity as well as sensitivity.[22] The subject typically keeps the record and records each day whether he feels better than the day before, the same, worse, or much worse. In Lawther's original work, the subjects were mostly men already suffering from emphysema who attended clinics in London on a regular basis. The diaries were periodically collected, coded, and correlated with the meteorologic and air pollution data for the corresponding period. This method has been extended and applied to other cities and to other types of subjects, in the United States and elsewhere.

Both daily mortality records and daily diaries have shortcomings. The additional deaths that occur during air pollution episodes are almost all among the very old or the very sick, whose condition may have nothing to do with air pollution itself; the pollution episodes may simply cause a bunching of deaths that would have occurred within a few days or weeks anyway. The diary method suffers from all the problems of self-operating experiments; interest in filling out the diary comes and goes, perhaps with time, seasons, or other things that may be correlated with air pollution; sensitivity to one's condition may be influenced by external events or by knowledge that one should feel worse on certain days.

Other measures of health effects, such as absences from school or work, visits to the doctor, athletic performance, age-specific death rates, and so on, have also been used, with varying degrees of success.

22. P.J. Lawther, R.E. Waller, and M. Henderson, "Air Pollution and Exacerbating of Bronchitis," *Thorax* 25 (1970): 525-39.

It is important that further work be done to measure health effects—
even though there never will be any single measure of health.

Establishing a Relationship Between Exposure and Effects

Once data, however crude, on exposure and effects are available,
statistical relationships can be sought. In extreme cases, such as the
1952 London episode, the raw data will convince most people that a
relationship does exist; but to quantify the relationship, or even to
demonstrate its existence in less extreme cases, requires sophisticated
analysis. The most common method of analysis is multiple regression,
in which equations are found "predicting" or "explaining" the measure of health effects in terms of the pollution and other variables.

The basic problem, beyond the data limitations, in establishing a
relationship between exposure and health effects is to adjust correctly
for factors other than pollution exposure that affect health. Cigarette
smoking and age are two obvious factors, but there are others, such
as climate and weather, diet, occupation, and past exposures. If data
on these other factors are available, they can be used in the statistical
analysis to try to separate the various effects on health. But if data
on these other factors are unavailable, and if any of the factors vary
systematically with pollution levels, then spurious relationships
between pollution and health effects may be found or real relationships masked. For example, there is a general urban-rural differential for air pollution, as well as for many diseases, including
nontuberculous lung disease and lung cancer; but many social economic, behavioral, and lifestyle factors affect these health differences,
so that it is difficult to say that the "urban factor" is a community
air pollution factor. Data on smoking habits of rural versus city
dwellers, for example, are essential to determine how much of the
difference in health may be due to smoking habits, but such data
are often unavailable.

Because of the many pollution, health, and other variables that
may enter into the relationship between pollution and health and the
lack of a generally accepted theory of the relationship, investigators
typically are forced to try many variables and mathematical forms in
their statistical equations. With modern computers, this can be done
quickly; literally hundreds of different equations can be fit to the
same data until a "good fit" is obtained, predicting "health" as a
function of "pollution" and other factors. But this process of fishing
for a "good fit" will often turn up many equally likely candidates,
some containing quite different variables. For example, of two equations predicting mortality equally well in terms of standard statistical

tests but using a different set of eight explanatary variables, one equation may indicate that the average sulfur dioxide level is and the average humidity is not statistically significant, while the other equation indicates just the opposite. Depending on how the investigator conducts his search for a "good fit" among the millions of possible equations, he may find one but not the other of these relationships; and even if he finds both, he may report only the result he regards as more plausible.

The problem of deciding which of several statistical relationships is the best arises whenever many variables are involved in a complex situation. We point it out here because it is one more reason for the great uncertainty and dispute about the health effects of air pollution, which is central to some important cases discussed below.

The Evidence Linking Health Effects to Air Pollution

Despite the problems of quantifying and identifying precise relationships, there is persuasive evidence that air pollution can have strong adverse effects on human health. Some of the principal studies indicating that it is not healthful to breathe dirty air are discussed briefly here.

Studies of Acute Air Pollution Episodes.   Acute episodes occur when pollutants build up under meteorologic conditions known as atmospheric inversion. Most episodes have occurred in river valleys, when the inversion has the effect of a lid, preventing the normal circulation of air and trapping the pollutants. When a severe episode hit the industrial valley town of Donora, Pennsylvania, in 1948, almost half of the population fell ill, and there were twenty deaths in the period of three or four days, compared to two or three deaths expected in that period. Most of those who died had preexisting disease of the heart or lungs. However, at that time atmospheric measurements were not being made, so that the precise levels and composition of the air pollution were not determined. Because of the presence in the area of a smelter and other sulfur-emitting industries, sulfur compounds were suspected, but were not known to have caused the health effects.

Four years later London, a much larger city, occupying the broad valley of the Thames in England, was subjected to an inversion that caused several thousand excess deaths, mostly during two weeks of December. Here the contributions of coal smoke and gases seemed clear, though measurements were not very satisfactory. The recorded maximum for smoke was 4,460 μg per cubic meter (μg/m^3), but this was clearly an underestimate, since all the measuring filters in central London were saturated. (Twenty-four hour average values are usually between 50 and 200 μg/m^3 in urban areas.) Sulfur dioxide levels were also very high, as one would expect, since coal was widely used for domestic heating, with grossly inefficient combustion in open fires, and was the principal fuel of power plants and industry.

The 1952 London disaster resulted in enactment of the British Clean Air Act of 1956, which, over the years, has produced remarkable improvement in the air of London and other British cities. The current daily levels of smoke in London are about 1 percent of what they were in the 1952 episode, and high concentrations are rare. British scientists consider that the air pollution has been reduced so effectively that few opportunities now exist for study of acute episodes, and reliance must be placed on results from earlier periods.

Effects on health in London have declined along with air pollution levels. During the 1960s it was observed that detectable excesses of mortality occurred when smoke and sulfur dioxide concentrations exceeded 750 μg/m^3; such high levels of air pollution and the associated excess deaths have diminished with time, and in recent years there has been no clear evidence of the effects of pollution on mortality after adjusting for other short-term influences on daily mortality rates, such as cold waves, heat waves, and influenza epidemics. The diary method studies also indicated that, during the 1960s, the symptoms of emphysema and chronic bronchitis patients worsened on days when smoke particles exceeded 250 μg/m^3 and sulfur dioxide was 500 μg/m^3 or higher; there are now few occasions on which this combination is reached. (These air pollution levels cannot be directly compared with U.S. data because the United States and the British use different methods for measuring suspended particulates; but for approximate comparison, the U.S. standard for twenty-four-hour averages are 260 μ/m^3 for particulates and 365 μg/m^3 for sulfur dioxide.)

It is significant that the British have focused their control efforts on black smoke (particulates) rather than on sulfur dioxide and, since 1956, have drastically lowered average particulate levels without much change in sulfur dioxide levels. Most British scientists and public health officials feel that they have largely eliminated the health hazard from air pollution without a direct attack on sulfur dioxide emissions or ambient levels. This is in dramatic contrast with the U.S. policy approach, where sulfur dioxide is viewed as the principal air pollutant associated with coal use.

Studies of Daily Mortality in New York City. Several investigators have studied the relationship of daily mortality in New York to air pollution variables and heat waves.[23] The studies by Buechley and associates in the New York–New Jersey metropolitan area help to put into perspective the influences of various environmental variables as well as to illustrate further the complexity of the air pollution-health problem.[24] The area used in the study had a population of 14.5 million people in 1970, about 7.1 percent of the U.S. population. With a population this large, the number of deaths each day can be used as a sensitive index of health status. Buechley used stepwise multiple regression to analyze deaths from 1962 through 1972, a period during which sulfur dioxide pollution was high for the first six years and was reduced to only one-tenth of those levels over the next five. A weakness in his study was that a single station on 125th Street, Manhattan, was used for all pollutant-monitoring data.

In order to isolate the effects of air pollution from other, more powerful influences on daily mortality, Buechley had to include other predictive variables. The annual cycle turned out to be the most important, with daily deaths going from a peak of 112 percent of "normal" at year end to a fifty-day low of 90 percent of normal during late July, August, and early September. Summer heat waves were a close second in importance; the highest spike recorded in the eleven year period was July 4, 1966, when there were 948 deaths (229 percent of normal) following a July 3 maximum temperature of 107°F. Influenza was the third most important factor, with three major epidemics in the eleven years. A meteorologic variable measuring temperature changes week to week was fourth in importance, predicting both the pollution variables and mortality. Eight other variables (such as day of week, day of month, holidays) made significant but questionably meaningful contributions to mortality. Three pollutant variables (SO_2, CO, total suspended particulates) were next in importance, all increasing mortality. Cold waves were the least influential, but still significant, variable.

Buechley's analysis of the New York data indicates that air pollution variables are significant predictors of daily mortality, although less important than other, natural variables such as heat. In particular, as the daily SO_2 varied from less than 30 $\mu g/m^3$ to above 500 $\mu g/m^3$,

23. See H. Schimmel and T.J. Murawski, "SO_2—Harmful Pollutant or Air Quality Indicator?" *Journal of the Air Pollution Control Association* 25 (1975): 738.

24. R.W. Buechley, W.B. Riggan, V. Hasselblad, and J.B. Van Bruggen, "SO_2 Levels and Perturbations in Mortality—A Study in the New York-New Jersey Metropolis," *Archives of Environmental Health* 27 (1973): 134.

daily mortality varied from about 1.5 percent less than normal to about 2 percent greater than normal (everything else equal). Significantly, the curve relating mortality to SO_2 is essentially linear over the range studied, with no suggestion of a safe or threshold level; the SO_2 variable remained a significant predictive variable even near the end of the study period, when average SO_2 concentrations had declined to one-tenth of their levels early in the study period. This fact suggests either that SO_2 is harmful at very low levels or (more likely) that the SO_2 variable is acting as a proxy for some other factors (such as sulfate levels) that vary on a daily basis with SO_2 but that may or may not have declined on average over the study period.

Epidemiologic Studies of Chronic Air Pollutant Exposures. If acute episodes of pollution can cause premature death of individuals having impaired lungs and hearts, it is reasonable to ask whether long-term exposure can cause chronic disease. A number of British studies have addressed this question, together with a few international comparisons and a few North American studies. In the United States, the most ambitious and best known have been the studies initiated by the U.S. Environmental Protection Agency called the CHESS program and the work of Lave and Seskin.

The CHESS Program. The Community Health and Environmental Surveillance System, or CHESS, was begun with a number of laudable objectives—to obtain baseline health information against which trends in pollution control could be measured, to elucidate the role of individual pollutants in causing health effects, and to strengthen the scientific base for existing air quality standards and possibly for new standards.[25] Field studies were started in communities across the country and were carried out intensively for several years by a large team of scientists. Pollution levels were measured, a variety of data on health effects was gathered, and statistical relationships were sought. It was an excellent opportunity to answer some basic questions about the health effects of air pollution.

Unfortunately, the CHESS program failed to accomplish its objectives. Although useful information was obtained, the program came under extensive criticism from the scientific community and from the Congress, which led to curtailment of the program before many of its results had been published. There were a number of reasons for

25. *The Environmental Protection Agency's Research Program with Primary Emphasis in the Community Health Environmental Surveillance System (CHESS),* An Investigative Report, U.S. Congress, Committee on Science and Technology (Washington, D.C. 1976).

this result. For example, CHESS was a pioneering scientific effort that was not given enough time to learn from its early mistakes. Moreover, there were political forces at play. Since the Clean Air Act had been written as though there were safe thresholds for air pollutants individually, CHESS had to be designed to find them—whether nature had put them there or not. The CHESS Investigative Report issued by the Congress was particularly harsh and explicit, observing that "technical errors in measurement, unresolved problems in statistical analysis, and inconsistency in data in the 1974 CHESS Monograph render it useless for determining what precise levels of specific pollutants represent a health hazard." There was no recognition of the possibility that "determining . . . precise levels . . . represent[ing] a health hazard" was a political misstatement of the scientific problem CHESS should have been trying to solve.

A program similar to CHESS, but one with improved methods, better quality control, and objectives defined by scientific realities rather than by legislation, is much needed. Even with its shortcomings, however, CHESS produced a number of important results. The CHESS studies reemphasized that air quality data from monitoring networks are of very little value to studies of health status and that more attention needs to be paid to indoor pollution, since indoor air is what most people breathe most of the time. The CHESS investigators felt that acid sulfate particles were more likely to be injurious than sulfur dioxide, a finding previously suggested by laboratory research. The CHESS findings abundantly confirmed the need to control for tobacco smoking and occupational exposure before assigning a health effect to air pollution. None of these findings made CHESS any more popular with those who wanted simple definitions of "safe" air. But they did lay the groundwork for later, improved studies, such as a current large-scale study of air pollution that is oriented primarily to products of fossil fuel combustion.[26]

The Lave and Seskin Study.   A major attempt to study the association between air pollution and health was conducted by Lave and Seskin and is summarized in their book *Air Pollution and Human Health* published in 1977.[27] Their primary objective was to provide a basis for cost-benefit analysis of air pollution effects and control. In order to justify the benefit calculations, it was necessary to demonstrate a causal relationship between measured air pollutant variables and mortality rates. They used mortality rates as a measure of health

26. "Air Pollutants and Health: An Epidemiological Approach," *Environmental Science and Technology* 11 (1977): 648.
27. Lave and Seskin.

effects because of the well-known inadequacy of attempts to measure morbidity on a wide and consistent scale. They carried out enormous numbers of analyses on the mortality rates of 117 Standard Metropolitan Statistical areas of the United States for 1960, 1961, and 1969. The air pollution variables were obtained from the National Air Sampling Network and for most of their work were limited to three measures of ambient sulfate levels and three measures of suspended particulates.

Lave and Seskin were aware of the severe limitations of the available data and made various attempts to control for the deficiencies. However, most epidemiologists with experience in this complex area of research do not believe that Lave and Seskin were successful. For example, cigarette smoking, known to be an important influence on mortality, could well be more prevalent in areas of high pollution and account for the effects attributed to air pollutants; social or income class, which Lave and Seskin used as a proxy for many life-style variables, is not a convincing control for this. Another very influential factor is age distribution, which Lave and Seskin represent by the fraction of the population over age sixty-five; chance variation in age structure, if associated with pollution levels, might account for all the effects attributed to pollution. Perhaps most important of all is the fact that several investigators have found the Lave and Seskin statistical model to be unstable, in the sense that reasonable changes in the independent variables may change pollutant effects from "significant" to "not significant"—that is, using the same data, it is possible to find relationships that "predict" mortality rates as well as do the Lave and Seskin relationships without using pollution as a significant predictive variable.

For all these reasons, it is not possible to place much scientific confidence in the quantitative relationships developed by Lave and Seskin—or by anyone else, for that matter. There is convincing evidence that air pollution is associated with mortality; but there is no reliable quantative information on the magnitude of the effect or on the number of lives that would be saved by a reduction in the level of any one or all air pollutants.

Conclusions Concerning Air Pollution and Health

The air pollutants of greatest concern in relation to increased use of coal over the next twenty years are sulfur dioxide (SO_2) and nitrogen dioxide (NO_2). These gases (especially NO_2) are of some danger themselves; but probably more serious are their oxidation and reaction products, such as sulfur and nitric acids, various sulfates and nitrates in the form of fine suspended particulates, and even photo-chemical oxidants. The sulfur comes primarily from coal, whereas the nitrogen is mostly from the atmosphere and is oxidized by the high temperatures in any combustion process; thus, NO_2 is more diverse in its sources. Trace metals from coal combustion and poly-nuclear organic materials, which could present problems in coal conversion, represent less concern to the general population, although they must be studied and controlled.

Our discussion has emphasized the complexity of atmospheric and biologic effects, the difficulty of studying low level effects of pollu-tants, and the lack of success thus far in identifying the role of individual pollutants in the complex and ever-changing mix of atmospheric pollutants that people breathe at home, at work, and out of doors. These problems show up in the disputes that arise in setting air quality standards and in deciding later whether any partic-ular standard is too lax or too stringent. The National Academy of Sciences and other groups, which have reviewed the new scientific information since 1970, have declared that there is no justification for lowering or raising the standard. Only the photochemical oxidant-ozone air quality standard has been changed, and then more because the old one was impossible to meet than because of any new evidence suggesting that oxidants were less dangerous than originally thought.

Government agencies can be and have been justly criticized for failure to develop health effects information that would be relevant for setting and changing standards. Although we have our own set of criticisms of government research planning and priorities, we think that the inability of scientists to find the health effects thresholds on which the air quality standards are supposed to be based has a more basic explanation: there are no meaningful thresholds. The sooner EPA and other agencies are freed from the responsibility of finding and defending things that do not exist, the sooner their research pro-grams can be reoriented toward discovering and quantifying the real relationships among pollutant emissions, exposure levels, and effects.

More epidemiologic studies along the lines of CHESS and the Lave and Seskin work need to be undertaken, with better methods, data, and quality control. Studies of an essentially descriptive nature will not fill the need; investments in analytic studies will be expensive and slow to yield results, but will be more valuable in the end.

Important opportunities have been lost from failure to measure changes in health status during marked temporal changes in air pol-lution; there should be plans to utilize such opportunities in the future as changes in air pollution occur. The National Center for Health Statistics collects and analyzes vast amounts of health information, but makes little effort to study environmental factors, including tobacco smoking, occupation, and air pollution, that play important roles in trends of illness and death rates.

Until better information is available from analytic epidemiologic studies and more effective use of routinely collected health statistics, we must continue to rely on laboratory studies. We need more information about sulfates and nitrates, not in general, but at levels realistic to urban environments, and not just to look for particular effects thresholds, but to obtain evidence on the public health implications of exposures.

In the present state of knowledge and ignorance, there is a huge range of uncertainty about the public health implications of expanded coal use. Although we have given two estimates of deaths attributable to air pollution—9,000 and 140,000 per year—we emphasize that reliable quantitative estimates of the overall health impact of air pollution do not exist. The proportion of deaths attributed to air pollution are estimated by various authors at between 0.1 percent and 10 percent, a range that includes the examples given. Similar variability characterizes the estimates of deaths attributable to a single coal-fired electric utility plant. Numerous reviews of the health impacts of specific air pollutants have been published in recent years, many of them conducted by the National Academy of Sciences—National Research Council at the request of the Congress. All objective reviews note the uncertainty of information about air pollutants and many recommend an increase of research in order to improve the scientific base for the very important and costly public policy decisions affecting air pollution control. For a variety of reasons, including the highly complex and dynamic nature of air pollution and a reluctance to fund careful long-term studies, scientific progress seems very slow and likely to continue to be slow.

Although there is no way now to quantify reliably the health effects of air pollution in general or to specific components of the pollutant mix, some limited information can be obtained by comparing U.S. and British policy and results over the past several decades. Before the air pollution control efforts of the past twenty years, there was unquestionably much sickness and many deaths from urban air pollution in British and other European cities and, to a lesser extent, in many U.S. cities. The British focused on the clean up of smoke rather than sulfur dioxide and have achieved much improvement in health status; the extent to which other factors contributed to the improvement is not known. In the United States, the low sulfur fuel policy led to much reduction of urban SO_2 levels, with little change in microparticulate sulfates. There is very little evidence that the reduction in SO_2 has improved health status, although it probably has in some places. Until more is known about just what is unhealthful about air pollution and how those agents are formed and interact, there remains a serious risk that large costs can be paid to obtain minimal benefits.

The Difficulty of Measuring Benefits

It is especially difficult to measure the value of the benefits derived from pollution control expenditures—that is, the damage prevented by such expenditures. Various studies have estimated the increased sickness that might result from a projected expansion of coal use to be from hundreds to hundreds of thousands of cases of asthma, bronchitis, and so on. Even excluding the difficulty of placing a monetary value on sickness or death, the wide range in the statistical findings themselves precludes a precise comparison of costs and benefits. Estimates have been made, however, of damage to health from all kinds of air pollution. One such estimate projects the national health benefits of air quality improvements expected by 1979 at $16.1 billion annually.[28] The dollar values are based on the imputed direct and indirect costs of early death and sickness presumably attributable to sulfates and particulates in the urban air in a number of cities. Unfortunately, as noted earlier, there are serious statistical, epidemiological, and measurement problems associated with this comparative urban study, as there are in the CHESS study of air pollution and the health of urban dwellers.

The same estimates of damage to health were included in a 1975 study by a committee of the National Research Council–National Academy of Sciences that estimated damage of $20.3 million per year as a result of a representative power plant emitting 96.5 million (10^6) pounds of sulfur per year in a metropolitan area of fifty million people. This damage figure represented a cost of 21 cents per pound of sulfur emitted or 46 cents per person per year. Roughly two-thirds of the cost was attributed to health effects ($13 million), 12 percent ($2.7 million) to damage to materials, 17 percent to aesthetics, and about 6 percent to acid rain. Damage to property and crops of billions of dollars per year have also been attributed to emissions from coal-fired power plants.[29]

Understandably, there is much debate over the usefulness of such damage estimates. Since few studies adequately delineate damage,

28. Lave and Seskin.
29. See William Ramsay, *Unpaid Costs of Electrical Energy: Health and Environmental Impacts from Coal and Nuclear Power* (Baltimore: Johns Hopkins University Press for Resources for the Future, 1979).

much extrapolation is needed to cover broad areas, and economists differ greatly about the dollar values to be attributed to health, aesthetic, and similar effects and about methods of calculating damage to agricultural and complex ecological systems. Nevertheless, while all estimates of damage are uncertain, many are very large, and none are trivial. The fact that costs of control are more readily calculated than benefits and are often large should not be interpreted as evidence that benefits must be smaller.

IMPLICATIONS FOR ENVIRONMENTAL POLICY

This brief review of the nature of the scientific information available on the behavior and the effects of a variety of air pollutants suggests some common threads and conclusions. These, in turn, suggest changes in the approach to environmental regulation and control to provide the flexibility in policy that is appropriate in light of the basic uncertainties about environmental processes, the effects of pollutants, the risks associated with such effects, and the benefits and costs of decisions about environmental controls.

Appropriate Environmental Objectives

Although there are differences of opinion regarding both the distribution and the magnitude of damages from air pollution, there is little doubt that damages are associated with emissions of pollutants such as sulfur and nitrogen oxides and particulates. No one argues that atmospheric pollutants are likely to be good for you.

While it is likely that damages increase with increasing emissions, what are uncertain and presently indeterminate are the shape and character of the relationships among a wide variety of pollutants emitted to the atmosphere, the quality of the air or level of pollution, and the observed damages. For example, there is apparently a statistical relationship between concentrations of sulfates and aerosols in the atmosphere and acid rain in some areas. A similar relationship is also suggested by some studies that relate mortality to ambient concentrations of sulfates and particulates. There is, however, no agreement about whether a reduction in sulfur dioxide emissions from specific power plants would either reduce sulfates in the desired localities or change the observed mortality figures. At a different scale, recent epidemiological studies indicate that children raised in homes where gas stoves are used are exposed to higher concentrations of nitrogen dioxide and appear to suffer reduced lung capacity compared to children raised in homes with electric ranges. This fact raises doubts about the wisdom of trying to reduce human exposure to nitrogen dioxide by controlling only power plant or mobile source emissions. It is just not clear how reducing emissions of a pollutant will actually reduce human exposures.

An additional finding suggests that for the entire population, including the young, the old, the sick, and the healthy, it is unlikely that there is a threshold exposure of any pollutant that is "safe" for everyone. Some in the population will always, through age or sickness, be close to a threshold at which an additional insult will be harmful or even fatal; long-term, low level exposures may eventually cause harm even to young, healthy people; and different people will react differently to the same stress. Whether or not there is a meaningful threshold for a single biologic function for a single individual, there will not be for the health of a population of diverse individuals. The best assumption is that more pollution is more harmful than less pollution, with the relationship between pollution and harm as likely to be linear as anything else. Similar "no threshold" concepts apply equally to most ecological systems. In the face of the great uncertainty about the behavior and effects of pollutants and the knowledge that there is no uniform threshold below which there are no effects, the setting of an ambient standard for individual pollutants in the atmosphere is a matter of judgment, involving a trade-off of the costs of reducing pollution levels for the benefits of doing so. Furthermore, there is no reason to expect the terms of this trade-off to be the same in different times and places. Where and when the benefits of cleaner air are higher and/or the costs of obtaining it are lower, the optimal air quality will be greater. Thus, a logical air pollution control program typically should result in air quality that improves at different rates at different times and places, not in a uniform march toward a uniform air quality goal to be reached by a certain uniform date.

Similarly, the emission levels from individual sources should not be determined primarily by the need to accomplish some predetermined ambient pollution level: although the goal of progressive improvement in air quality in many areas may be desirable, there is no logical way to choose a target ambient level without consideration of the cost of accomplishing it; there is no reliable way to determine how emissions from a particular source will affect ambient levels at a given point; and there is no clear relationship between accomplishing some ambient level and reducing damages (for example, spreading pollution more thinly over a wider area will not necessarily reduce overall damages). Absolute ambient standards set without regard to the cost of accomplishing them and then used to establish rigid control regulations in effect try to force the expenditure of unknown but possibly very large amounts of money to achieve a level of air quality that may or may not yield significant benefits.

These facts and general considerations convince us that an appropriate objective for air pollution control policy is to reduce emissions in a cost-effective way over time, rather than to accomplish arbitrary ambient standards by arbitrary deadlines. By cost-effective emission reduction over time we mean emission reductions where and when they are relatively less expensive (per unit of emission reduction accomplished), so that whatever economic resources are devoted to pollution control accomplish as much emission reduction as possible. The rate of emission reduction should depend on the interaction between social judgments about how much it is worth overall to reduce emissions further and the cost of the emission reduction at a specific time and place; the required social judgments involve the difficult problem of assessing benefits and costs, discussed above.

Effective implementation of such an objective requires innovative policy approaches, some of which are discussed in the next chapter. But we are convinced that such an approach is entirely consistent with, indeed virtually required by, the nature of the air pollution problem and other environmental problems. What are required are processes permitting trade-offs between the rate of reduction in environmental pollutants and the rate of development of energy resources and other activities to be made effectively and efficiently. We believe that this can be accomplished if the uncertainties inherent in environmental problems are explicitly recognized and a more flexible approach to environmental policy is adopted—an approach directed toward reducing environmental impacts over time while allowing a wide variety of adaptations to be made in response to changing conditions, such as the need to move from heavy reliance on oil to other energy sources. In moving to a flexible approach, it becomes increasingly important to monitor the quality of the environment, to improve understanding of environmental processes, and to determine more accurately the ways in which human beings are affected by changes in pollution and other environmental variables—instead of somehow choosing a fictitious "safe" and absolute goal, it is necessary to understand, evaluate, and weigh effects of various kinds.

The Prognosis for Energy and the Environment

With respect to the environment and energy, uncertainties about the future impacts of current choices suggest that what are acceptable or even good choices for the immediate future may prove less satisfactory for the distant future. Obviously this could be true of social and economic choices as well, but the scale of effects as well as the reversibility and the time required to reverse some environmental impacts may be quite long.

As our analysis makes clear, we believe that coal can be used increasingly over the next twenty years without significant deterioration of human health or the earth's natural environment. The threat of increasing carbon dioxide as a result of the burning of fossil fuels does not demand that we cease using fossil fuels now since time is available to permit the accumulation of a better base for decision-making and then to change course, if necessary. But we have also pointed out that the large-scale use of coal with present technology— even with significant environmental regulation of extraction, transport, burning, and waste disposal—poses the prospect of increasing damage to man and to the environment. In addition to the problems posed by sulfur in the atmosphere and in water, coal contains many trace metals, including radioactive nuclides, that are emitted to the air, land, or water under present methods of burning. For these reasons it is not difficult to suggest that a "coal economy," meaning the dominant use of coal for both the generation of electricity and for many other uses, may not be welcomed in the twenty-first century.

The pessimistic view of coal economy can, of course, be countered with a suggestion that new technology, such as burning of coal in a fluidized bed or new processes of cleaning coal and removing sulfur before burning, may well be developed for use by early in the twenty-first century. We propose in Chapter 15 that research in these areas be actively pursued because of the large quantity of coal available in the world and in the United States and because the prospect of a clean coal is enticing. We do not, however, hazard a prediction as to whether, or when, such technological innovations might become both effective and economical. Such uncertainties compel us to suggest that the potential cumulative impact of too heavy reliance on coal in the long future warrants strong efforts to look elsewhere both for a large energy resource and for alternative mixes of many energy resources.

In contrast to coal, the energy content per unit of radioactive energy resources is very large, and with breeders, and perhaps fusion reactors, the supply is even more abundant. Here again, we take the view that work should continue toward the development and use of nuclear energy as an important alternative to overdependence on coal or other hydrocarbons. As in the case of coal, however, one must rely on estimates (some would say speculation) about future technology in dealing both with the design of nuclear reactors and in the disposal of radioactive wastes. We believe that the technological problems surrounding the handling and disposal of wastes from commercial reactors are amenable to solution. Again, however, as in the case of coal, this estimate is a matter of judgment, and in this instance, repeated suggestions that solutions to the disposal problem were just around the corner have lessened confidence and credibility in renewed reiterations of this expectation. We recognize the even more difficult problems of control and management of the entire nuclear fuel cycle, including the potential proliferation of nuclear weapons as a result of increased availability of plutonium.

Given that both coal and nuclear energy conceivably pose unacceptable health and environmental problems for the future, one is led to look for more benign alternatives. As noted elsewhere in this book, all energy options are flawed; none are without some adverse environmental effects. Yet problems comparable to the longevity of nuclear wastes and the possibility of global heating from the burning of fossil fuels may not be associated with the conversion and application of solar energy, at least in some forms. There is reason to believe that the characteristics of the environmental dislocations associated with potential development of solar energy may be less destructive than those noted earlier for hydrocarbons and radioactive materials.

Ultimately one must be concerned with all projections of growth that presume equivalent or nearly equivalent growth in the use of energy. Even with considerably increased efficiency, the heat rejected in the generation of electricity by the major conventional fuels is large and places a heavy demand on local resources of air and water. While utilizing coal as an energy source requires the digging, transport, and disposal of very massive volumes of material, reliance on nuclear energy requires management of a complex technology with a finite probability of accidents of devastating magnitude. One need not be afraid of the future to stress the necessity not only of seeking new solar and other options, but of developing policies that will assure that energy is efficiently used and that in the long run, conservation will be looked upon as a desirable objective on grounds aside from efficiency in the event that optimistic predictions about the handling of nuclear material or the cleaning up of coal prove unwarranted. It is not necessary to invoke the stationary state in discussing energy and environment, but it is necessary to be aware that growth involving large quantities of material resources and energy carries with it the necessity for dealing with an enlarging potential environmental disruption accompanying such growth. Increasing complexity and manipulation of global resources requires sophisticated management to assure that the resilience of global systems is not exceeded and that potentially irreversible disruptions are avoided. It is impossible to say that this cannot be done and equally impossible to say that it will be. There is harm in panic, but perhaps there is no harm in running scared.

Managing Air Pollution

11

As demonstrated in the previous two chapters, the conflicts that are likely to arise during the next twenty years between efforts to expand the use of energy and efforts to conserve environmental quality will be especially sharp in one area—the processing and use of fossil fuels, especially coal, in stationary facilities. There will, of course, be other areas of conflict, but none is likely to be as pervasive and as complex. Depending on the outcome, efforts to reduce oil imports by expanding domestic fuel production could be sidetracked, or attempts to maintain and improve air quality could suffer.

In our view, however, much of the conflict between these two apparently competing goals can be avoided by broadening the terms in which the issues are debated. In the present air pollution control program, each specific standard, deadline, and regulation is attacked by those who perceive it as impeding energy development and is defended by those who fear that any compromise will open the door to a wholesale sellout of the environment in the name of energy. So far, the result of this conflict seems to be that air quality barriers to energy projects, particularly those involving coal use, are becoming more rather than less formidable. But the long-term outcome of this debate over narrow issues is not likely to serve society's need for either energy or clean air. Thus, in this chapter we outline an approach toward air pollution control that would put less emphasis on specific, narrow targets and deadlines and would instead pursue broader objectives in a cost-effective way over time.

This section reviews the history and current status of the U.S. air pollution control policy and describes the most serious energy-related problems. Our conclusion is that the Clean Air Act, despite major modifications since 1970, still defines national air quality objectives in a way that is too narrow and absolute in light of the fundamental uncertainties and variabilities involved and that does not allow enough desirable trade-offs to be made. We also conclude that the way the Act has been implemented is too inflexible and lacks incentives to encourage efficient and effective progress over time. As a result, progress toward the more basic objectives of both energy policy and air pollution policy may be slowed.

To improve this situation, we suggest using economic and market forces to motivate progress toward cleaner air. Although a few encouraging steps in this direction are already being taken, much more can be accomplished in the long run by adopting broader, scientifically supportable views of air quality objectives and by implementing innovative policies that let market forces rather than governmental regulators determine more of the details of technology, degree, and timing of control.

U.S. AIR POLLUTION CONTROL POLICY TODAY

Until 1970, the federal government's role in air pollution control was limited to providing information and technical assistance to states, trying ineffectively to deal with interstate pollution problems, and setting undemanding emission standards for new automobiles. The 1970 Clean Air Act (CAA) changed this situation dramatically and gave to the U.S. Environmental Protection Agency (EPA) the dominant role in U.S. air pollution policy.

The 1970 Clean Air Act

This first federal venture into air pollution policy was launched at the peak of the "environmental crisis," before the complexities and difficulties of the air pollution problem were generally appreciated. As a result, the CAA took a straightforward and simple approach to the problem: EPA would establish National Ambient Air Quality Standards (NAAQS) defining the "safe" levels of air pollution; the states (or EPA, if the states failed to act) would develop plans specifying what must be done to accomplish these safe levels; and all the required actions would be taken by 1975, or by 1977 at the latest, with the threat of state and EPA legal action providing the motivating force. In addition, the CAA specified precise limitations on the levels of emissions that new automobiles could produce; directed EPA to set nationally uniform emission standards for major new stationary sources of air pollution and for sources of "hazardous" air pollutants; and contained language that the courts quickly interpreted as requiring EPA to prevent "significant deterioration" of air quality everywhere.

National Ambient Air Quality Standards. The 1970 CAA required the EPA administrator to establish air quality standards for nitrogen oxides, hydrocarbons, carbon monoxide, sulfur dioxide, and oxidants. The "primary" air quality standards were supposed to protect human health (including that of people already suffering from serious diseases) from all adverse health effects of the pollutant in the ambient air "with an adequate margin of safety" and were to be met everywhere by 1975, or by 1977 at the latest. The "secondary" standards were supposed to "protect human welfare" against all "known or anticipated" effects and were to be met in "a reasonable time." The cost of meeting the standards was not to enter into the administrator's decision about where to set them.

As discussed in the preceding chapter, there is little reason to think that the overall health effects of air pollution on diverse human populations exhibit any meaningful "thresholds" or "safe" levels. In general, there is no scientific way to say, for example, that an exposure to a pollutant in a concentration of 100 parts per million for twenty-four hours is "safe" while exposures above that level are "dangerous." There is not even any way to say that effects of exposure become rapidly more serious at or near some particular exposure level. At best, it will be possible to find an exposure level at which no experiments have yet demonstrated adverse effects and slightly above which only slight effects have been observed. This could then be called the "safe" level—at least until more sensitive experiments are done. But the resulting arbitrary "standard" cannot logically be used to make fine distinctions between "safe" and "unsafe" air, independent of all other factors, if important resource allocation decisions turn on this distinction.

Despite the logical and sceintific weakness of the concept, the five air quality standards required by the CAA were established in 1971 and have not been significantly changed since. EPA quickly withdrew the secondary standard for sulfur dioxide when challenged in court to justify it. And the primary oxidant standard was relaxed in 1978—more because it was impossible to comply with the old standard than because of any new health effects information. Occasional reviews by the National Academy of Sciences and others have found no reason to suggest changing the existing standards and insufficient information to establish any more, such as for sulfates or lead.

At present, everybody seems to understand (but not to acknowl-edge) that the standards are only administrative fictions, and there is no reason to think that the EPA administrator did not use his "best judgment" in choosing them initially—which is all he was required by law to do. To point out now that the existing standards are not really "safe" levels would only make the air quality management problem more intractable, which is in the interests of neither the EPA nor those subject to its regulations.

Initially the air quality standards were the principal driving force behind the CAA strategy, since all emissions that contributed to a violation of them were to be eliminated by 1975-1977. This did not happen, of course, and the CAA has been amended several times to extend the deadlines, thereby reducing the "absolute" nature of the standards. However, the standards continue to play an important role in air pollution policy, because different procedural requirements apply in a region depending on whether it is "clean" or "dirty" as defined by the standards. As detailed below, it is our view that the standards should be deemphasized even more, in light of their limited value as an administrative goad and their questionable scientific and logical validity.

State Implementation Plans. The 1970 Clean Air Act required the states to develop, within a few months, State Implementation Plans (SIPs) specifying precisely what would be done to accomplish the primary National Ambient Air Quality Standards by 1975. If state-developed SIPs were inadequate or did not promise accomplish-ment of all standards everywhere by 1975-1977, the EPA would pro-duce its own plan. When California refused to submit an impossible plan, EPA was forced to promulgate a plan calling for restricting automobile traffic in Los Angeles by 70 to 80 percent in 1975. Other states submitted, and EPA approved, SIPs that had no realistic chance of being implemented.

Now, after the passage of several years and several sets of amend-ments to the Act, the SIPs have evolved into general statements that "too much" air pollution is illegal, and the states and EPA have entered into lengthy negotiations with each source of pollution to find workable definitions of "too much," of "compliance," and of "on schedule." EPA reports that as of September 30, 1976, 21,731 major pollution sources had been identified, of which 20,010 were complying with standards or were on "enforceable compliance schedules" (i.e., were taking promised actions to reach compliance eventually), 1,450 were violating standards or schedules, and 271 were in unknown compliance status. The 1,450 violators were primarily large sources such as power plants and steel mills—that is, the serious air polluters. EPA estimates that about 130,000 smaller sources also will have to be controlled in areas where standards are not met.

In this source-by-source negotiating process, the deadlines in the CAA have not played an important role. The principal effect of the air quality standards has been to reduce pressure for control on sources in "clean" areas. Technology and economics, not the requirements of the standards, have set the upper limits of control in dirty areas. Congress has repeatedly weakened EPA's ability to impose controls on transportation, facilities siting, or land use, even where these are clearly required to bring about compliance with ambient standards. Of course, this inability or unwillingness to impose costly, onerous measures to meet arbitrary standards by an arbitrary deadline is evidence more of the basic good sense of the U.S. system than it is of failure. But the result has been to repeatedly move back the deadlines for meeting the standards.

The 1977 amendments to the CAA provide specific regulatory mechanisms for dealing with "nonattainment" regions—that is, air quality control regions where standards were not met by 1977—and set new "absolute" deadlines. According to these amendments, states must submit by 1979 revised SIPs that promise to attain all air quality standards by 1982—or by 1987 for automobile-related pollutants. There is not much reason to expect that these new deadlines will be more effective in forcing control on existing sources than the original 1975-1977 deadlines were; the processes of regulation available to EPA and the states are not very good at forcing costly action on reluctant emitters, deadline or no. However, these regulatory and administrative processes are very good at preventing things from being done, and hence the other principal mechanism for dealing with nonattainment regions—the limitations on new sources—may be quite effective.

A major new emission source, such as a large fuel-using or fuel-processing plant, will be able to locate in a nonattainment region— that is, in almost any of the major industrial areas of the country —only if it installs the best available control technology for all air pollutants and only if offsetting emission reduction can be obtained at existing sources. As will be seen presently, this emission offset requirement is particularly interesting because it begins to establish marketlike mechanisms in air pollution control, which we strongly favor. However, the restrictive nature of the offset requirement, and the difficulties and uncertainties involved in setting technology-based emission limitations, suggest that it may be difficult to build any large new industrial facility in a nonattainment region.

New Source Performance Standards. The 1970 Clean Air Act required EPA to set nationally uniform emission standards for all new or substantially modified stationary sources in broad categories. These standards were to reflect the best available control technology and were to take into account the costs of meeting them. The resulting New Source Performance Standards (NSPS) were originally stated in terms of required performance. That is, they specified how many pounds of a pollutant could be emitted per unit of plant input or output, leaving it to each source to decide what combination of technological measures and fuel changes to use to meet the NSPS. The 1977 CAA amendments, however, directed EPA to take a new approach to setting NSPS for three pollutants from fossil fuel combustion—sulfur dioxide, nitrogen oxide, and particulates. Under the new method, emissions are to be reduced by a specified percentage below what they would be in the absence of technological control measures, regardless of how high or low this no control level would be. Thus, for example, use of low-sulfur coal is no longer an accepted way to reduce sulfur emissions from new sources.

There is no simple logic behind the NSPS as they have evolved. The intent of the 1970 Act seemed to be to prevent states from competing for new industry by setting undemanding emission standards in their SIPs. As policymakers recognized the value of controlling emissions even where the original 1975–1977 standards have been met, the NSPS came to be viewed as a way to reduce total emissions in a reasonably cost-effective fashion. The idea is that emissions should be controlled wherever it is cheap to do so, and it is relatively cheap to design emission control into new sources. The 1977 CAA amendments implicitly seek to install control technology for its own sake, even if the result is higher costs and higher emissions (as it may be if the average sulfur content of coal burned increases enough). In fact, the real political intent of the new NSPS has little to do with air quality or economic efficiency, but is basically intended to encourage use of high-sulfur eastern coal over low-sulfur western coal; another provision of the 1977 CAA amendments allows EPA to require use of local coal in order to protect local jobs.

One of the superficially appealing features of emissions standards for new sources is that they should be relatively easy to set, because all new sources in a category are more or less the same, and there is much flexibility in the design and construction phases. However, modern industrial processes are complex and varied and are becoming more so. Even within a narrowly defined industrial category, different sources in different regions will use different processes and raw materials to provide different product mixes and will have to use different methods with different costs to meet any specified NSPS. Where emissions control is inexpensive and effective, it may be possible to set an NSPS that can be easily met by all sources in a category. But in other cases, more typical of fossil fuel sources, control methods are poorly developed, expensive, highly varied, and of uncertain effectiveness. In such cases, the standard setters must choose between setting a lax NSPS that neither forces improvements in technology nor reduces emissions effectively or setting a demanding NSPS that must be defended and modified on a case-by-case basis in the face of the determined, and often quite justified, opposition of each source subject to it.

The principal argument for setting a demanding NSPS is that it will "force technology": faced with an inflexible performance standard, industry will have no choice but to try to meet it, and the technology will become better and cheaper in the process. There are, however, several problems with this approach. The most important is that it may have just the opposite effect. To demonstrate the feasibility of its chosen NSPS, EPA must point to a particular technology that can be used to meet the standards. Then anyone wanting to try something different must demonstrate that the new approach is or will be as good—not in overall economic or environmental desirability, but purely in terms of meeting the NSPS for a single pollutant.

A new technology seldom performs predictably in any specific application, and there is always variability in any industrial process even after it is built. Furthermore, there is uncertainty about how EPA eventually will define compliance for a complex, variable process. Thus, despite all efforts to comply, the builder of a plant subject to a demanding NSPS will always run some risk that the EPA will find the plant to be out of compliance and will impose a severe penalty, such as a shutdown. The best way to minimize this risk is to avoid experimentation and innovation and to use the method that EPA seems to prefer. No matter how this method performs, the plant owner is less likely to be severely penalized than if he tried something new that, whatever its other advantages, might fall short of EPA's expectations in accomplishing the NSPS.

The most important case in which EPA has used NSPS to "force technology" has been flue gas desulfurization (FGD) in power plants. By setting an NSPS for sulfur dioxide that could be met without

FGD only for the lowest sulfur coal, EPA provided what they thought was a strong incentive for development of FGD. EPA and the utilities have argued in the courts and in the political arena for years about whether FGD was a proven, reliable, economical technology. Progress in improving the technology has been slow, and it has not been widely applied. As of 1979, several dozen systems were operating or being built, most of them small demonstration units, or governmentally funded units, or units being built under protest as an experimental part of a utility's overall compliance agreement with EPA. Meanwhile, little has been done to develop other emission control systems, such as fluidized bed combustion, that may not initially be as effective in controlling sulfur dioxide, but may have more long-term potential and other environmental advantages.

The 1977 amendments requirement that EPA set an NSPS for each pollutant, specifying a uniform percentage by which emissions must be reduced, is particularly likely to slow the development of new technology. With the "old" NSPS, stated in terms of pounds of emissions per Btu of coal burned, there was at least the option of using low-sulfur coal in the event that the control system did not work as well as hoped. Now this fallback position is less effective, increasing the risk that a new plant or technology will fail to comply with NSPS despite best efforts.

Furthermore, EPA no longer has the option of redefining the NSPS categories as a means of encouraging development of new technologies that might be better in terms of cost, overall environmental impact, or energy efficiency. With a uniform percentage removal requirement for each individual pollutant, applicable to all sources burning fossil fuels, there is no way to make trade-offs: fluidized bed combustion, solvent refining of coal, coal gasification, FGD, and any other new coal-using technology must all meet precisely the same quantitative percentage reduction standards, for each pollutant individually, or be unsalable in the United States.

The debate about precisely where to set the percentage removal NSPS for fossil fuel sources is underway and can be expected to continue indefinitely. EPA's principal antagonists have been the utility industry and the Department of Energy. In the case of the sulfur dioxide NSPS, EPA originally considered a 90 percent removal requirement, but then proposed an 85 percent removal requirement, on a daily basis. DOE argued that 85 percent removal on a monthly basis would be more realistic in light of the inherent variability of these processes and the EPA's insistence on a daily standard would result in significantly higher costs and widespread noncompliance. DOE also proposed that some minimum sulfur emission factor (0.5 or 0.8 pounds of sulfur dioxide per million Btus, compared to a current NSPS of 1.2) be specified, so that very clean coals need not be treated to the same extent as higher sulfur coals. This would also provide a fallback means of compliance for sources whose control devices failed to work as planned. According to DOE, its proposal would reduce the present value of control costs by $24 billion, or 33 percent, from $73 to $49 billion and would increase sulfur dioxide emissions by only 1 million tons per year, or 5 percent, from 19 to 20 million tons per year in 1990. These figures suggest that on an annualized basis it would cost more than $2 to eliminate each pound of sulfur dioxide beyond DOE's suggested level of control.

DOE was also concerned that the requirement to eliminate 85 percent of sulfur dioxide would inhibit development of the most promising new coal-using technologies, particularly fluidized bed combustion and solvent refining of coal. Although these could remove 80 percent of sulfur dioxide and may have compensating environmental and economic advantages, it is not certain that they could remove 85 percent of sulfur dioxide. As a result, they might never be developed. A somewhat different set of problems surrounds the new NSPS for nitrogen oxides, where certain boiler manufacturers may be eliminated from the market if the NSPS is set at 0.6 rather than 0.7 pounds of nitrogen oxides per million Btu and where much more stringent NSPS levels (such as 0.2 pounds per million Btu) are necessary if the development of dramatically better technologies is to be forced.

No matter how these issues are eventually decided, it is clear that the setting of absolute emission standards for each pollutant confronts energy and environmental policymakers with many unpleasant dilemmas. These are not just matters of trading off cleaner air for cheaper electricity. Rather, these dilemmas involve dead losses: nobody gains if a new technology, which might be cheaper and environmentally better overall, remains undeveloped for fear that it will control only 89 percent instead of the required 90 percent of one pollutant. The real economic, energy, and environmental impacts of all this are difficult to measure, even after the fact. But it is likely that, as the 1977 requirements begin to have their impacts, the conflict over these standards will increase, and further legislative action will be considered. Unfortunately, instead of taking a new approach, the future amendments will probably only relax the numbers, so that the result will be delay and high cost in the short run and disappointing improvements in air quality in the long run.

Prevention of Significant Deterioration.   The 1970 CAA was ambiguous about how to protect air quality in regions already meeting the National Ambient Air Quality Standards. However, a series of court cases quickly established that clean air should be prevented from deteriorating "significantly," and EPA spent several years trying to decide what "prevention of significant deterioration" (PSD) should mean. In the 1977 CAA amendments, Congress reasserted its commitment to the principle of PSD and wrote into law some of the concepts and procedures proposed by EPA as regulations earlier.

Under the PSD program defined in the 1977 amendments, each region in the country that already complies with the air quality standards is to be put in one of three classes. Large national parks and wilderness areas are automatically Class I areas, where very little deterioration of air quality (defined in terms of increments in ambient concentrations) is allowed. All other areas are Class II areas, where moderate increases in ambient concentrations are allowed. A state governor may reclassify a Class II area in his state as Class I or as Class III, where larger increments in ambient concentration are allowed, but no increments are allowed that would lead to violation of an air quality standard.

To establish a new source in any PSD area, the owner must apply for a permit. To obtain this permit, the applicant must demonstrate that the emissions will not cause any violations of the ambient concentration increments allowed in the area or in any other area. He must perform baseline studies of ambient air concentrations surrounding the proposed plant for at least one year. And he must agree to apply "best available control technology" for all pollutants, whatever the classification of the area and whether or not this is necessary to meet the allowable increments. Some exceptions can be made if the applicant can demonstrate to the satisfaction of state and federal officials, and after public hearings, that emissions will have no adverse effects and that the plant cannot be built without the variance. In addition, EPA is to issue regulations to protect visibility in Class I areas.

As with nonattainment area regulations restricting new sources, the PSD regulations put the environmental authorities in a strong position in negotiating with potential emitters. A plant cannot be built until the necessary governmental procedures are carried out, analyses conducted, determinations made, and permits granted. In addition, private intervenors have any number of opportunities to challenge the applicant's analyses and the agency's determinations. Thus, the PSD program will probably be "successful" in that firms that want to establish large, fossil-fuel-using facilities in an important PSD region will find it hard to do so. However, there are several potential difficulties with the PSD policy that may require legislative changes as the effects of the restrictions begin to be felt.

Among the potential problems with the PSD policy is that absolute protection of visibility in Class I areas may require impractical levels of control on, or total elimination of, all large pollution sources over very large areas, especially in the Western states. The fine particulates, sulfates, and nitrates produced by a coal-burning power plant, even with "best available technology," can affect visibility hundreds of miles from the source. Whether, or to what extent, a specific plant will affect visibility in a distant Class I region cannot be determined with certainty until the plant is built and operated, if then. The debates over the predictions of visibility impacts and over the allocation of the blame if the predictions are perceived to have been wrong will be lively. Similar problems exist in deciding whether or not the arbitrary allowable PSD increments in ambient concentrations will be or have been met and what to do if the prediction turns out to be wrong.

Even if there were no uncertainty about what the air quality and visibility impacts would be for a specified level of emission, there would remain the problem of selecting the emission level defining "best available control technology" for each pollutant from each source. And if several applicants are competing for an allowable air quality increment or visibility impact or if a new source wants to compete with an existing source for access to the air resource in a PSD region, there is no allocation or rationing mechanism—not even a "trade-off" policy similar to the one allowed in nonattainment regions.

THE MAJOR ENERGY-RELATED
WEAKNESSES OF THE CLEAN AIR ACT

As a result of the many ad hoc changes made in the CAA since 1970, the basic strategy of the U.S. air pollution control program has been changed significantly—in practice if not in rhetoric. Some of these changes are desirable in that they reflect recognition that the air pollution problem is more difficult than originally assumed and that more flexibility and discrimination are required. But discrimination and flexibility were consciously minimized in the original version of the Act for some good reasons: history and logic demonstrate that administrative and regulatory procedures become increasingly complex, lengthy, uncertain, and ultimately ineffective as they are required to discriminate and show flexibility on the basis of fine distinctions. Thus, although the U.S. air pollution control program seems to be evolving, if only unconsciously, toward a better understanding of the complexity of the air pollution control problem, it is still a long way from having a set of policy instruments that can deal with this complex problem effectively and efficiently.

The most fundamental change in the air pollution control program since 1970 is its increased reliance on a strategy of reducing emissions everywhere and of considering the overall impact of a new source, as opposed to a strategy of meeting air quality standards everywhere. This change is more real than may be apparent because the CAA continues to distinguish between regions on the basis of whether or not they meet standards—a distinction that is often difficult to make. But the implications of this distinction may turn out to be relatively unimportant. New sources locating anywhere will have to use best available control technology for all air pollutants under one name or another (BACT, NSPS, LAER); in practice, these will all be defined about the same, but only after lengthy negotiations in each case. To locate in a PSD region, a source will have to demonstrate nonviolation of allowable ambient concentration increments, while in a nonattainment region, it will have to find offsetting emission reductions. In either case, there will be uncertainty about whether or not the requirements will be met or have been met. Much room for negotiation, redefinition, and accommodation will exist if there are strong economic reasons for locating the plant at the proposed site, and much room for opposition, challenge, and confrontation will exist if there are strong environmental reasons for not locating the plant at the proposed site. Whether or not National Ambient Air Quality Standards are met may turn out to be a relatively unimportant factor in determining the outcome.

Despite (or perhaps because of) the deemphasis of ambient standards in the CAA control strategy, it is clear that the decision processes are becoming more complex, especially where new sources are concerned. The cost, delay, and uncertainty involved in obtaining approval for a new source will certainly discourage some potential applicants. Restrictive interpretations of policies and inefficient implementation of procedures could make it virtually impossible to build a major new facility that uses or processes fossil fuels anywhere in the United States—either because the air is too dirty or because it is too clean.

It is too early to tell just how severe the impact of the PSD and nonattainment programs will be on construction of new energy facilities. But it is not too early to begin thinking about ways that the existing programs might be improved, so that the level of unnecessary delay and uncertainty is reduced, for the benefit of both energy and air quality.

AIR POLLUTION POLICY FOR THE NEXT TWENTY YEARS

Despite the problems and shortcomings of U.S. air pollution policy, progress has been made in reducing the level of the common air pollutants in many of the most populated areas. Where stationary fossil fuel sources are involved, however, most ot this progress has been due to the replacement of coal by oil and to the removal of pollutants to less-populated areas, both by locating new plants (especially powerplants) in remote areas and by using tall stacks to diffuse the pollution over wider areas. These were probably desirable air quality control methods during the past twenty years and may still be useful in certain cases. But the oil crisis and resulting emphasis on coal use, better appreciation of the effects of pollution over long distances and at low concentration levels, and the perceived value of maintaining high quality air where it now exists all suggest that use of naturally clean fuels and remote location of plants cannot be the principal methods of protecting air quality over the next twenty years.

The evolution of the national energy system over the next twenty years is impossible to predict in any but the most general form. It will almost certainly involve increased use of coal, although the exact future levels cannot be predicted. The only thing that is certain is that there are many complex possibilities, many of which should be developed and used for particular purposes. Coal can be used in many forms and for many purposes; electricity from coal or other sources can be substituted for direct coal use and for other fuels; synthetic fuels can be produced in many forms by many different methods. Many different types of energy conservation technologies and renewable energy sources can reduce the need for coal use. Each of the various technologies, methods, and products has its own mix of positive and negative impacts on air quality, oil imports, and energy costs that cannot be reliably predicted. In short, the number of possible outcomes and combinations is huge.

We are no better able than anyone else to predict the outcome in this complex situation. We do know, however, that the traditional processes of governmental regulation are hopelessly inadequate to the tasks of choosing and generating the right mix of technologies, of coordinating investment plans, and of matching supply and demand. Only if market forces are allowed a major role in stimulating innovation, encouraging diversity, and coordinating individual plans and decisions will a satisfactory outcome result. This presumption that an effective energy policy must rely strongly on market forces is central to our overall theme. But it also applies directly to air pollution control policy, where crude regulatory devices threaten to distort the desirable outcome significantly—to the overall detriment of the energy system, the economy, and the environment—and where there are feasible, if untried, market alternatives.

Market Mechanisms in Pollution Control

The problem of controlling air pollution is, among other things, a massive resource allocation problem. It is not fundamentally different from other resource allocation problems that are handled in the United States with passable success by letting markets do much of the work, within a general framework of property rights, law, and regulation. Air pollution control is probably the only area of U.S. public policy that uses brute force regulation to try to solve a resource allocation problem of such magnitude and complexity. The result has not been remarkably successful, even in terms of air pollution control itself, let alone in terms of the broader interests of society. Yet there are ways in which market forces could be used to help solve this difficult resource allocation problem.

The idea of using marketlike mechanisms as instruments of pollution control policy is not new, but is relatively untried, especially in the United States. A large literature exists developing the theoretical economic bases of the idea and documenting its application, in limited ways, to situations as diverse as river basin management in Germany, air pollution control and compensation of victims in Japan, and municipal sewer systems in some U.S. cities. A few halting steps toward using market forces are being tried under the 1977 CAA amendments, as will be discussed below.

There are two basic "pure" types of market systems that might be applied to pollution control. The first uses effluent fees or emissions charges to impose a specified charge (in, for example, dollars per pound) on each unit of a specific pollutant discharged in a given area and time interval, and each discharger is then free to adjust his discharges as he sees fit in light of his individual economic and technological situation. The second allows emissions only in quantities, at times, and at locations specified in permits or rights defined and enforced by the environmental authorities, and these rights are then marketed among dischargers under specified conditions. In principle, there is little difference between these two types of systems, since with perfect knowledge, either charges can be set to accomplish any desired total emission levels or marketable rights can be defined that will have any desired market prices. In practice, however, either type of system will be implemented in the face of great uncertainty and limited ability to monitor and enforce, and hence it may make a good deal of difference whether charges or marketable rights are used.

Because marketable emission right (MER) systems specify the total level of emissions in a region and time period, the total or incremental costs that will be incurred in meeting the objective cannot be predicted precisely, and the total emission levels will not depend on costs and technology; thus, MER systems are best suited to situations in which the optimal level of total emissions can be selected rather precisely beforehand (perhaps on a declining schedule over time) and will not need to be reconsidered as more is learned about costs or as conditions change. Emission charge systems, on the other hand, put an upper limit on the costs that will be incurred to reduce emissions one more unit and hence result in emission levels that cannot be definitely predicted and will vary as conditions change. Thus, charge systems are best suited for situations in which the "best" total emission levels and rates of progress toward them cannot be determined beforehand, but depend on unpredictable and variable costs and technological factors. Emission charge and MER systems can be combined in various ways to get some of the advantages of each.

In many air pollution control situations, there is a general idea of what should be accomplished and by when, but there is enough uncertainty about costs and technology that some flexibility should be incorporated. In such cases, the best combination of charges and MERs may involve applying emission charges only to emissions above some allowable level for each source. The initial allowable levels may be relatively undemanding, but can decline over time to reflect the progress society would like to see. If the allowable emission levels can be traded among the emitters, and if they are fixed in total number so that nobody can get more fuel by, for example, building a new plant, then they become a form of marketable emission rights. The resulting market in MERs will allow and encourage each source to develop its own plan for reducing emissions over time, encourage phaseout of uneconomic emissions, and provide a mechanism to allow economic growth. The monetary penalties for emissions above the maximums specified in the MERs can be set at levels high enough to discourage "violations," but not so high that the entire schedule need be reconsidered if the job turns out to be more costly and difficult than originally thought. Variations in the timing and sources of emissions, due to valid causes that would be grounds for variances under a regulatory program, can be accommodated automatically. The application of such systems within particular programs in the CAA is discussed below.

Much of the debate about market systems applied to pollution

388 Coal: An Abundant Resource—with Problems

control seems to deal with technical questions: How can emissions be monitored accurately? How does one choose the number of market-able emission rights? What criteria should be used to set the level of the monetary charge on excess emissions? These are all difficult and important questions that can be (but seldom are) asked about tradi-tional regulatory programs: How does one define "best available technology?" How should an absolute deadline be chosen, and when should it be relaxed? How does one know "reasonable progress" and "good faith" efforts when one sees them? The answers are essentially the same, whether or not market systems are involved: estimates and judgments must be made.

The need for estimates and judgments in dealing with air pollution control is taken for granted in regulatory programs, but seems to be harder to accept in thinking about market systems. Ironically, one reason for this may be that there is a well-developed, logical theory of how market systems should be applied to pollution problems to obtain the economists' "ideal" solution. Because there is always a large gap between theory and practice (and because even theorists find it easier to demonstrate how hard the problem is than to suggest practical solutions), discussions of market devices tend to highlight how far any proposed practical system is from the ideal. By way of contrast, regulatory programs are expected to be pragmatic and ad hoc, guided by no neat theory of what should be accomplished sub-stantively: as long as the proper procedures are followed and all the right factors are weighed in some unspecified fashion, a regulatory program cannot easily be criticized for failure to accomplish the (undefined) ideal.

This asymmetry in the way policy alternatives are viewed leads to some peculiar results. For example, people who are quite willing to set infeasible deadlines to "hold the rascals' feet to the fire" criticize emission charge proposals on the grounds that the marginal social damages caused by pollution cannot be measured precisely and hence there is no "rational" way to set the charge. And it is often asserted that the inability to measure emissions accurately and continuously makes an emission charge impractical—as though the estimating and compromising common in regulatory programs must be totally ex-cluded as soon as market devices are introduced. But policy alterna-tives should be compared to each other in terms of the same set of criteria, not by holding just one of them up to an unattainable ideal.

Looked at in this way, even the problem of estimating emissions—often cited as the main technical obstacle to market systems—loses much of its force as an argument against market devices in air pollu-tion control. Any logical and effective pollution control program must estimate and control the quantity and timing of mass emissions from each of many sources. Even common air pollution regulations that do not explicitly control quantity—such as regulations enforced by an inspector who walks around looking for violations of "good practice"—must, in important cases, be translated into mass emission numbers. Emission estimates must be made by some combination of such things as direct stack monitoring, measurement of fuel composition, engineering estimates of equipment capability, spot checks, and operating logs. There is much room for error and proving a violation of a specific standard in court is so difficult that it is seldom attempted—especially since the alleged violator will claim that (1) he didn't do it; (2) if he did, it wasn't his fault; and (3) even if it were his fault, the proposed penalty is out of proportion to the crime. The question is whether market devices necessarily make these difficult problems more difficult.

Presumably, the problem of monitoring and enforcing individual mass emission limitations is made no more difficult by allowing the individual sources to trade allowable emission rights among themselves in a regulated market. It makes no difference whether source A got his emission limitation of X tons per week by buying all or some of it from source B or by persuading the authorities that he "needed" it. In fact, the emitter is more likely to avoid successful prosecution for a violation if there is no market, because then he has the defense that the level allowed by the authorities is technically or economically unreasonable. Furthermore, no matter how the permitted levels are set, if emissions above these levels are something to be paid for, like water or electricity, instead of being potentially the subject of legal prosecution, the whole question of estimating emissions takes on a different significance. Instead of lawyers arguing over what is admissible evidence, whether emissions were 99 percent or 101 percent of the standard, and who is really to blame, the engineers and accountants can sit down and reach a negotiated or, if necessary, arbitrated agreement on what emissions were, plus or minus 10 percent or 50 percent, and the bill to be paid. It may be well that the difficulty of monitoring emissions should be regarded as a further argument for, not against, putting air pollution problems into the market rather than into the courts.

Considerations of this kind convince us that there are no insurmountable technical obstacles to the use of market devices in air pollution control programs, as long as the objective is pragmatic improvement more than accomplishment of the economists' ideal. But a number of political and ideological objections to the use of markets have proven difficult to overcome, even though we do not find them persuasive. Perhaps the most basic is the general aversion to the idea of selling a "license to pollute." This emotion-laden phrase obscures the fact that any pollution control program that does not ban all pollution immediately is, in effect, granting "licenses to pollute" to someone, for some reason, for some length of time. It is our view that the best way to see that these licenses are limited in size and duration and go where they are really the most valuable is for a government agency not to give them away free but to sell them to those willing to pay high prices for them. Both environmental and energy-economic objectives are likely to benefit from such an approach to allocating the inevitable "licenses to pollute."

Other objections to the use of market devices in air pollution control are more narrowly political. Environmentalists, when they think the political winds are blowing in their favor, do not want to give the polluters the option of paying prices—even very high prices—as an alternative to investing in costly control devices. Polluters fear that market devices will be used merely to force on them cost-ineffective control measures, not as a means of striking a reasonable balance between controlling and living with pollution. Those who will have to live with the effects of pollution in each case are not convinced that they would benefit as much from monetary charges as from expenditures on control—and because disposition of the proceeds is a tricky subject, they may be right. So there are real and difficult political obstacles to adoption of market systems in pollution control, most of which cannot be discussed in detail here. Nevertheless, we are convinced that the economic and administrative advantages are significant and that the concept should be applied to the maximum extent feasible.

In applying market devices to air pollution control, their fundamental advantage should be kept in mind, so that it is not lost in the details of implementation. Market devices allow the processes of government to concentrate on "external" matters that only they can handle—such as determining where emissions come from, where they go, what harm they do, and how much it is worth to society to reduce them. Individual emitters are given the incentive and the opportunity to evaluate and respond to the "internal" technological and economic details that each can best deal with individually. This is a natural and efficient division of responsibility of the type that is essential for successful management of any complex and dynamic system, whether a modern corporation or an economy. To the extent that market devices of various kinds can effect this kind of division, they can help solve difficult environmental management problems.

An example of a system that has some of the trappings of a market mechanism, but essentially leaves regulators with the responsibility of deciding the details of individual responses, is the "delayed compliance penalty" in the 1977 Clean Air Act amendments. This provision requires EPA to define not only what a particular source must do and when, but with what technology and at what cost. Then, if compliance does not result, EPA can, at its discretion, collect an amount of money equal to costs saved by the source as a result of its noncompliance—unless there are good reasons, somehow defined, for noncompliance. This is an innovative enforcement device for which some success is claimed in the Connecticut state air pollution control program, and we hope that experiments with it prove fruitful. However, we do not expect it to provide the needed fundamental improvements in flexibility and incentives, because it does not let market forces determine the details of the response.

Of course, letting market forces help with some of the details means that the authorities cannot predict and determine the outcome precisely. Plant owners will have their own ideas about which technology to develop, when to install it, and so on, given the strength of the market incentive and their individual situations. But if the experience with the Clean Air Act has demonstrated anything, it must be that the future cannot be predicted and determined precisely, no matter how tough the language in a law. Furthermore, as more is learned about the complex relationships among policy, technology, economics, environment, and health, it becomes more clear that there is no simple way to decide in detail what should be accomplished and on what schedule. Instead of setting arbitrary, detailed, short-term goals and deadlines and then overregulating with an expensive and ultimately ineffective program, it might be preferable to set general objectives, use market devices to provide the proper overall incentives, and then let market forces determine the detailed who, when, where, and how of emission controls. The potential benefits to both environmental and energy-economic values in the long run are great enough that such an approach should at least be considered and tried with some enthusiasm.

Applications of Market Mechanisms Within the CAA

In this section we discuss some ways that market devices might be incorporated by legislative action into the existing basic structure of the Clean Air Act to help make it more flexible in the short run and more effective in the long run. Our intent here is not to be definitive or exclusive in our suggestions, to debate all the pros and cons of markets versus regulations, or to provide ultimate solutions. Rather, we intend only to express support for the basic idea of using market devices, to show the kinds of systems that we think could be workable and helpful within the existing programs, and to recommend that these ideas be given more serious consideration than they have so far received. A later section will recommend some more fundamental changes in the philosophy of the Clear Air Act.

Control of New Sources. We agree with the basic idea behind New Source Performance Standards—that new potential sources of the important air pollutants, wherever they are built, should take some minimum steps to reduce emissions. We take this position for a specific reason: in light of the uncertainties and complexities involved in determining the ultimate impact of pollution, emissions should be reduced whenever and however it is relatively cheap and easy to do so, such as in the design and construction of new facilities. Our objective is cost-effective reduction of emissions over time, not the installation of control equipment for its own sake, protection of eastern coal mining, or anything else.

To reduce emissions over time in a cost-effective way, it is not enough for new sources to incorporate the control devices that the authorities are able to identify and require. Even where new sources are concerned, at least three other requirements must be met:

1. Improved control technology and improved combinations of fuel, processes, and controls must be continually sought, demonstrated, and applied.
2. Once built, new sources must be operated, maintained, and periodically modified.
3. Existing sources must be phased out or upgraded continually.

In our view, the NSPS provisions of the Clean Air Act do not meet these three requirements. They are ineffective, even counterproductive, as stimulants to the discovery, development, trial, and application of new technology or fuels—especially when stated in terms of required uniform percentage reductions as mandated by the 1977 CAA amendments. They do little to encourage continual attention to emissions once a new source is built. And they tend to discourage modernization of a polluting industry, since the existing sources must surrender their valuable "rights to pollute" if they are retired or modified.

The most obvious way to introduce market devices into the NSPS program to help overcome some of these deficiencies is to replace the set of "absolute" new source performance standards with a set of allowable levels of emissions of each pollutant and then to impose per unit charges on emissions above the allowable levels. If we set the initial allowable levels at or—better—below the existing or proposed NSPS for each pollutant separately and set the charges high enough so that each new source will make serious efforts to reduce emissions to or near the allowable levels, we reduce a number of problems:

- All potential new sources (within some broadly defined class) would be subject to the same, certain rules, reducing the need for case-by-case negotiation.
- The precise level allowed for each pollutant is less critical, because no one would expect all sources to meet all required levels simultaneously and because the penalties for small failures would be small.
- Each source would be free to find its own best solution, involving some combination of clean fuels, control devices, careful operation, and so on. Any source wanting to try a new technology or new fuel would be free to do so without negotiating a special arrangement with EPA.
- Temporary malfunctions, interruptions in fuel supply, the high emission levels unavoidably associated with startup and shutdown of a plant (which are now simply ignored), and the like would be subject to a predictable but manageable penalty.

A set of allowable emission levels of this sort could be established for the same source categories used in the NSPS, where distinctions are made on the basis of fuel type (oil, bituminous coal, lignite, etc.), size of facility, boiler design, and so on. However, distinctions of this type, which are necessary when one is setting absolute performance standards, are neither necessary nor desirable in market type systems. Considerations of technical feasibility and cost effectiveness may dictate that small or coal-fed boilers emit more pollutants per million Btu of fuel used than large or oil-fired boilers. But the best way to determine to what extent this is true and whether the alleged economic advantages of these dirtier technologies justify their higher emissions is to require them to pay the same emission charges, down to the same allowable levels, as the cleaner sources pay.

   In fact, strict economic logic requires that the "allowable" or costless level of emissions be zero, so that zero emission alternatives—such as nuclear power or energy conservation as a substitute for a coal-fired plant—are given proper encouragement. Any system that gives to a new source some free emission rights in the form of an allowable emission level provides some noneconomic "subsidy" to new pollution sources. By keeping the allowable levels small or by charging a high price for them, this problem can be reduced or eliminated; and in any case, such a system would still be a great improvement over the present one, in which all emissions are allowed free of charge unless the regulators can get a conviction in court. But, strictly speaking, there should be no allowable emission rights made available free of charge to anyone who wants to start polluting. Thus, even if many distinctions are made in setting allowable emission levels initially, these distinctions not only can be but should be reduced over time, perhaps by decreasing all allowable levels proportionally until they approach zero.

Although we support the eventual application of this concept to all potentially significant sources, it would be particularly useful for energy policy purposes if some such system were instituted for fossil fuel facilities as soon as possible. For example, let us say that all fossil fuels were given an allowable emission level of 0.3 pounds of sulfur per million Btu of heat value at the mine mouth, wellhead, or point of import, decreasing by, let us say, 10 percent per year. If all emissions of sulfur from fossil fuel used in new sources were also subject to an emission charge of 50 cents per pound, increasing with inflation, then the continual problem of setting and periodically modifying the sulfur NSPS for various types of fossil fuel sources would essentially be solved. A plant operator would find it uneconomical to burn eastern coal with a 3 percent sulfur content in a new source with no control because the excess emission charge would be $1.05 per million Btus, comparable to the cost of mining.[1] However, a power plant with a flue gas desulfurization (FGD) unit that removed 85 percent of the sulfur would pay only 3 cents per million Btus in excess emission charges. If the efficiency of the FGD unit slipped to 75 percent, the excess emission charge would increase to 15 cents per million Btus—a powerful incentive to keep the unit

---

1. The emission charge in this example is calculated as follows: The amount of sulfur (S) per unit of heat value is

$$\frac{1 \text{ ton coal}}{25 \times 10^6 \text{ Btu}} \times \frac{0.03 \text{ ton } S}{\text{ton coal}} \times \frac{2{,}000 \text{ lb.}}{\text{ton}} = \frac{2.4 \text{ lb. } S}{10^6 \text{ Btu}}.$$

Therefore, assuming an allowable emission level of 0.3 pound of sulfur per million Btu and an emission charge of 50 cents per pound of sulfur, the excess emission charge is

$$(2.4 - 0.3)\frac{\text{lb. } S}{10^6 \text{ Btu}} \times \frac{\$0.50}{\text{lb. } S} = \frac{\$1.05}{10^6 \text{ Btu}}.$$

operating efficiently, but not a total disaster if the poor efficiency is due to disappointing performance of a lower cost or experimental FGD unit. By using 2 percent sulfur coal until the performance of the FGD unit was improved, the plant could reduce the excess emission charge to 5 cents per million Btus.[2]

With such a system of allowable emissions and an excess emission charge, the developers of new processes such as solvent-refined coal or fluidized bed combustion would not have to negotiate special deals with EPA for fear that they would be excluded from the market for failing to meet the NSPS precisely. If their systems are almost as good as those of competitors at controlling sulfur and have compensating advantages, they will find a market. And if a similar system is used for nitrogen oxide emissions (with an allowable emission level of, say, 0.5 pounds per million Btus, decreasing by 5 percent per year) technologies such as fluidized bed combustion, which are relatively good at nitrogen oxide control, could even trade off sulfur dioxide for nitrogen oxides. In short, a lot of desirable flexibility would be introduced.

The only problem not addressed by this sulfur dioxide new source emission charge system is the protection of eastern coal mining. Although we do not favor distorting the concept of NSPS in order to further such objectives, there is apparently political support for doing so. If it must be done, the system outlined above could be modified to do so by granting sulfur emission rights on the basis of the amount of sulfur in the fuel, rather than on the basis of the

---

2. The arithmetic supporting the estimate of excess emission charges in this hypothetical case is as follows:

a. With 85 percent sulfur removal on 3 percent sulfur coal with 2.4 lb. S per million Btus, the excess emission charge is

$$\left[\left(\frac{2.4 \text{ lb. } S}{10^6 \text{ Btu}} \times (1{-}0.85)\right) - \frac{0.3 \text{ lb. } S}{10^6 \text{ Btu}}\right] \times \frac{\$0.50}{\text{lb. } S} = \frac{\$0.03}{10^6 \text{ Btu}}.$$

b. With 75 percent sulfur removal on 2.4 lb. S per million Btus coal, the charge is

$$\left[\left(\frac{2.4 \text{ lb. } S}{10^6 \text{ Btu}} \times (1{-}0.75)\right) - \frac{0.3 \text{ lb. } S}{10^6 \text{ Btu}}\right] \times \frac{\$0.50}{\text{lb. } S} = \frac{\$0.15}{10^6 \text{ Btu}}.$$

c. With 75 percent sulfur removal on 2 percent sulfur coal with 1.6 lb. S per million Btu, the excess emission charge is

$$\left[\left(\frac{1.6 \text{ lb. } S}{10^6 \text{ Btu}} \times (1{-}0.75)\right) - \frac{0.3 \text{ lb. } S}{10^6 \text{ Btu}}\right] \times \frac{\$0.50}{\text{lb. } S} = \frac{\$0.05}{10^6 \text{ Btu}}.$$

amount of fuel (Btus) in the fuel. Any desired political objectives can be incorporated into this system as well as into the NSPS.

Nonattainment Regions. The most important provisions of the Clean Air Act in nonattainment regions are the requirements on new sources: "lowest achievable emission rates" (LAER) must be met, emission "offsets" must be obtained, and existing sources under the same ownership must be in compliance. For existing sources, there is little more than the traditional regulatory pressure to do one's best to reduce emissions if it is not too costly.

For new sources in nonattainment regions, a system of allowables and excess emission charges, similar or identical to the one outlined above for NSPS, could be used to implement the LAER requirements. The offset requirement already introduces some rudimentary market-like arrangements. For example, to get permission to build a pipeline terminal in Long Beach, California, Standard Oil of Ohio (Sohio) agreed to pay for emission control by three large dry-cleaning establishments and an electric utility. If this project goes forward, the result will be a net reduction in emissions of hydrocarbons and sulfur dioxide, as well as a step toward reaching the region's and the nation's economic and energy objectives. Similar arrangements have allowed approval of a Volkswagen plant in Pennsylvania, a General Motors plant in Oklahoma, and an oil refinery in Virginia.

We fully support the general concept of allowing sources to find (and buy) offsetting emission reductions and recommend that it be broadened and made more flexible. In particular, it should be remembered that the problem in many nonattainment regions is that it has been, and may long remain, unreasonably costly to meet the national air quality standards. If allowable emission levels are defined as those necessary to meet these standards, continuing "excess emissions" must be expected, in total and for important individual sources, no matter what kind of policy is used. Thus, an effective policy must simultaneously legitimize these excess emissions (because there is no other choice) and apply continuing pressure to reduce them (because that is the only way to make progress), for both new sources and existing sources. And it should not be presumed that the excess emissions, in total or for individual sources, are best eliminated according to some legislatively or administratively determined schedule.

An effective way to allow something while discouraging it is to make it expensive. The offset policy is effective because it makes emissions expensive both for the new source (which must purchase emission rights from others) and for the existing sources who sell

their emission rights (and would lose the sales revenue if they continued their emissions as before). However, by requiring only new
sources to find emission offsets and by requiring owners of new
plants to obtain offsets equal to several times the expected emission
levels of the new source, the present offset policy strongly discriminates against new sources. The reasons for this discrimination are
clear enough: potential plants do not have the immediate economic
and political importance that existing employers have. Moreover, the
threat to deny a construction permit to a potential new source is a
powerful weapon to force emission reductions that might otherwise
be impossible. But by making it difficult for new sources to enter
and by insulating existing sources from both the opportunities and
the pressures that market forces can provide, the existing offset
policy may produce rates and patterns of emission reductions and of
economic growth that are less beneficial than they could be.

One way to expand the offset concept is to set individual emission
rights for existing sources equal to those emission levels that the state
implementation plan (SIP) promises to enforce to accomplish the
national air quality standards; to allow these rights to be transferred
on a one-for-one basis from existing to new sources and among existing sources; and to apply to emissions in excess of the number of
emission rights uniform monetary charges that start at low levels and
increase with time. Under such a system, a potential new source
would be required to acquire emission rights equal to its expected
emissions and to pay an additional charge to the extent that its emissions turned out to be higher than expected. Existing sources that
find it relatively expensive to reduce emissions to the levels required
by the SIP and that would be in a good position to postpone taking
action under the regulatory system, will find it cheaper to purchase
emission rights from sources that can reduce emissions relatively
cheaply. As competition among new and "noncomplying" existing
sources makes the emission rights more valuable, all sources will
make greater efforts to reduce emissions, but in a cost-effective way
over time, so that both economic and environmental objectives can
be advanced.

A potential obstacle to implementing this concept is that the SIPs
do not, in fact, define specific emission levels for each source, but
rather contain regulations stated in terms of allowed emissions per
unit of input or output, good engineering practice, and the like; this
is especially true for smaller sources. As a result, the "allowed" emission levels for a particular source may depend on variables over
which the air pollution control agency has no direct control, such as
scale or hours of operation. To the extent that this is the case in a

region, no emission control program can effectively force emission reductions. An offset policy that allows emission rights to be sold only after they are clearly defined in terms of a maximum quantity of emissions per hour or day may encourage existing sources to negotiate such clear definitions, because only then will the rights have any market value.

To move toward a system along the lines that we suggest for nonattainment regions, a first step is to allow a large industrial facility under the control of a single manager to allocate emission levels among the various sources in the facility as the manager chooses, so long as the overall emission limitation is attained. This "bubble concept" has been used sparingly to allow internal expansion of some emitting process in a facility, if emissions from other sources in the facility are reduced by offsetting amounts. That this should be controversial at all is evidence of how far some would push the regulatory process into the internal affairs of others.

Once the "bubble concept" for large industrial facilities and the offset policy for new sources are accepted, the evolution of a more comprehensive system of marketable emission rights should not be far behind. The "bubbles" can begin trading among themselves and with new sources; all of these will be searching for existing emissions rights that can be defined and purchased.

In the long run, any number of variations and refinements can be made in this general approach, while maintaining most of the advantages. For example, the initial emission rights could be based on current emission levels, decline over time according to some schedule, and be enforced with a high excess emissions charge. Emission rights could be purchased and retired by private citizens or public agencies who want to improve air quality. Brokers could purchase and hold emission rights for resale to new sources. Transfers of rights from one time period to another could be allowed on some basis, so that, for example, a plant could trade future for present emission rights to allow itself time to make required changes.

The offset policy is an important and encouraging step in the right direction; but there are more steps in this direction that can be taken to further both environmental and economic goals in nonattainment regions. We recommend that these steps be considered and taken with less timidity.

Prevention of Significant Deterioration

Every region that is not a nonattainment region (where the air is dirty) is a prevention of significant deterioration region (where the air is clean). However, because definitions of nonattainment and PSD are set for each pollutant separately, a region may be a nonattainment region for some pollutants and a PSD region for others. How the two programs will interact with each other in such cases is only one of the many uncertainties in the Clean Air Act.

In any PSD region, a new source must apply "best available control technology" for all air pollutants. This requirement, like the requirement that "lowest achievable emission rates" be attained by new sources in nonattainment regions, has the same strengths and weaknesses as the new source performance standards. One of these weaknesses of particular importance in clean air regions is the inability of performance standards to prevent emissions from a plant during periods of startup, shutdown, or malfunction, when emission rates may be ten or a hundred times as large as during normal operation of a plant with effective control equipment. Since regulations cannot prevent equipment malfunctions or control the operating cycle of individual plants, they typically specify emissions only under "normal" conditions. The "unavoidable" emissions during nonnormal periods are left uncontrolled and can easily become the principal problem in regions with a few large new plants. Our general recommendation here is that all new sources, anywhere, be subject to a system of charges on emissions in excess of low and declining allowable levels, as discussed in the earlier section on NSPS. Such a system need not ignore nonnormal periods and, by making malfunctions and plant cycling expensive, will discourage both.

The more difficult parts of the PSD provisions are those that assign each PSD region to one of three classes, according to the value of maintaining air quality in each region, and then establish procedures for deciding whether or not to allow a new emission source into a region. The basic concept here is that, even where national air quality standards are met, increases in air pollution should be tolerated only if there are "good reasons" for doing so. In most cases, the "good reasons" will be economic: it is too costly to put the new power plant somewhere else, to install more effective control equipment, to use another fuel, or to conserve further on the use of energy. Thus, the basic problem in PSD regions is to decide whether the alleged economic benefits of some proposed project are large enough to justify the air pollution costs. This problem is made more difficult by the fact that the benefits often do not accrue to the same people who bear the costs. Complications arise if several sources want to come into the same region, if future growth is anticipated, and if existing sources could be reduced to allow new ones.

The problem of deciding what it is worth to control emissions in specific regions is inherently a political and administrative one and is not our immediate concern here. The three way regional classification procedure established by the 1977 CAA amendments is reasonable enough, at least if used in an overall PSD program more flexible than the existing one. The balance of the problems that must be dealt with in a PSD program, however, are economic ones, not fundamentally different from the kind of problems handled routinely by markets. And marketlike mechanisms can help here. The best way to estimate the economic value to be gained by accepting dirtier air is to find out how much the potential emitters are willing to pay for the privilege of dirtying it. To allocate costs and benefits more equitably, those who benefit should make payments for the account of those who bear the costs. If several existing or potential emitters want to use the same air resource, let them bid or trade among themselves, so that the one with the most valuable use for the resource gets it. To the extent that market processes can help solve some of the difficult economic problems involved in implementing a PSD policy, they should be used.

The most direct way to introduce market policies into the present PSD program is to translate the allowable ambient air quality increments in a region into allowable emission levels (perhaps stated in terms of emission density) and then to establish markets in these emission rights. Although it is not a simple matter to determine what levels of emissions, in what locations, will result in what maximum increases in ambient concentrations, it is no more arbitrary to set maximum emission increments than to set maximum ambient concentration increments; and, since it is emissions that must ultimately be controlled, there are important administrative advantages to stating the goals in terms of emissions and not of ambient levels. There is no technical reason this cannot be done—although it might require new legislative authority. There is no apparent reason why, on average, the emission increments should be more or less protective of regional air quality than the ambient increments, although clearly the effect will not be the same in every region.

Once the allowable emission increments in a PSD region are determined, the problem of deciding whether a proposed new source would meet the allowed increments would be greatly simplified, removing a major source of delay and uncertainty in the PSD program. These emission rights could then be sold—to potential sources wanting to use them immediately; to brokers or speculators who want to hold them for future resale; to environmental groups who are able to raise money to protect air quality. A new source willing to pay high prices for these rights would be demonstrating, in a concrete way, that there really are sound economic reasons why society should agree to allow the air to be degraded. The revenues could be used for environmental or other worthy purposes, to compensate for the lowered air quality. If several sources want entree into the region, they could bid for the rights; and as economic and technical conditions change, the rights could change hands, so that they continue to be put to good economic use, without further degrading air quality.

A further refinement would involve applying charges on emissions in excess of the number of rights held by a source. This would help solve one of the most serious problems with the PSD program: what to do if a new source fails to accomplish the emission levels promised in the construction permit. It would also provide an automatic escape valve for when it turns out to be particularly valuable to allow economic activity in a region, even though this leads to emissions in excess of the specified PSD increments. And by letting the level of emission rights decline over time, continuing pressure to reduce emissions can be maintained, even in regions where effective competition for the rights is weak.

If some market devices along these lines are introduced into the PSD program, in such a way that they reduce the administrative and procedural costs and delays and introduce some flexibility, the number of new sources admitted to PSD regions in the short run would probably be greater than otherwise. To those who view trade-offs between energy and environmental concerns individually and in the short run, it will be obvious that environmental values have lost to energy and economic values. However, from a longer and broader perspective, it is not clear that environmental values would lose from adoption of a policy that reduces bureaucratic roadblocks to new sources while encouraging cost-effective emission controls. More new sources in some PSD regions may mean fewer new sources elsewhere and, perhaps more important, more rapid phaseout of older sources everywhere. A system that provides continuing incentives for technological development, good operation and maintenance, and periodic upgrading, may result in both less pollution and more economic growth even within a single PSD region.

Even in the short run, emissions for an individual source may be less as a result of these incentives. The closest thing there is to a "free lunch" in economic affairs occurs when efficient programs replace inefficient ones; and the PSD program as it now stands is as good a place as any to look for such a "free lunch."

Perhaps equally important is the likelihood that the PSD program will not continue unchanged very long. It is so cumbersome and restrictive in its present form that, once its economic costs become clear, it will probably be modified—and not in ways that favor clean air. If this is correct, then the real choice may not be between the present system and one that allows economic growth while minimizing the adverse impact on broader and longer term environmental values, but between a system of this latter type and a greatly weakened, ineffective, and inefficient system trying to protect clean air areas.

Beyond the Clean Air Act

The preceding suggestions for applying market concepts are all within the narrow confines of the Clean Air Act, as amended in 1977. Over the next twenty years, however, this Act will continue to evolve, not just in its particular mechanisms of implementation, but in basic philosophy. Following are some of the directions in which we would encourage evolution, sooner rather than later.

Deemphasize National Ambient Air Quality Standards. These standards have already been deemphasized since 1970, but continue to play a role in the administration of the Act that is inappropriate in light of the scientific and logical weakness of the whole "threshold" concept and of the arbitrary nature of the particular levels selected. In particular, to the extent that important resource allocation decisions turn on whether a region is classified as a nonattainment or a PSD region, the national standards are too important. Whether an imprecise estimate of pollutant concentration in one or several places is more or less than the arbitrary standard should be a minor consideration in most significant resource allocation decisions.

This is not to say that measurements of air quality and continued research on the relationships between pollutants and human health should be regarded as less important or less necessary. Indeed, one of the principal reasons for deemphasizing the national standards as policy instruments is to encourage more diligent, long-term scientific effort devoted to improving these measurements and defining these relationships. When important decisions turn on the precise level of a measurement or standard, there is no way to avoid letting policy considerations influence scientific judgments, as when prestigious scientific groups deny the significance of evidence of low level effects because to do otherwise may result in establishment of clearly absurd policy goals. By reducing the policy role of the standards in ways that we will shortly suggest, we would encourage scientific research to follow the evidence where it leads, with less concern about the immediate policy implications.

If the policy role of the national standards is reduced, the objectives of air quality measurements and of scientific research will change. No longer will the problem be to find and defend "safe" ambient concentration levels or to make yes-no decisions about whether emissions do or will cause violations of ambient standards somewhere or other. Rather, the questions will concern how much damage will ultimately be caused by increased emissions of pollutant A at point X, at least relative to pollutant B at point X or to pollutant A at point Y, and whether costly control measures will lead to reductions in air pollution damages. These are obviously difficult questions, requiring more, not less, scientific research on where air pollution goes, what happens to it along the way, and what it does when it gets there.

Recognize Long Distance and Low Level Effects. The basic strategy of the 1970 Clean Air Act was to manage emissions within regions of relatively small size—metropolitan areas or several rural counties—in order to bring about "safe" levels of air pollution everywhere. Over the years this strategy has become increasingly inappropriate, and individual regulatory decisions and legislative amendments have, in effect, rejected its basic assumptions. The increased emphasis on "best available control technology," the rejection of tall stacks as an acceptable control device, and the PSD requirements are all justified partially by the fact that pollution may travel long distances and be harmful at low levels. However, these various ad hoc provisions, added on top of the basic State Implementation Plan and NAAQS strategies, do not add up to a logical policy for dealing with long distance transport and low level effects of pollutants.

The case of sulfur dioxide and sulfates is clearly one in which managing on the basis of "safe" air quality standards and control regions is inappropriate. Sulfur dioxide emitted in the industrial Midwest can be converted to metallic and acid sulfates while being transported hundreds of miles and then be harmful to health at low levels in the populated Northeast or be harmful to fish life in mountain lakes. The sulfur dioxide standard is of no direct relevance even if a no damage standard could be established for sulfates themselves (scientific groups say there is "insufficient information," although the real problem is probably that there is no threshold), it would never be attainable until sulfur emissions were virtually eliminated over much of the country. And managing on a region-by-region or state-by-state basis makes little sense when sources hundreds of miles away may be the cause of a local problem. Similar processes and problems are being recognized where nitrogen oxides and oxidants are involved.

Obviously, situations of this type are so complex and involve so much fundamental uncertainty that highly sophisticated management systems, based on precise objectives and fine distinctions, are inappropriate. However, basic considerations of cost effectiveness do suggest the outlines of a logical control strategy. For example, a workable system for sulfur might be constructed by combining information on the general relationship between sulfates and health effects (e.g., effects are linear in exposure), with atmospheric chemistry and transport models to construct a crude index of average relative health effects of sulfur emissions from different locations. This index would be higher where prevailing winds carried sulfates into densely populated areas. It could perhaps be adjusted to reflect other impacts of emissions, such as impaired visibility and acidified lakes. A reasonable case can even be made that, when all the uncertainties, the diversity of effects, and the political and administrative factors are considered, the impact index should not vary at all with location, at least over areas as large as "the industrial Midwest."

Be that as it may, once an impact index is attained for each state or county or air quality control region, control measures would be implemented with varying degrees of stringency, depending on the index. The simplest way to do this would be to impose an emission charge proportional to the index. The absolute level of these relative charges would have to be determined by a political and administrative process that compares the costs of control to the benefits obtained.

Whether it is worth going to the trouble of instituting this or any other management system to deal with long distance, low level effects depends on whether doing so would make any real difference. Even good management is costly and should not be expanded without good reason. If the existing programs remain in place, perhaps modified along the lines we suggest, and have some effect on total emission levels, they will reduce the long distance, low level problems even though they are not designed for this purpose. Thus, it is necessary to decide that a particular long distance, low level problem is serious enough to worry about and will not be handled adequately by existing programs before deciding to do something about it. Then it must be decided whether relatively minor changes in existing programs, additional programs, or replacement programs are called for.

Our reading of the scientific information on sulfur dioxide, sulfates, and long distance transport convinces us that sulfates and not sulfur dioxide are the threat to health from sulfur emissions and that the long distance, low level problem is serious enough that, if we were starting from scratch in designing an air pollution control program for sulfur, we would adopt an extensive regional strategy rather than the localized strategy of the Clean Air Act. The same is true for most other pollutants, for that matter. However, the existing programs are in place and are not likely to be replaced soon. Thus, we conclude that the best way to deal with the long distance, low level problems is through changes in existing programs, along the lines we suggest above and in the next section.

Simplify and Decentralize Implementation Programs. There is a natural tendency to think that as more is learned about a difficult public policy problem such as air pollution and its control, it will be possible and desirable to manage the problem more tightly and centrally. The objectives will become more precise, the methods of accomplishing them better understood, the fine distinctions easier to make, and the social consensus more solid. If so, then the agency charged with managing the problem can be given more authority to make distinctions, can use a range of different programs for dealing with special situations, and can decide more of the details of the solution.

There may be important public policy areas that meet this description, but air pollution is not one of them. For this reason, we favor less effort by centralized administrative processes or agencies to set precise, quantitative objectives, make absolute distinctions on the basis of poor knowledge, and decide and enforce the details of the solution. The policies we favor may use centralized administrative processes to establish guidelines, incentives, and the general framework; to monitor performance and recommend needed changes; and to gather the basic information required. But they will make distinctions only when necessary and when the bases for the distinctions are clear; and they will use continuous penalty-reward mechanisms, so that the effects of small differences are small. Then they will let local agencies, private firms, and individuals make their own best detailed adjustments.

The specific policy steps that can be taken to move in the recommended direction within the narrow confines of the Clean Air Act have been already discussed in general. The most immediately useful step beyond these confines would be to combine the "best available control technology" requirements for PSD regions, the "lowest achievable emission rate" requirement for nonattainment regions, and the NSPS requirements into a single set of allowable emission levels and excess emission charges, as outlined above in the discussion of controlling new sources. Beyond that, gradual expansion of the use of the offset and "bubble" concepts in nonattainment regions and of the marketable emission increments in PSD regions would lead to simplification and uniformity in the methods of administering the national air pollution control program.

Eventually, the CAA could evolve into a system with a fixed number of marketable emission rights in each region, with emission charges applied to emissions not covered by rights. The value of the emission rights in each region would depend upon the demand for them relative to the supply and the method and strictness of enforcement. In some nonattainment regions where "violations" are, for good reasons, the rule rather than the exception, an emission charge large enough to discourage emissions but not so large as to cause economic chaos might be used. In critical PSD regions, a high excess emissions charge, intended to make any violation virtually unthinkable, might be used. Even standard regulatory enforcement methods might be applied, with varying degrees of toughness. But the existence and marketability of the clearly defined emission rights and the uniform treatment of technology for new sources would simplify the processes of planning, setting standards, taking enforcement actions, and negotiating with existing or potential new sources. Both the control agencies and the emitters would benefit. The detailed technical and economic decisions would be made by the emitters themselves in a decentralized, market-guided manner.

Within this overall structure, determined by federal policy, each state or region could be allowed as much decentralized control as desired. Initially, the number of emission rights might be determined in the nonattainment regions by the requirement that the NAAQS should someday be met and in PSD regions by the three way classification scheme (in which states already play a significant role) and the federally determined allowed ambient increments. The federal EPA might also establish minimum levels for the excess emission charges applied to all new sources, as well as to all sources in nonattainment and PSD regions, and might impose surcharges in some regions to deal with the long distance, low level problem. But the state or localities could apply surcharges to excess emissions or to all emissions and perhaps should even be allowed to reduce the rates below federally set levels to stimulate growth—at some cost to their own revenues. Eventually, state and local governments might be allowed to vary the number of emission rights within a region (presumably by buying or selling them) to accommodate their individual situations.

With a system along these general lines, emissions would be decreased (or increased) in each region to extents and at rates that depend on many things—the cost of emission control, the local economic conditions, the preferences of the citizens, the perceived

value of clean air, and the extent to which local emissions have effects elsewhere, for example. Evidence concerning the impact of
emissions on air quality, visibility, ecological systems, human health,
and materials, both locally and at distance, would be important, even
though the precise relationship between some variable measure of
concentration and some arbitrary standard might be of limited
interest. There would be reasonable assurance that there was good
economic justification for the emissions that remained.

CONCLUSIONS

The processes involved in air pollution and its control are complex,
poorly understood, and inherently variable; there is no realistic prospect that either air pollution itself or the need to take costly
measures to keep it under control will go away soon. These facts convince us of the need for long-term air pollution policies that pursue
general, flexible objectives by providing information and incentives
and by allowing decentralized processes to determine the details of
the response over time.

As we have noted, the problem of controlling air pollution is not
fundamentally different from other resource allocation problems
that are handled with passable success in most of U.S. society by
letting markets do much of the work, within a general framework of
property rights, law, and regulation. Our general recommendation is
that air pollution policy take the hint from other, more successful
parts of U.S. society and use market forces to do those things that
they do rather well.

Within the structure of the Clean Air Act as it now exists, a
number of steps can be taken to begin the evolution toward a long-
run air pollution policy that strikes a more appropriate mix between
governmental regulation and market processes. An obvious and
important step would be to replace the difficult and overlapping
"best technology" standards for new sources with a single, unified
system of allowable emission levels and monetary charges on excess
emissions. The "emission offset policy" being used to allow new
sources in nonattainment regions should be expanded toward a full-
fledged market in emission rights, with increasing charges on excess
emissions to put existing sources on notice that failure to make
progress will eventually become expensive. And in Prevention of
Significant Deterioration regions, the best way to assure that clean
air is not degraded for insignificant economic gains is to make incremental emissions costly, as can be done with marketable emission
increment rights and excess emission charges.

Over time, these administratively similar programs can be integrated into a unified but decentralized system in which governments make the kind of general policy judgments that are appropriate under the prevailing conditions of uncertainty and variability, while local agencies and private firms and individuals decide their own detailed responses. Unfortunately, the trend in policy is (with minor exceptions, such as the offset policy) in the opposite direction. Each succeeding amendment to the Clean Air Act seems to impose more absolute limitations, add more sources of confusion and uncertainty, and give EPA more authority and responsibility to control technical and economic details. It is too early to tell where this trend is heading, but if the constraints on economic growth and valuable energy projects become costly enough, it will be reversed. We only hope that if and when this reversal occurs, the result will be improved administrative and economic efficiency, rather than simple relaxation of existing regulations, so that both air quality values and energy-economic values can be advanced.

Alternatives to Fossil Fuels

IV

Introduction

There is no dearth of energy sources. Oil and gas, extracted from their more or less conventional environments, can be available for the balance of the century and some distance into the next; coal exists in abundance, here and abroad; close substitutes for conventional oil and gas are widely distributed and could be made available at higher costs; and there are a great many unconventional resources, some very large, but depletable, like geothermal, some very large and nondepletable (or more commonly lumped under the heading "renewable"). We have dealt with oil and gas as resources in Chapter 7 and with policy questions in Chapters 4, 5, and 6 and on and off throughout the book, while coal has formed the focus for Part III. In this part we have selected two sources of energy supply—nuclear and solar—that merit consideration separate from the oil-gas-coal trinity.

Nuclear now supplies some 13 percent of U.S. electricity and varying percentages abroad. It could conceivably be expanded in its present technological guise to supply much more, and the breeder has the potential for delivering practically unlimited amounts of energy. But nuclear energy is beset by extremely serious problems that make its very survival a question. Chapter 12 is not a comprehensive report on nuclear energy. Rather, it focuses on a limited number of issues that need to be resolved or at least managed satisfactorily if the nuclear option is to survive.

The chapter begins with a brief description of the role of nuclear energy, its technological features, competitive setting, and so on. Among the critical issues it addresses in detail are the problems and management of nuclear waste material, the weapons linkage, safety and other hazards, and the nature and extent of differences in the approach to nuclear energy taken by different countries. These are issues that form the background for such policy decisions facing this and other countries as breeder development, reprocessing and recycling of nuclear fuel, spent fuel storage or disposal, and management of uranium mill tailings. Over all of these specifics there hovers, now more than ever, the question of the viability of this source of energy, a viability that is important both because diversification of energy sources is a desirable objective and because the breeder remains a technology that would relegate fuel resource limitations to a far horizon.

By contrast, Chapter 13 deals with the global, durable, and, by and large, environmentally least objectionable source, solar energy, which offers a way out of having to rely on the two flawed sources— coal and nuclear. At the same time, it is our least exploited source. It can be tapped in a vast variety of ways, directly or indirectly, for heating, cooling, electricity generation, to raise crops, drive engines, and so on, but its costs are as yet relatively high and its performance uncertain in most applications. We do not favor slowing down the pace of learning and developing when we suggest that solar energy's greatest potential would seem to be as a very large and perhaps critical source in the somewhat more distant future, a role that could easily be jeopardized by rushing into narrow, premature selection of a small number of ways of utilizing solar energy and bringing them to quick commercial use. Consequently, the chapter we have prepared takes a somewhat different tack from most of the current writing on the subject. It attempts to present in the briefest possible space and with the least feasible reliance on technical language the basic nature of solar energy: what it can and what it cannot be expected to do, what its advantages and disadvantages are, what some of the misconceptions are (for example, it is obviously not "free" since substantial equipment and other efforts are called for to capture it, and it does have environmentally adverse features), how one can usefully put its cost in perspective and finally, what research directions are especially worth pursuing. It does not inform the reader (because we do not believe it is feasible) what share of total energy will be furnished by the sun at what future date, nor what amount of government funds will produce how much commercial technology. It will, we hope, convey to the reader a useful notion of what solar energy is all about and why we should develop it along a broad front of opportunities.

12

Nuclear Power

INTRODUCTION

Nuclear generating capacity in the United States has been growing at over 20 percent a year during the past five years and in 1979 accounts for about 13 percent of the electricity generated. It is an even larger fraction in some other countries. Yet throughout the world, the industry is in serious trouble. The construction of new plants has been opposed, and in some instances halted, because of generalized opposition to "hard" technologies—those characterized by bigness, complexity, and centralized decisionmaking—and specific concerns about safety, waste disposal, and environmental damage. These latter concerns have been reflected in frequently changing and increasingly stringent regulations, which have been a major factor in the doubling since the 1960s of the time required to bring a nuclear plant on stream. Higher standards and accompanying delays have in turn contributed to the escalation of construction costs, which have risen at more than twice the general rate of inflation. Uncertainty both about future demand for electricity and about the acceptability of nuclear power has increased. As a result of these developments, cancellations of existing orders for nuclear units have exceeded new orders in recent years, and a substantial excess in manufacturing capacity now exists in both the United States and Europe. Projections of the future growth of nuclear generating capacity have been falling steadily over the past few years. In the United States the estimate for the year 2000 has been reduced from 1,200 gigawatts electric (GWe), made in 1972, to about 300 GWe.

The changing outlook for nuclear energy was clearly evident by 1976, and then the industry was struck by another serious blow—one that many nuclear proponents considered to be at best unreasonable and unwarranted and, at worst, likely to have effects the opposite of those intended. The future of nuclear power became entangled with United States policy toward the proliferation of nuclear weapons.

The issue of proliferation became an important concern, especially in the United States and Canada, following the 1974 detonation of a nuclear explosive device by India. By the time of the 1976 U.S. election campaign, both presidential candidates had pledged to try to prevent nuclear power programs being exploited to produce nuclear weapons. Of particular concern was the possibility of the development of a "plutonium economy."

Here it should be pointed out that if plutonium, which is produced whenever uranium is irradiated in a reactor, is separated from the fission products that are also produced in a reactor, it can be used either to make nuclear weapons or as a reactor fuel. Prior to reprocessing, plutonium is "protected" from easy access because of the intense radioactivity of the fission products with which it is mixed in spent fuel. Plutonium separation has been occurring since World War II, and it continues as part of the weapons programs of the United States and the other nuclear weapons powers. It had long been assumed in the industry, both in the United States and abroad, that reprocessing of spent fuel from commercial reactors would also occur, so that plutonium could be used in reactor fuel to supplement uranium. However, following his election, President Carter announced that the United States would defer indefinitely the reprocessing of spent commercial fuel and would otherwise seek to prevent plutonium from becoming an item in world commerce, in order to reduce the likelihood of its becoming available for weapons manufacture. Efforts were also initiated to secure agreement among the major industrial nations to forgo supplying others with reprocessing equipment and to reverse decisions already made by France and West Germany to do so. Finally, the U.S. Congress passed the Nuclear Non-Proliferation Act of 1978 (the NNPA), a major feature of which is to give the United States a veto on reprocessing of fuel manufactured from uranium that had been produced or enriched in the United States.

The issues implicit in President Carter's decision and in the NNPA are complex but of great importance in any consideration of the future of nuclear power. They are discussed at some length subsequently. Suffice it to say here that the reaction in the nuclear industry and among many other governments was decidedly negative. In the United States the decisions were widely interpreted as biased against nuclear power generally and more specifically biased against "breeder" reactors—advanced reactors that are especially efficient in producing plutonium and that make no sense except in a situation where reprocessing occurs. Abroad, the decisions were widely interpreted as an effort to achieve commercial advantage for the United States by cutting off developments in areas of technology to which it was not as committed and where it was not as far along toward commercialization as others.

The latter concerns about U.S. intentions have been largely allayed, but this does not mean acceptance of the U.S. position on the desirability of delaying the reprocessing of spent reactor fuel and the introduction of breeder reactors. It is widely believed abroad, and by many in the United States as well, that the hope of limiting nuclear weapons proliferation by constraining nuclear power developments is illusory—an attempt to solve what is fundamentally a political problem with a "technical fix." It is argued that any nation likely to have a nuclear power program can produce weapons independently of its power program or will soon be able to; indeed, that could well be the preferred route. Second, others see reprocessing and breeders as essential for the solution of their energy problems, even though the United States may view delay as acceptable—indeed, possibly desirable—on economic grounds. The United States appears to have enough uranium to fuel all of its light water reactors (LWR)—the kind that are now being built—until well into the next century, even if it does not reprocess spent fuel to recover plutonium and unconsumed uranium. Other countries with major commitments to nuclear power, notably Japan and those of Western Europe, are not individually so well endowed with uranium (or other energy sources). They, therefore, find reprocessing, which can increase the efficiency of use of uranium in present generation reactors by 35 percent and by sixty- to one hundredfold in breeders, very attractive. In varying degrees, other nations have also viewed reprocessing as a desirable precursor to the disposal of waste products from the nuclear fuel cycle—a viewpoint with which our study differs (see below).

In these circumstances speculation about the future of nuclear power is necessarily fraught with great uncertainty—an uncertainty heightened by the Three Mile Island accident. The growth of the industry will depend on the regulatory environment (and not only that affecting nuclear power directly but also on that affecting its competitors), on local attitudes, on the resolution of such questions as spent fuel storage and waste disposal, on the constraints imposed relating to nonproliferation objectives, and presumably not least on economic considerations. We now elaborate on some of these questions, turning later to the specific policy questions about which there is debate.

NUCLEAR POWER AND ENERGY REQUIREMENTS

Costs of Light Water Reactors and Fossil Fuel Plants

In adding new electrical generating capacity, the choice between nuclear and fossil fuel plants has been increasingly dominated by uncertainty about future demand for electricity. Projection of future demand has been very uncertain since 1974, and the electrical utility industry has consequently been disposed to delay commitments as long as possible. This uncertainty has weighed heavily against the nuclear choice, because licensing and construction time has been much longer and capital costs larger for nuclear units. It has been entirely rational for utilities to accept the possibility of higher fuel costs if a plant must be built (the extreme case obtaining with gas turbines) in preference to a very early commitment to large capital costs for a nuclear plant for which there might not be adequate demand by the time it becomes operational. The points are made in Table 12-1, which gives some illustrative estimates, which would have been appropriate for the United States two or three years ago, of construction time, capital costs, and fuel costs for alternative generating options.

The total time for construction and licensing of nuclear plants has probably increased by a couple of years since then and that for fossil fuel plants by even more—perhaps by as much as four years—with the passage of the 1977 amendments to the Clean Air Act and the requirements for "best available control technology" (for sulfur dioxide emissions) and "prevention of significant (air quality) deterioration" (see Chapter 11). These considerations, and uncertainty about public

Table 12-1. Estimated Costs of Adding Electrical Generating Plants.

	Nuclear Units	Coal with Flue Gas Desulfurization	Oil, Steam Generation	Oil, Combined Cycle	Oil, Gas Turbine
Construction and Licensing Time	10 years	6	5	5	3
Capital Cost per kWe (1976 dollars)	$700	600	450	370	200
Fuel Cost Mils per kWh (1976 dollars)	6	12	21	16	24

acceptance of nuclear power, disposal of nuclear wastes, the future regulatory environment for both nuclear and fossil fuel power plants, and future demand for electricity, make the planning of additional generating capacity extremely difficult.

In other countries, additional considerations may militate against a nuclear choice. Among these are foreign exchange problems, the fact that nuclear units tend to be too large for small grids, and uncertainty about access to fuel.

In the light of these considerations, simple comparisons of generating costs are not likely to be controlling in decisions about new generating capacity even where they are made, as in the United States, by utilities. Nonetheless, cost comparisons will presumably be at least an important factor.

The costs of constructing nuclear plants have generally greatly exceeded estimates, and the capacity factors (the ratio of electricity produced to that that could be provided, assuming operating 100 percent of the time at maximum capacity) have on the average been substantially lower than estimates. Since capital charges are directly proportional to the former and inversely proportional to the latter, and since such charges are about two-thirds of the total cost of generating electricity with a nuclear plant, costs per kilowatt hour have been very much higher than estimates. Nevertheless, costs of electricity from nuclear power have been lower than for coal generation for many American utilities. Since the dramatic increases in oil prices following the 1973 embargo, they have been much less than for oil generation as well.[1] Some of the relatively low cost nuclear plants have been "turnkey" units purchased from the vendors at substantially less than cost.[2] However, even with nonturnkey plants, costs per kilowatt hour have been, and may continue to be, lower than for fossil fuel plants finished at the same time for most of the United States. The major exceptions are in the northern Rocky Mountain and western North Central states where low-sulfur coal can be mined at low cost. We reach this judgment believing that

- There is now enough experience with nuclear plants so that it may be reasonable to assume that capacity factors will be around 60 to

1. In response to the Atomic Industrial Forum survey for 1977, three-fourths of the utilities that reported costs for both nuclear and coal generation indicated the latter were higher. All those that reported costs for nuclear and oil generation indicated oil was more costly. The weighted average for nuclear-generated power was 1.5 cents/kWh; for coal 2.0 cents/kWh; and for oil 3.5 cents/kWh.

2. The vendors of these plants agreed to provide a ready to operate plant at an agreed price; they, rather than the utilities, accepted the increase in costs resulting from unforeseen contingencies and construction cost escalation.

65 percent—substantially lower than many earlier estimates but perhaps little, if any, poorer than for coal plants with sulfur removal equipment.

- Increasingly, uncertainties relating to nuclear power will be those relating to the back end of the fuel cycle, where the cost implications will be small, rather than to plant construction and operating conditions where the cost implications are large.[3] (The point is illustrated by comparing capital costs and back end costs in Table 12-2 and is discussed subsequently.)
- Costs for coal-generated power may escalate as rapidly as for nuclear power, assuming the imposition and enforcement of requirements to severely limit sulfur dioxide and possibly other emissions.

Because of high capital costs of nuclear plants, the costs and difficulties associated with sulfur removal in the case of coal, and the environmental problems with both, oil and gas will continue to be attractive alternatives in some countries, notwithstanding high fuel costs and concerns in many countries and, indeed, restrictions in some, relating to the use of these fuels for electricity generation.

In Table 12-2 we give estimates for electrical generating costs for nuclear and coal plants that might come on stream in the United States in 1990. These are bus-bar costs, which exclude transmission costs. For nuclear plants the estimates imply commitment now. For coal, a delay of perhaps two years would probably be possible. Even without further escalation in oil prices, base load generating costs with new oil plants would likely be higher throughout the United States than for either of the other alternatives. All costs are in constant 1978 dollars. It will be noted that uncertainties in cost projections exceed the differences between the estimates for the two alternatives. Thus, the fact that the cost estimate for the nuclear case is lower than for coal should be interpreted as suggesting only that it will *probably* be the preferred option on narrow economic grounds in most, not all, of the United States. The assumptions on which the table is based are given in the footnotes, and some are discussed subsequently.

One, the capital charge rate, perhaps merits discussion here. Costs in the table are calculated based on a rate of 12 percent, a rate that would be appropriate for use by utilities in making procurement decisions in a constant dollar—that is, inflation-free—economy. It

3. This assumption could well be vitiated if, as a consequence of the Three Mile Island accident, there is a movement to change standards and licensing and regulatory requirements.

Table 12-2. Projected Base Load Generating Costs for Plants Coming on Stream in 1990 (mils/kWh, 1978 dollars).

Cost Factors	Nuclear		Coal	
Capital Charges[a]	15.4	$\pm 3.3^e$	12.9	$\pm 2.9^e$
Operations and Maintenance	2.0	± 0.6	3.5	$+ 2.5$
				$- 1.5$
Fuel[b]				
U_3O_8	3.2	$+ 3.0$	12.0	± 3.0
		$- 1.5$		
Conversion to UF_6	0.1	± 0.02		
Enrichment	2.0	$+ 0.5$		
		$- 1.5$		
Fabrication	0.5	± 0.1		
"Back end" costs (i.e., spent fuel storage and waste management and, in the event of reprocessing, including credit for recovered Pu and U)	0.8	± 1.5		
Decommissioning[c]	0.02	$+ 0.10$		
		$- 0.015$		
Total[d] (U.S. average)[f]	24	$+ 5$	28	± 5
		$- 4$		

[a]Based on costs of $730/kWe and $610/kWe for nuclear and coal generation respectively, assuming in the nuclear case dual 1,150 MWe units and in the coal case triple 800 MWe units with flue gas desulfurization; capacity factors of 0.65 for both coal and nuclear; and a capital charge rate of 12 percent.

[b]Based in the nuclear case on tails assay of 0.20 percent U-235; no reprocessing of spent fuel and costs of $40/lb. for U_3O_8, $3.50/kg of U for conversion of U_3O_8 to UF_6, $85/kg of separative work for enrichment, $115/kg for fuel fabrication, and $230/kg for "back end" costs. Cost of coal is assumed to be $1.25/$10^6$ Btu. All costs in 1978 dollars.

[c]Based on a plant life of forty years and a cost of $70/kWe.

[d]These are bus-bar costs. The charge to the consumer would be 50 to 200 percent higher because of distribution costs, the need for high cost peak power, and so forth.

[e]Some comments on estimates of uncertainties in cost components: those for capital charges are based on an assumption of an uncertainty of ± 20 percent in the cost of plant (both kinds) and of ± 7 percent in capacity factor for nuclear and ± 10 percent for coal, the higher figure being used for coal because of the limited (and generally poor) experience with large scale flue gas desulfurization. The 20 percent uncertainty in cost of plant reflects simply the variation in estimates (adjusted) of a number of utilities and architect-engineering firms. The upper bounds could well be exceeded if environmental and regulatory constraints become more restrictive. Uncertainties in U_3O_8 and enrichment costs are not independent. If enrichment costs drop, as they might with changing technology, tails assay would presumably be reduced, and with that, U_3O_8 consumption would decrease.

[f]Regional variations in generating costs are likely to be substantial: for nuclear power, ranging from $+ 7$ percent (for the Northeast) to $- 7$ percent (for the Southeast), and for coal from $+ 15$ percent (for the Northeast) to $- 25$ percent for Montana, Wyoming, and the Dakotas.

includes allowance for the cost of capital—both equity and debt—amortization of investment, taxes and insurance.[4]

For many countries decisions about new generating plants are taken at the national level rather than by utilities, and for them a lower capital charge rate will be appropriate. It should presumably be based on the discount rate and depreciation of investment over the true expected life of the plant, with no allowance for taxes (except for those directly related to providing plant protection and insurance). For modern industrial societies a capital charge rate of something like 6 percent, corresponding to a discount rate (in constant dollars) of 5 percent and depreciation over forty years, might be appropriate. This lower rate, which will tend to favor nuclear—compared with coal—and particularly oil-generated power, will also, of course, be the appropriate one for countries like the United States, Japan, and Germany if the issue is nuclear power versus the alternatives from a national perspective rather than from that of utilities.

The impact of differences in choice of capital charge rate on the attractiveness of different generating alternatives is sufficiently great that we thought it useful to recompute the figures from Table 12-2 for the lower rate of 6 percent and also for 18 percent (see footnote 4). The results are displayed in Table 12-3.

From a consideration of Tables 12-2 and 12-3 it will be apparent that the most critical factors affecting the economic acceptability of nuclear power are likely to be capital costs and fixed charge rates. Capacity factors and U_3O_8 costs will also be important. Uncertainties about operating and maintenance costs, decommissioning, and other

---

4. The actual capital charge rates used by American utilities tend to be in the range 17-20 percent, these higher rates being a reflection of the high cost of capital in *current* dollars in an economy where inflation has averaged, and presumably can be expected to continue at, 6-7 percent per year, at least. A 12 percent rate in constant dollars, of course, equates to about 18 percent in current dollars, assuming inflation constant at 6 percent. It will make no difference whether a rate appropriate to a constant dollar calculation or one appropriate to a current—that is, inflating—dollar calculation is used in making comparison of alternatives if the total discounted cost of generating equal amounts of power is used as the basis for comparison. If, however, a high rate that reflects inflation, say 18 percent, is used and comparison is made on the basis of "levelized costs," there will be a peculiar distorting effect. Use of the inflated rate implies accelerated amortization. Unfortunately, all too often one sees comparisons made on this basis. Since capital costs comprise a much larger fraction of total generating costs for nuclear power than for the alternatives, the use of the higher rate makes nuclear power *appear* to be relatively less attractive (see Table 12-3). If a capital charge rate appropriate to a constant dollar calculation is used, comparison of alternatives on the basis of levelized costs is equivalent to comparison based on total discounted costs, and there will be no distortion in favor of the less capital-intensive alternatives.

Table 12-3. Projected Base Load Generating Costs (mils/kWh, 1978 dollars).

Capital Charge Rate (percent)	Nuclear	Coal
6	16	22
12	24	28
18	32	34

components of the fuel cycle, including back end costs, are likely to be relatively much less significant.

If one believes that capital costs can be prevented from escalating substantially, nuclear power should continue to be attractive relative to the principal alternatives—provided that it is acceptable considering safety, environmental, and nuclear weapons proliferation concerns and provided that reliance on large, centralized plants remains socially acceptable.

Projections of the Growth of Nuclear Power

During the 1950s and 1960s electricity consumption and generating capacity grew at a steady compound rate of about 7 percent a year in the United States. Many utilities based their plans for expansion on a continuation of the trend, including the remarkable stability in the rate of increase. Thus, there appeared to be a solid basis for growth of nuclear power. This was reinforced, not only in the United States but worldwide, by the dramatic increases in oil prices following the embargo of 1973. Official projections envisaged growth in U.S. nuclear capacity from about 20 GWe at the end of 1973 to 1,100 to 1,200 GWe at the turn of the century. In 1973, a record of forty-five new orders were placed for nuclear steam supply systems from U.S. vendors, and it appeared that demands for fuel might exceed supply capacity for both uranium and enrichment service at some time during the 1980s.

But then with increasing licensing time, costs, and environmental opposition—and with electricity consumption actually dropping rather than continuing to increase at the 7 percent rate—the situation changed dramatically. By 1976, cancellations of existing orders for nuclear units exceeded new orders ten to six. One of the five American reactor manufacturers had effectively withdrawn from the market, and U.S. Department of Energy projections for nuclear power for the United States had dropped by over 50 percent. They have dropped another third since then and now range from 158 to 193 GWe for 1990 and from 256 to 396 GWe for the year 2000.

Now, contrary to some earlier estimates, it appears that there will be no problem whatever, at least for the United States, in meeting demand for both uranium and enrichment service at reasonable prices for at least the rest of the century. Indeed, the United States is now operating its enrichment plants well below capacity.

Other countries have been less willing to reduce estimates of future growth as prospects for nuclear power have dimmed. Pronuclear "establishments" have been concerned in part about the possibly self-fulfilling implications of acknowledging poor prospects for the industry. However, there have been some dramatic cutbacks in plans, mostly notably in Iran and Spain, and some of the other larger programs will almost certainly fall far short of recent "official" expectations. Cutbacks are likely in Italy and Brazil, too, in large measure because of financing problems, and in West Germany because of political pressures, opposition of environmentalists and other groups, and consequent litigation.

In these circumstances, estimating the future growth of the industry can hardly amount to more than bounded speculation. Yet estimates will be of importance not only in considering matters directly relating to the industry but more generally in considering energy policy and nuclear proliferation policy.

Until the Three Mile Island accident, and notwithstanding a referendum in Austria that resulted in a completed plant not being allowed to operate, it had seemed unlikely that there would be many instances of completed plants actually being shut down permanently. Discounting this possibility, one gets as a lower bound for future capacity simply that now on stream. This is given in Table 12–4. It is

Table 12–4. Nuclear Generating Capacity, Late 1978 (excluding centrally planned economies) (GWe).

United States	54.1
Japan	11.5
United Kingdom	8.8
France	7.4
Germany (Federal Republic)	6.4
Canada	4.8
Sweden	3.9
Belgium	1.7
Italy	1.4
Taiwan	1.3
Switzerland	1.1
Spain	1.1
All others	2.7
Total	106.2

relevant to later discussion to note how few countries actually have a significant nuclear generating capacity at this time. The picture will not change qualitatively until the mid-1980s when plants in some developing countries will be completed.

Since it will be virtually impossible to bring a plant on stream in less than about seven years (probably twelve for the United States), one can obtain a reasonable upper bound for the late 1980s by simply adding in the capacity of plants on which some commitment has been made. The result is an upper bound for world generating capacity, excluding that for the centrally planned economies, of about 300 GWe. Taking account of possible cancellations and deferrals, 200 GWe seems a more reasonable upper bound.

In attempting to look beyond 1990, we have no good basis for concluding that lower bounds on nuclear generating capacity for the world, or for that matter the United States, will be higher than our estimates for the late 1980s. Although further growth seems likely, much depends on public attitudes and political considerations, which cannot be predicted for a decade or longer with any confidence. As an upper bound it is perhaps reasonable to assume that as much as 60 percent of generating capacity added between the late 1980s and 2000 might be nuclear in countries that have indicated a clear interest in the nuclear option. (The U.S. Department of Energy's high estimate for the United States is consistent with this and a generous assumption of a two and a half-fold increase in total U.S. electricity consumption between now and 2000.) This suggests an upper bound for the world, excluding the centrally planned economies, of about 800 GWe for the year 2000.

Uncertainties in projections of nuclear power growth have several implications. For the United States, the difference between DOE's high and low figures for 2000—140 GWe; (396–256)—is equal to about 15 percent of the country's electrical generating capacity projected for that time. Assuming the alternative to nuclear power to be coal, the uncertainty in the nuclear projection equates to a requirement for about 350 million tons per year—about 60 percent of present production. If the country were to burn this much more coal at the turn of the century because nuclear power was curtailed as a response to environmental or other objections, the cost would be slightly over 0.1 percent of GNP—of the order of $5 billion in 1978 dollars (assuming an average difference in generating costs for nuclear and coal of 6 mils/kWh, using the figures in Table 12-3 for a capital charge rate of 6 percent). This would be out of a GNP that will presumably have grown by that time to something like $4 trillion ($10^{12}$) in 1978 dollars. In reality, because demand for electricity will be somewhat elastic, less than total substitution of coal-generated power for forgone nuclear power would be expected. Therefore, the reduction in GNP would be considerably smaller. Even if the burning of coal were restricted because of environmental concerns and expansion of oil (and gas) use were restricted because of concerns about dependence on imports, the loss to U.S. GNP of forgoing 140 GWe of nuclear capacity would be but a fraction of 1 percent. It is only in the period after 2000 that the effects on GNP of restricting nuclear growth would begin to be substantial, and then only if the burning of fossil fuels were curtailed for reasons other than direct cost. Still, in the absence of such restrictions, curtailing nuclear growth would imply a substantial increase even in this century in coal production or oil imports, or both, above what would otherwise obtain.

Curtailment of growth in nuclear capacity in other industrialized countries that might result from environmental or other objections would likely be made up largely by greater use of oil rather than coal, depending on the rate of increase in the price of oil. With the cost differential between nuclear- and oil-generated power likely to be greater than we have estimated for the differential between United States nuclear and coal, the effects of curtailing nuclear power in other countries, notably Japan and some in Western Europe, would be greater than in the United States—but it would still not be large in this century. Nevertheless, if a substantial fraction of any shortfalls in nuclear capacity for countries other than the United States were made up through oil imports, the effect on world oil trade could be quite large. The uncertainties in the year 2000 projections of world nuclear capacity, excluding the United States and the centrally planned economies, suggest that this shortfall might be of the order of 200 to 300 GWe capacity. This would equate to 5 to 7.5 million barrels of oil per day—a quantity that is large enough to suggest that world oil prices could depend substantially on the extent to which Western European and Japanese nuclear aspirations are met.

The economies of developing countries, with the exception of those of some of the largest such as India and Brazil, are not likely to be very adversely affected in a direct way by frustration of whatever nuclear aspirations they might have. This is because the alternative of nuclear power is likely to be less advantageous to them than to large industrial nations. There are two principal reasons for this: (1) the capital (and foreign exchange) cost of nuclear units is likely to be high because much of the technology will have to be imported and because the electrical grids of many utilities are not likely to be large enough to justify installation of nuclear units large enough to permit realization of economies of scale; and (2) the discount rate appropriate to developing, capital-short economies is likely to be substantially higher than the rate for industrial countries.

It should be noted that developing countries that do not export oil may suffer directly, and perhaps quite substantially, if world oil prices are driven up because of radical cutbacks in the nuclear expansion programs of Japan and the major European countries, or for any other reason.

ALTERNATIVE FUEL CYCLES AND URANIUM CONSUMPTION

We have noted earlier that the possible exploitation of nuclear power programs to produce nuclear weapons has become a major issue of public policy. The problem arises in large measure because many countries, owing in part to their concern about dependence on others for the uranium needed to fuel reactors, have sought to reduce uranium consumption by means that increase concern about nuclear weapons proliferation. The options that have commanded the most attention involve extracting plutonium and unconsumed uranium from spent reactor fuel and then using these materials for new fuel, either for reactors of the kind now in operation or for "breeders." Once plutonium is extracted from spent fuel, it can be fabricated into weapons without the need to take extraordinary measures to prevent those involved from being exposed to high levels of radioactivity. With the reprocessing of spent fuel, the plutonium becomes accessible.

One would expect that the motivation to reprocess spent fuel would be reduced if access to uranium at reasonable prices could be assured or if alternatives for reducing uranium consumption that did not require reprocessing could be developed. These beliefs, as well as possible economic advantage, underlie much of the current interest in alternative nuclear fuel cycles.

In addition, concern about nuclear proliferation has focused attention on alternative reprocessing and enrichment schemes involving both different technologies and institutional arrangements that might be more "proliferation resistant." We reserve comment on the effectiveness of these alternatives in achieving nonproliferation objectives until later, limiting our discussion at this point to a description of alternatives to current fuel cycles and comments on cost implications.

Reprocessing and Recycling of Spent Fuel
From LWRs

All but one of the commercial power reactors in the United States and the great majority of those in the rest of the world are light water "converters" or "burners." They are fueled with uranium in which the U-235 content has been enriched to about 3 percent from its normal concentration of 0.711 percent, the other 99.3 percent of natural uranium being U-238. During normal operations, 70 to 75 percent of the U-235 is fissioned and about 2.5 percent of the U-238 is converted to plutonium, up to two-thirds of which then also fissions. It is the fissioning of U-235 and plutonium (principally the isotope 239, but also 241) that releases most of the energy that is used ultimately to produce electricity.

It will be apparent from the above that the spent fuel from a light water reactor contains a large amount of unconsumed U-238 and also small amounts of U-235 and plutonium. If the uranium and plutonium are separated from the fission products with which they are mixed in spent fuel, either or both can then be used in the fabrication of new fuel that can then be used in reactors of the same kind. If only the uranium is recycled, uranium requirements will be reduced by about 20 percent; if both uranium and plutonium are recycled the reduction will be about 35 percent, and there will be a reduction of about 15 percent in enrichment requirements as well.

For many years it was widely believed that recycling of uranium would be economically advantageous and that recycling of plutonium would be, too, but for the fact that it might be put to better use in "breeder" reactors. However, such conclusions were predicated on reprocessing costs of the order of $50 to $100/kg of spent fuel. It now seems likely that the cost will be almost an order of magnitude greater. The increase is largely attributable to more stringent health and safety standards. It also seems probable that the cost of fabricating mixed oxide fuel (that is, fuel containing plutonium as well as uranium) will be higher than was heretofore thought likely. Aside from uncertainties in these costs, the economic advantages of reprocessing and recycling will depend heavily on the cost of uranium and to a lesser degree on the cost of enrichment and on how plutonium is valued—at its worth if recycled in light water reactors or at its worth if used in breeder reactors where it can be used more efficiently. (In the latter case separation should be delayed if delay is economically preferable.) Discounting the last possibility—that is, assuming recycling of plutonium as well as uranium in LWRs and assuming present costs for uranium and enrichment—the net economic benefit of reprocessing and recycling appears to be about zero but with there being an uncertainty of perhaps ± 1 mil/kWh.

It is plausible that the plutonium produced in French converter reactors will just about meet the initial fuel loading needs for French breeder reactors during the rest of this century. For the world as a whole, however, and for all countries other than France, plutonium in spent fuel from LWRs (and other converter reactors) will greatly exceed requirements for breeders. Thus, except possibly in France, economic arguments for large-scale recycling and reprocessing must depend very much on assuming that uranium prices will increase substantially in the next couple of decades.

There appears to be substantial interest in recycling of plutonium in LWRs only in Belgium, West Germany, and Japan; but West Germany and Japan are interested not so much in reducing LWR fuel cycle costs as in waste disposal and the desire to develop capabilities in anticipation of breeder deployment later.

Breeder Reactors

The United States has had an active breeder R&D program for many years. As in other countries the major emphasis has been on the liquid metal fast breeder reactor (LMFBR). The United States will continue to have a breeder program even if the Congress goes along with President Carter's decision against completing the Clinch River demonstration breeder. Clearly, though, there is not the sense of urgency about early deployment in the United States that there was several years ago—and that there now is elsewhere, notably in France, Germany, Japan, and the Soviet Union. The change in the American view reflects concern about the proliferation of nuclear weapons, but that aside, it can be rationalized as a response to changed conditions—the greatly reduced projections for nuclear power growth and demand for uranium and increases in estimates of the uranium resource base.

It is generally believed that the capital cost of breeder reactors will be substantially higher than for light water reactors (and higher than that for some other converters as well), but that fuel cycle costs will be much less because breeders use uranium (or thorium) fifty to one hundred times more efficiently. Thus, it is usual to discuss the economics of breeder deployment in terms of a trade-off between the differential in capital costs for breeders and LWRs and the price of uranium. This is illustrated in Table 12-5.

With uranium costs now in the $40 per pound range and with little reason to expect dramatic increases in cost (in constant dollars) con-

Table 12–5. Uranium Indifference Costs for LWRs (throwaway fuel cycle) and FBRs.

FBR/LWR Capital Cost Differential $/kWe Capacity	Uranium Costs $/lb. of U_3O_8 (1978 dollars)
200	55–65
350	80–90
500	100–115
650	125–145

Note: These indifference costs are computed assuming a 5 percent discount rate (6 percent capital charge rate)—see p. 421. For LWRs, the other assumptions are the same as those for Table 12-2, except that fuel burnup is assumed to be 40 MWd/kg rather than 30, the rationale being that by the time breeders might compete with LWRs, the higher burnup for fuel for the latter is likely to have been realized. The breeder is assumed to be a "first generation" liquid metal fast breeder (breeding ratio 1.21) with the same capacity factor (0.65) as the LWR. The sum of fuel fabrication, O and M, and back end costs for the breeder is assumed to exceed that for the LWR by 4 mils/kWh. The low and high indifference costs for each capital cost differential are a reflection of the value assigned to surplus plutonium produced by the breeder. The low figures would be appropriate to a breeder program expanding at a rate such that it would be necessary to use some enriched uranium to supply enough fissionable material to the breeders. The high figures would be appropriate where plutonium production by breeders would be in excess of their requirements, the plutonium then being valued at its worth as an alternative to enriched uranium as a fuel for LWRs. (The breeding ratio is the number of fissile atoms produced per atom fissioned. If the ratio is greater than one, a reactor is a "breeder.")

sidering the much reduced projections of demand, it will be apparent from Table 12-5 that it is hard to make a case for early breeder deployment on narrow economic grounds, even assuming no further RD&D would be required. Recomputation of Table 12-5 including amortization of RD&D over a reasonable pattern of breeder deployment would result in much higher uranium breakeven costs. So, too, would the use of higher capital charge rates, such as might be appropriate to utility (rather than national) decisions, or very much lower enrichment costs, such as might be realized with laser enrichment.

Reducing Uranium Consumption in LWRs by Other Means

Not surprisingly, minimization of total cost rather than minimization of uranium consumption has dominated both design and operations of LWRs heretofore. However, if economy in the use of uranium assumes importance aside from considerations of price or if uranium becomes much more expensive, there may be a case for trying to use fuel more efficiently in LWRs. The attractiveness of available options will depend on uranium and enrichment prices.

One possibility, which requires no change in either reactor design or operation, is changing the "tails" assay in the enrichment process. The question is simply one of a trade-off. In producing uranium enriched in U-235, a stream depleted in U-235 is also produced. With the expenditure of increasing amounts of separative work, the amount of U-235 left in the discharge stream—that is, in the tails—can be reduced so that less uranium ore is needed to produce a given amount of product. The optimal tails assay at present uranium and enrichment prices is about 0.2 percent, and the U.S. Department of Energy requires customers to deliver uranium in amounts that would be consistent with such a tails assay. By going from 0.20 percent to 0.05 percent tails assay, uranium requirements would be reduced by 19 percent but separative work requirements would increase by 82 percent. At present prices the cost of product—namely, enriched uranium—would be increased by 18 percent. It is likely that the ratio of enrichment cost to uranium cost will drop with time, especially if laser enrichment technology works out, and in that case it may be that a tails assay of 0.05 percent U-235 or less will be economically optimal.

Uranium requirements will also be reduced if a larger fraction of the uranium in fuel rods can be made to fission. This can be accomplished by moving fuel rods within a reactor more frequently to achieve more even burning at the price of shutting the reactor down more frequently and hence reducing the capacity factor. A more interesting possibility lies in extending the time of fuel exposure beyond the present design values of about 30 MWd(t)/kg (megawatt days of heat generated per kilogram of fuel) perhaps to as much as 40 to 50 MWd(t)/kg. The savings in uranium would be 6 to 12 percent, rather less than these figures might at first suggest, because a higher degree of enrichment will be required for fuel with increased burnup. There will be virtually no saving in enrichment requirements. Fuel cycle costs might be reduced by 5 to 10 percent.

More efficient use of the neutrons produced in fission offers a third possibility for reducing uranium consumption substantially. Savings of 12 to 15 percent can be achieved through operating the reactor in a spectral shift control mode. Heavy water is partially substituted for light water during the early part of the operating cycle. This shifts the neutron spectrum to higher energies, thereby leading to the production of more plutonium than is produced when an LWR is operated in a normal mode. Some of the plutonium later fissions to produce energy.[5] In contrast, in the normal LWR operating mode, poisons are used during the early part of the operating cycle to absorb excess neutrons, but this is unproductive in generating power. Operation in a spectral shift mode requires the installation of equipment to concentrate heavy water (it is diluted during each operating cycle), and there are other considerations that may require some changes in plant. It will be important to limit releases of tritium because of its radioactivity and to limit losses of heavy water because of its high value. Because of these changes in plant and operating procedure and the cost of heavy water, the spectral shift control option cannot be expected to be economically attractive from the perspective of utilities at uranium prices of less than about $100/lb, although some savings may be achievable with optimum operations of a pair of reactors. There are a number of other ways of achieving more modest reductions in uranium consumption in LWRs through changes in design or operating procedures.

Alternative Converter Reactors

Almost 90 percent of the world's nuclear power is produced in light water reactors, which will probably be of comparable importance for the remainder of the century. Their dominance is due no doubt in part to heavy government subsidization during development—LWRs derive from ship propulsion reactors—but also to their favorable economics. The fact that they use uranium inefficiently compared with the principal competitors is offset by relatively low capital costs.

Competitors to the LWR have been developed and have achieved market penetration, not because they use uranium efficiently, but primarily for other reasons: ability to operate with natural rather than enriched uranium, an important attribute of the Canadian pressurized heavy water reactor, the CANDU; higher operating temperature and therefore greater thermodynamic efficiency, attributes of high temperature gas-cooled reactors; and in some cases, a desire to exploit an indigenous technology rather than to import an American or other design. With the expectation of rising uranium prices and interest in more efficient use of uranium, the CANDU, the high temperature gas reactor (HTGR), and some other designs that use ura-

---

5. This need not involve reprocessing. In any reactor, some of the plutonium that is produced fissions before the fuel is removed. In LWRs operated commercially the fraction may be, as noted earlier, as high as two-thirds.

nium very efficiently are commanding attention. However, in the light of much diminished expectations for the growth of nuclear power, and with breeders being developed as they are, it is somewhat doubtful whether any of these, other than the well-established heavy water reactors (HWRs), will be developed beyond the prototype stage. They do not offer enough of a cost advantage, if any, over LWRs, assuming no reprocessing and recycling, or over breeders, if reprocessing or recycling is assumed.

HWRs probably will continue to command 8 or 10 percent of the market for the remainder of the century. Consideration has been given to their introduction in the United States, but this prospect seems unlikely because of the entrenched—even though newly furnished—position of LWRs, the licensing problems with a new reactor, and the fact that access to enrichment service is unlikely to be a problem for American utilities. Were a CANDU or similar reactor to be introduced, capital costs would almost certainly be higher than for LWRs, particularly if the cost of the initial inventory of heavy water were included, as it should be. Much lower fuel cycle costs and possibly higher capacity factors would compensate partially at least for higher capital cost. As it is presently operated, the CANDU requires no enrichment and uses about 20 percent less uranium than LWRs on a once through fuel cycle. With 1 percent enriched fuel it would use 40 percent less uranium than LWRs.

Using Thorium in LWRs and CANDUs

Current generation converter reactors could be fueled initially with oxides of thorium and enriched uranium. With such fuel, U-233 would be produced (along with plutonium). Like U-235 and plutonium, U-233 can be used as a reactor fuel (or for bombs). If spent fuel of this kind were processed, the uranium, including the U-233, could then be recycled (and the plutonium, too, if so desired). Assuming breeding ratios of less than 1.0, it would be necessary to provide makeup fissionable material in addition to the U-233 (and plutonium) that had been produced. This would likely be done in the form of uranium enriched in U-235 (although plutonium and/or U-233 from sources other than the reactor in question could be used). As Table 12-6 shows, the use of such fuel cycles would require larger amounts of uranium for initial fuel loading but could result in substantial reductions in equilibrium and total uranium requirements. For comparative purposes the table also shows the effects of LWR uranium recycle and of using slightly enriched uranium in CANDUs.

Table 12-6.   U$_3$O$_8$ Requirements (short tons/GWe).

	LWR			CANDU		
	Once Through	U Recycle	Th cycle with 93 Percent U-235	Once Through Natural U	Once Through 1 Percent Enriched	Th Cycle with 93 Percent U-235
Initial Fuel Load	417	417	539	164	257	548
Equilibrium Requirement (per yr at 75 percent capacity)	192	150	96	156	114	48
Thirty Year Requirement (at 75 percent capacity)	5985	4767	3323	4688	3563	1331

Source: Y.I. Chang et al., *Alternative Fuel Cycle Options: Performance Characteristics and Impact on Nuclear Power Growth Potential* (Argonne, Ill. ANL-77-70, Argonne National Laboratory, 1977).

NUCLEAR POWER AND WEAPONS PROLIFERATION

We have noted earlier that much of the controversy about nuclear power, particularly in the international arena, is related to the possibility that nuclear materials might be used to produce weapons. There are potential problems with both the "front end" and the "back end" of the fuel cycle for light water reactors, and some of the problems arise with other reactor fuel cycles as well. Three materials are of concern—uranium containing high concentrations of the isotope 235, plutonium, and U-233.

The first of these, highly enriched uranium, is produced through enrichment of normal uranium by isotopic separation techniques, the same processes that are used in producing fuel for light water reactors. For weapons use, however, the degree of enrichment must be much greater. Normally the U-235 content in bombs is greater than 90 percent compared with about 3 percent for light water reactor fuel (the principal other isotope in both cases being U-238). Highly enriched uranium is particularly worrisome because it is possible to build weapons based on simply bringing two pieces of metal together rapidly (the Hiroshima weapon was of this design), something not possible with plutonium. These weapons in particular, and uranium weapons in general, require less sophisticated technology and can be built with greater predictability of performance than those using plutonium. On the other hand, the amounts of uranium required are much larger—tens of kilograms for primitive uranium weapons compared with about one-fifth as much for plutonium weapons of similar design.

Plutonium is produced as a result of neutron capture by the U-238 that is contained in reactor fuel. If it is to be used for weapons (or for recycling in reactors), it must be separated by chemical means from the highly radioactive materials, mostly fission products, that are produced along with it. This requires remote handling, but such separation has generally been regarded as easier than the isotopic separation that is required to produce highly enriched uranium. For this reason, most of the concern about weapons proliferation has focused on plutonium rather than uranium possibilities.

In producing plutonium for weapons, the fuel is subjected to only brief exposure in a reactor. As a result, the plutonium will then contain 90 percent or more of the isotope 239. If the exposure is longer, as is normal in commercial reactors where cost minimization is important, the fractions of the heavier plutonium isotopes, particularly Pu-240, are increased as a result of multiple neutron capture.

This is highly undesirable from a weapons design perspective. Presence of the isotope 240 complicates bomb design and may result in less predictable performance. Because the levels of radioactivity in spent fuel from commercial reactors are higher than for fuel that has been exposed only briefly, processing is also more difficult.

Should a nation wish to use its commercial reactors to produce plutonium for weapons, it could obtain "better" plutonium by removing fuel from its reactors when it had been subjected to only a fraction of the normal exposure to neutrons. In this case though, there could be some reduction in power generation as a result of the need for refueling more frequently than is normal, although removing fuel after one year would incur minimum penalties.

As they are normally operated, light water reactors of the 1,000 MWe class produce enough plutonium for fifteen to forty weapons per year. Heavy water reactors of the same size produce about twice as much.

Very little U-233 has been produced either commercially or for weapons. It is, however, conceptually attractive for the latter purpose and, therefore, worrisome from a weapons proliferation perspective. It is produced by neutron capture in a reactor if the fuel contains thorium. Like plutonium, it must be separated from the highly radioactive components of spent fuel.

There arises a question as to whether any nation interested in making nuclear weapons would use all or parts of a commercial fuel cycle for that purpose or whether it might prefer to build special purpose facilities. A mixed verdict is in order. A country without a power program could acquire weapons more quickly and much less expensively through a special facility than through the development of a full power program, including, as would be required for a weapons capability, either an enrichment or a reprocessing plant. A large research or special purpose reactor that could produce enough plutonium for several weapons per year would cost tens of millions of dollars and could be built in perhaps three to six years, whereas a power reactor would cost hundreds of millions and would take perhaps twice as long to build. For most countries that might be interested in producing enough material for a few weapons, the case for building small enrichment or reprocessing plants or both, rather than building commercial facilities, will be equally strong. This is because economy of scale arguments suggest that enrichment and reprocessing will be commercially attractive only for plants costing a billion dollars or more and capable of servicing dozens of reactors. Small plants for a modest weapons program would, in contrast, cost perhaps a tenth as much.

Having said this, we should add that a nation that had, or was building, the elements of a power program might elect to use them for weapons purposes. It would likely turn to special purpose reactors only as a way of evading safeguards and disclosure of weapons acquisition intent or if it decided that it wanted to base its weapons program on "weapons grade" plutonium and was unwilling to accept the inconvenience and cost to its power program of producing such plutonium in power reactors.

Similar arguments apply to reprocessing. Facilities designed to handle the high burnup (long exposed) fuel from a commercial reactor would be more than adequate to handle low burnup fuel such as would be preferred for weapons purposes. (It would not matter whether that material had been exposed in special purpose reactors or removed from commercial reactors after less than normal exposure.) Building a special purpose reprocessing plant in addition to a commercial plant would seem to make sense only to avoid disclosure or if the capacity of the commercial plant was fully committed to reprocessing fuel for recycling purposes.

Two additional points can be made with respect to reprocessing: a nation committed to it for commercial purposes would probably build a pilot scale plant first, and this could well be adequate for a small explosives program. India provides an example. Second, a large plant, even if under safeguards, would be worrisome because of the diversion potential. Such a plant would likely have the capacity to separate several thousands of kilograms of plutonium per year from spent fuel, and with materials accountability likely to be limited to the order of 1 percent, at least for the near future, there would be the potential for undetected diversion of enough plutonium for several weapons per year.

Using commercial enrichment plants for the production of highly enriched uranium raises different kinds of problems. A commercial plant designed to produce, say, the 3 percent enriched fuel required for reactors could not be easily and immediately used to produce weapons-suitable material because only a very slight degree of enrichment can be achieved in a single operation. (There may be an exception to this in the case of laser enrichment.) Thus, any process (lasers possibly excepted) must be repeated many times in a "cascade" of stages. A plant will normally be built with just enough stages to produce whatever degree of enrichment is desired. Diffusion plants require on the order of 1,000 stages for 3 percent enrichment but 4,000 for enrichment to 93 percent, a typical value for weapons production. With centrifuges, the number of stages may be only one-hundredth as great.

Assuming a commercial plant built to produce a 3 percent product, there are two obvious options available for producing highly enriched uranium. Either the 3 percent product might be used as feedstock for a small additional facility designed to produce the highly enriched product or the 3 percent product might be fed back through the commercial facility in a batch-processing mode to accomplish the same objective. The latter is not a very attractive option for gaseous diffusion plants because of the large gas inventory and long time required for the cascade to come to equilibrium. Batch processing would be much more feasible in gas centrifuge enrichment plants. Also, with centrifuge plants, in contrast to gaseous diffusion plants, there exists a third option for obtaining highly enriched fuel. This is the possibility of increasing the number of stages in a plant at the expense of reducing the capacity per stage, simply by changing the plumbing between centrifuges. It would then be possible to operate the plant in a continuous mode to produce highly enriched uranium. The greater flexibility of centrifuge plants obviously makes them more worrisome from a weapons proliferation perspective. A third enrichment technique has been developed to the pilot plant stage— the aerodynamic nozzle process, one version of which is now operating in South Africa, with another planned for Brazil. How adaptable these, and laser, plants will be to the production of highly enriched uranium when designed to produce low enriched material is not yet clear.

OTHER RISKS OF NUCLEAR POWER

An earlier study, in which a number of those involved in this study participated, included separate chapters on health, environmental, safety and waste disposal aspects of nuclear power.[6] We have reviewed that and more recent work and conclude that for the most part the findings of that study remain valid. They are summarized, with some additional comments, in this section.

Even assuming normal operations of nuclear facilities, the public and those working in the industry will be exposed to nuclear radiation that can be expected to produce somatic and genetic damage. Considerable uncertainty attaches to estimates of the magnitude of these effects because of imperfect understanding of the effects of low levels of radiation. However, we believe that the total number of fatalities from all causes, except possibly the release of radon from

6. Report of the Nuclear Energy Policy Study Group, *Nuclear Power: Issues and Choices* (Cambridge, Mass.: Ballinger, 1977).

mill tailings, will probably be less than one per year per 1,000 MWe plant and almost certainly less than ten per year. This suggests that fatalities from nuclear power are probably less than those associated with the use of coal. The adverse effects on the environment can also be expected to be much less than in the case of the use of coal, the major exception arising because present generation nuclear plants produce about 50 percent more thermal "pollution" than modern fossil fuel plants.

The worst conceivable accidents with nuclear power, which would involve a meltdown of a reactor core and massive release of fission products to the environment, could be very serious indeed, possibly producing tens of thousands of fatalities, mostly from long-delayed effects, such as leukemia developing years after exposure to radiation that might have been released. Estimates have been made suggesting that accidents of this degree of seriousness would be extremely unlikely—one chance in 200 million per reactor year (WASH-1400). However, reviews of WASH-1400 suggest that the uncertainties surrounding such probability estimates are enormous, partly because of the impossibility of identifying all possible accident sequences and partly because of the considerable subjective judgment that is required in assigning probabilities to individual events. As a result, the Nuclear Regulatory Commission (the agency that sponsored WASH-1400) has concluded that WASH-1400's numerical estimates of the overall risks of reactor accidents are not reliable. The point has been strikingly illustrated in both of the two worst accidents so far experienced with light water reactors—the Brown's Ferry fire of 1975 and Three Mile Island accident in 1979. Nevertheless, approximate bounding estimates can be made, and the aforementioned *Nuclear Power: Issues and Choices* study suggests that one in twenty thousand might be a reasonable upper bound on the probability (per reactor year) of the worst accident. It concludes that almost certainly the likelihood of extremely serious accidents will be within the range of other catastrophes incident to man's activities that society accepts and that in consideration of fatalities from all causes, nuclear power will compare favorably with coal.

We believe that with careful attention to plant design, operator training, and enforcement of rules and standards, the probability of a "worst case" accident can probably be made this small or smaller. Preliminary findings with respect to the Three Mile Island accident are indicative, however, of serious problems in all of these areas, which may well be generic, at least as regards the United States. This accident also revealed serious inadequacies in contingency planning at both the state and the federal level and an apparent lack of responsibility on the part of the utility company. Finally, it illustrated the problem of comparing strikingly different kinds of accidents. That is, it is easy in concept but extremely difficult in practice as well as in public perception to compare the small but continuously occurring respiratory ailments caused by coal combustion with the dramatic but infrequent types of accidents that reactors can cause or the disruption potentially created by carbon dioxide accumulation in the atmosphere—let alone the ever-present chance of military conflict with incalculable consequences inherent in continued pressure on Middle East oil resources.

Pending further investigation, drawing more specific conclusions and making specific recommendations seems premature. Even if the chances of a worst case accident can be contained to the levels mentioned above, we realize that there is at least the possibility that the difficulties and costs of implementing measures to improve safety will be so great, and the residual uncertainties so large, that the nuclear option will no longer be attractive to utilities or acceptable to the public.

Additional long-term hazards from nuclear power arise from waste disposal at both ends of the fuel cycle—in mining and milling of uranium and in the disposal of spent fuel or the wastes therefrom. These hazards are viewed by some as particularly worrisome because of the very long time periods involved—thousands to hundreds of thousands of years. They raise questions of the responsibility of this generation to those that follow—that is, of intergenerational trade-offs of benefits and risks or costs.

The production of uranium for reactor use (or for weapons) leads to the accumulation of very large quantities of mill tailings—so far, about 140 million tons. Unfortunately, this results in the release of radioactivity to the environment, primarily because of the presence in the tailings of thorium-230, which decays with a half-life of 78,000 years to a radioactive chemically inert gas, radon-222. Because radon-222 has a relatively short half-life (3.8 days), it normally decays before it can diffuse to the surface from an underground uranium deposit, but after the ore has been mined and milled into fine particles and left in tailings piles exposed to the air as it has been in the past, a significant percentage of the radon escapes into the atmosphere.

Since thorium is present in surface soil and rock, radon-222 is already in the atmosphere at a "natural background" level. At much higher concentrations it has been found that radon-222 and its radioactive decay products produce excess lung cancers among uranium miners. If it is assumed that the same lung cancer risk per unit radiation dose applies at low exposure levels, then on the order of 2,000 lung cancer deaths might be resulting in the United States each year from exposure to the natural background level of radon-222 and its decay products. The added increment from mill tailings is very small. An upper bound would probably be on the order of 0.1 deaths per century per GWe year of power generated—a very low figure compared with others associated with power generation, whether by nuclear or other means. However, if instead of accepting a century as a reasonable period of concern, one sums up fatalities over many millenia, the number of fatalities per GWe year of electricity generated is two or three orders of magnitude greater.

Recently the long-term hazard represented by uranium mill tailings piles has come to be an issue in the national debate about nuclear power, and the Nuclear Regulatory Commission is now moving to establish regulations concerning how they must be covered. Simply leaving the piles on the surface and covering them over is unlikely to be a permanent solution since wind and water erosion could easily remove a covering of a few meters of soil in a time short relative to the hundreds of thousands of years during which the piles would be generating radon at high rates. If a permanent solution is desired, therefore, it is likely that the piles will have to be buried—possibly in part at least in the original mines.

If it were decided to bury the tailings piles, it would probably be advisable to mix the tailings with cement and let them solidify so as to reduce any potential water pollution problem. In that way the hazard to man could be reduced to approximately the level that obtained before the original ore was disturbed. According to a recent Oak Ridge National Laboratory report, the cost of such disposal should only amount to $1 to $2 per pound of U_3O_8 recovered. The corresponding increase in the cost of light-water-reactor-generated electricity would be about 0.01 cents/kWh—a relatively trivial amount.

If managed with some care, wastes that accumulate in spent fuel will almost certainly pose much less of a hazard than those from uranium production. However, estimation of the magnitude of risk is much more difficult, and historically, waste disposal has been mismanaged, at least in the United States. It is probably the combination of these two factors that has pushed the waste disposal problem to the point where it is now ranked with the risk of reactor accidents as a major impediment to the acceptability of nuclear power.

As with mill tailings, the long-term nature of the problem arises because the very long half-lives of some of the constituents of spent reactor fuel. The problem is more complex because of the much greater variety of radioactive materials present and because, at least for many years, the level of radioactivity will be much higher than that arising from mill tailings.

For the first few centuries after removal of spent fuel from a reactor, the dominant contributors to radioactivity will be fission products, particularly strontium-90 and cesium-137. However, after about 1,000 years, most of the fission products will have decayed, and the dominant hazard will increasingly become that from the transuranic elements, which typically have very much longer half-lives. Were it possible to remove all of these elements before waste disposal, the problem of isolation would be much simplified. One would need only to emplace the remaining non-transuranic wastes in geological formations whose stability over some thousand years or so could be assured. However, even if spent fuel is reprocessed and most of the uranium and plutonium is separated out, some of it will remain with the fission products, along with the other actinides. The result is that the hazard is reduced only by a factor of five to ten during the period from a thousand to ten million years. The weight to be attached to reductions of this magnitude is of some significance because there are those who argue, particularly in Germany, that the removal of plutonium from spent fuel is so important that reprocessing is virtually a prerequisite to the disposal of high level wastes.

As it is, then, there is much feeling that these wastes must be isolated for hundreds of thousands of years. A consensus has developed, at least in the United States, that manmade containers cannot be relied on and that the primary barrier to the release of wastes to the environment must be the geological formations in which they are placed. (A Swedish study has given more weight to multiple manmade barriers, in addition to geological barriers, to contain high level wastes.) The principal concern is protection from dissolution and transport by groundwater, but erosion and rediscovery through drilling or mining are other possibilities. A number of geological media have been considered, particularly salt (favored in Germany and the United States), granite, shale, clay, and deep sea sediments. Unfortunately, it is not yet possible to demonstrate the superiority of any of these media over any of the others. It is likely that the acceptability of sites will be dependent on detailed characterization of geology.

Notwithstanding our inability to make specific recommendations at this time, we are optimistic about ultimate waste disposal. In thinking about the problem of what we must do it can be helpful to consider what nature has done in immobilizing radioactive materials, particularly radium-226, within the earth's crustal rock. Since strontium-90 is the dominant fission product hazard over the period of one to several hundred years, and since it behaves chemically like radium-226, a comparison of these two isotopes is of special interest. Measured by the standard international limits of annual intakes by ingestion, one finds that the toxicity of the strontium-90 of fresh spent fuel is approximately 1,000 times that of the radium in the uranium ore that was originally mined to manufacture it. Offsetting this enhanced toxicity, however, is the fact that the amount of uranium in the crustal rock of the United States down to 500 meters is at least 1,000 times the amount that is likely to be mined for light water reactors. Since the radiation dose to bone lining cells from ingested radium is probably causing only about ten bone cancers per year in the United States, it would appear that even for a nuclear power economy many times the present size, the hazard from buried spent fuel can be made very low if waste can be placed several hundred meters below the surface and immobilized as well as is radium. The hazard is likely to be of a few percent, at most, of that arising from other parts of the nuclear fuel cycle. Since the most troublesome fission products, strontium-90 and cesium-137, have half-lives of about thirty years, the levels of radioactivity will decay rapidly compared with those incident to mining and milling.

Many problems must be dealt with, however, before we can be confident that the technology for waste disposal and our ability to select acceptable sites are adequate.

- The waste form may be subject to the synergistic effects of radioactive heating, radiation damage, high pressure hot water, and chemical attack from radiation activated species. It is not clear that under these conditions the waste form can be given the longevity that most rock has under ordinary subsurface conditions.
- The heat deposited in the geologic medium surrounding a waste depository by the radioactive decay of the wastes would cause the rock and trapped water to expand significantly with the potential for creating cracks, a problem particularly in media such as granite. Flow through cracks would be much more rapid for a specified "permeability" than the flow through homogenous porous media that are often still postulated in discussions of water flow around aquifers. In fractured media there is also the possibility of thermal

convection, with cold water flowing into the repository from the sides being warmed and rising to the surface above on a relatively short time scale.[7]

- Human activities such as exploratory drilling for petroleum or the inadequate sealing of the mine shafts used in emplacing the radioactive wastes may create channels of weakness for the flow of water into previously dry geological media such as salt deposits that might be used as waste repositories.

- While it is ordinarily assumed that the toxic radioactive ions such as strontium, cesium, and plutonium will, if leached out of wastes, move from 10^2 to 10^4 times more slowly than the ground-water due to exchange with ions in the rock, the possibility cannot be excluded that these ions may react with other chemicals present in the waste or the geologic formation and change into species that migrate with little or no retardation at all.

- It is possible that the uptake by the food chain and the absorption through the human intestinal wall of the long-lived trans-uranic elements may increase by several orders of magnitude as a result of changes over time in their chemistry from the simple inorganic forms currently being assumed in risk analyses. This would mean that the near term hazard would be increased some-what, but the principal effect would be to increase the longer (greater than several hundred years) hazard rather substantially.

INTERNATIONAL DIFFERENCES OVER NUCLEAR POWER AND PROLIFERATION

Since the end of World War II there has been a conflict between permitting, and indeed at times encouraging, the exploitation of nuclear energy for peaceful purposes and preventing the spread of nuclear weapons. With the signing of the Nuclear Non-Proliferation Treaty (NPT) in 1968, it appeared that the United States and the Soviet Union had resolved the dilemma: exploitation for peaceful purposes was to be essentially unrestricted provided that states

7. The rate of heat generation in spent fuel or high level waste diminishes by a factor of about nine between one and ten years after discharge from a reactor and by another factor of six to nine during the following ninety years. Heat-related waste disposal problems can accordingly be much mitigated if spent fuel or high level wastes are kept in above ground storage for some decades before burial. The Canadians and the Swedes, particularly, have favored this approach. Opponents argue that such above ground storage may be unacceptable on environmental grounds and, in the case of spent fuel particularly, because of potential access to plutonium.

that did not have nuclear weapons would renounce an interest in weapons acquisition and would accept safeguards on their nuclear programs designed to detect diversion of materials that might be used for weapons. Other countries objected to the treaty because it explicitly recognized the division of nations into two classes—the nuclear weapons states and the non-nuclear weapons states—and seemed to accept that division as permanent. They argued that the treaty was one sided in imposing restraints on the latter class of nations but not on the former. By including an article (Article VI) calling on the nuclear weapons states to disarm and others (Articles IV and V) that made it explicit that the peaceful exploitation of nuclear energy was not to be interfered with but, indeed, to be facilitated, the assent of most of the world community was secured. There were exceptions, however, notably Argentina, Brazil, China, France, India, Israel, Pakistan, South Africa, and Spain.

Nonetheless, it seemed clear that most of the world, the United States included, would follow a similar nuclear energy policy. Converter, mostly light water, reactors would be deployed; the spent fuel from them would be reprocessed; the recovered uranium would be recycled, as would the plutonium (either in converter reactors or to fuel breeders); and the high level wastes from reprocessing would then be disposed of, probably underground. The safeguards program, which had been developed to monitor reactors, would have to be extended to cover reprocessing plants, plutonium storage facilities, fuel fabrication plants that would incorporate plutonium in fuel, and any enrichment plants in nonweapons states. Parties to the treaty might sell such facilities to others or assist in their construction, provided arrangements were made to apply safeguards to them.

It was widely recognized that such a system would not prevent power programs from being exploited to produce weapons. First, it seemed quite likely that, notwithstanding safeguards, diversion of modest amounts of weapons-usable fissionable material could occur without detection, particularly from reprocessing plants. Second, once such material had been produced, the time interval between diversion or abrogation of safeguarding arrangements and weapons manufacture could be quite short—probably too short for diplomacy or sanctions to be effective in preventing weapons acquisition.

It was these considerations, particularly the latter, that led to the aforementioned changes in U.S. policy during the period 1976–1978. The United States decided that "timely warning" of the attainment of a weapons capability would be an essential requirement of U.S. policy and that the very existence of reprocessing or enrichment plants or of significant amounts of plutonium or highly enriched uranium in nonweapons states would be inconsistent with this criterion.

During the period when the Carter administration's nuclear policy was evolving and being enunciated there clearly was hope in administration circles and in the Congress that American views would be accepted and that the evolution of a world plutonium economy could be avoided, or at least delayed, while mechanisms for its management could be developed that would be more likely to provide a "timely warning" than those envisaged in the NPT. It is now clear that at least the first hope is not going to be fully realized. The U.S. position has been widely represented as inconsistent with Article IV of the NPT. France, the United Kingdom, and the Soviet Union seem virtually certain to proceed with the reprocessing of spent commercial fuel on a large scale; India, Japan, and Germany will continue on a small scale; Japan is likely to build a large plant, and Germany will too, if not prevented by domestic opposition; Belgium is likely to reactivate and increase the capacity of the Eurochemic plant, a plant that was formerly operated as an international facility; and there are other prospects, notably in Spain, Argentina, and Brazil. Moreover, the Netherlands, Germany, Japan, Brazil, and South Africa have been moving toward the acquisition of indigenous enrichment capabilities. At issue is not whether reprocessing of spent fuel and enrichment of uranium will occur in states that do not have nuclear weapons, but rather, how much and under what conditions. Of special interest is the disposition of separated plutonium in cases where reprocessing will be done in one country for another.

While the United States can probably temporize until early 1980 in facing up to major policy changes, the coincidence of the termination of the International Nuclear Fuel Cycle Evaluation effort and the second quinquennial review of the Nuclear Non-Proliferation Treaty will make difficult policy choices unavoidable.[8] Aside from the immediate and cumulative effects, both negative and positive, of a continuation of present policy, there is a strong argument for early consideration of possible changes. Effecting them is likely to be time consuming and fraught with great difficulty in the light of the intensity of commitment of members of Congress to particular

8. The International Nuclear Fuel Cycle Evaluation is a roughly fifty nation effort to evaluate technological and institutional alternatives for nuclear power, with particular attention being given to "proliferation resistance" as well as to technical and economic feasibility. It is scheduled to end in February 1980.

points of view and the likely necessity of changes in the Nuclear Non-Proliferation Act of 1978.

It should be noted here that as a condition for provision of enrichment service or the export of uranium or nuclear-energy-related equipment by the United States, an "agreement for cooperation" must have been concluded with the recipient nation. The Act requires inter alia that new agreements give the United States a veto on reprocessing of spent fuel, uranium enrichment, and the storage of plutonium. Also required is a right of veto on the transfer to third parties of equipment or material of U.S. origin and the right to insist on safeguards on all nuclear activities on non-weapon-state recipients. In addition, the Act stipulates that existing agreements be renegotiated by next spring to reflect these requirements. Of special significance is the fact that there is little evidence as yet that the Euratom nations are willing to accept the U.S. conditions in renegotiation of the agreement with that organization. Morever, Japan is likely to insist, as it has heretofore, on equal treatment with Euratom.

The Act also includes provisions for U.S. cooperation in nuclear-energy-related activities such as setting up a "fuel bank." However, the unilateral aspects of the Act, and particularly the requirement for a unilaterally determined renegotiation of existing agreements, have been the center of attention in the comments of other governments about U.S. policy.

The NNPA includes provisions for waiver or conditional approval with respect to its requirements; and so far Japan has been authorized by the president to reprocess spent fuel of U.S. origin on a limited scale and to ship spent fuel to Europe for reprocessing. The shipment of enriched fuel to India has also been allowed notwithstanding the refusal of India to accept safeguards on all its nuclear activities. The decisions encountered objections from members of Congress and from the Nuclear Regulatory Commission.

Whether or not the United States exercises its veto on reprocessing, transfer of spent fuel to third parties, return of plutonium to producers, and so on, the existence of the Act, and the implications of unpredictability about future U.S. behavior, are bound to cause friction between the United States and other nations with which it is directly involved in nuclear commerce and also between the United States and nations like the United Kingdom and France that wish to provide others with services over which the United States might exercise a veto.

Moreover, there is the possibility that U.S. policy generally and the passage of the Act in particular may have effects on nuclear weapons proliferation just the opposite of those intended. The specter of U.S. denial of uranium and enrichment service and of an American veto of reprocessing by third parties may serve as an inducement to the development of indigenous enrichment and reprocessing capabilities and premature interest in breeders.

The most severe critics of U.S. nuclear export policy and of the Act argue that the only certain effects have been to increase doubt about the United States as a reliable trading partner and to exacerbate relations with a number of nations that arouse little concern on the issue of weapons proliferation. In other nations the actual effects on weapons proliferation potential remain indeterminate, though the policy probably serves to stimulate rather than reduce interest of these nations in reprocessing and enrichment technologies.

Such criticisms fail to take account of the positive effects, and the potential for more, that can be traced to U.S. statements and efforts of the last two years. It does seem clear that in other nations, particularly in the "supplier" states, there has developed an increased concern about nuclear weapons proliferation, there has been substantial convergence on the undesirability of exporting reprocessing and enrichment technologies, and more attention is being paid than would otherwise be the case to the degree of "proliferation resistance" of alternative fuel cycle technologies.

Still, it would appear to be timely to consider whether changes in U.S. policy might not be in order, especially since most of the other states with a substantial interest in nuclear power are prepared to follow the United States lead only to a limited degree, and some perhaps not at all.

In the absence of change, and particularly if the United States should apply the NNPA strictly, a two (or more) tier structure in nuclear trade could well emerge, with one group of nations accepting U.S. conditions while others do not. Buyers in the latter group would presumably pay more for uranium and technology as the price for perhaps less encompassing safeguards and greater independence with respect to acquiring and operating indigenous reprocessing and enrichment facilities.

POLICY ISSUES FOR THE UNITED STATES

Hardly any major questions about American nuclear energy policy can be treated without considering their impact on the nuclear power programs of other countries and on our international relations. Discussion is, however, perhaps simplified by first considering a number of issues from a largely domestic perspective, recognizing that international considerations may modify conclusions.

The Breeder Question

The low and high projections for growth in U.S. nuclear generating capacity (256 and 396 GWe) for the year 2000 imply a demand of 1.6 to 2.5 million tons of U_3O_8 for all U.S. reactors deployed by then, assuming a once through fuel cycle, no improvements in uranium utilization, expected reactor life of forty years, and an average capacity factor of 0.65. These amounts of uranium are likely to be producible at costs below the midrange of those given in Table 12-5. This suggests that it would not be economically advantageous from a national perspective to bring commercial fast breeder reactors on stream before 2000. They would be even less desirable from a utility perspective, considering the higher capital charge rates that utilities are likely to believe appropriate. When one takes this last point into account, along with the possibility of a 15 to 30 percent improvement in uranium utilization in LWRs by 2000, it seems probable that the breeder would, in fact, not be competitive before about 2010 and very likely not before 2020.

This being the case, we see little merit in the RD&D program of the United States being oriented to a demonstration of a commercial type reactor before the year 2000. We do not accept the argument that the United States should accelerate its own breeder program beyond what is needed to meet its own domestic requirements. That other countries have breeder programs to meet what they may perceive as more urgent needs should play little or no role in a U.S. decision on the breeder. In fact, the decisions of France and the Soviet Union to move ahead on a more accelerated time scale only reinforce our belief that the United States need not do so. This is a technology where we can well afford to let others take the lead, taking advantage later, if need be, of their developments through licensing or other arrangements. Moderation in our efforts is particularly warranted since there is some possibility that conservation and the development of other technologies may persuade us to forgo breeder deployment entirely.

Still, we believe that there is enough of a likelihood that we will need breeders that we should continue to have a broad and substantial RD&D program. Since others are focusing their efforts almost exclusively on liquid metal fast breeders, more U.S. attention to alternatives such as the gas-cooled and molten salt breeder concepts may be warranted. With projections for growth of nuclear power radically lower than a few years ago, we need not be as concerned as we have been about future breeders having short doubling times,[9] and our R&D program should reflect this.

Specifically, the United States needs a vigorous R&D program focused on providing candidates for a decision whether to build one or two large-scale (but not commercial) steam-generating breeder reactors. The target decision date, given the possibility that a commercial breeder may need to be deployed by 2010 or so, should be 1985–1990, and the decision criteria should include, importantly, low capital cost (including low fissile inventory) and the nature and cost of the fuel cycle facilities. In accord with the principles that we have proposed to guide RDD&D in general, foreign LMFBR technology should be adopted if the results of the American R&D program at the time do not promise a breeder economically superior to the LMFBRs that will have been demonstrated abroad—the Super Phénix and others.

We believe that we should continue to cooperate with others in breeder R&D, particularly in information exchange, but would in most instances oppose participation in joint development and demonstration programs, mainly because of concern about premature commitment to commercialization and possible difficulty in disengagement if termination or redirection seems warranted.

Recycling of Uranium and Plutonium in LWRs

Recycling of uranium in LWRs would probably be desirable from an economic perspective if reprocessing of spent fuel were judged to be desirable as a precursor to waste disposal or to obtaining plutonium for use in breeder reactors. It is, however, not needed for the first purpose and is premature for the second, in our view.

Whether the benefits of recycling *both* uranium and plutonium in LWRs would justify reprocessing and recycling on economic grounds at this time is unclear. The net benefit, if any, would certainly be small even if one considered the investment already made in the large uncompleted U.S. plant at Barnwell, South Carolina, as "sunk"—that is, even if in calculating costs of reprocessing and recycling, one counted only costs yet to be incurred.

In short, economic arguments and concerns about weapons proliferation indicate that it would be unwise from a narrow national perspective for the United States to begin commercial re-

---

9. The doubling time for a breeder is the time required for it to produce enough plutonium (or U-233) in excess of its own refueling requirements to provide the initial fuel load (both in and out of core) for another breeder of the same kind.

processing and recycling at this time. When account is taken of the fact that others will be reprocessing spent fuel whether the United States does or not, what the United States should do may be less clear cut.

Other Measures to Reduce Uranium Consumption

Reductions in U.S. uranium consumption of 15 to 30 percent can probably be realized relatively quickly, and with economic benefit, through such measures as improving burnup of fuel in LWRs and development of the laster isotope separation technology described earlier. Such reductions might delay the need for breeders, and hence for reprocessing, by a decade or more, possibly even obviating the need for breeders totally, depending on progress in developing alternatives. Accordingly, we believe that federally supported R&D on such techniques is well justified.

We are much more skeptical about the wisdom and political realism of government efforts to reduce uranium consumption through measures that require much longer to implement and that are less likely to be advantageous economically. We have in mind efforts to stimulate investment in other reactor types such as HTGRs or HWRs, even though these might be operated on once through cycles with considerable reductions in uranium consumption.

We see even less merit in the allocation of government resources to concepts involving both alternative converter reactor technology and reprocessing and recycling. Breeders would seem more attractive from a cost-benefit perspective and little, if any, worse as regards proliferation potential.

Interim Spent Fuel Storage

All of the U.S. nuclear power plants now operating and under construction were designed when there was a general expectation that spent fuel from the reactors would be held in large water-filled pools at the plant for about six months while most of the short-lived fission products would decay. At the end of that period, the level of radioactivity of the fuel elements is still so high that remote handling and massive shielding is required. However, it was contemplated that fuel elements would then be sent to reprocessing plants, where they would be stored for an additional period, again under water, and then be reprocessed.

Normally reactors are refueled only about once a year, and then only about one-third or one-fourth of a core is removed. This implied that the storage pools at plants would never contain large numbers of fuel elements, although they have been sized with reserve margins for delay in shipment of spent fuel and so that a full reactor core could be removed if necessary.

However, with the deferral of commercial reprocessing of spent fuel in the United States, the storage capacity at some plants will probably be exceeded soon. In the absence of an expansion of capacity, there will be a problem on a national scale by about 1985 if the criterion is the maintenance of enough capacity at each reactor site to permit normal discharge—that is, discharge of about one-third of a full core. If reserve capacity large enough to discharge a full core is the criterion, additional storage capacity will be required no later than 1983. The need has been mitigated somewhat by construction of storage basins at the three commercial reprocessing plants. Of course, none of these plants is operating, but some spent fuel is now being stored at these facilities. While there is room for about 1,000 tons more, it may be that only about half that much capacity will be made available by the owners.

In any case, there is a clear need for very large amounts of additional storage capacity, considering that the rate of discharge of spent fuel is now about 1,000 tons per year and will triple within a decade unless many plants under construction are now canceled. The problem is compounded by virtue of the United States having offered to store limited amounts of spent fuel from other countries' reactors.

A decision has been made that the government will offer to take title to spent fuel, but there remain questions about the nature of the storage facilities to be built, the charges to be levied for storage, and the disposition of recovered uranium and plutonium in the event that reprocessing is allowed at a later date. These questions (and that of storage of spent fuel of foreign origin) are the subjects of recent draft environmental impact statements.

Storage capacity can be expanded most economically, up to a point, by "reracking" or "densification"—that is, by putting more fuel elements in a storage basin than originally planned. This requires introducing neutron-absorbing materials into the reactor basin so that a criticality accident will not occur. (In the absence of such materials, and if too many fuel elements are placed in the basin, a chain reaction will occur, just as in a reactor.) We see little objection to further expansion of the reactor capacity, either by "reracking" or construction of additional storage basins. However, particularly if reprocessing is to be long deferred, waste disposal and retrievable storage of spent fuel is probably best handled by the federal government. This, economies of scale, safeguarding considerations, and the importance of the offer to accept foreign spent fuel argue for the construction of a very limited number of large, away from the reactor, federally operated, spent fuel storage facilities.

Finally, there is the question of cost recovery. Should the government base its charges on full recovery of costs, or should it subsidize services? From a domestic perspective we would argue for the former. However, the purpose of offering to accept foreign fuel storage is to reduce the risks of nuclear proliferation. If subsidization would contribute substantially to the realization of this objective, we would favor it—recognizing that it would be politically difficult, probably impossible, to accept foreign fuel at charges lower than those applying to domestic fuel. As it is, the question of charges is likely to be a secondary consideration in other nations' consideration of any U.S. offer, and so we would incline toward full recovery of costs. What is likely to be critical is to be able to implement the offer to accept foreign fuel fairly soon.

We believe that utilities should be given the option of paying an annual charge while retaining title to the fuel or of paying a one time charge immediately, or at any time thereafter, with the government then taking title. If it chose the first option, the utility would then have the right to have the fuel reprocessed at any time after reprocessing is allowed and would then have title to the recovered uranium and plutonium and the responsibility to pay for permanent disposal of the wastes. If the government takes title to the fuel, it could, of course, reprocess at its option and sell the recovered uranium and plutonium. However, it would also bear the risk of having to assume any costs that might subsequently emerge.

There is an overarching issue with respect to spent fuel storage that also arises in connection with a number of other questions relating to nuclear power—that is, the reconciliation of national needs with state and local interests and the preeminence of federal law. This, of course, arises with respect to virtually all other aspects of energy policy and is treated in Chapter 14.

Waste Disposal

In the past four years the issue of radioactive wastes from civilian nuclear power plants has provided a major arena for those debating the future of nuclear power. Indeed, the lack of arrangements for the ultimate disposal of radioactive wastes has been used as the reason for delaying further commitments to nuclear power in a number of European countries and U.S. states. The nuclear industry is therefore urging a rapid "demonstration" of a "solution" to the radioactive waste problem. Since uncertainties and potential trade-offs are involved in any of the "solutions" so far proposed, other participants in the discussion argue for more research before a commitment is made to a particular waste disposal strategy.

Uncertainties concerning both the best strategy for disposal of radioactive waste and the ultimate use of the plutonium and uranium in spent LWR fuel lead us to conclude that decisions regarding "ultimate" disposal must be deferred until at least the 1990s. This does not mean that we favor a continuation of the policies of "benign neglect" that obtained during the thirty years following World War II. Far from it. There must be a vigorous program to resolve, insofar as possible, uncertainties about preferred means of disposal and to characterize possible sites. Even if decisions should be taken to abandon the nuclear option, the waste disposal problem will not go away. There are already large amounts of wastes from both military and commercial programs that will require attention.

For at least the next decade there is no realistic option for the United States for dealing with wastes from the nuclear industry other than interim storage above ground. In looking beyond that time, it has been proposed that we begin emplacing spent fuel in underground vaults, with the main shaft left open for a number of years so that the spent fuel can be removed relatively easily in the event that unexpected problems arise.

With the current U.S. emphasis on forgoing reprocessing, the possibility of retrieval to recover plutonium and unconsumed uranium has not been given much, if any, weight. This is probably a mistake. Many of the other countries of the world are not likely to accept irretrievable disposal of wastes containing significant amounts of plutonium and uranium, and the U.S. position could change as well a few years hence.

This suggests to us that in considering the "back end" of the fuel cycle, we should be looking at three alternatives: (1) interim above-ground storage of spent fuel; (2) underground storage for spent fuel, which could provide a greater degree of protection from both natural catastrophes and undesired access with an expectation of retrieval; and (3) underground disposal of either spent fuel or high level wastes, with no expectation of retrieval, but with a possibility of retrieval at reasonable cost, at least for some years. No decision need be made for many years as to whether the wastes are to be in the form of spent fuel or high level wastes from reprocessing since site selection and designs of geological repositories will be quite indifferent to that question.

Mill Tailings

We have noted earlier that the hazards from mining and milling of uranium are likely to be more troublesome than those arising from the disposal of spent fuel or high level wastes. The institutional problems may also be greater. The utilities still retain their high level wastes in the form of spent fuel, but responsibility for existing mill tailings has been less clear. It is impossible to distinguish between tailings associated with military programs and those produced in connection with civil programs. Moreover, some of the corporations that generated tailings are no longer in business, leaving open to dispute the sharing of responsibility between the states and the federal government.

Since we in general favor internalization of costs of energy production, we believe that in the future, the mining and milling companies should be made to bear the burden of protecting the public from the radioactive emissions from the tailings insofar as this can reasonably be done through the setting of standards by the federal authorities. However, there will be a residual responsibility because of the long times involved that should be borne by the federal government.

Maintaining the Viability of the Nuclear Power Option

The lower and upper DOE projections of growth in nuclear capacity for the United States equate to 10 and 20 GWe per year during the 1990s. These rates of growth are far lower than those projected several years ago, and there must be a real question as to whether demand will continue to support four U.S. vendors of nuclear steam supply systems, particularly if it is at or near the lower end of the range of estimates. To make matters worse for U.S. industry, competition in export markets, particularly from Germany, France, Canada, and the Soviet Union, is likely to be intense.

A question arises as to whether special government efforts should be made to assure the survival of the industry. There may be some case for federal help, considering the possible superiority of the nuclear option to coal on environmental grounds and because all fossil fuels may be troublesome if the carbon dioxide problem turns out to be severe. Nevertheless, we would not single out nuclear power for special subsidization at this time. Its superiority to other options is not yet so clear nor the necessity of a clear choice so imminent that we would argue for federal support of other than

R&D and regulation. Support for nuclear energy should be provided on roughly the same grounds it is provided to other energy sources—namely, the likelihood that public benefit would not be realized by simply letting the market work.

INTERNATIONAL POLICY QUESTIONS

We take it as given that the United States ought not to interfere with the efforts of other nations to exploit nuclear energy for peaceful purposes, provided that it can be done safely. Indeed, it may be desirable to encourage nuclear energy if it is environmentally more acceptable than alternatives or if it reduces the rate of exploitation of key depletable resources, especially oil and gas. On the other hand, it is U.S. policy, with which we agree, to try to prevent nuclear power programs from being used to facilitate weapons acquisition. As noted earlier, until 1976 the United States sought to achieve this latter objective largely by emphasizing the application of safeguarding arrangements to "sensitive facilities" and critical materials, particularly in nonweapons states. More recently it has decided to try—by example, through positive measures and through threats of denial of access to U.S. uranium, enrichment service, and technology —to induce others to forgo acquiring such facilities and materials. So far the emphasis, or at least the perception, has been on the denial aspects of U.S. policy.

We believe that to the degree that there has been emphasis on denial, it has been largely misguided on three counts:

1. It is based on an exaggerated view of the efficacy of controlling nuclear proliferation through limiting nuclear power developments. The costs and technological requirements of a program dedicated to the acquisition of modest amounts of materials for weapons will be much less than those required for a power program (see p. 434). Any nation able to manage a power program is likely to be able to implement a weapons program without great difficulty. Thus, denying others reprocessing or enrichment capabilities for commercial power programs will not be very effective in preventing those who are determined to acquire weapons from doing so and will not be necessary for those with no interest in weapons. Such denial might be effective in lengthening the lead time between a decision to acquire weapons and the attainment of a capability, during which time a decision might be reversed; or it might affect marginal decisions. There is a question, however, considering the rather limited impact on weapons proliferation, of how large a price should be paid.

2. It is based on the erroneous view that the development of "plutonium economies" can be prevented by a U.S. policy of denial. Since at least some nonweapons states, notably Japan and some in Europe, are going ahead with the separation of plutonium and with indigenous enrichment, the United States will have to choose between enforcing such a policy indiscriminately, thereby greatly exacerbating relations with some of its most important allies while probably not inducing them to change their nuclear policies; not enforcing the policy against them while doing so against others, with relations with the latter then being adversely affected; or forswearing the policy. As regards the second alternative, we are concerned not only about the generally adverse effects on foreign relations but also about the likelihood of actually inducing nations to acquire indigenous sensitive facilities, very likely without safeguards.

3. Emphasis on denial seems to be based on an exaggerated view of the ability of the United States to effect denial unilaterally. While the United States has been quite successful so far in inducing other supplier nations to forswear the sale of enrichment and reprocessing technology to developing countries, it is not likely to be able to prevent for long all sales by an increasing number of technologically capable vendors or the construction of such facilities based substantially on indigenous capabilities. It is particularly noteworthy that with a world surplus in enrichment capacity, the United States cannot exercise the leverage that had been anticipated as recently as 1976 when projections of growth in nuclear power were much higher than they now are.

Having made these points, we would not argue that denial be entirely eschewed as a policy instrument. With perhaps rare exceptions, states that refuse to accept IAEA safeguards on nuclear fuel and facilities that could be used to produce weapons should probably not be supplied with uranium, enrichment service, or other nuclear technology or services by the United States or other supplier nations. While denial even in these circumstances will be resented, unwillingness to accept safeguards can be taken as such a strong indicator that facilities or materials will be used for a weapons program that there is probably less risk in denial than in providing the facilities or materials—even though a likely consequence of denial may well be an indigenous unsafeguarded reprocessing or enrichment program. Moreover, there may be instances, even where there is a willingness to accept safeguards, where it will be undesirable to provide nations with sensitive facilities or critical materials because there is strong reason to believe that they would be misused. The executive branch should continue to have the authority to prevent such sales. The use of authority and diplomacy, including the threats of sanctions not limited to the nuclear sector, will be preferable to legislatively mandated denial.

We conclude that even if nations acquire enrichment or reprocessing plants, retain plutonium or highly enriched uranium in their possession, or assist others in the development of such facilities or the acquisition of such materials, they should in general not be subject to discriminatory treatment, provided that they are willing to accept safeguards on such facilities and materials and to insist on others accepting them as a condition for sale or technical assistance.

This is not to say that we view with equanimity the possibility of widespread proliferation of reprocessing and enrichment plants and a growing commerce in plutonium and highly enriched uranium, with these materials being stored in widely dispersed sites. Not at all. But we feel that the best hope of prevention, particularly in the longer term, lies not so much in denial as in persuading others that such developments may not be necessary in their particular circumstances and in offering better alternatives.

A substantial effort to improve information and analysis is certainly justified. Much of the interest in indigenous reprocessing, recycling, and early commitment to breeders has been based on unrealistic estimates. Many estimates of cost and technical difficulties of various reactor and fuel cycle programs have been unrealistically low, and estimates of demand for nuclear power have been unrealistically high. The effect of these errors has been to make breeders and recycling of uranium and plutonium in LWRs seem more attractive than the facts warrant. Moreover, much of the interest in moving in these directions appears to have its basis in aspirations for energy independence without any realistic appreciation of the fact that such movement can have no substantial impact until well into the next century. The advantages of reprocessing as a precursor to waste disposal have probably also been overestimated. Analytical efforts and argumentation may lead to more soundly based decisions.

But there is more to be done to respond to interests that motivate nations to get into enrichment, reprocessing, and breeder programs. Putting aside considerations of prestige and possible weapons development, we focus attention on the other major interests— solution of waste management problems, reducing dependence on others for fuel, and reducing vulnerability to interruption of supply.

The first of these needs is conceptually much the easiest to respond to. It is especially important in the case of West Germany and

Japan. Both have an interest in reprocessing because they believe in its desirability as a precursor to waste disposal.

One of their alternatives is to ship spent fuel to France, the United Kingdom, or both for reprocessing, and agreements have been reached to have this done. The questions of the disposal of the resulting high level wastes and the disposition of the separated uranium and plutonium remain open. We turn to these questions later.

The other alternative now open to these countries is to reprocess at home. Again, there arise the problems of disposal of high level wastes and of what to do with the plutonium that is separated in excess of immediate needs. Waste disposal is particularly worrisome to Japan since it has not identified acceptable disposal sites and in West Germany because of domestic opposition to the development of the site that has been selected. The plutonium problem is troublesome for both countries, for other countries that might also reprocess, and because of the proliferation potential, for the world community.

A third possibility would be for the United States, and possibly other countries, to offer to accept fuel from West Germany, Japan, and other countries for retrievable storage. If the storage costs were low and the conditions relating to later return of plutonium and disposal of wastes acceptable, the provision of such a service could serve as a strong disincentive to reprocess, to store plutonium, and to recycle plutonium in LWRs in Japan, West Germany, and other countries with similar views about waste management. The attractiveness of such an offer would depend very much on the conditions attached. The most that might reasonably be expected is that the United States would offer to store spent fuel at cost, with the country of origin having the right to have it returned at any time or sent to a third party for reprocessing or for permanent disposal without reprocessing, subject only to the requirement of IAEA safeguards.

It is, of course, impossible for the United States to implement such an offer until it has its own interim fuel storage problems in hand. Moreover, the NNPA would have to be repealed or amended. Finally, offering to accept unlimited or even large amounts of fuel for temporary storage could be politically difficult, although it should be noted that the amount from Japan and West Germany would not be likely to exceed 40 percent of that arising from our own program.

It may be that, as in the case of some American utilities, some foreign utilities or governments would prefer to have their spent fuel disposed of without reprocessing, either immediately after shipment from reactor sites or after a period of interim storage elsewhere. Offers to accept spent fuel for such disposal might also reduce reprocessing incentives. We believe that the United States should offer to accept spent fuel for permanent disposal at cost as soon as it has resolved its own waste disposal problem, and it should try to persuade other countries with suitable sites to do so as well.

Probably few, if any, countries, however, want to get into reprocessing primarily to help solve the waste disposal problem. The main interest is rather in more efficient use of uranium than is possible in a once through reactor fuel cycle. Interest is focused on the breeder, which offers the hope of using uranium (or thorium) so efficiently that even nations that have little, if any, indigenous uranium will be able to achieve independence with respect to nuclear fuel. This can be realized with breeders only by nations that can also carry out their own reprocessing and fuel fabrication.

As we pointed out earlier, the breeder is not likely to be economically competitive with LWRs (or some other converter reactors) unless uranium prices get very much higher—higher than seems likely in this century, considering the worldwide supply and demand situation. With the breeder at a substantial economic disadvantage, other nations may be induced to forgo it, and the reprocessing that goes with it, for quite a number of years, provided that they can be assured of a supply of uranium at reasonable prices adequate to the needs of whatever other reactors they might buy. Nations electing to buy LWRs will have an interest in building indigenous enrichment plants, unless the supply of enrichment service, as well as of uranium, can otherwise be assured at reasonable cost.

Of great interest, then, from an antiproliferation perspective are measures that might be taken to reduce insecurity about supply of uranium and enrichment. We identify five alternatives that merit comment.

First, there is the question of establishing—in some cases, reestablishing—the credibility of individual suppliers. This is particularly a problem for the United States because in 1974 it announced that it would not supply additional enrichment service to customers who had come to rely on it. Moreover, with the passage of NNPA, others saw their supply of both enrichment service and uranium subject to interruption by the United States, notwithstanding existing contracts. Confidence was also undermined by diffusion of authority among the president, other officers in the executive branch, the Congress, and the Nuclear Regulatory Commission. There appear to be three alternatives for dealing with this problem— although probably none can completely undo the damage done—

establishing a record over time of consistent and, from the point of view of buyer nations, acceptable behavior; repeal or amendment of the NNPA, leaving the executive branch with authority to disapprove sales that are judged not to be in the U.S. national interest; or treaty arrangements. The second alternative is probably to be preferred, but the first may be the most satisfactory that can be reasonably expected in the near term.

Second, there is the possibility of encouraging the development of a multiplicity of suppliers. There are now four major uranium suppliers—the United States, Canada, Australia, and South Africa. Canada and Australia are perhaps regarded by buyers as even less reliable than the United States because of the stringent conditions that they have imposed on sales of uranium, differences between political parties in Australia on uranium export policy, a history of moratoria on exports imposed unilaterally by Canada, and allegations that both have engaged in cartel activity. South Africa's reliability is questionable because of its political instability. There is some hope of increasing access to uranium through exploration programs in other countries. Regarding enrichment, there is perhaps more promise and concern. With the growth of EURODIF (a French-dominated consortium) and URENCO (a British-Dutch-German organization) and with the Soviet Union providing services to Europe, buyers can now turn to three suppliers other than the United States. Each has a somewhat different view about constraints on service. In addition, South Africa and Japan may develop substantial enrichment capacities. Some of the plants are, or will be, in non-nuclear weapons states, but there is some comfort to be drawn from the fact that diversity and surplus capacity will reduce the economic incentive for still more countries, particularly small ones, to develop indigenous capacity. We believe that the United States should certainly encourage the development of additional sources of uranium. We would not take the same position as regards construction of enrichment capacity, particularly in nonweapons states, but we do not believe that the United States should as a general policy try to prevent such developments by sanctions against the countries in question, those who might assist them, or those who might turn to them for enrichment service. While the United States should be a ready supplier of enrichment service, we believe that any effort to regain its monopoly position would be unwise. The United States should in general oppose the imposition of conditions on sales of uranium and enrichment service other than the application of safeguards to nuclear programs, and it should oppose any cartel developments in these areas.

Third, there is the much discussed proposal for the establishment of a "fuel bank," from which nations could obtain uranium, enriched if necessary, in the event of an interruption of supply. Whether such a bank would be effective in reducing nations' incentives to recover and recycle plutonium would depend on the amounts of fuel that could be made available and on the conditions of sale. While it might be desirable to be able to provide small amounts of fuel on short notice to cover failures of delivery due to strikes, natural catastrophes, and so on, hedging against such developments is not presumably what motivates nations to get into reprocessing and breeders. Far more important will be hedging against being denied access to uranium at reasonable prices and under otherwise acceptable conditions over a period of many years. Thus, if a fuel bank or authority is to affect motivations, it will have to be able to meet those kinds of needs. The resources of the "bank" would have to be large, even to meet the needs of countries with modest nuclear power programs. There would have to be a clear understanding on conditions under which uranium would be made available. These should probably be limited to a willingness to pay a reasonable price and acceptance of safeguards on the facilities that might use or process the uranium. The utility of the bank could be seriously compromised if nations were required to forswear acquiring enrichment or reprocessing facilities as a condition for participation. Suppliers of uranium and enrichment services could certainly not have a veto on release of fuel, and the stocks would presumably have to be held outside the supplier nations, perhaps on the soil of likely recipients. Questions on the role of the authority in stabilizing prices and the coverage of carrying charges would require a resolution. On balance, we see the concept as possibly desirable but not likely to be *the* definitive answer to the problem of assurance of access to uranium supplies.

Fourth, national stockpiling is probably a more promising possibility. It has been less discussed but is much more realistic than stockpiling of oil or any other fuel. In fact, some nuclear fuel stockpiles exist. In contrast to the situation with oil, few, if any, nations' nuclear power programs would be critically affected by an interruption of supply for a year or so.

Stockpiling as a hedge against an interruption of enrichment service seems particularly attractive, since a nation would need only a six to ten year stockpile of enriched fuel—to cover the time required to construct indigenous enrichment facilities in the event of service being unavailable elsewhere. The carrying charge on such a stockpile would increase generating costs by about 10 percent at present cost. Such a stockpile is likely to be less expensive than acquiring an indigenous enrichment plant, because with enrichment technology developing rapidly, costs are likely to drop. Thus, there is a strong argument against investment in new plants at this time.

It is also possible to stockpile as a hedge against interruption of uranium supply. However, the stockpile would have to be large if it is to be able to supply reactors for their full expected lives, and carrying charges would be great. Still, at prices that are likely to prevail for the rest of this century, it will probably be less costly (from a national perspective) to rely on converter reactors with such stockpiles than to generate the same amount of electricity with breeders.

Stockpiling natural uranium, or even low-enriched uranium, would not add much to the feasibility of a modest weapons program (although it could be important in a large one), and it might be seen by some nations as an alternative to reprocessing and breeders. Therefore, we believe it should be encouraged or at least permitted without restrictions other than acceptance of safeguards on nuclear facilities. So far, the United States has not encouraged the stockpiling of enriched uranium except in the case of West Germany, where a major motivation was balance of payments considerations.

Fifth, there is the possibility of reducing uranium consumption through alternative reactors and fuel cycles and modification in existing cycles. If the development of plutonium economies is to be avoided, one is practically limited to those measures not involving reprocessing and recycling. We have noted earlier that exploiting at least some such possibilities would seem attractive for the United States. The conclusion depends substantially on the likelihood that uranium that can be produced at reasonable prices will be able to meet demand for American LWRs that might become operational at least until the year 2000 but probably not beyond 2050. In few, if any, other countries is the uranium resource base either as well known or apparently as closely matched to medium term demand (twenty-five to fifty years). For several countries, including Canada, Australia, and South Africa, uranium resources will so exceed domestic demand that there will be no question about assurance of adequate supplies. In these countries, reductions in consumption will be important only to the extent that total costs can be reduced. For other countries, including most with substantial commitments to nuclear power, resources exploitable at reasonable costs will fall so far short of projected demand that reductions in consumption of, say, 10 to 50 percent are not likely to affect decisions about breeders. In France and the Soviet Union, where resources and projected demand might be sufficiently in balance so that uranium savings could be important in reducing dependence on others, the commitment to reprocessing and breeders is probably irreversible. One concludes that for most countries other than the United States, improvements in the efficiency of uranium consumption in converter reactors are likely to be important in decisions about reprocessing and breeder deployment only if coupled with stockpiling of uranium as a hedge against interruption of supply.

It will be apparent from the foregoing that suasion, denial, and positive inducements collectively may not suffice to induce other nations to forgo acquiring indigenous enrichment plants and to delay reprocessing and commitments to breeders. It is, therefore, important to consider whether the dangers of weapons proliferation might be mitigated by technical or institutional arrangements, or both.

We are somewhat skeptical about the efficacy of most of the "technical fixes" that are being considered. The rate of advance of weapons-relevant technology tends to be underestimated in political circles. Moreover, in some political and scientific circles, as distinct from industrial circles, the time required to introduce new reactor types and fuel cycles under commercial conditions is also underestimated. As a consequence, fuel cycles that could not be commercially important before perhaps 2010 to 2020 command support of those concerned about weapons proliferation on the assumption that reprocessing and isotopic separation will continue to be so difficult as to be significant impediments to national weapons programs.

The attention being accorded to the thorium U-233 fuel cycle is an example. It is argued that it may be attractive from a nonproliferation perspective because the U-233 can be "denatured" with U-238 and because U-233 will normally be contaminated with highly radioactive daughter products of U-232, thereby making weapons manufacture (and fuel fabrication) difficult. But by the time thorium U-233 systems could be deployed to a significant extent, obtaining U-233 in a relatively clean form, in amounts sufficient for weapons purposes, is not likely to be much of a problem.

In considering the more developed uranium-plutonium fuel cycle, most of the search for "technical fixes" has been focused on reprocessing of spent fuel and the fabrication of plutonium-containing fuel. The hope is again to make the weapons-usable material—in this case, plutonium—relatively inaccessible.

Contaminating it with radioactive materials so that remote handling techniques would be required is one possibility. This can be accomplished by incompletely separating plutonium from fission products during reprocessing, by adding radioactive materials to separated plutonium, or by producing radioactive fission products in plutonium-containing fuel elements by slightly irradiating them before shipment from fuel fabrication plants. The more difficult the removal of contaminants, the better from an antiproliferation perspective, but there will presumably be little point in making decontamination more difficult than separation of plutonium from old spent fuel elements. Unfortunately, any effective level of radio-active contamination will complicate handling of fuel for commercial purposes and, in some schemes, for reprocessing and fuel fabrication as well. The result will be higher fuel cycle costs and probably higher risks of exposure to workers to radiation. Since the separation of plutonium for weapons purposes is not likely to be difficult or, if carried out on a small scale, expensive, there arises the question of whether denaturization of plutonium would be worthwhile, not to mention the question of who would pay the costs.

Co-processing—that is, incomplete separation of plutonium from uranium during reprocessing—and locating facilities for fabricating plutonium-containing fuels at reprocessing plants are other ap-proaches to the "back end" of the fuel cycle that are being con-sidered. While they offer only limited protection against plutonium diversion they avoid many of the problems inherent in "spiking" plutonium with radioactivity. Although some nations might prefer to do their own fuel fabrication for balance of payments or other reasons while having reprocessing done elsewhere, we believe that co-location of reprocessing and fuel fabrication should be encour-aged in general. The case for co-processing is less clear.

In turning to enrichment, there arises the question of whether it might be feasible and desirable to try to induce nations to opt for one technology in preference to others on the grounds that it is less susceptible than others to being modified to produce weapons-grade material. Gas centrifuge plants will be much more susceptible to such adaptation than gaseous diffusion plants. But the centrifuge option is likely to be less costly, particularly for nations with modest nuclear power programs. Since it seems unrealistic to expect most nations to pay the higher costs of diffusion if lower cost options are available, inducing them to buy the higher cost options will presumably require subsidy from the international community or those nations (such as the United States) most strongly committed to nonproliferation. We question the realism of such subsidization. There may be more promise in the laser option in that it may prove to be less costly but more proliferation-resistant than centrifuges. The ready availability of low cost enrichment services, together with the encouragement of stockpiling low-enriched uranium, may be the best option.

The development of international facilities might also meet national demands for assurance of service while reducing weapons proliferation potential. We have already noted examples in enrichment —in URENCO, where each of the three partners is, or will be, engaged in all aspects of the activity; and in EURODIF. In the latter, which is basically under French control, the other partners, although assured of a share of the output in exchange for provision of capital, are not involved in all aspects of the technology and plant operations.

Such concepts might be extended to other stages of the fuel cycle. In general, the acceptability to participants is likely to depend on the nature of their involvement in the design, operation, and governance of the facility and also on their sensitivity to possible interruption of the service. Presumably, something approaching a URENCO type structure would be most acceptable to junior partners or less technologically advanced countries, but such an arrangement could be troublesome from a proliferation perspective because of the possible transfer of technology—a matter of substantial concern with respect to enrichment and of some, but less, concern in reprocessing. Because the technology transfer problems will be of little concern with respect to fuel storage and waste disposal, multinational facilities of these kinds may be easily established at this time. However, they cannot be expected to serve as much of a disincentive for others to engage in indigenous reprocessing.

There remains the possibility of the spread of reprocessing, storage of weapons-usable materials, fabrication of fuels involving them, and enrichment being limited as a result of such services being provided on acceptable terms by a small number of "safe" countries. We have referred to this earlier in discussing enrichment and reprocessing. Natural evolution in this direction is likely to occur with respect to all aspects of the fuel cycle because of economies of scale. It will be facilitated if prices are low and if, in the case of reprocessing, the nations providing the service can be induced to dispose of high level wastes.

Any advantage of reprocessing from an antiproliferation perspective would be practically nullified if pure plutonium, U-233, or undenatured fuel containing them were to be returned to the country of origin without restriction on amount or use. Ideally, one would like to see a regime in which low-enriched uranium of equivalent energy content would be provided in exchange for these materials, which would be stored or used in the country where reprocessing occurred. Agreement might be reached that such plutonium and

U-233 as might be returned would be under safeguards and would be limited in amount to what was immediately needed for R&D and, later, for actual deployment of breeders.

Since the United Kingdom and France are going ahead with plans to offer reprocessing services, they should be encouraged to do so on terms that will make it attractive to others not to acquire their own reprocessing plants.

If French and British reprocessing charges should continue to be very high or if, when pressed, they refuse to accept wastes, their service may do little to dissuade others from building their own facilities. While the United States might consider using the unfinished reprocessing facility at Barnwell, South Carolina, to provide such services, it would then be politically difficult, if not impossible, to deny service to American utilities. It might be better, at least for some years, to offer the alternatives, discussed earlier, of retrievable storage or permanent disposal of spent fuel without reprocessing. If necessary to dissuade others from developing indigenous reprocessing capability, the United States could offer to provide them with plutonium from DOE processing plants to meet R&D requirements.

To sum up, there appear to be no practicable ways of insuring absolutely that nuclear power programs cannot be used to facilitate weapons manufacture. Yet much can be done to reduce the likelihood of this happening. We are skeptical about the efficacy of "technical fixes" and about policies of denial as means of preventing access to sensitive facilities and weapons-usable materials. We feel more positive about the application of safeguards to the sensitive parts of the fuel cycle, about the development of international institutional arrangements, and about other measures that could reduce the incentives of nations to acquire individual reprocessing and enrichment capabilities.

Solar Energy

13

INTRODUCTION

Solar energy is everybody's great hope for the future—infinitely renewable, environmentally benign, available everywhere. However, judgments differ about the extent and timing of the contributions solar can make. To some, solar is a practical alternative now on a large scale, that can make increased use of fossil and nuclear energy sources unnecessary over the next decade or two and can allow these less-favored sources to be phased out early in the next century. Others view the prospect of large-scale use of solar energy as a seductive but elusive dream that is even counterproductive because it delays commitments to the nuclear and fossil technologies that are the real hope for the next decades. But everybody agrees that solar energy is a "good thing"—if it is used correctly.

Our own position is that solar energy should and will become an increasingly important part of the energy picture within the next twenty years, even though it is impossible now to quantify its likely impact. We recommend policies that will encourage the economical development and use of solar energy in the short run; and we are optimistic about the probability that, with proper policies, technical and economic developments over the next decade or less will fundamentally change the economics of solar energy, making it much more attractive than it is now perceived to be. Thus, we are basically optimistic about the contribution that solar energy can actually make to energy supply within our twenty year period.

Whether or not our optimism about the short-run future of solar energy turns out to have been warranted, however, there is another, perhaps more important, role for solar as an insurance policy for the longer run. Solar is potentially a fundamental alternative in the event that nuclear and fossil fuel sources are found to be partly or wholly unacceptable. Thus, we support policies that will encourage the early use of solar energy by lowering its real costs and by reducing the artificial price advantages (consisting of both direct subsidies and unpaid environmental costs) enjoyed by competing sources. We especially urge greater attention to the basic science, technology, and economics of solar energy systems to increase the odds, that the citizens of the next century will have a wider range of economical energy supply options open to them.

This emphasis on enhancing the long-term potential of solar energy is a result of our basic view that, for a time much longer than our twenty year perspective, physical scarcity of nonsolar energy is not the problem. The real problem is the high cost of obtaining clean, reliable energy; and except in limited applications, solar energy is not now significantly cheaper than other, equally reliable, energy forms and hence cannot do much to solve the latter problem. The extent to which solar will become an economically viable option on a large scale within the next twenty years is uncertain. Over the longer term, however, there is a real possibility that environmental or political considerations may impose constraints on or greatly increase the cost of energy from nuclear and fossil fuel sources, while new scientific discoveries or systems concepts may make it economically practical to supply most of the world's energy from solar sources. In our view, maximizing the probability of this latter event in view of the possibility of the former is of much greater importance than pushing hard for a few extra solar roof panels, wood burners, or uneconomic technology demonstrations in the short run.

Implicit in this view of solar energy is an international perspective on energy problems and on solar's role in the solutions. Unlike fossil fuels or uranium, solar energy is accessible in large quantities everywhere. Local progress in the development of solar energy may advance local energy self-sufficiency, thereby helping to reduce the international tensions produced by world dependence on scarce and unevenly distributed supplies of energy. Furthermore, the huge variety of climatic, economic, demographic settings and the many potential ways to adapt solar energy to local conditions are already engendering a rich diversity of concepts from which to choose in deciding how best to use solar energy for meeting energy demands in a particular setting. Thus, solar energy is inherently an international resource and must be analyzed as such.

Our emphasis on solar energy as a fundamental, international option has determined the contents and organization of this chapter. We do not describe the various existing or near-term solar energy technologies in any detail;[1] neither do we try to guess which of the several promising longer term concepts will turn out to be winners, in the United States or elsewhere. Instead, we concentrate on some of the basic physical and economic considerations that will help determine the ultimate role of solar energy. These basic considerations delimit the general size of the solar resource, reveal some of the practical and environmental problems confronting various solar concepts, and help us to identify a number of promising but often overlooked areas for further study and experimentation.

BASIC CONSIDERATIONS: QUANTITY AND QUALITY OF SOLAR ENERGY

It is much easier to invent new solar energy concepts than it is to find ways intelligently to assess new or old ones. The number of conceivable different ways to collect, store, convert, distribute, and finally use solar energy is immense. So is the number of types of settings in which solar energy might be used, not only worldwide, but even within regions of the United States. The performance, economic attractiveness, and impacts—environmental, social, legal, and political—of a particular solar technology can differ considerably in different settings. For example, small methane generators may be appropriate for small family farms in China or India if they can be built with local materials, but not otherwise. Principles that work for small digestors may not be appropriate for converting large volumes of animal wastes from commercial feedlots. Electricity from solar cells may become economically attractive for refining aluminum in sunny tropical regions but not elsewhere. A solar hot water and space heating system that is best for central Boston is likely to be much different from the system that works best in suburban Boston or rural Massachusetts, not to mention Texas.

Because of this great diversity of opportunities and constraints related to wide-scale use of solar energy technology, we have chosen to focus our analysis on a few basic considerations that apply to broad classes of solar energy technology rather than on specific technologies now in use or under development. We begin this section by discussing the most basic question of all: What is solar energy and

1. This has been done recently and extensively, for example, in Congress of the United States, Office of Technology Assessment, *Application of Solar Technology to Today's Energy Needs* (Washington, D.C.: OTA, 1978).

how can its contribution be measured and projected? We then describe some of the basic physical characteristics of solar energy that will inevitably influence its development.

The "Quantity" of Solar Energy

Solar energy can be defined to include all forms of energy renewably derived from solar radiation and used for human purposes. This definition includes not only "direct" solar energy derived from solar radiation that is intercepted by collectors deployed by humans (e.g., solar cells, flat plate collectors), but also "indirect" solar energy intercepted by humans at some time in a natural cycle after solar radiation strikes the earth (e.g., hydropower, wind, ocean thermal energy, fuels from plants). Broadly construed, this definition also includes the chemical energy stored in food, fodder, and nonfuel plant products (nearly 100 quads per year worldwide, compared with about 200 quads per year from other energy sources) and even the energy in air used to dry materials, to ventilate by natural convection, or to heat space in warm climates.

As these last examples illustrate, it is easy to define solar energy so broadly that the concept has little operational meaning. Solar energy is everywhere and affects everything humans do, in ways that are seldom recognized or measurable, let alone measured. The earth and its natural systems are what they are largely because the sun is what it is, and there is no unambiguous way to say what is due to "solar energy" and what is not. In particular, there is no analytically useful way to say that solar energy is now providing a total of X quads of Btus of energy used by or of value to humans or that this number will increase by Y quads over the next twenty years. About the best one can say for analytical purposes is that certain activities more or less related to the sun affect the demand for some other energy forms (usually oil or fossil fuels generally).

Treating solar energy as a way of reducing demand for scarcer energy forms is not just a practical necessity, but is the analytically appropriate approach. The energy problem, after all, is not a shortage of sunlight or of physical Btus of heat, but rather the scarcity of particular, convenient sources of high quality energy and the high cost of getting high quality energy from other, more abundant sources, such as solar, geothermal, or nuclear. Thus, the analytically relevant measure of the contributions of these abundant sources is not how much or what kind of energy they really contain, in some basic physical sense, but rather the extent to which they allow humans to satisfy their wants and desires with less of the scarce and valuable forms of energy, such as oil and natural gas.

Because solar energy is so diffuse and ubiquitous, and because its contribution is measured by its impact on demand for more conventional energy forms, it is often difficult to specify the dividing line between energy conservation and solar energy in some of its forms. For example, homes can be designed to be cooled by summer breezes and warmed by winter sunlight; clothes can be hung outside to dry instead of being put in a dryer; ice can be produced in large quantities in winter and used in the summer to improve the efficiency of steam turbines or air-conditioning systems; population can shift from colder to warmer areas. Any of these actions will reduce demand for electricity and/or fossil fuels compared with what it otherwise would be. But there is no logical way to decide how much of the savings should be attributed to increased use of solar energy and how much to energy conservation or to fundamental changes in energy demand. Thus, different analysts, agreeing on the basic facts but using different accounting conventions, may come to quite different judgments about energy demand and about the role of solar in meeting that demand now or in the future.

A more important source of confusion and disagreement about the future role of solar energy is the tendency to label as "solar" certain activities or technologies that, while they may be necessary for solar, can be used just as well with other primary energy sources. For example, a home designed to use locally collected solar energy as its primary heat source must be well insulated and carefully constructed; solar electricity systems must have good energy storage capabilities; widespread use of solar energy may require development of such things as efficient heat pumps, engines capable of operating on small temperature differences, energy-efficient transportation systems, and colocation of facilities so that low temperature heat can be used. Plausible scenarios of the future can be developed in which these methods and technologies reduce total energy demand to the point where it can be satisfied almost entirely with solar energy. But while this is good news for those who earnestly desire an all-solar future and even for those who, with us, regard it as important that a largely solar option be available in case other options develop badly, it does not demonstrate that a largely solar future is either the most likely or the best of the available options.

Writers of scenarios of an all-solar future often forget that, once we do all we can to conserve energy and put into place the new energy-efficient technologies and systems that will allow solar to meet most of the energy demand, we could just as well use oil or coal or nuclear power. Moreover, at the lower level of use, the conventional energy sources might be cheaper and just as acceptable environmentally. Many of the actions or technologies appropriate for a solar-dominated future will become economically and socially acceptable as the cost of nonsolar energy increases. But some of these could be used in conjunction with conventional primary energy sources before they are used with solar on a large scale: storage and distribution of waste heat from fossil-fueled or nuclear power plants to district heating systems can use much of the same technology required for solar district heating systems and may be economical before solar systems are. A solar rooftop collector may be an expensive way to satisfy the small remaining heating load in a well-designed building. Once a Pakistani farmer learns how to build a more energy-efficient cooking stove, he may be able to afford to cook with kerosene instead of going miles for firewood or using dung for fuel instead of for fertilizer. Basic advances in technology have wide applicability, and efforts to improve the absolute economics of solar energy may or may not improve its relative economics compared with competing energy forms.

Throughout this study we have avoided making quantitative projections on the ground that one of the most fundamental characteristics of energy problems is the great uncertainty about future events and outcomes and, hence, that policy should definitely not be tied to any specific quantitative forecasts or targets. This general proposition is particularly valid where solar energy is concerned, for the reasons suggested above: the contribution of solar to total energy supply cannot even be defined unambiguously, let alone measured accurately; solar's future will be critically affected by relative rates of technical advance and cost increase that are even harder to predict than averages; and the most important role for solar energy is as insurance in the event that nuclear and fossil energy sources both turn out to be unacceptable options for the long run.

The basic considerations discussed in this chapter lead us to some general policy recommendations intended primarily to advance solar energy in its various forms at economically appropriate rates in the short run and to develop it as a fundamental option for the long run—even though we do not and cannot know what solar's quantitative contribution will turn out to be. But our policy recommendations should be valid whether one is bullish or bearish about the actual role of solar energy within the next twenty years.

The "Quality" of Solar Energy

Not all Btus of energy are the same: some have more value, are more useful, or have a "higher quality" than others. Physicists generally define the "quality" of energy by the temperature at which it is available, because the laws of thermodynamics say that a Btu of high temperature heat can do more work than a Btu of low temperature heat. But other characteristics of an energy source are also important in determining how useful it is to humans. These include its concentration, reliability, cleanliness, and ease of transport. Here we discuss some of the factors influencing the "quality" of energy in a general sense, with emphasis on solar energy.

Direct Solar Energy. Unlike all other forms of energy available to humans in large quantities, solar radiation is pure energy that is not associated with any material. As a result, some types of solar energy systems, once constructed, can maintain a constant inventory of facilities and materials, releasing nothing but useful and waste energy at the same rate as the absorbed solar radiation. For example, operating solar electric cells convert some of the incident solar radiation directly into electrical energy and reradiate, reflect, or discharge the rest. The process produces no waste materials that must be accumulated or released to the environment. Even solar electric plants that require cooling water or solar power systems that use plants for fuel could, in principle, recycle all associated materials within the system, releasing only useful energy and waste heat. In contrast, use of fossil fuels or any type of nuclear energy requires the addition of mass (fuel) to an energy supply system and either the accumulation of increasing stockpiles of waste products (such as ashes, sulfur, carbon dioxide, or radioactive isotopes) within the system or their release to the environment.

This potential advantage of direct solar energy systems may or may not result in less harmful overall environmental impact in any specific application than would result from fossil fuel or nuclear energy systems supplying the same types and quantities of energy for end use. Even solar energy systems that are completely closed with respect to materials can change the flow of solar radiation, water (e.g., rainfall on collectors), or air in ways that could have adverse environmental effects. And potential pollutants are produced in the materials processing, manufacturing, and construction phases and as a result of wear, replacement, or decommissioning of components after a system has been assembled. These issues are discussed more fully in a later section dealing with environmental impacts.

In terms of its capacity to generate high temperatures, direct solar radiation is roughly of the same quality as fossil fuels. Nearly perfect focusing of sunlight can produce temperatures above $5,000°C$ $(9,000°F)$, about what can be produced by complete combustion of pure carbon (derived, for example, from coal) in pure oxygen. Temperatures higher than 100 million degrees Centigrade are achieved inside fission and thermonuclear explosives and would be maintained in confined plasmas in fusion reactors. The highest temperatures actually produced in fuel in any existing fission reactors are less than 2,000°C for electric power production and about 3,000°C in space propulsion test reactors; gas core reactors that might achieve 10,000°C have been seriously proposed but never built. Even these high fuel temperatures now yield useful heat at temperatures of only 450°C (in light water reactors) to 800°C (in gas-cooled reactors).

The temperature of the heat produced by an energy source is an important measure of energy quality when the heat is to be used to do mechanical work or (equivalently) to generate electricity. The laws of thermodynamics say that more work can be done by a Btu of heat in a heat engine the higher the temperature of the heat source relative to that of the heat "sink" used to cool the engine. For example, if the heat source is hot water at a temperature of only 90°C and the heat sink is a natural water body at 20°C, the maximum theoretical (Carnot) efficiency of a heat engine is 20 percent. But if the temperature of the heat source is raised to 600°C (1,112°F, typical steam temperatures in modern power plants), the ultimate Carnot efficiency is 60 percent, so that more than three times as much useful work is obtained from each Btu of heat. Increasing the steam temperature to 1,200°C (2,192°F, beyond the range commonly used today) would increase the ultimate Carnot efficiency to 80 percent, adding only 20 percent more useful work per Btu of heat used. Thus, increasing the temperature of the heat source from that of hot water (90°C) to that of high pressure steam (600°C) greatly increases the theoretical efficiency, but going well beyond that temperature (to 2,200°C) adds relatively little. And of course, the efficiency can never exceed 100 percent. Thus, less and less is gained by going to higher and higher (and more expensive) temperatures in converting heat to useful work.

We stress these technical points only to put into perspective the concept of "thermodynamic potential," which is often used incorrectly. Temperatures within fission or fusion reactions, or perfect carbon flame temperatures, are no more relevant thermodynamically than are temperatures (some 10 million degrees Centigrade) inside the sun: none of them is yet used to do work by humans, and given the increasing costs and decreasing returns associated with going to higher temperatures, they will probably never be fully used. There are expensive ways to use these extreme temperatures to heat working fluids to the thermodynamically useful upper limit (several thousand degrees Centigrade) by burning fossil fuels carefully, by controlling nuclear reactions, or by focusing sunlight. There are also cheaper ways of using any of these energy sources to get lower temperatures, such as heating water to 60°C. As an economic matter it may be "wasteful" to use one or the other of these energy sources to heat bath water, depending on how valuable the energy form is and how difficult it is to use in a particular situation; but as a thermodynamic matter, there is little difference between a low temperature nuclear reactor, an open gas flame, or a nonfocusing solar collector— all fail to achieve the full thermodynamic potential of the energy source if they simply heat bath water.

The situation is somewhat different when high temperature steam from any source produces electricity that is then used to provide low temperature resistance heat, because here the thermodynamic potential of the steam is never fully realized even as heat. Complex, multistep systems of this type, whatever the primary energy source, tend to be costly and hence uneconomic except in special cases. Even in this case, however, it may be economically efficient to "waste" thermodynamic potential (e.g., in order to get transportable energy), just as it may be economically efficient to focus sunlight to high temperatures in the process of heating a home (e.g., to drive a heat pump). Economics and not thermodynamics must ultimately decide how any energy source is best used.

A particularly attractive feature of direct solar radiation is its form as photons or "bundles" of electromagnetic energy that are capable of stimulating physical effects having nothing to do with Btus of heat energy as such. Nearly 50 percent of solar radiation energy is in photons energetic enough to activate the eight or so steps required for photosynthesis or to transfer electrons in some semiconductors, such as silicon, to states that allow generation of direct electrical currents. These are the reactions that account for solar energy's role in plant growth and in direct production of electricity. But in addition, many of the photons in sunlight are sufficiently energetic to split water into hydrogen and oxygen, in two steps. Thus, solar radiation may someday be usable for directly producing a wide variety of fuels or chemical feedstocks, with theoretical efficiencies as high as 50 percent. This technical field is especially rich in possibilities, many of which have yet to be explored.

Indirect Solar Energy. Unlike direct solar radiation, energy derived indirectly from solar radiation has been intercepted by some natural collector (such as the ocean or the atmosphere) and hence is associated with material (water, air, or even soil) that, like conventional fuels, must enter an energy conversion system. In principle (and except for biomass conversion systems), only the physical heat content and not the chemical form of the material that passes through an indirect solar energy system need be changed—a much different process than that in a nuclear or fossil fuel plant. However, capturing and releasing the medium that contains the solar energy (such as gravitational energy in water behind a dam or kinetic or heat energy in wind) involves altering the flows of large quantities of water or air that contain the solar energy. This process may have undesirable environmental effects.

The "qualities" of solar energy in its many indirect forms are widely variable and difficult to compare. Nonetheless, interesting results can be obtained from some comparisons. For example, Table 13-1 lists the heat and theoretical mechanical energy potentially extractable from a metric ton of various natural carriers of solar energy under commonly occurring conditions. The work theoretically obtainable (that is, a perfect Carnot engine) is less than the heat because any engine for turning heat into work must lose some of the heat in the process, the more so as the temperature of the heat source declines. These energy densities (in million Btus per metric ton) measure one aspect of energy "quality," in the sense that a

Table 13-1. Heat and Work Theoretically Available per Unit of Mass of Various Natural Carriers of Solar Energy.[a]

	Energy per Unit Mass (million Btu/metric ton)	
Energy Carrier	*As Heat*	*As Work*
1. Solar radiation	"infinite"	"infinite"
2. Dry biomass[b]	16	9.0
3. Heat of fusion of ice[c]	0.32	0.022
4. Thermal energy of water[d]	0.040	0.0014
5. Thermal energy of dry soil[d]	0.012	0.00041
6. Thermal energy of air[d]	0.011	0.00038
7. Freshwater/seawater osmotic pressure[e]	0.0021	0.0021
8. Water behind 100 meter dam	0.00095	0.00095
Kinetic energy of air or water at:		
9. Velocity of 10 m/sec (21 mph)	0.000046	0.000046
10. Velocity at 3 m/sec (6.3 mph)	0.0000043	0.0000043

[a]Mechanical energy converted to heat at 100 percent efficiency. Heat energy converted to work at Carnot efficiency under stated conditions.
[b]Dried cellulose. Work produced from 400° C steam, cold leg at 20° C.
[c]Work calculated assuming ice used to cool low temperature end of a steam engine from 20° C to 0° C.
[d]Assuming temperature of 20° C lowered to 10° C.
[e]See text.

system extracting solar energy from a high energy density carrier can obtain a unit of useful heat or work by processing less material. Thus, there is a presumption that solar technologies using carriers with higher energy densities should be cheaper, less disruptive environmentally, and smaller—perhaps even transportable.

Even apart from direct solar radiation (which has an "infinite" energy density per unit mass because it has no mass), the energy densities per unit mass of the carriers listed in Table 13-1 vary by factors as great as four million. Nonetheless, all of them have been used or seriously proposed for practical application.

Biomass (line 2 in Table 13-1) is considered an indirect form of solar energy for our purposes. It is by far the highest quality form of indirect solar energy with a potential release of oxidation energy (by combustion) of 16 million Btu/ton, about 60 percent that of high grade coal. Furthermore, because its energy can be released at flame temperatures of several thousand degrees centigrade, both its theoretical (Carnot) efficiency and its practical efficiency in doing mechanical work or generating electricity are also high; if biomass is used to raise steam at even $400°C$, as assumed in Table 13-1, the Carnot efficiency is 56 percent. Even higher quality fuels than dried biomass (considered to be primarily cellulose), such as methane and hydrogen produced by natural ecosystems or by processing biomass, could have been included in the table. They have not been included because the natural abundance of such energy forms is much smaller than that of any others in the table. The high relative energy density of biomass suggests that it has great potential as a solar energy carrier, limited only by its availability—that is, it is the cost of getting, not using, the biomass that limits use of this energy carrier.

The heat of fusion of natural ice (line 3 in Table 13-1) is the heat (0.32 million Btus per metric ton) that is absorbed as ice at $0°C$ is melted to water at $0°C$. This "heat sink" attribute of ice has been used since antiquity for food preservation and, to some extent, for space cooling. Serious engineering and economic studies have been undertaken to explore the feasibility of towing icebergs from the Arctic to arid regions, such as the Middle East, and of making and storing large quantities of ice in regions with cold winters. These studies suggest that using "harvested" natural or stored manufactured ice as heat sinks for refrigeration and air conditioning could be practical and economically attractive under some conditions—especially where the resulting fresh water would be valuable. In fact, ice could be used to lower the discharge temperature of heat engines, thereby improving their thermal efficiency. The Table 13-1 estimate that 0.022 million Btus of useful work could be obtained from the heat of fusion of 1 metric ton of ice is based on lowering the low temperature end of a steam engine from 20°C to 0°C; the increased thermal efficiency would add 0.022 million Btus of work for each ton of ice melted whatever the high temperature end of the engine.

The kinetic energy per unit mass of water or air moving at velocities that commonly occur in nature (lines 9 and 10 in Table 13-1) is the energy that drives the worldwide multitude of windmills, sailcraft, and water wheels (without dammed reservoirs to build up a significant head) and that can substantially enhance (or retard) the progress of aircraft in the jet stream or ships navigating ocean currents. Power-generating windmills were available three decades ago and are again being built for demonstration purposes; and proposals have been made to capture the kinetic energy in ocean currents by mooring submerged turbines to the ocean bottom in regions with relatively strong currents, such as the Gulf Stream. Yet the density of this kinetic energy is much less than that of other forms of indirect solar energy, including that available from even the thermal energy that can be extracted from air, water, or soil (lines 4 to 6) if its temperature can be lowered a few degrees.

The relatively high values of the thermal energy densities in natural materials suggest that, in principle it should be possible to extract much energy from natural bodies of water or even (at somewhat lower densities) air or soil. Externally driven heat pumps using natural masses as sources of or sinks for heat can be economically applied to space heating and/or cooling. The thermal energy in wind, even at very low speeds (e.g., 3 m/sec), is often a more effective means of drying than direct exposure, without wind, to bright sunshine. Differences in temperature within or between natural bodies of water or currents of air can also be used to do work and operate electric power generators by using engines that operate between two temperatures that are not far apart, although the low efficiencies (3-10 percent Carnot) at these small temperatures make the work per unit mass far less than the heat per unit mass. One example is the ocean thermal energy converter (OTEC) concept for using the 10-15°C difference in temperature between tropical ocean water near the surface and deep subsurface water to operate engines using working fluids (such as ammonia or propane) that vaporize at low temperatures.

Dissolving salt in water or diluting a salt solution with fresh water releases some of the energy stored in the chemical bonds of the salt. At the mouths of rivers where fresh water is flowing into seawater, this "heat of solution" is potentially extractable by using a semipermeable membrane to prevent mixing. Osmotic pressure will then raise the level of the salt water relative to the level of the fresh water, theoretically to a height as great as 200 meters, allowing mechanical work to be done by the resulting water column. The potential energy per unit mass indicated in line 7 of Table 13-1 is a theoretical limit to the potentially extractable mechanical work from a ton of roughly equal parts of seawater and fresh water. It is interesting to note that this energy density corresponds to about twice the potential energy of water that can fall 100 meters, the height of a large hydroelectric dam. Thus, the energy density of osmotic energy is of a quality comparable to that of hydropower, which is now the principal form of indirect solar energy.

Another way to compare the "quality" of various forms of solar energy is in terms of the energy "flux," or energy per unit time (power) that crosses a unit area of "collector." This measure is an important factor in energy quality because some kind of fabricated structure or other material (including growing plants, in some cases) must be used to intercept or channel all forms of solar energy, and the larger the energy flux, the smaller the collector area needed for a specified yield. The numbers listed in Table 13-2 indicate the absolute maximum rate of energy extraction per square meter of energy interceptor for various types of solar energy.

All the indicated energy fluxes generally vary considerably during a day, a season, or a year. The exceptions are solar radiation intercepted in space outside the earth's shadow and mechanical or thermal energy extracted from large water reservoirs such as seawater in the tropics or steady currents such as the Gulf Stream. In particular, annual average intensities of solar radiation on the earth's surface are much smaller than maximum intensity at noon on a clear day and vary considerably from place to place on the earth, as illustrated by the numbers in lines 2 through 4 of Table 13-2.

The most striking features of the comparative flux values in Table 13-2 are their wide range and the fact that the solar systems most used or talked about tend to be those with the lowest energy fluxes. For example, hydroelectric dams are now the principal source of solar energy; yet the power ratings of most large hydroelectric systems, per unit area of the reservoir behind a dam, are surprisingly low. The figures shown in line 6 of Table 13-2 are approximately the maximum and minimum values for the U.S. Tennessee Valley Authority's hydroelectric systems. Of course, the kinetic energy of a solid column of water with a 1 m^2 cross section that has dropped 100 meters from the top of a high dam is very high (about 45 MW/m^2), so that the size and cost per unit power output of the direct collectors—the turbines in the power plant—can be low. But the reservoir area needed to sustain such a power flux at the tur-

Table 13-2. Heat and Work Theoretically Available per Unit Time (Power) per Unit of Collector Area, for Various Forms of Solar Energy.[a]

	Power per Unit Area (kW/m^2)	
Form of Solar Energy	*As Heat*	*As Work*
Direct solar radiation[b]		
1. In earth orbit	1.4	0.14
2. Peak at sea level	1.0	0.10
3. At Marigat, Kenya[c]	0.31	0.031
4. At Hornsund, Norway[c]	0.056	0.0056
5. Five meter ocean waves	~10	~10
6. Hydropower dam[d]	$10^{-2} - 10^{-4}$	$10^{-2} - 10^{-4}$
Water flowing at 3 m/sec (6.3 mph)		
7. Thermal energy[e]	1.25×10^5	4.1×10^3
8. Mechanical energy	13.5	13.5
Wind blowing at 10 m/sec (21 mph)		
9. Thermal energy[e]	120	4.0
10. Mechanical energy	0.6	0.6
Wind blowing at 3 m/sec (6.3 mph)		
11. Thermal energy[e]	36	1.2
12. Mechanical energy	0.016	0.016

[a]Mechanical energy converted to heat at 100 percent efficiency. Heat energy converted to work at Carnot efficiency under stated conditions.
[b]Work calculated assuming 10 percent collector efficiency.
[c]Annual average on horizontal surface.
[d]Energy extracted per unit of reservoir area.
[e]Assuming temperature at $30°$ C lowered to $20°$ C.

bine is typically very large. Thus, when other proposed forms of solar energy are criticized for their large land requirements, the hydroelectric case should be kept in mind.

The large thermal energy potential in air and water at near ambient temperatures is indicated by the comparative power fluxes in lines 7 through 12 of Table 13-2. The power flux associated with the mechanical energy of a 10 m/sec wind (21 mph, approximately the velocity below which many wind generators cut off) is one two-hundredth of the thermal energy associated with a $10°$C difference in temperature of the air and some other thermal reservoir. Hence, drying crops and other materials is often a more effective use of wind, especially at low speeds, than generating electric or mechanical power by using its kinetic energy. Even more mechanical energy could be produced by running a heat engine that operates on a temperature difference of $10°$C (using a cold reservoir $10°$C cooler than the wind) than by using the wind's kinetic energy directly. We have found no reference to such possibilities in the technical literature.

The thermal flux associated with flowing water moving at a speed as low as 0.1 m/sec (a very sluggish stream) and 10°C colder or hotter than some other heat reservoir represents a potential thermal power flux that is several thousand times greater than peak solar insolation at sea level. Conversion of this thermal energy to work, at an efficiency equal to 50 percent of the maximum theoretical (Carnot) efficiency for an engine working between water at 20°C and a reservoir 10°C cooler, corresponds to mechanical or electric power of about 70 kW/m² of intercepted warm water. This illustrates the huge potential of using natural temperature differences for warming and cooling and even for doing mechanical work or producing electric power. Use of stored ice made in regions with cold winter climates can accentuate the usefulness of thermal energy in natural flowing water or air.

The potential power density associated with vertical ocean wave motion is also very high. The value of 10 kW/m² in Table 13-2 for 5 meter waves is associated with the oscillatory vertical motion of a volume of water that is a 1 m² in cross section and 5 meters high, rising and falling with a period of ten seconds. This is a not uncommon situation in the open ocean. Extraction of this energy would require displacing the water with a buoyant float and mechanically connecting the float with some object that remains relatively fixed with respect to the surface. (For example, cables could be attached to the ocean floor or to a relatively large, submerged, neutrally buoyant structure.)

Conclusions

These basic considerations concerning the quantity and quality of solar energy lead us to several general conclusions about the role of solar energy in the future.

- Direct solar radiation and the natural systems of the earth that are driven by solar energy provide, in principle, many opportunities for reducing reliance on fossil and nuclear energy sources—even though, as we have pointed out above, there is inherently no way to draw a precise line between energy conservation and solar (or more generally, "renewable") energy supply.
- The thermodynamic quality of direct solar radiation, in terms of its capacity to produce useful work, is at least as great as that of fossil fuels and is higher than that of the thermal forms of energy that are accessible in present commercial nuclear power plants.

Solar radiation is of considerably lower quality than the nuclear or thermal energy produced internally in very high temperature nuclear (fission or fusion) power plants that have been proposed or are under development; but these high temperatures are, with present technology, thermodynamically inaccessible and may never be of any economic relevance to energy production. Thus, economic and not thermodynamic considerations will (and should) determine which solar and other energy forms are used.

- Of the indirect forms of solar energy, those associated with the potentially usable thermal energy content of water or air have much higher "quality," measured by energy density or flux, than those associated with the kinetic energy of water or air. Thus, systems that take advantage of natural temperature differentials or that capture natural warmth or coolness and store it from one season to another (e.g., solar-heated ponds or ice reservoirs large enough to be insulated economically) could be important sources of air conditioning and refrigeration (especially in association with heat pumps). They could even provide electrical power and mechanical work, with heat engines that operate at low temperature differentials. Although this possibility is receiving serious attention for water (ocean thermal energy conversion), uses of thermal energy in air appear to have received much less attention than conversion of its kinetic energy to mechanical or electric power.

- Land use requirements for indirect solar energy systems for which the overall conversion efficiency is greater than about 1 percent, such as the cultivation of high yield crops for fuel, are generally much less than the reservoir area required for hydropower—the most extensively used form of solar energy, broadly defined, today.

- Some indirect solar energy forms that are now unused but that could produce electrical power or mechanical work rather directly have high energy qualities as measured by energy mass density or power flux. Examples are the potential energy in waves and in the mixing of fresh water with seawater. This fact suggests that these energy forms could be of great value someday—but not until new concepts for their exploitation are developed.

- Widespread use of solar energy on a large scale requires the development of generic technologies, such as low temperature heat engines and long-term storage, that would also benefit other forms of primary energy, such as fossil fuels or nuclear. Thus, while basic considerations convince us that a largely solar future is not only inherently plausible but should be made available as an option, it cannot be determined now how soon it will be necessary or desirable compared with a fossil or a nuclear future improved by technologies now associated primarily with solar concepts.

PRACTICAL CONSIDERATIONS: TECHNOLOGY, ECONOMICS, AND ENVIRONMENTAL FACTORS

Basic considerations can take one only so far in deciding when, where, and how solar energy will be used. Practical considerations, such as where and when the energy is available, what it costs to use it, and what its environmental effects may be, will ultimately decide solar's fate. In this section we discuss some of these practical questions—although still at a rather abstract level, rather than in terms of specific machines or systems.

Solar Energy: Where, When, and in What Form

Solar energy may be available in some form everywhere, but is not available in all forms at all times anywhere on earth. The precise conditions and forms of availability are all-important in determining whether and how solar will be viable at a specific location.

Direct Solar Radiation. The intensity of solar radiation is measured in terms of the energy per unit time (power) crossing a unit area; this measure of "insolation" is expressed here in watts per square meter (W/m^2). In the earth's elliptical orbit around the sun, outside the earth's atmosphere, the insolation on a surface perpendicular (or normal) to the sun's rays varies from a minimum of 1,345 W/m^2 on July 4 (when the earth is farthest from the sun) to a maximum of 1,438 W/m^2 on January 2, with an annual average of 1,350 W/m^2. The earth's atmosphere reflects and absorbs some of the solar radiation, so that total insolation at noon on a clear summer day is perhaps 1,300 W/m^2 at the top of Mt. Everest and is about 1,000 W/m^2 on the desert at sea level.

The total insolation on a horizontal surface on the earth, averaged over some specified period, is called the "global" insolation. It differs from peak perpendicular or normal insolation because the sun is not usually directly overhead and the sky is often cloudy (both of which tend to decrease insolation) and because scattered light comes in from the entire sky (which tends to increase insolation). Actual values of global insolation depend on latitude, time of day or year, the character and extent of cloud cover and other atmospheric conditions, altitude, the local topography (including buildings), and (under some conditions) the reflection and absorption properties of the surrounding surfaces.

Because the earth has a tilted axis and a slightly elliptical orbit, the highest value of the global insolation during a clear period of twenty-four hours is at the South Pole in December, when the sun is above the horizon all day and the earth is closest to the sun. The twenty-four hour global insolation on clear days is also higher in midsummer at latitudes about 40° north and south than on any day at the equator, because of the longer daylight period at higher latitudes in summer. But winter daylight hours are shorter at higher latitudes, and hence the difference between daily average insolation in summer and in winter increases with latitude.

In terms of the collector area (not the horizontal land area) required to intercept a specified amount of solar radiation, the effects of increasing latitude or of changes of the sun's angle above the horizon during a day can be partly compensated for by using fixed, tilted collectors or movable tracking collectors that always point at the sun. However, because the atmosphere absorbs and scatters more solar radiation as the sun approaches the horizon, a tilted or tracking collector has less to collect and misses more of the total radiation as it tilts toward the horizon. Thus, fully tracking or permanently tilted collectors are less than fully effective in compensating for low sun angles and have even less advantage where there are more clouds. Focusing collectors collect primarily direct radiation from the sun's disk when it is unobscured by clouds and hence work best when the "direct normal" insolation is highest.

Some selected insolation measurements and estimates are presented in Table 13-3. The locations were chosen to exhibit the large differences between sunny and cloudy areas within each of three latitude bands—near the equator, at midlatitudes, and at high latitudes. The values for average global insolation rates are actual measurements averaged over several consecutive years. All other figures are estimated, using geometric calculations and empirical data relating direct to global (which includes direct and scattered) insolation. The estimates of direct horizontal and direct normal insolation, especially in the three cloudy areas, could be wrong by as much as a factor of two.

Near the equator, the minimum monthly average insolation occurs during the cloudiest part of the season, not necessarily in winter. At higher latitudes the minimum monthly average is in midwinter, and the maximum is in midsummer. The maximum daily average figures, at all locations, are for unusually clear days during the period of

Table 13-3. Insolation at Various Locations.

Insolation Data	Equatorial		Midlatitude		High Latitude	
	Sunny Marigat, Kenya (0°35'N)	Cloudy Benin City, Nigeria (6°33'N)	Sunny Adrar, Algeria (27°52'N)	Cloudy Tomei, Japan (32°37'N)	Sunny Washington, USA (46°10'N)	Cloudy Bochum, FR Germany (51°29'N)
Total Horizontal (Global)						
Annual average (W/m²)	312	170	260	110	194	91
Minimum monthly average (W/m²)	251	139	160	61	50	14.5
Maximum monthly average (W/m²)	338	194	339	144	343	173
Lowest 4 month total/annual total	0.31	0.30	0.24	0.26	0.12	0.090
Maximum daily average (W/m²)	342	342	380	385	384	380
Total, Fixed Tilt at Latitude Angle						
Annual average (W/m²)	312	170	265	102	220	85
Minimum monthly average (W/m²)	251	140	217	67	96	20
Maximum monthly average (W/m²)	338	200	286	144	325	170
Lowest 4 month total/annual total	0.31	0.30	0.28	0.30	0.16	0.13
Maximum daily average (W/m²)	342	342	310	320	330	340
Direct Horizontal						
Annual average (W/m²)	258	77	194	37.4	139	33.5
Minimum monthly average (W/m²)	175	52	116	17.5	26.1	3.3
Maximum monthly average (W/m²)	287	97	261	73.4	280	67.6
Lowest 4 month total/annual total direct	0.30	0.26	0.23	0.27	0.089	0.060
Maximum daily average (W/m²)	290	290	320	330	330	320
Direct Normal						
Annual average (W/m²)	360	110	310	62	240	60
Minimum monthly average (W/m²)	200	60	230	40	100	17
Maximum monthly average (W/m²)	380	130	350	120	430	100
Lowest 4 month total/annual total direct	0.26	0.22	0.25	0.35	0.15	0.12
Maximum daily average (W/m²)	385	385	430	460	510	510

highest insolation. The insolation conditions in the United States range approximately between those similar to Adrar, Nigeria (in the southwestern U.S. deserts) to those near Bochum, Germany (in the cloudiest parts of coastal Washington state). The annual average global insolation in the continental United States is about 190 W/m^2.

Perusal of Table 13-3 suggests the following rather broad conclusions:

- Differences in latitude and in average cloudiness contribute about equally to differences in annual insolation. A noncloudy location at high latitude (e.g., eastern Washington State) can have a higher annual average global insolation than a cloudy equatorial location (e.g., Benin City, Nigeria). Thus, direct solar energy is not always or only for the tropics.
- At high latitudes, even where cloudiness is not a major factor and the annual average global insolation is relatively high, the summer-winter differential is large. For example, the two seasons differ by a factor of seven in eastern Washington State and about twice that if one takes into account the lower efficiency of the collector at the lower insolation level. Thus, the capacity to store solar energy for six months or more for delivery in winter would dramatically increase the total annual useful energy collected per unit area of collector and may well be the key to making solar generally economic in these areas. In equatorial regions, insolation varies from season to season by only a factor of two or so, and hence long-term storage is generally not needed.
- The relative advantage (or disadvantage) of tilted or tracking collectors over fixed horizontal collectors, averaged over the year, depends greatly on average cloudiness but not much on latitude. In clear regions the direct normal insolation is only 15-25 percent more than the global insolation on an annual average basis, and in cloudy regions the direct normal insolation is 35-45 percent less than the global insolation. Fixed, tilted collectors, even in clear, high latitude regions, have no more than a 15 percent advantage over horizontal collectors on an annual average basis. However, tilted and tracking collectors can increase winter insolation (per unit of collector area) by factors of about two in clear, high latitude regions. Thus, if tilted collectors can be spaced far enough apart that they do not shade each other, they can help to reduce the critical winter problem at high latitudes.

These general conclusions highlight the common-sense notion that the best system for collecting and storing direct solar energy will differ widely, depending on local climate and other conditions. But they also indicate that direct solar may be economically feasible under a wide range of latitude and climate conditions; it is not necessarily just for the sun belt.

Proposals have been made to put into earth orbit large arrays of solar electric cells, transmitting power back to earth with microwaves or lasers. As a purely technical matter, such a scheme is feasible and could, in principle, provide as much power as desired. Its advantages over earth-based solar cell arrays would be its uninterrupted exposure to sunlight and (perhaps)the smaller land area required for the receiving equipment. But the energy beamed back to earth could have serious adverse environmental consequences and would have to be collected on large areas away from population centers. The losses in collection, conversion, transmission, and then distribution on earth would be large. The costs of constructing and maintaining such a system would clearly be immense—and anybody with the capacity to orbit a payload of a few hundred pounds could put a large energy source out of service within a few hours. Although solar power from satellites cannot be ruled out for the distant future, it is not something that should be counted on or even invested in heavily within the next twenty years.

Indirect Solar Energy.   The geographical distribution of most potential forms of indirect solar energy is not as well known as is the distribution of direct solar radiation. An exception is hydropower, the potential for which has been surveyed in most parts of the world. Theoretical upper limits to the quantities of energy potentially extractable from wind, fresh and sea water, and plants can be calculated for a particular region, but may have little relation to practical upper limits—especially because fluctuations in available wind or water energy are as important in determining what is practical as is the total amount of energy available annually. Nonetheless, estimates of the maximum energy potentially available in some particular form of indirect solar energy in some region can be helpful in putting the various forms into some perspective.

Hydropower. The estimated maximum worldwide potential for water power is the equivalent of about 3 million megawatts of installed electrical capacity—equivalent to 3,000 large nuclear powerplants—of which a little over 10 percent has already been developed. In the United States, the present hydroelectric capacity of nearly 70,000 MW represents about 40 percent of the estimated potential. Worldwide, hydropower now accounts for 23 percent of total electric power production.

Hydropower, because it is available directly as mechanical rather than as thermal energy, is naturally suited for producing electricity or doing mechanical work directly (e.g., for driving machinery or compressing air or lifting water that can be used to produce electricity or to drive machinery). And it is one of the more easily storable forms of solar energy, which can, within limits, be used as needed. This accounts for the fact that, both for the United States and the world as a whole, the average load factors for hydroelectric plants are a relatively high 50 percent.

For all these reasons, hydropower is a highly desirable form of solar energy. It has great promise, especially in developing countries where the potential is not so fully developed and where small-scale hydropower may be usable without expensive power transportation and distribution grids. However, full development of most of the world's hydropower potential, whether in the form of very large power plants or of much larger numbers of "minihydro" plants with capacities of 10 MW or less (and little storage capability), might have major environmental impacts. The upper limit to potential water power in environmentally acceptable, practical installations is surely much lower than that represented by the above estimates.

Windpower. Windpower, like direct solar energy, is potentially available everywhere on earth. It now plays a significant role in all countries in drying crops and materials, although this role is impossible to define quantitatively. Like hydroelectric power, wind energy is available directly as mechanical energy, but is unevenly distributed in terms of its practical availability for conversion to mechanical or electrical energy. Unlike hydropower, wind energy cannot be stored and used as needed.

The World Meteorological Organization has estimated that the best land sites worldwide might provide about 20 million megawatts of installed capacity or about seven times the world's potential for hydropower. The electrical-mechanical energy potential in wind is much greater even than this, in principle, if all the earth's wind kinetic energy within about 50 meters of the surface were included. However, even with good sites for wind power, variations in wind speed generally make it unrealistic to assume an effective annual average load factor greater than about 30 percent, and protracted calm periods can result in zero capacity; some kind of energy storage system or backup is required in order for wind to be used as a major energy source.

There is little doubt that, as a physical matter, a substantial fraction of the U.S. and world demand for electricity or mechanical work could be met with windpower, even if consumption should rise several times above present consumption levels. Cost (especially for storage) and environmental considerations, however, will limit the use of wind energy to considerably less than its physical potential.

Ocean Thermal Energy. The energy potentially extractable from ocean thermal gradients worldwide is huge. Even at an overall thermal-to-work conversion efficiency of 3 percent, the average electric power or mechanical work theoretically extractable from one million square miles of tropical ocean is more than 10 million megawatts. The large thermal capacity of the ocean and the existence of such ocean currents as the Gulf Stream make ocean thermal energy available year round in locations far from where the solar energy is actually absorbed by the oceans. Again, cost, environmental, and power transmission considerations are much more likely to limit the use of this source than its potential availability.

Osmotic Energy. The osmotic pressure differences between fresh and salt water, could, in principle, be used at the mouths of rivers to produce mechanical or electric energy. Roughly 100,000 MW of electric power might be extracted from this source at the mouths of all major rivers in the United States, which is significantly less than is available from other indirect solar forms. Much more work needs to be done on technical and economic feasibility, as well as on understanding the environmental impact of this source, before its practicality can be demonstrated.

Ocean Waves. The mechanical power associated with the rising and falling of ocean swells is huge but, like the wind energy that drives the swells, is unevenly distributed geographically and in time. Costs and environmental impacts are highly uncertain.

Economic Factors Affecting Solar Energy

To a certain extent, solar energy produces a unique product—free, clean, non-nuclear, decentralized energy. Some people will work hard, spend extra money, change their habits, or live in strange-looking houses for the privilege of obtaining this special product. To other people, however, solar energy is essentially just a way to heat water and space, to provide light, and to do work, with little advantage over other ways of doing these mundane things. It may be possible to demonstrate that the environmental and other advantages of solar energy are large enough (or with rising energy consumption, will become large enough) that it is worth some inconvenience and added cost to use it or even that a solar "lifestyle" is somehow better. If so, then private or public subsidies may be justified to stimulate the early adoption of solar energy. But ultimately, the success of solar energy will depend on its being cheaper—all costs considered— or at least not too much more expensive than the other ways of heating water or doing work.

The problem of comparing the costs of solar and nonsolar energy is complicated by many factors. For one, the production and use of nonsolar energy often imposes large "external" costs on society, particularly in the form of environmental damages and dependence on foreign oil exporters, which are not automatically reflected in market prices. Other complications are that public energy utilities are regulated in ways that distort the prices paid by energy users; future costs of nonsolar energy cannot be predicted with certainty; and the use of solar energy requires changes in established ways of doing things that are resisted because they are costly or simply because they are different. For all these reasons and others, there will seldom be a clear-cut answer to the question of whether solar energy in some particular application is economically justified, either for an individual or for society as a whole. Nonetheless, the question must be continually asked; and until enough people answer in the affirmative, solar energy will not become a major factor in the overall energy picture.

In this section we do not try to deal with all the technical or institutional complications involved in comparing the costs of solar energy to the costs of other energy forms. Instead, we discuss some of the factors that strongly influence the costs of solar energy, in order to make two main points: it is easy to make mistakes in estimating and comparing costs, and there are some technical problems that, if solved, could make solar energy (as well as some of its competitors) much cheaper. We do not attempt to judge whether avoiding the analytical mistakes and solving the technical problems will make solar energy in general look more or less economical compared with its competitors.

Capital Costs versus Operating Costs. As an economic matter, a solar energy system—whether it is a device installed on the roof, a large central power plant, or a redesign of a building for better passive heating and cooling—is a substitute for some other energy source. In some cases, such as some passive home designs, the initial cost of the solar energy system may be no greater than the initial cost of the alternative, so that subsequent savings in fuel costs are obtained at no cost as long as they are not offset by higher maintenance and operating costs or lower performance of the solar system.

In the more typical and difficult cases, however, the solar energy system will have higher initial costs and (because it needs no fuel) lower operating costs than the conventional alternative. The economic comparison then turns on whether the savings in future fuel costs are worth the higher initial capital costs.

In general, the comparison of initial capital costs to a stream of future fuel cost savings requires a complete "present value" calculation, in which costs and benefits at different times are adjusted for inflation and then discounted to reflect the real "time value" of money. As in all calculations of future costs and benefits, discounting is necessary because there are other investment opportunities that are expected to yield future benefits in excess of present costs. Unless the discounted cost savings from the solar project are at least as great as its capital cost (also calculated taking into account the time value of money), larger future benefits could be obtained with other investments, such as insulating homes, exploring for oil, building fast food restaurants, and the like.

A complete present value calculation is difficult because it involves estimating future energy costs, allowing for the durability and mode of operation of the solar devices, considering alternative investments, and so on, but it is the only way to make the comparisons properly. Most short-cut methods are wrong or at least misleading, the more so the more they try to simplify an inherently complex question.

One of the most misleading ways to look at the relative cost question is to compare the capital cost per unit of peak capacity for different energy devices. For example, the cost of building a solar electric cell or a windmill that can generate 1 kW of electrical power under ideal sunshine or wind conditions is sometimes compared to the capital cost per kilowatt of peak capacity for a nuclear or coal-fired power plant. But the solar device, unless it has expensive storage capacity, cannot be used in the same way the other sources can—that is, to provide power continuously and as needed. Moreover, the solar device may or may not last as long. Comparing capital cost per unit of peak capacity is meaningless unless the devices compared have very nearly the same operating and durability characteristics.

A somewhat better way to analyze the economics of solar energy compared with alternatives is to calculate the annual capital charges (including both interest and depreciation) on the solar investment, per unit of solar energy produced or other energy form displaced each year. This capital charge per unit of solar energy (for example, $/million Btus or ¢/kilowatt hour) can then be compared to the cost or price of nonsolar energy in an equivalent form.

While the annual capital charge rate includes depreciation and thus takes into account, if only imperfectly, the durability of capital, it can still be misleading in any but the most simple cases—that is, one in which the quantity or the value of the energy displaced does not vary with time. In the contrary case, if, for example, the cost of the fuel displaced is expected to increase over the life of a solar investment, comparing unit fuel costs in the first year to first year capital charges per unit of fuel saved will understate the value of the solar energy device. If the solar device produces or displaces electricity, the cost of which depends on whether or not the utility system is operating with slack capacity, comparing capital charges per unit of solar electricity produced to some average cost of nonsolar electricity may either overstate or understate solar's value. In such cases, a valid economic comparison requires careful attention to the type, timing, and value of energy purchased or displaced by the solar investment, so that a complete present value calculation can be done; comparing a single estimate of capital charges per million Btus or per kilowatt hour of solar energy to some cost or price of nonsolar energy will not result in meaningful economic comparisons in any but the very simplest cases.

Solar Energy in Integrated Systems.   When a single solar energy investment is being considered as part of a discrete project, it is relatively easy to define its capital cost: it is the sum total of the extra costs of land, labor, materials, engineering, testing, and so on necessary to incorporate some solar energy features into a project corrected for the effects of inflation and the time value of money during the construction period. This incremental capital cost attributed to solar energy is then compared with the present value of the expected future fuel cost savings, as discussed above. However, if solar energy is to become a significant share of total energy supply, its economics cannot be analyzed simply in terms of independent, incremental projects: solar energy facilities will have to be integrated into larger energy supply systems, where their economic value will depend on whether they complement or conflict with other elements in these systems.

Solar energy devices have a number of characteristics that make them awkward to use in combination with conventional energy supply systems such as electric utilities. They typically have high capital cost–operating cost ratios, so that once constructed, they should be used as much as possible. However, unlike nuclear and coal-fired power plants, which also have (relatively) high capital–operating cost ratios, solar facilities without storage cannot be counted on to meet the base load of the system. Nor, because peak demand does not necessarily concide with peak solar energy availability, can they be counted on to meet peak demand, which is ordinarily met by low capital cost, high operating cost equipment such as diesel generators or gas turbines. Solar energy without storage is a capital-intensive way to save some fuel costs at unpredictable times. Unless capital costs of solar power devices come way down, as they could with a radical improvement in direct photoconversion technology, they will seldom be an economical addition to an integrated energy system.

An "integrated energy system" in this sense is not necessarily a large electric utility. A single house or factory is part of an integrated system if it relies on the utility for backup energy, and it is itself an integrated system if it provides its own backup capability in such forms as on-site diesel generators or fuel-burning heaters. Either way, without energy storage it is often difficult to justify the incremental capital cost of the solar system by the energy savings it yields.

In principle, solar energy can be applied without energy storage, in nonintegrated, intermittent processes. For example, a solar-powered aluminum smelter or fuel (e.g., hydrogen) generator might be designed to operate only when the sun shines or the wind blows. But the capital required for the additional processes will then also be used only intermittently, raising the capital cost per unit of energy used. Such intermittent process systems are most usefully thought of as solar energy systems that store energy in the form of refined product or fuel.

Energy storage, then, is critical to making solar energy economic on a large scale. Where the solar energy is to be used on site for space or water heat, energy storage is already technologically and often economically practical, in the form of simple hot water or hot rock storage. Where mechanical or electrical energy is called for, however, economical energy storage has yet to be demonstrated. Some concepts for storing mechanical energy are promising, such as pumped hydropower or compressed air. And it may be possible to store hot water for long periods in large ponds, extracting the energy as work with engines capable of operating on low temperature differentials. But any of these add significantly to the capital costs of solar energy systems.

It should be noted that widespread use of thermal heat storage in forms now technologically and (under some conditions) economically practical would be useful whatever the primary energy source. A house heated and cooled by water that circulates through a rock-filled heat storage reservoir can use solar heat to warm the rocks during the day when the sun shines, but it can also use a heat pump or even resistance heating to heat the rocks with electricity used during off-peak hours. What combination will be optimal depends on the costs of electricity, of solar panels, of storage capacity, and so on. Energy storage may be critical to lowering the cost of solar energy, but it does not guarantee that solar will become relatively more economical than alternatives. The point made here has led the study group to stress the important role of basic general research (see Chapter 15).

Illustrative Capital Cost Calculations. The difficulties involved in estimating the cost of energy from solar energy systems can be illustrated with some simple calculations. Because capital costs are such a large share of the cost of solar energy, we deal only with the capital component of energy costs, using the information presented previously to compare capital costs in different locations. We also concentrate on direct solar energy systems, typically but not exclusively for the production of electrical power. These calculations and comparisons are not meant to be statements about the absolute values of capital costs for actual solar energy devices, but only to illustrate the importance of certain critical characteristics of solar energy.

The total capital cost of a solar energy installation includes the cost of all land, construction, equipment, and components for collecting, storing, and converting the energy, and then delivering it to the end user or to the distribution system. The capital cost can be stated in terms of either the total capital cost per unit of peak power capacity (dollars per megawatt: $/MW) or the total capital cost per unit of collector area (dollars per square meter: $/m^2). Given either of these capital cost figures, along with data on local solar energy availability and system energy demand at various times, the capital cost contribution to initial unit energy cost can be calculated.

In simplest algebraic terms, the capital cost contribution (in real, or inflation-free, terms) to initial unit solar energy cost can be calculated from the following relationship:

$$C_E \ (\text{¢/kWh}) = 11.4 \ r \times \frac{C_p}{F} = 11.4 \ r \times \frac{C_c}{Ie},$$

where:

C_E = capital cost contribution to initial cost/unit output energy;
r = capital charge rate, at percent/year of capital costs;
C_p = capital cost, in dollars per peak watt of output;
F = load factor = annual average-peak power output;

C_c = capital cost, in dollars/m² of collector area;
I = incident solar energy in watts/m² , annual average;
e = efficiency of collectors;
11.4 = factor to convert \$/watt year to ¢/kilowatt hour.

The capital charge rate r depends on the real rate of return available elsewhere in the economy, the durability of the solar energy equipment, and the expected future costs of competing energy forms. If the real costs of other energy forms are not expected to change over the life of the solar installation, then r is (roughly) the inflation-free interest rate plus a depreciation factor and would be about 10 percent a year for an installation lasting twenty years or more. However, if the costs of other energy forms are expected to increase, then r should be reduced initially to reflect the fact that more of the capital costs of the solar installation should be allocated to the more valuable energy produced later; roughly speaking, the inflation-free interest rate should be reduced by the rate at which the real cost of other energy forms is expected to increase.

C_p is the capital cost per unit of peak power capacity or per unit of energy the system would produce if it operated for a year at peak power. F is the ratio of the energy produced during a year to what could be produced if conditions were always ideal; it is less than unity both because the sun is not always shining brightly and because, even when it is, there might not be a need for the power. The factor C/F (\$/watt) is, then, the capital cost per unit of energy actually produced in a year.

Dividing C_c, the capital cost per unit area of collector (\$/m²) by I, the average annual incident solar energy (watts/m²), yields the capital cost per unit of solar energy (\$/watt) incident on the collector during the year. The efficiency factor e is less than unity both because solar systems convert only about 3–15 percent of incident solar energy into useful energy and because the energy that could be produced is not always needed. The factor $C_c/(Ie)$ (\$/watt) is then the capital cost per unit of energy actually produced in a year, just as is C/F, above.

When the capital cost per unit of energy actually produced during a year is multiplied by the initial annual capital charge rate r, the result is the capital cost component of initial unit energy costs in dollars per watt year. The factor 11.4 in the above formula simply converts this to the more common unit of cents per kilowatt hour (¢/kWh).

The relative simplicity of these algebraic formulas hides some complications. The problem of determining the proper capital charge rate r when the cost of other energy forms is expected to change over the life of the solar energy installation has already been mentioned. F and e cannot be determined without knowing the characteristics of the energy system into which the solar installation must fit; strictly speaking, it is not even enough to know how much energy the solar energy installation will actually produce, since variations in the value of the energy over the daily and annual cycle should be taken into account. Nonetheless, these simple formulas can be used to illustrate the importance of some of the variables involved.

Consider, for example, a complete photovoltaic system that meets the DOE 1986 cost goal of $1/watt of peak power, including not only the costs of the collector array, but all storage and power conditioning equipment. For illustrative purposes, we use a capital charge rate of 10 percent per year and assume that the system is designed to produce the maximum energy possible per year, without regard to any particular load demand. Since the peak power output is produced when insolation is at its maximum of about 1,000 W/m^2, for a horizontal collector array the annual load factor F would be about 0.312 in Marigat, Kenya (where, according to Table 13-3, annual average insolation on a horizontal surface is 312 W/m^2 and only about 0.091 in Bochum, Germany (where annual average horizontal insolation is 91 W/m^2). Thus, the capital cost contribution to unit energy costs would be 11.4 $\times$ 0.1 $\times$ 1/0.312 = 3.65¢/ kWh in Marigat, Kenya, and 11.4 $\times$ 0.1 $\times$ 1/0.0981 = 12.5¢/kWh in Bochum, Germany.

Focusing and tracking collectors, if they cost the same $1 per peak watt, would decrease the capital cost contribution to 3.17¢/kWh in Marigat, where the annual average direct normal insolation is 360 W/m^2; but in Bochum, where the annual average direct normal insolation is only 60 W/m^2, focusing and tracking collectors would actually increase the capital contribution to unit cost to 19¢/kWh. This illustrates again the disadvantages of focusing collectors in cloudy areas, unless they can be built more cheaply than can horizontal, flat collectors—which may be possible, since focusing collectors do not need such large active areas.

These illustrative calculations suggest that, under favorable conditions, the capital cost per unit of electricity from a solar energy installation might be about 3–4¢/kWh. This would be somewhat more than for nuclear or coal-fired power plants that operate continuously. But capital costs per unit this low require that the solar energy system have a long life (otherwise the capital charge rate will be greater than the assumed 10 percent/year) and that it have enough energy storage or backup power capacity so that all the energy producible is usable. And some of the cost of this backup or storage capacity—whether batteries or peaking turbines or coal-fired power plants operated only when the sun does not shine or an aluminum smelter operated only when the sun does shine—must be included in the capital cost of the solar energy system.

In the absence of significant long-term storage, a solar energy facility might be designed to provide each month the amount of energy available during the least sunny month, using storage or backup power only for short periods. In Marigat, Kenya, where the minimum monthly average insolation on a horizontal surface is 251 W/m^2, the capital cost component of unit cost for such a system (still assuming a \$1/peak watt capital cost, 10 percent/year capital charge, etc.) would be $11.4 \times 0.1 \times 1/0.251 = 4.5¢/kWh$, only 24 percent higher than for the horizontal collector system with perfect storage. In Bochum, Germany, however, where horizontal insolation is only 14.5 W/m^2 in the worst month, the capital cost component of unit cost would be $11.4 \times 0.1 \times 1/0.0145 = 78.6¢/kWh$, over six times as great as for the system with perfect storage.

This calculation illustrates the point that, in higher latitudes where there are large differences in seasonal insolation, long-term energy storage is critical if solar energy is to become an economical source of electricity on a large scale. Storage of electric energy directly for periods of more than several weeks at most is likely to be uneconomical for the forseeable future. Some other method for producing solar electric power, which allows economical long-term storage (as in large hot water reservoirs or stockpiles of chemical fuel) of some intermediate form of energy from which electric power can be derived, is therefore required to obtain economically competitive electric power from solar energy at high latitudes.

The importance of energy storage and the difficulty of storing electricity add support to the contention that careful matching of energy sources to end uses can do much to make solar energy more economical. If use of electrical energy is limited to those jobs for which it is uniquely suited—lighting, electronics, motor drives, some process heat—then solar energy can be produced, stored, and used in its most economical forms to provide heat. The key to using solar energy on a large scale is not simply to force solar energy to do the same things that oil or nuclear power do, but to adjust the way energy is used so that solar energy becomes the most economical source in more and more of these uses.

Solar Energy and the Environment

Solar energy participates continuously in innumerable processes within the biosphere, as it has for billions of years. This is in contrast to energy from the combustion of fossil fuels and from nuclear fission, which is now released on a large scale only as the result of human activities (although spontaneous oxidation of fossil fuels certainly has, and even nuclear fission may have, occurred in the geologic past). Today, the human-caused energy flows are less than 0.1 percent of the solar energy absorbed from the sun on a global, annual basis.

The fact that solar energy is "natural" while other forms of energy are not is often cited as a reason for expecting the environmental effects of using solar energy to be relatively benign. No such generalization is warranted, however. Indeed, on a priori grounds one could just as easily draw the opposite conclusion—that interfering with the large, natural flows of solar energy could cause more disruption than carefully adding small amounts of energy from nonsolar sources. The relative environmental effects of solar and nonsolar energy technologies can be determined only by careful comparisons in specific cases, not by generalized, a priori reasoning. Nonetheless, some order of magnitude calculations can be useful in putting into perspective some of the potential environmental impacts of solar energy.

The Land Requirements for Solar Energy.   Because solar energy is so diffuse, "large" surface areas are required to collect it. These areas do not have to be covered with manmade collectors, however, because indirect solar systems can, in a small area, tap flows of solar energy collected naturally over very large areas. For example, an ocean thermal energy converter could, in a single location, draw on some of the solar energy absorbed over a million square miles of ocean. Similarly, a line of windmills, taking up relatively little land area, could capture some of the solar energy that falls on thousands of square miles, and a single wood-burning power plant can draw its fuel from hundreds of square miles of forest land. Thus, there is no simple way to say that a unit of solar energy capacity must "use" a certain amount of surface area or that its "use" of an area to collect solar energy is necessarily incompatible with other uses.

Although there is no simple connection between surface area used and solar energy generated, a simple calculation can put into perspective the "large" area needs of solar energy. Suppose total world energy demand were to be supplied by a solar energy system on land devoted exclusively to the purpose. This solar system could provide an appropriate mix of different types of energy:

- Direct heat with efficiencies of about 50 percent, that is, water and space heat and perhaps focused industrial process heat.
- Electrical and mechanical energy with efficiencies of 5–10 percent, such as solar electric cells.

• Fuels for transportation, with efficiencies of 3–5 percent, such as those derived from cultivated biomass or solar-driven electrochemical processes.

Taking into account losses in storage and transportation of energy and assuming a high load factor, the system might attain an efficiency of 5 percent overall in converting incident solar energy to useful energy at the point of use. With an overall world annual average insolation rate of 160 W/m² and 5 percent efficiency, this system would produce an annual average of 8 W of useful power for each square meter or 8 MW per square kilometer (MW/km²) of collector surface; a solar power plant with 1,000 MW capacity (equivalent to one typical nuclear or modern fossil power plant) would require 125 km² of collector surface area.

With an average power output of 8 MW/km², an all-solar system would produce useful energy at the rate of about 2.4×10^{-4} quads/year/km², on average. Approximately 700,000 km² of collector surface would be required to satisfy the present world energy consumption of 170 quads/year (at point of use, including noncommercial fuels such as firewood, but excluding food and nonenergy plants). Of course, 700,000 km² (270,000 square miles) is a lot of land—about the size of Texas. But it is less than 0.5 percent of the total land area in the world and only 2–3 percent of the land area devoted to growing cultivated crops. If limited amounts of fossil or nuclear fuels were used to satisfy those energy demands for which solar is least appropriate and if the collectors were concentrated more heavily where insolation is higher than average, the land requirements could be reduced significantly.

Using averages and current values of changing variables can be misleading, of course. If world energy use doubles while population increases to, say, eight billion people, the required land area increases to 1.4 km² or 175 m² per person. In Singapore or Manhattan, a doubled population simply would not have 175 m² of land per capita for all uses. In the Netherlands, with a population density sixteen times greater and average insolation 20 percent less than the United States, about 18 percent of the total land area would be required to meet total energy demand. If solar energy is to be the dominant source in large, densely populated metropolitan areas, it will have to be transported, either in forms that can be used there (e.g., electricity, fuels, compressed air) or embodied in manufactured goods.

These simple calculations say nothing about the economics of using solar energy to supply most primary energy. Land is costly, especially near population centers, while converting solar energy to transportable forms gets one back into the problem of "wasting" high quality energy in low temperature end uses. Effective (and costly) energy storage devices would be needed to get the high collector load factors assumed in the above calculations. The relocation, equipment conversions, and other steps necessary to use solar where and in the forms in which it is most economic would be costly. But the physical availability of land, somewhere, is not the real issue; land availability (like thermodynamics) is an important influence, but only one among many, on the economics of solar energy in each particular application.

Solar Energy and Heat Flows. The solar radiation striking the earth is partially reflected directly back into space (primarily by light-colored areas such as clouds, snow and ice, sand) and partially absorbed by the atmosphere, oceans, darker rocks and soil, and vegetation. The absorbed radiation, along with the fossil, nuclear, and geologic energy generated on the earth, may be stored for a while in various forms (e.g., plants, wind), but ultimately it heats the earth so that it radiates energy into space, mostly as low temperature, infrared radiation. The earth's equilibrium temperature is reached when the radiated energy equals the absorbed solar energy, both averaged over the year.

The amount of fossil and nuclear energy added to these natural flows, on a global basis, is and will long remain insignificant—although local heating and climatic effects, due to high concentrations of energy use or transformation, may be important in some cases. The more critical question regarding energy balances is whether human activities could alter the reflective and radiative characteristics of the earth enough to change the huge natural flows of solar energy and to significantly affect climate or the biosphere, globally or regionally.

Solar collectors, to the extent that they replaced highly reflective surfaces such as snow or white sand, could increase the absorption of solar energy by a factor of about four on the area actually covered by the collectors. Also, large-scale heat storage systems such as hot water ponds or ice reservoirs could change local patterns of absorption and radiation over the diurnal or annual cycle. Moreover, the earth's heat balance may already have been affected by human use of solar energy in the form of agriculture, which often replaces sparse, slow-growing natural vegetation with high densities of rapidly growing crops. The fact that humans have lived with such effects for a long time does not demonstrate that the effects are either small, benign, or desirable or that adding fuel crops to food crops would not increase these unknown effects.

As demonstrated above, however, the maximum total land area requirements for solar systems are so small (a fraction of a percent of the earth's land surface) that it is difficult to imagine how the differences between natural and solar collector reflectance could have significant effects. If it turns out to be necessary, collectors could be distributed so that there would be no net change in the global heat balance, since most plants and many soil surfaces have about the same reflective properties as do fabricated collectors. Artificial reflecting surfaces could even be used to offset the global, if not the local, effects of increased absorption by solar collectors.

These general considerations demonstrate that the direct effects of solar and other forms of energy on natural energy flows are not likely to be large or uncontrollable on a global basis. The most important energy-related potential threat to global heat balances is due to fossil fuel combustion, which is increasing the carbon dioxide (CO_2) concentration of the atmosphere (discussed in Chapter 10). This could conceivably raise the earth's surface temperature several degrees on average within the next fifty years or so, with unpredictable effects on world climate. Solar energy can mitigate these still speculative effects by reducing the use of fossil fuels and hence the net addition of CO_2 to the atmosphere. Even solar biomass systems recycle atmospheric carbon over the seasons or decades without adding to it on balance. This possible indirect effect far outweighs any of the direct effects of solar energy on global heat balances.

Local or regional effects of solar collectors or storage systems on the earth's heat balance could be important, as is now often the case with energy from fossil or nuclear sources. Present annual energy consumption in New Jersey, for example, equals about 4 percent of the total solar radiation incident on the state each year, with the percentage rising to about 10 percent in midwinter; analogous percentages for New York City are more than ten times as large. The local effects of such heat are sometimes substantial, often resulting in average temperatures in cities that are more than 3°C (6°F) higher than in the surrounding countryside, along with substantially less snowfall and different precipitation and fog patterns. Whether such effects are desirable or undesirable is debatable; but there is no doubt that they exist and will continue wherever energy conversion or use takes place on a large, concentrated scale—whatever the primary source.

Waste heat from solar electric power systems can have more or less environmental impact (for the same amount of electrical power) than waste heat from fossil or nuclear fueled plants. Air-cooled photovoltaic systems can have nearly the same reflectivity as the ground they cover and hence could actually withdraw (in the form of electricity for export) as much as 15 percent of the incident solar energy from the local environment. On the other hand, water-cooled photovoltaic systems may release as much as 60 percent of the incident solar energy as heat in warm or hot water, while solar thermal plants can have significantly bigger waste heat disposal problems per unit of electrical power because of their lower thermal efficiency.

The discharged heat for a solar electric plant operating at relatively high input temperatures (i.e., several hundred degrees Centigrade) could be used for space or water heating rather than simply being wasted; exactly the same is true, of course, for the waste heat from a large nuclear plant or a small diesel generator. Large district heating systems and widespread use of small cogeneration plants are becoming increasingly economical ways to conserve expensive primary energy of all types, and their use can be expected to increase. But there is no real technical or economic reason to regard waste heat recovery as a specifically "solar" technique—except in the sense that, if other energy forms are so expensive or become so unacceptable that now contemplated solar electric systems become economical, expensive systems for using waste heat will also be economical.

Materials in the Environment. Solar radiation may be "pure" energy, but the systems necessary to apply it to human use are not. These systems contain modern industrial materials and chemicals that are not necessarily any safer, cleaner, or otherwise more benign environmentally than those used in nonsolar energy systems. Because solar systems are often so massive (per unit of energy produced) relative to traditional energy systems, even small "leak" rates can result in large absolute discharges. Furthermore, interfering with large natural flows of air or water may redistribute naturally occurring materials in environmentally damaging ways. Building, operating, and eventually dismantling a few large or many small solar devices, whether done commercially or by do it yourselfers, could result in the wrong materials getting to the wrong places at the wrong times.

The best illustration of this point comes from the largest existing solar energy systems used by humans—the growing and harvesting of food, fiber, timber, or animals. Chemical fertilizers, pesticides, and herbicides are among the most serious environmental threats produced by humans. Soil erosion, wilderness destruction, and species extinction are important social costs. All of these effects can only increase if solar energy, in the form of biomass production, becomes a large-scale enterprise at the same time that food production is increasing to feed a richer, larger global population.

Even collectors for capturing direct solar energy on a large or small scale use materials that may damage the environment during their manufacture, use, or disposal. Chemicals must be added to water in solar devices to reduce corrosion and growth of fouling or health-threatening organisms. Some solar systems operate with large quantities of antifreeze type chemicals (e.g., ethylene glycol) that are not necessarily totally benign in the environment. Heat pumps and thermal engines operating at low temperatures use chloro-fluorocarbon refrigerants such as Freon that, although confined to "closed" systems, may damage the earth's ozone layer if released. About 2,000 tons of Freon would be required for every 1,000 MW of electrical output capacity in plants operating with hot water input temperatures of 85°C.

Systems that extract solar heat from natural materials often must process massive quantities of these materials, with possibly serious environmental effects. For example, ocean thermal energy converters move water from the deep ocean to the surface. The deeper water is relatively rich in nutrients and in dissolved carbon dioxide, and bringing it to the surface might stimulate biological activity and release CO_2 to the atmosphere. Although increased biological activity would be of some potential value in food production, it could change the reflectance of the oceans; and the CO_2 might contribute more to the greenhouse effect than would the burning of coal to produce the same useful energy. Similarly, osmotic power systems would change the distribution of fresh water and salt water and the flow of nutrients in estuaries, with effects on marine organisms. There is now no way to predict what all these effects would be, but there is certainly no reason to expect them to be small or benign.

Solar electric cells and electrical storage systems may contain large quantities of dangerous metals and exotic chemicals. Cadmium sulfide solar cells are likely to require about 100 tons of cadmium for every 1,000 MW of average capacity; the safe disposal of lead-acid and mercury batteries is already cause for some concern. A society in which homeowners and automobile owners—the same people who now overapply fertilizers and pesticides to their lawns and who dump waste crankcase oil down storm drains—are responsible for their own individual rooftop solar cells and bank of automobile batteries is not necessarily an environmental paradise.

Conclusions on Environmental Impacts. Whether using solar energy is environmentally preferable to using oil or gas or coal or uranium cannot be determined in the abstract. Solar energy technologies, like other energy technologies, have environmental impacts that may be severe and that are different in different situations. Even careful study in a particular case may not yield unambiguous answers, because of uncertainty about likely effects or disagreements about which effects are more serious.

Some order of magnitude calculations do help to put into perspective some of the environmental effects of solar energy. Land requirements for solar systems are large enough that the cost and availability of land in particular cases will be an important element in economic calculations; the local environmental and land use implications of large solar systems may also be important. But on a U.S. or global basis, physical availability of land or the environmental impacts of covering land with collector systems are not likely to become serious constraints on the use of solar energy.

Human use of energy adds insignificantly to global heat balances. But large-scale energy production could affect the earth's temperature and climate by changing the natural systems that determine the absorptive and reflective properties of the earth. The most plausible mechanism by which this could occur is the "greenhouse" effect of increased carbon dioxide in the atmosphere. Although it is possible that ocean thermal energy systems could add to the atmospheric CO_2 levels by bringing to the surface CO_2-rich water from the ocean depths, on the whole, solar energy acts as a substitute for fossil fuel use and hence tends to reduce the rate at which CO_2 is released by energy-producing activities.

Solar energy systems may contain dangerous substances such as heavy metals or exotic gases or may change the distribution of naturally occurring materials such as salt or nutrients in the environment. Careful selection and control of the substances used in solar devices and careful analysis of the impact of large-scale solar systems are essential. Perhaps most important, however, is the fact that solar energy appears to be free of the severe to critical hazards that make both coal and nuclear energy such flawed options.

CONCLUSIONS AND RECOMMENDATIONS

The preceding sections illustrate the diversity of possibilities for using solar energy, as well as the complexity of any effort to assess specific types of solar energy in detail. The possibilities extend far beyond the specific technologies now commercially available or under intensive development. Even the new concepts being reported in the technical literature at an accelerating rate are only a subset of the most interesting possibilities for the long run. We have, therefore, not undertaken a detailed survey of solar energy technology—a survey that would undoubtedly be seriously out of date soon after publication of this book. Instead, we have used the basic characteristics of the broad types of solar energy technology to help us select a set of guiding principles that we recommend be incorporated in solar energy policy during the next twenty years. These principles are set forth briefly below and are illustrated by reference to specific approaches that appear to us to be especially consistent with or in conflict with them:

1. A primary objective of energy policy must be to put solar energy and energy conservation on an equivalent economic basis with energy derived from fossil or nuclear fuels, within an overall energy policy framework that relies primarily on market forces to determine when, where, and how particular energy sources are used. Public opinion appears strongly to favor the use of solar energy and energy conservation over fossil fuels or nuclear energy, but this potential public demand for solar energy is at least partly thwarted by policies that prevent prices from reflecting the true marginal costs of fossil fuels and nuclear energy. Thus, as we have stressed throughout this study, we urge that all energy be priced at its true economic value, through decontrol of producer prices, reform of utility pricing practices, and implementation of environmental policies that cause environmental costs to be reflected in product prices. Only then can solar energy and energy conservation proceed at economic rates.

2. Government R&D programs should be reoriented away from direct participation in demonstration programs for solar energy technology, especially those that are clearly not yet economically competitive on a broad scale. Instead, emphasis should be placed on basic and applied research related to major categories of solar energy technology. Examples of research that might further this goal and that would also lower the costs of other energy forms include:

- Short- and long-term storage of various forms of energy, whether or not derived from solar radiation;
- Conversion of biomass to chemical fuels;
- Photochemistry and photogalvanic processes;
- Concepts for low cost transmission and distribution of heat;
- Basic new approaches to converting solar radiation to electricity;
- Methods for the efficient long distance transport of electricity;
- Low temperature and low velocity heat transfer and fluid dynamics, related to low temperature solar-thermal and to wind energy conversion systems; and

- Basic concepts for collecting and storing solar heat for use in passive solar heating systems for buildings.

3. Government and privately supported solar energy R&D programs should identify basic reasons for present high capital or operating costs of components of solar energy systems and then should try to find ways to reduce these costs as much as possible. Some examples of approaches that look promising are:

- Sharp reductions in the mass per unit area of fabricated materials used for solar collectors, perhaps by using air-supported or inflated plastic support or protective structures;
- Long-term storage of hot water or ice in reservoirs that are sufficiently large to reduce seasonal heat losses to acceptable levels, perhaps coupled with long distance transmission and district distribution of hot water, chilled water, or ice slurry and engine-generators that can operate economically at small temperature differentials;
- Further development of continuous, automated techniques for fabricating, doping, and annealing photovoltaic cells and then depositing electrical conduction channels on their surfaces; and
- New approaches to the production of fuels from biomass, such as the hydrolysis of wood and other forms of cellulose to starch or sugar with subsequent fermentation to ethanol and the pyrolysis of biomass with solar heat to produce chemical fuels.

4. Government should support basic and applied solar energy research programs specifically designed to provide useful technical information to foreign institutions concerned with solar energy development, especially in the less-developed countries. Such U.S. programs should emphasize research that requires specialists, facilities, and other resources that are not available in the other countries. Examples of such programs are:

- Basic theoretical and laboratory investigations of the biological and chemical processes that are involved in the digestion of organic materials to produce methane; and
- Detailed investigations of physical phenomena that govern heat transfer rates between air and various types of substances relevant to solar-heated air drying of crops.

5. Government should generate, assemble, and make widely known data and information on solar energy technologies and the

likely impact of their deployment under various conditions to assist those people searching for types of solar energy of interest to them for specific settings.

6. Government and private investors should be encouraged to seek out opportunities to apply solar energy where it is cost-competitive and otherwise appropriate.

As we have said repeatedly, we do not know whether adoption of these principles and suggestions would lead to much or little solar energy actually being used in the year 2000 or 2020; that will depend on many factors we cannot now predict, such as technological discoveries, the social acceptance of nuclear power, or the nature of the global carbon dioxide cycle. Nor are we prepared to judge whether a government research and development budget of $500 or $1,500 million is the "right" amount to devote to solar. These amounts should arise as the aggregate of the costs of useful objectives, not as percentage increases or as a result of comparison with other research budgets. But we are confident that the principles we suggest for guiding solar energy policy over the next twenty years will help to position solar energy technologically and economically so that it is ready to take the place appropriate for it, both in the short and in the long run. That is as much as any set of policy principles can reasonably hope to accomplish under conditions as complex and uncertain as those surrounding solar energy.

V

Improving the Process

Introduction

It is possible to set down the proper objectives of energy policy and be animated by the right spirit to attain them. Yet, if the process by which they are to be achieved is ineffective, inefficient, cumbersome, things will go wrong. Chapter 11 illustrates this point with regard to air pollution management (and could have been placed in this part as well as in Part III). This final part of our study addresses the process. It does so in two areas and, as throughout the book, is less than comprehensive. Specialists in political science, government, or administration, for example, may find meager pickings, while others will question the widsom or even appropriateness of including consideration of decisionmaking, a subject that has wide validity for governance generally, in a volume considering energy. Perhaps so; but we believe that the two areas we have selected have vital implications for the energy future and need to be considered.

Chapter 14 deals critically with the process of reaching energy decisions in general and in some detail in the matter of power plant siting. In doing so, it illuminates the relationship between federal and state roles, the influence of the judiciary, and the role of the interested citizen—specifically the "intervenor." We are cautious in offering recommendations, but confident that the important questions raised in this chapter can help the reader understand what the debate is all about. It considers, too, what kinds of remedies might be available or fashioned to affirm legitimate rights and discard fictional ones.

Chapter 15 is, we believe, rightly placed at the end of our study. It represents the bridge to the future—that is to say, to the world beyond the twenty years that form the time frame of this undertaking. Research, development, demonstration, and deployment are the stages that move an idea from conception to commercial use, some in a matter of a few years, others over decades. Decisions need to be made, are being made now, and will continue to be made during the next twenty years that will greatly affect the shape of the energy future. An analysis of R&D, to use the conventional term, cannot be omitted, even if its fruits fall beyond our time horizon.

The chapter does two things. First, it analyzes the respective roles of government and the private sector in performing the different phases of R&D and identifies what characterizes energy R&D as contrasted with, say, defense or space R&D and what follows from these differences for the allocation of tasks. It concludes, generally, that government's strength and the need for its participation lie in the initial stage, basic research and development. It should see to it that neglected areas do not stay neglected and that research be broad and provide for the unfolding of competitive ideas. Industry's strength lies in bringing ideas to fruition in an economically efficient way, but with greatest possible freedom of choice as to the specific path and with competitive viability always in the forefront.

Second, in a detailed appendix, the chapter identifies a number of possible energy sources, processes, and technologies. In little more than thumbnail sketches, it characterizes the choice for attention and their prospects for success. While it is far from comprehensive in exposition, the appendix serves to illustrate the guiding principles evolved in the body of the chapter and to convey at least a loose consensus of the group's judgment on priorities.

14

Jurisdiction, Regulation, and Decisionmaking

One perspective on the energy problem that contains more than a grain of truth is that it is as much an institutional as it is an economic problem. The institutional dimension is especially striking for particular energy sources and regulatory systems. A good example is coal, where most of the constraints on both demand and supply stem less from the economics of coal or even from the energy-environmental trade-off as such than from institutional uncertainties and impediments (see Chapters 10 and 11).

Because it is not possible in a volume of this scope to discuss the wide range of institutional problems encountered in the production and distribution of all forms of energy, we have chosen the field of electricity generation as an example of the conflicts of jurisdiction among levels of government and among agencies at the same level of government. This example also illustrates the conflicting statutory and regulatory mandates typical of many institutional problems of energy. Even within electricity generation we have limited ourselves largely to a single problem—the licensing of power plants—and even there we are concerned primarily with the licensing of nuclear plants. One justification for this specialization is the widely held perception that whatever the economics of nuclear power, the future of the industry will be decided in courts, agencies, and legislatures. Hence, there is a special value in taking a close look at one of the prime battlefields—the licensing of nuclear plants—more than ever under scrutiny following the Three Mile Island accident in March 1979. However, in the process we shall also look briefly at coal-fired plant licensing and at some institutional aspects of the waste disposal question.

POWER PLANT SITING: THE LAST TEN YEARS

Before proceeding to a discussion of the causes of delay in siting decisions today, we should ask what has changed in the last ten years. Perhaps the largest change stemmed from the passage of the National Environmental Policy Act of 1969 (NEPA). This act required all federal agencies to file an environmental impact statement with "every recommendation or report on proposals for legislation and other major Federal actions significantly affecting the quality of the human environment." The Act does not specifically provide for judicial review, and "the legislative history is virtually silent on the possibility of judicial enforcement of the Act."[1] Nevertheless, the courts found no difficulty in concluding that judicial review was proper. Indeed, some early court decisions manifested a startling enthusiasm for plunging into environmental issues.[2]

A number of judicial interpretations of the Act have underscored the importance of the 1970 legislation. First, it was held, as indeed NEPA clearly provided, that agencies were required to consider environmental issues even though their statutory powers would not otherwise enable them to entertain such issues. For example, in the 1971 *Calvert Cliffs'* case, a federal court of appeals held that the Atomic Energy Commission (AEC) must consider environmental issues in licensing proceedings, even where no party or intervenor raised such issues, even though before NEPA the AEC had no power to consider environmental factors in its licensing decisions.[3]

Second, some courts found that NEPA not only imposes upon agencies certain procedural requirements but that it also permits the courts to determine whether the agencies were "arbitrary and capricious" in their weighing of environmental values. In particular, courts have been willing to determine whether "the actual balance of costs

1. Frederick R. Anderson assisted by Robert H. Daniels, *NEPA in the Courts. A Legal Analysis of the National Environmental Policy Act* (Baltimore: Johns Hopkins University Press for Resources for the Future, 1973), p. 13
2. See particularly Judge J. Skelly Wright's opinion in the Calvert Cliffs' case where he spoke of the "judicial role" of making the "promise of this legislation become a reality." "Our duty, in short, is to see that important legislative purposes, heralded in the halls of Congress, are not lost or misdirected in the vast hallways of the federal bureaucracy." Calvert Cliffs' Coordinating Com. v. AEC, 449 F. 2d 1109, 1111 (D.C. Cir. 1971).
3. *Id.*

and benefits that was struck [by the agency] was arbitrary or clearly gave insufficient weight to environmental values."[4]

Third, the Act's requirement that the environmental impact statement include "alternatives to the proposed action" has been construed to mean that the agency must consider alternatives (including the environmental effect of those alternatives), even though it would have no statutory power to implement such alternatives. For example, a landmark case held that the Department of Interior, in approving a lease sale for offshore oil and gas, was required to consider the alternatives of elimination of oil import quotas, increased onshore exploration and development, and increased nuclear energy development.[5] On the other hand, the court ruled that the agency need not consider more remote alternatives such as development of oil shale and coal liquefaction, but only because such alternatives would not be available in what the court considered the relevant time period. The court conceded that the discussion of the environmental effects of alternatives need not be "exhaustive" but only provide "information sufficient to permit a reasoned choice of alternatives so far as environmental aspects are concerned."[6] Nevertheless, the alternative rule requires energy agencies to consider at some length even alternatives that they consider "nonstarters." They cannot be sure that the courts will not have more expansive or romantic notions about alternative energy sources. On the other hand, some courts have been reluctant to review the substance of an agency's NEPA statement and have confined their scrutiny to a determination of whether or not the agency has taken a "hard look" at environmental factors.[7]

It is difficult to assess the overall impact of NEPA. It has clearly led to better agency decisions in some instances. However, some agencies have made decisions and then covered themselves by writing supporting impact statements, much as a lawyer writes a brief to support a predetermined conclusion. Further, a known cost of NEPA is that agency action is sometimes delayed, first in the agency itself during the drafting of the statement, then in the courts as the review that almost always accompanies controversial decisions occurs, and then again in the agency if the agency failed to predict what the judiciary would require in the particular situation.

4. Environmental Defense Fund, Inc. v. Corps of Engineers, 470 F. 2d 289, 300 (8th Cir. 1972).

5. Natural Resources Defense Council, Inc. v. Morton, 458 F. 2d 827 (D.C. Cir. 1972).

6. *Id.* at 836.

7. See, e.g., New England Coalition on Nuclear Pollution v. NRC, 582 F.2d 87 (1st Cir. 1978).

STANDING

"Standing" is a legal concept concerning the circumstances under which a private party may challenge government action in the courts. Traditionally, standing to obtain review of agency action was limited to those having a direct economic stake in the outcome of the agency proceeding. Standing was, however, greatly expanded in the first half of the 1970s. In keeping with that general trend, NEPA was interpreted to permit almost any group that alleges that its members have an interest in an agency decision to seek review of that decision.

After the Supreme Court held in its 1972 *Sierra Club* case that that organization did not have standing to attack the proposed development of the Mineral King valley,[8] the Sierra Club and other organizations established standing by including in their amended complaints allegations that their members had a personal interest in the agency decision through actual use of the area for recreational and other purposes. Similarly, a group of law students, having formed a group called Students Challenging Regulatory Agency Procedures (SCRAP), attacked an Interstate Commerce Commission interim order allowing an across the board 2.5 percent rail freight surcharge. That order perpetuated a rate structure allegedly favoring shipment of virgin metal ore over recyclable metal scrap. The student group argued that the ICC had failed to file a NEPA impact statement. The organization's allegation that its members used the forests, streams, mountains, and other resources in the Washington area was held to establish standing.[9]

The courts' willingness to entertain appeals by citizens' groups about energy facility siting decisions dates back to an earlier period than decisions involving national environmental groups. In a 1965 decision the second circuit allowed the Scenic Hudson Preservation Conference, a local conservationist organization, to appeal a Federal Power Commission order granting a license for the construction of a pumped storage hydroelectric project.[10] In the *Scenic Hudson* case the conference alleged that the commission, in licensing the project, had failed to comply with a statutory requirement of a comprehensive plan for waterway projects that would include "beneficial public uses, including recreational purposes." In fact, most licensing statutes provide a similar substantive basis under which any local group can find a basis for intervention in the agency proceedings and for sub-

8. Sierra Club v. Morton, 405 U.S. 727 (1972).
9. United States v. SCRAP, 412 U.S. 669 (1973).
10. Scenic Hudson Preservation Conference v. FPC, 354 F.2d 608 (2d Cir. 1965).

sequent appeal to the courts. What the NEPA adds is, first, another ground on which to attack agency action—namely, the failure to prepare an adequate impact statement—and, second, a basis for court attack on agency decisions that, because they do not involve hearings, do not involve an opportunity for intervention.

JUDICIAL REVIEW OF AGENCY PROCEDURES

An increasingly important source of lengthening lead times for nuclear power plants has been the willingness of courts to set aside administrative agency decisions on the ground that the agency followed the wrong procedure. The District of Columbia Court of Appeals, to which under current law a large share of energy decisions is appealed, has been particularly active in supervising agency procedures. In most of the cases where the agency decision has been set aside for procedural reasons, the ground has usually been either that the agency failed to hold a trial type hearing (with oral testimony and cross-examination) on a particular factual issue or that it failed to make an independent inquiry into some issue that the agency felt was tangential but that the court decided was sufficiently central to the agency's responsibilities to require more deliberate attention. In either type of situation, the result of reversal by the courts may be to require a more time-consuming procedure. Indeed, because agencies are not able to determine in advance when courts may require the additional procedural steps, agencies have often decided to undertake trial type adjudicatory hearings and to make supplementary inquiries in situations where the courts might well not have required the additional steps but where a reversal on appeal would have been a major setback for the agency's program. The strategy parallels defensive medicine practiced by physicians fearing malpractice suits. Resulting delay cannot, of course, be ascribed directly to particular court decisions.

This process came to a head in the nuclear power plant licensing arena in *Vermont Yankee Nuclear Power Corp. v. Natural Resources Defense Council, Inc.*[11] The Supreme Court, in an outspoken opinion, severely criticized the willingness of the District of Columbia Court of Appeals to set aside decisions of the Nuclear Regulatory Commission on what the Supreme Court regarded as unjustified procedural grounds. In the *Vermont Yankee* case, the court of appeals had set aside an NRC decision granting an operating license for a

11. 435 U.S. 519 (1978).

nuclear power plant. It had done so on the ground that the NRC had not considered the nuclear waste disposal issue in the licensing proceeding itself but rather in a separate rulemaking proceeding in which it had issued a rule specifying numerical values to be used in determining the overall cost-benefit balance for each operating license. Although the court of appeals approved the use of rulemaking proceedings to address the waste management issues, it found the rulemaking procedures inadequate. This rulemaking proceeding, which involved both written comments and a public hearing, was held to have generated an inadequate record; and the court of appeals implied that a separate trial type hearing on the waste disposal issue should have been held. The Supreme Court reversed the court of appeals' decision, which it characterized as "Monday morning quarterbacking."

In the same opinion the Supreme Court reviewed an NRC decision involving the grant of a construction permit to Consumers Power Company for two nuclear plants. The District of Columbia Court of Appeals in that case had held that the NRC's Licensing Board should have considered energy conservation as an alternative to new plant construction in carrying out its duties under NEPA, even though the intervenor who raised the energy conservation issue had not provided any evidence that conservation provided a realistic alternative. The appeals court had also held that the licensing board should have returned a report of the NRC's Advisory Committee on Reactor Safeguards (ACRS, a group of outside experts) to ACRS for a more complete report that would be understandable to a layman so that a "concerned citizen" would be able to assess the safety of the reactor design. The Supreme Court was especially scathing in reversing the appeals court on the ACRS report issue. The language used by the Supreme Court, which seldom lectures lower federal courts in this manner, is especially relevant to the licensing delay issue:

> To say that the Court of Appeals' final reason for remanding is insubstantial at best is a gross understatement. Consumers Power first applied in 1969 for a construction permit—not even an operating license, just a construction permit. The proposed plant underwent an incredibly extensive review. The reports filed and reviewed literally fill books. The proceedings took years. The actual hearings themselves over two weeks. To then nullify that effort seven years later because one report refers to other problems, which problems admittedly have been discussed at length in other reports available to the public, borders on the Kafkaesque. Nuclear energy may some day be a cheap, safe source of power or it may not. But Congress has made a choice to at least try nuclear energy, establishing a reasonable

review process in which courts are to play only a limited role. The fundamental policy questions appropriately resolved in Congress and in the state legislatures are *not* subject to reexamination in the federal courts under the guise of judicial review of agency action. Time may prove wrong the decision to develop nuclear energy, but it is Congress or the States within their appropriate agencies which must eventually make that judgment. In the meantime courts should perform their appointed function.[12]

The fact that the Supreme Court was unanimous makes this language even more obviously a calculated and telling reprimand to the District of Columbia Circuit. The Supreme Court decision is likely to have a sobering effect on court review of nuclear licensing decisions.

FRAGMENTING OF
FEDERAL AUTHORITY

With an increase in federal legislation governing the environment has come a fragmenting of federal agency responsibility. For example, although the NRC must prepare an environmental impact statement, the Federal Water Pollution Control Act (FWPCA) amendments of 1972 require the Environmental Protection Agency to regulate all nonradioactive liquid effluents from electric power plants as well as the plants' water intake. As a result, each nuclear plant must not only receive a construction permit and an operating license from the NRC, but must also obtain an EPA National Pollution Discharge Elimination System (NPDES) permit. Some thirty-two states have been delegated the power to issue NPDES permits.

Other federal agencies play a role in plant licensing in particular situations. The most common problems involve the need for a permit from the U.S. Army Corps of Engineers for the cooling water intake and discharge facilities of power plants. All relevant agencies are able to comment on other agencies' draft environmental impact statements.

A host of other relatively new statutes may play a role in energy facility construction, as the recent incident with snail darters under the Endangered Species Act of 1973 demonstrates. Pacific Gas and Electric Company has listed twenty-one federal agencies from which approvals or permits must be obtained to construct a nuclear power plant.[13] In addition, the 1977 Clean Air Act amendments create the

12. *Vermont Yankee Nuclear Power Corporation vs. Natural Resources Defense Council, Inc.* 98 S. Ct. 1197 (1978).

13. Cited in Arthur W. Murphy et al., *The Licensing of Power Plants in the United States* (Mt. Kisco, N.Y.: Seven Springs Center Farm, January 1978), pp. 35–36.

possibility that the NRC may lose its exclusive competence to deal with radioactive emissions from nuclear power plants. Those amendments require the EPA to determine whether emissions of radioactive pollutants endanger public health. If so, the EPA, and perhaps the fifty states as well, would be able to establish regulatory standards and to that extent undertake their own regulation of nuclear power plant construction and operation.

If the only interest at stake were the construction of nuclear plants, one might be led to recommend that all powers under the various statutes should be delegated to the NRC so that only one agency could hold up plant construction and so that the same agency would have an incentive to expedite the overall plant licensing process. Like most government reorganization proposals, however, such a recommendation only makes sense from one point of view. Those whose primary interest is in the maintenance of water quality and environmental purity not only would be loath to transfer FWPCA powers from EPA to NRC but would make the sound point that EPA is likely to maintain a more uniform and consistent system of water quality than would result from the cumulative efforts of agencies whose prime functions were in a variety of other fields.

Therefore, it is probably a mistake to believe that the problem of fragmentation of federal agency responsibility can be solved by "moving boxes around" in some simple kind of government reorganization. The root of the problem lies in the proliferation of broad, general, and overlapping environmental and safety statutes (each passed by Congress without any consideration of the cumulative impact of all such statutes on particular economic activities like electric power generation), as well as in the tradition of delegating enforcement responsibilities to an agency with one value responsibilities and legalistic procedures. The NRC itself is such an organization. It is mandated to rank nuclear plant safety as a number one value, and its decisions are subject to judicial review to assure that other values do not cause it to downgrade safety.

A logical argument can be made for consolidating all of the federal environmental and energy agencies in a single department, so that all energy-environmental trade-offs could be made by a single set of politically responsible officials, but such a recommendation runs counter to the political tendency to have regulatory programs run by the advocates of the interests served by the regulation, does not mesh well with legalistic procedures established by Congress for most regulatory programs, and appears utopian in view of the Three Mile Island accident.

Indeed, such an approach would leave open the question of how to deal with nonenergy industries that create pollution problems.

Moreover, it is far from clear that consolidation necessarily promotes efficiency. Finally, some of the current fragmentation stems from a preference for countervailing sets of power centers, as manifested in the division of the Atomic Energy Commission into a safety agency (NRC) and a promotional agency (formerly the Energy Research and Development Agency, now the Department of Energy).

The best one can hope for, therefore, is better coordination of the timing of decisions on particular energy facilities among various government agencies. Although coordination is desirable and in fact already practiced to a degree between NRC and EPA, it would be a mistake to expect too much from such an approach. Coordination among equals, even under the aegis of a lead agency, is unlikely to lead to better or even to more consistent decisions. Even if it does so, the result will be achieved by a lowest common denominator process. This does not mean, however, that governmentwide time targets could not be negotiated for particular kinds of projects.

STATE-FEDERAL CONFLICTS

Seen from a national perspective, the role of the states in the licensing of power plants can be viewed as a source of delay. From such a perspective, NRC proceedings, so long as they proceed reasonably expeditiously, are not thought of as causing delay because, by consensus, safety licensing is necessary. Whenever particular states require additional procedures, such requirements are (from a national viewpoint) parochial, and therefore, any additional time required for the state proceedings necessarily extends lead times. Viewed from this national perspective, the simplest way to eliminate an important source of delay in the nuclear energy field would be to preempt all state legislation insofar as it had an impact on nuclear power plant construction.

For all important state legislation, preemption raises no significant constitutional issues. All that is required to preempt state law is an intention by the Congress expressed in a federal statute. The reason is that any nuclear power plant, even if it produces power consumed within a single state, has such a large effect on interstate commerce that Congress clearly has the power to regulate all aspects of its design, construction, and operation, including its impact on local health and safety.[14] The vast amount of litigation on preemption has been

14. See, e.g., *Northern States Power Co. v. Minnesota* 447 F.2d 1143 (8th Cir. 1971), *affirmed per curiam*, 405 U.S. 1035 (1972), upholding federal preemption of state authority to regulate radioactive effluents of nuclear plants. See also *Ray v. Atlantic Richfield Co.*, 98 S. Ct. 988 (1978). Other grounds for sustaining congressional authority exist, but the interstate commerce ground is the most inclusive.

generated by the failure of Congress, when it enacts federal regulatory legislation, to specify precisely whether and to what extent state regulation would be preempted.

Recent legislative trends have been toward giving more rather than less scope to state authorities. A number of states have assumed authority under a delegaton from EPA to issue NPDES permits for water discharges. (Such delegation is possible only because the 1972 FWPCA legislation specifically provides for it.) In additon, at least thirty states and some local governments have adopted their own NEPA statutes (usually called SEPAs). Even if research for a federal impact statement may suffice for state purposes, the federal requirement does not preempt state requirements. Moreover, state environmental reviews may be required at several stages in the state regulatory process, including determination of the need for power, selection of the site, and approval of the actual plant. The need for coordination of federal and state requirements has been recognized by the Commission on Federal Paperwork and the Council on Environmental Quality.

Some twenty-five states now have legislation dealing specifically with energy facility siting. Of those twenty-five statutes, twenty-three purport to be "one stop" statutes in which a single designated state agency has been delegated power to act on behalf of all state authorities in granting or denying a permit. Despite the one stop form of legislation, it is in fact necessary in some states to deal with some other state and local authorities to operate the plant, even though the siting permit is granted by a single agency. Hence, one stop siting legislation does not necessarily mean that some other state agency cannot effectively delay the start of operations of a new plant. These considerations led to the conclusion in the report of the Western Interstate Nuclear Board that one stop approval "remains more of a hope than a reality."[15]

Whether or not a state has specific energy facility siting legislation, the state public utilities commission may de facto regulate new plant construction either directly by a traditional certificate of public convenience and necessity or indirectly. One state with an indirect approach is Oklahoma, which regulates new securities issues that would be required to finance a plant.[16] A central issue in the state public utilities commission proceeding, whatever its form, will be what is

15. Quoted in Murphy, et al., p. 65.

16. See display of state requirements in John B. Noble and John B. Hemphill, *Need for Power: Determinants in the State Decisionmaking Processes* (Washington, D.C.: Nuclear Regulatory Commission NUREG/CR-0022, March 1978), p. 11. (Table 1).

called the "need for power." Thirty-nine states require some form of formal demonstration of a need for power.[17] Indeed, even under SEPAs, the need for power tends to be a prime regulatory issue because the state impact statement must balance degradation of the environment against the need for power. Because of NEPA, the need for power is a central issue in NRC licensing proceedings. Hence, the need for power is a key issue on which there is overlap between federal and state proceedings. Because the inquiry into the need for power is an uncertain process involving forecasts rather than hard evidence, the issue can easily be resolved in different ways in different proceedings. Moreover, the issue is especially open to delaying tactics by intervenors.

Many people argue that the need for power is an issue more appropriate for state consideration than for federal consideration and therefore that the issue should be left to the states. Indeed, there seems to be a broad consensus to this effect in the federal and state governments. Nonetheless, two offsetting considerations should be borne in mind.

First, it is difficult to see how the federal environmental decision that the NRC has to take under NEPA can avoid consideration of the need for power where the plant would add to existing capacity. (Need for power issues can be avoided where the new plant would substitute for older and more polluting plants.) If environmental effects are to be measured without regard to the need for power, then the harm to the environment will by definition outweigh benefits to be derived from any particular nuclear plant. Hence, it would not be practicable to leave the need for power issues exclusively to the states unless the state decision was made binding on the NRC in connection with its environmental review.

Second, states may not always be in the best position to make responsible judgments as to the need for power. Electric power generated by a particular plant is often sold in more than one state, and even where it is not, it may displace power imported from outside the state. Indeed, there is something incongruous about determining the need for power, which because of interconnections involves a regional and even an increasingly national market in the context of local siting decisions. The mechanisms for regional determination of the need for power either do not exist or are of questionable efficiency. At the same time, preemption would be inconsistent with current state regulatory policies and would imply a measure of eco-

17. M.J. Scarpa, M.S. Fertel, and E.F. Dul, *Licensing Coal-Fired Plants. Current and Emerging Requirements* (New York: Envirosphere Company, October 1978), p. 22.

nomic deregulation to the extent that state review of utilities' construction programs would thereby be restricted. Moreover, the need for power is often so tied to local conditions that to make a federal level determination of the need for power binding on the local utilities commission might work poorly in practice.

Although the need for power in a general sense becomes part of any environmental inquiry, the importance that should be attached to the issue in economic regulatory proceedings is open to question. One may doubt, for example, that a regulatory agency is necessarily in a better position to judge whether a plant is needed than the utility. It is often argued that, where regulations allow a firm to earn on new investment more than its cost of capital but less than it could make in an unconstrained market, the firm will tend to overinvest. According to this argument, the regulatory authority cannot therefore trust the utility's judgment on investment matters, but must itself make the capital investment decision, which it does under the rubric of the need for power. The validity of the argument depends, however, on whether the assumptions behind the proregulation arguments are valid.

In contrast to those assumptions, it is frequently asserted that under today's conditions, utilities are not enabled to earn the marginal cost of new capital and may not even be able to earn the average cost of old and new capital. They cannot earn the marginal cost because the state utility commissions are unwilling to permit them to earn the new capital rate on the entire rate base when the embedded cost of old capital is far below current market rates. Moreover, because of delays in the ratemaking process, the unwillingness to permit a return on a utility's investment in construction work that is still in progress may mean that the overall rate of return that the utility actually earns may be less than the average cost of capital in a period when replacement costs and market interest rates are both rising rapidly. Under such circumstances utilities would tend, if anything, to underinvest. For example, they might favor coal-fired plants over nuclear plants because of the lower capital cost relative to fuel cost, or they might prefer to buy power from neighboring utilities rather than become self-sufficient.

Ironically, state utility commissions have been increasing the intensity of their need for power reviews at the very time when the objective basis for any fear of overinvestment has been dissipating as a consequence of changing cost conditions. One's view of the desirability of preempting the state's role in need for power issues or, at the other extreme, of delegating the NRC's role to the states depends in part on the care with which the NRC and particular states discharge their need for power reviews.

From the foregoing review, it is clear that the preemption issue raises the larger question of the desirability and scope of state economic regulation of utilities. Moreover, it raises large political questions as to the desirability of maintaining a role for state governments in politically sensitive issues involving the electric power industry. Finally, it can be reasonably expected that the Three Mile Island accident will bring to prominence other reasons for a greatly enhanced federal role, both because the event has drawn attention to the fact that releases of radioactive substances do not respect state borders and because the capacities of neither the utility nor the local and state authorities are apt to suffice for managing emergencies. Pending the outcome of the deliberations of the presidential commission appointed to investigate the incident and make recommendations, it is premature to go beyond these observations.

NUCLEAR WASTE MANAGEMENT

The issue of federal preemption versus delegation to the states arises in a particularly controversial way in the debate over nuclear waste management. Some states have taken steps, either by statute or by regulatory measures, to bar or limit storage of wastes within their territory. If this state "veto movement" were to continue, it would render untenable the administration's present effort (expressed in the October 1978 report to the president by the Interagency Review Group on Nuclear Waste Management) to use a "consultation and concurrence" approach in which federal agencies and state and local governments would cooperate in selecting storage sites.

Under these circumstances the public policy case for federal preemption would be strong. The national interest in developing permanent storage facilities for wastes is unchallengeable in view of the large volume of wastes already generated by both defense and nondefense reactors and in view of the limited space available for on-site storage at existing nuclear power plants. Much of the burden, particularly that arising from public anxiety about health and safety consequences, is borne by state and local governments. If federal preemption legislation were passed, it would be desirable (and no doubt politically unavoidable) to provide federal subsidies to those state and local governmental units in places where the wastes would be stored, not simply to cover the direct costs imposed on them but also to overcome local political reluctance. Such payments could take the form of a transfer to local governments of some of the fees paid by storing utilities. Many states and localities would find the prospects of tax relief through federal payments a powerful incentive to cooperate in selection and approval of sites, so one could expect that competition among them would lead to a reasonable limit on the federal sums required.

A precedent for federal fiscal transfers to states in connection with state cooperation in programs of overriding national interest may be found in the Coastal Zone Management Act Amendments of 1976. That legislation created a Coastal Energy Facility Impact Fund that provides money to coastal states for planning and for programs to reduce the "net adverse impact" of offshore oil development.

Another kind of precedent is educational impact aid, which transfers federal money to local districts in which federal facilities are located. The theory behind this program is that the federal facilities take real estate off the tax rolls while at the same time placing greater strains on local educational facilities. This theory would apply especially to defense bases having families who send their children to local public schools. Educational impact aid, however, illustrates the fiscal danger of any such intergovernmental transfer program. Impact aid has become a general grant in aid program for federal support of education and is no longer targeted solely at communities bearing special burdens. Any waste management transfer program should be designed to reach only those state and local governments actually providing nuclear waste facilities as part of a national program.

A somewhat different federal-state problem has arisen in connection with the transportation of nuclear materials. Some municipalities (notably New York City) have attempted to prohibit such transportation, a move that can have a serious impact where transportation through the city is necessary to service a nuclear facility. More recently, South Carolina officials acted to turn back trucks carrying nuclear materials from Three Mile Island intended for disposal in that state. If these local prohibitions were to spread, a crisis could arise. Nuclear materials would not be able to reach some nuclear plants, or some nuclear plants would have to close down because their operators could not transport wastes from already filled on-site facilities to away from reactor storage facilities. Here the federal government already has the legislative authority to preempt local regulation by itself regulating such transportation, but it has not exercised that power. A current rulemaking proceeding in the Department of Transportation, which has authority over such transportation under the Hazardous Materials Transportation Act, may provide the means for resolving this crucial problem.

NRC LICENSING PROCEEDINGS

Aside from problems of federal fragmentation and federal-state relations discussed in the last two sections, the NRC licensing process itself raises major policy issues. That process involves two stages in which two separate licenses are granted—a construction permit and an operating license. As the names convey, the construction permit precedes construction (with limited exceptions), and the operating license is to precede the commencement of operations. The NRC must hold formal adjudicatory hearings with oral testimony, cross-examination, and so forth before it can issue a construction permit. Although such a hearing must be held for an operating license only if one is demanded by some member of the public, demands for hearings are typically forthcoming. As discussed previously, the prime issues for the NRC are radiological health and security, but the NRC also becomes involved in environmental and need for power issues. In addition, physical security (safeguards) and antitrust issues are also confronted in the NRC process.

As a point of departure, one can estimate that it takes about ten or twelve years for the operator of a nuclear power plant to obtain all necessary licenses.[18] However, the time seems to be lengthening. NRC had calculated the median for plants commencing operations in 1976 at 12.3 years. What may be happening is that "unusual" delays are becoming usual.

From the point where the utility decides to build a new nuclear plant, it normally takes about two years of "site-specific" planning by the utility before the NRC construction permit review can begin. The NRC review takes about two and a half years (up from about twenty months in 1970), ending in the issuance of a construction permit.[19] Construction then requires about five and a half years or more. After about the first two and a half years of the construction period, the NRC operating license review begins. The operating license is not actually issued until construction is completed, but hearings are commenced far enough in advance so that the operating license hearings are not normally "on the critical path" (that is, they

18. The following description of the cycle is based on Comptroller General, *Reducing Nuclear Powerplant Leadtimes: Many Obstacles Remain* Report to the Congress, EMD–77–15 (Washington, D.C.: G.A.O. March 1977).

19. Murphy, p. 23.

occur at the same time that construction is proceeding and hence do not delay completion). Nonetheless, if there is intervention coupled with a lengthy adjudicatory hearing, there may possibly be a delay before the completed plant can be placed in service. The NRC has developed an "immediate effectiveness" rule, which provides that once the NRC has granted a construction permit or an operating license, the plant may begin construction or operations even while the appeal is pending. Of course, nothing prevents the appellate court from staying construction or operations while it reaches its decision, and the commission itself may stay the effectiveness of an initial decision pending review.

Most of the proposals for improving the NRC licensing process itself have centered on the construction permit requirement. The heart of the problem is that the construction permit proceedings, as opposed to the operating license hearings, are "on the critical path" for a considerable period of time. Whatever can be done to reduce this time will reduce the total time from date of conception to date of operation. At present the utility can seek a Limited Work Authorization (LWA) in the course of the construction permit proceeding. An LWA takes the construction permit proceeding off the critical path until the utility completes site preparation and preliminary construction. But the utility takes the risk that it may lose the investment it makes if the NRC later denies the actual construction permit. And the LWA proceeding itself, which is part of the construction permit proceeding, remains on the critical path.

One approach to reducing lead times would be for the NRC to consider the plant design and the site before the utility is prepared to file the application for the construction permit. With respect to plant design, such a reform is possible only if standardized plant designs are considered. Although nuclear plants once were one of a kind designs, a number of designs, albeit apparently only for portions of plants and not for a plant as a whole, have already been considered by the NRC under a standardized design procedure that it has introduced administratively. With respect to site selection, the notion is that sites could be reviewed in advance and then "banked" for future use. Although the NRC will grant early review of sites upon the filing of a construction permit application by a utility, legislation would be required to permit the NRC to "bank" sites upon application by states or other public bodies as well as utilities.

Such efforts to speed up the first phase of the licensing cycle cannot affect completed plants until the late 1980s, but that fact does not make these reforms less desirable. Nevertheless, the question of binding the NRC at a later date when plants and sites are "married" may raise difficulties, in part because the nuclear industry still needs to resolve a number of safety and operational issues. Indeed, even within the regular construction permit and operating license cycle, NRC has found it difficult to deal with the central question of whether proposals under review—and, indeed, plants under construction (or even already completed)—should be required to conform to new safety standards and requirements. In a significant number of cases, the regulatory response has been to impose the new requirements, even though to do so delays the start of operations and adds the cost of "backfitting."

One of the major complaints of the utility industry has been "ratcheting" of requirements—that is, backfitting for marginal safety gains. The industry contends that this practice directly affects costs and introduces unpredictability into the regulatory process. In the industry's view, once a plant or component has been approved, it ought not to be subsequently disapproved after the process of design implementation has begun. There is, however, the competing notion that new knowledge pointing to a safer way of doing something ought not to be ignored simply because costs will be higher, in view of the very much larger costs that may be anticipated in the event of a nuclear plant accident. This view is likely to receive strong support as a result of the Three Mile Island affair.

The ostensible balance that has been struck so far is that the NRC has required retroactive modifications only where they would lead to substantial additional protection to public health and safety. However, the complex and lengthy licensing process affords many occasions for NRC staff to achieve such modifications by informal means, including the implicit threat of time-consuming and costly regulatory delays involved in litigating the need for the retroactive modification. More recently, attempts have been made to introduce cost–benefit analysis in the backfitting decision. The ratcheting-backfitting process currently constitutes a major disincentive to utilities contemplating the installation of new nuclear capacity. Because the bureaucratic element plays such a large role in safety-cost trade-offs in a safety-oriented agency like the NRC, it seems unlikely that this problem could be "solved" by substantive legislation.

If attempts to shorten lead times were not successful before the reactor mishap at Three Mile Island in March 1979, they seem even less likely to be successful after it, even though logically a long review process does not necessarily have anything to do with the quality of the outcome. In short, the Three Mile Island accident has not changed our view that simplification is needed; it has only made that objective less likely to be reached.

COAL-FIRED PLANTS

Coal-fired plants, as well as oil-fired and gas-fired plants, have been regulated by the states. There has been nothing at the federal level equivalent to NRC review. Federal review has been largely limited to EPA and Corps of Engineers regulation, and for many states, most or all EPA review has been delegated to the states. As a result, coal plants have not been confronted with the major sources of delay facing nuclear plants.

Consequently, it is sometimes said that environmental and licensing regulations do not have any significant effect on the time required to construct coal-fired plants. Certainly some of the concerns with nuclear plants involving safety, radiological hazards, and disposal of highly radioactive wastes have no direct counterpart in coal-fired plants. Nevertheless, it is a rare case in which a federal environmental impact statement will not be necessary for a coal-fired plant, either because a Corps of Engineers permit is needed or because EPA has not delegated its NPDES permit procedures. The application for these federal permits triggers the requirement for an impact statement. As noted earlier, there are a host of different state regulatory requirements and procedures.

One estimate is that (1) where there are no significant state licensing procedures, the process of site selection, field study, conceptual engineering, and federal licensing activities (EPA, NPDES, and Corps of Engineers permits) consumes thirty-five months; (2) where there is a state one stop licensing statute, this period stretches to thirty-nine months; and (3) where the state has licensing requirements but no one stop statute, the period becomes forty-two months.[20] At the same time, this study points to a proceeding before the "one stop" New York State Board on Electric Generating Siting and the Environment. In that case, hearings began in September 1976 (following several years of company preparatory studies), the utilities' testimony and cross-examination continued for twelve months, and the state's testimony was expected to last another seven months. After that, there would be arguments, culminating in a final agency decision expected in June 1979.[21]

What seems to be meant by those who say that there are few delays in coal-fired plant siting is that there is little private intervention in coal-fired plant proceedings, including few challenges to

20. Envirosphere Company, *Study Presenting Federal and State Licensing Requirements Affecting Coal-Fired Electric Generating Units* (New York: December 1977), pp. 46–50.
    21. Ibid., pp. 63–64.

federal environmental impact statements. This relative freedom from
the regulatory delays that have affected nuclear plant construction
has been changed by the Clean Air Act amendments of 1977. As a
result of those amendments, delays in building coal-fired plants may
come to equal average delays for nuclear plants.

Perhaps the principal reason for anticipating increased delay is the
prevention of significant deterioration provisions as detailed in Chapter 11. Under the 1977 amendments, a coal-fired plant will have to
obtain a PSD permit. The permit cannot be issued unless "a public
hearing has been held with opportunity for interested persons including representatives of the Administrator [of EPA] to appear and
submit written or oral presentations on the air quality impact of such
source, alternatives thereto, control technology requirements, and
other appropriate consideration." Although the permit procedure is
to be administered by the states upon delegation by EPA, this
language would seem to mandate an extensive hearing procedure—
though not necessarily a full-fledged adjudicatory hearing—with
opportunity for judicial review.

Among the issues to be resolved in the permit proceeding are the
following:

1. Whether the facility will comply with new source performance
   standards;
2. Whether the facility will interfere with implementation plans for
   the attainment of the national ambient air quality standards in
   the relevant air quality control region;
3. Whether the facility will use best available control technology; and
4. Whether certain tests have been made and, in that connection,
   whether the utility agrees to undertake whatever monitoring the
   permit-issuing agency thinks is necessary.

Just what effect these substantive requirements may have on the
availability of permits will depend on many circumstances, not the
least of which is the severity of the new NSPS and visibility standards.
Whatever the substantive impact, the procedures will be a major
source of uncertainty. Moreover, they will lengthen baseline lead
time for coal-fired plants and create a procedural setting in which
additional delays may be generated.

The statute provides that the permit must be issued or denied
within one year of the filing of a "completed application," but of
course there may be serious questions as to when an application is
"completed," and the statute presumably does not foreclose judicial
review by intervenors. Moreover, for some situations, there is a

special "variance" procedure involving state governors where the permit is initially turned down. It may turn out that many plants will be built under variance, so this procedure too may be viewed as a potential source of delay.

Finally, the statute specifically requires the collection of data "at the site . . . and in the area" of "ambient air quality, climate and meteorology, terrain, soils and vegetation, and visibility . . . for each pollutant" for a period of one year before the submission of the application. Exemptions from this requirement may be given by the EPA, which appears to prefer modeling to data collection in any one year. The studies based on such data could easily be a major source of litigation delay if adjudicatory procedures are required and there is intervenor opposition to a plant. The adequacy of the collection procedures provides fertile ground for oral testimony and cross-examination. The statute requires a decision by EPA within one year, but nothing controls the time that may be consumed in judicial review.

The foregoing discussion has been limited to PSD permits, which are required in attainment areas—that is, the areas where national ambient air quality standards are being met. A separate permit procedure is established by the 1977 Clean Air Act amendments for new power plants in nonattainment areas. It is possible that any given plant will have to run the gauntlet of both PSD and nonattainment area permit procedures, because whether an area is an attainment or nonattainment area is determined separately for each pollutant.

Another new regulatory system that will have an important impact on coal-fired plants and that might conceivably provide an opportunity for delay is the Resource Conservation and Recovery Act of 1976, which establishes a permit procedure for the treatment, storage, or disposal of solid waste. Power plant ash and certain sludges are expected to be considered hazardous solid wastes. The permit would probably trigger a further, or expanded, NEPA statement.

A CLOSER LOOK AT DELAY

Current discussions of power plant siting tend to use the concept of delay. There is no such thing as absolute delay; rather, one should think of "optimum delay." For example, a requirement that NRC act on all applications within twenty-four hours would certainly cut delay, but few would consider that period optimal. Therefore, in reducing lead times, the public policy goal should be to achieve some optimum delay.

A second preliminary consideration is that delay can be viewed both ex ante and ex post. What appears an unwarranted delay when the process is just starting (ex ante) may not be so once the process has produced results (ex post). For example, a lengthy delay caused by the site review process may not seem unwarranted if it results in the discovery that a proposed nuclear plant would have created a grave risk to public safety or health.

Finally, a distinction should be made between anticipated and un-anticipated delay. If a utility anticipates the period from application to final operating license approval, it can plan accordingly and minimize the cost of delay. The costs of unanticipated delay, on the other hand, are likely to be greater because of interest, wage, and other expenditures that could have been avoided if the delay had been foreseen.

In making dollar estimates of the cost of delay, the most important variable is the state of construction of the plant at the time the delay occurs. A delay caused, say, by requiring a utility to undertake certain environmental tests before applying for a construction permit will be less costly than an equal delay lengthening the time between the end of construction and the start of operations. Such a delay in the startup of a nuclear power plant may lead the utility to use a peak-load-generating plant that has higher fuel and total operating costs. Alternatively, the utility may have to use an obsolescent plant with high running costs or purchase high-priced replacement power from another utility. Under any of these three circumstances, the delay is likely to be costly to the utility and to the economy as a whole.

In trying to estimate delay, analysts have used differing concepts of cost, often without any clear distinction between (1) fully allocated versus out of pocket costs; (2) costs measured in current or nominal dollars (which leads to high figures for delay as inflation raises the price level) versus constant dollars (which better measures the use of real resources); and (3) pretax versus posttax costs. Hence, no two estimates of the costs of delay are likely to be consistent. Nor is it easy to place different estimates on a comparable basis. With these qualifications, a review of some estimates will nonetheless suggest the magnitude of costs at issue.

The NRC has estimated that a delay in starting construction costs, on average, about $9 million per month and that a delay in starting operations averages between $8.5 million and $13 million per month.[22] The Congressional Budget Office uses nearly identical

22. Marcus A. Rowden, "Licensing of Nuclear Power Plants," *Regulation* 40 (January-February 1978): 42, referring to NRC estimates.

figures: $8.9 million per month for delays during the construction permit licensing phase and $10.6 million per month for delays thereafter.[23] Commonwealth Edison has estimated the cost for a delay at $50 million per year at the early end of the cycle and $200 million per year at the end of the construction process.[24] The American Nuclear Energy Council has stated that if it were possible to reduce the time to build a nuclear plant by one-third, the saving would be from $200 to $300 million.[25]

Even if one has economically relevant definitions of costs and careful estimates using those definitions, assigning causes to any particular delay is a tricky and subjective undertaking. Analysts can assign causes to delays in one plant and then average the cost or the length of the delay attributed to each cause across several plants. However, the underlying data appear to be defective because this process often involves assigning several causes for a single period of delay. Even if the data were complete and internally consistent, assigning several causes to a single period of delay in a particular plant would still be subjective.

One reason for such subjectivity is that since 1973, utilities have been reassessing their investment plans, frequently canceling or deferring new plants. Utility executives may, for example, be ambivalent about certain regulatory delays and therefore be unwilling to compromise issues that they otherwise would have compromised. Another problem is that the causes of delay are mutually reinforcing. Many utilities have been unable to obtain financing (at least on terms they like) for new plants, but the utilities insist, perhaps correctly, that the cause is the uncertainty created by unstable regulatory requirements and delays. Under these circumstances, it is not surprising that utilities blame regulatory procedures and the escalation in regulatory requirements for the lengthening construction periods, whereas others cite the financing problems themselves as the cause.

Moreover, it is not clear that data from the period since 1973 can be projected into the future. A large portion of the delay during the period is attributable to the unanticipated higher prices for primary energy, which reduced the anticipated demand for electrical energy and forced utilities to reconsider projects that were on the drawing boards in 1973. If future energy prices prove to be correctly anticipated, then there should be few such "demand delays" and thereby

23. Congressional Budget Office, *Delays in Nuclear Reactor Licensing and Construction: The Possibilities for Reform* (Washington, D.C., March 1979), p. 29 (Table 5).
24. Murphy, p. 16.
25. Statement dated July 12, 1978.

also probably fewer "financing delays" in the future. Hence, calculations such as those by the Congressional Budget Office to the effect that demand and financing delays have caused 46 percent of all delays may be no guide to the future.[26]

PRIVATE INTERVENTION AND DUE PROCESS

The term "due process" is used in the plant-siting field to mean something different from what it means in the lexicon of constitutional lawyers. To the latter, due process refers to constitutional rights—that is, rights to fair procedure (and sometimes substantive rights) rooted in the Constitution itself.

To lawyers whose attention is fixed on plant sites, however, due process refers to the practice of private intervention in plant-siting proceedings and to the purported undesirability and perhaps political impossibility of restricting private intervention. (Intervention is often characterized as "public," but the adjective "private" better describes what is involved, particularly where the intervenors espouse a particular political position toward nuclear power.)

The term "due process" thus generates confusion. And the confusion is not without consequences. Invoking a constitutional term suggests that intervention is somehow rooted in the Constitution. Nothing could be further from the truth. Congress has the power to eliminate all private intervention in federal siting proceedings. Congress could, furthermore, eliminate judicial review of NEPA impact statements. It could probably even eliminate third party review of NRC decisions. Moreover, such an outcome would merely align NRC proceedings with analogous proceedings, such as Federal Aviation Administration aircraft certification proceedings, where private intervention does not exist.

Rather, private intervention is a by-product of the growth of regulation, particularly environmental regulation. Moreover, intervention is not a necessary by-product of the underlying regulation. If Congress chose, it could—as indicated above—eliminate all intervention both in agency proceedings and in the form of third party appeal from agency decisions. (Under general constitutional principles, the utility might have a constitutional right to some kind of hearing before an application could be denied.[27]) Indeed, the courts and agencies have been able to expand intervention only because Con-

26. Congressional Budget Office, p. 24 (Table 3).
27. See *Goldberg v. Kelly*, 347 U.S. 254 (1970); *Willner v. Committee on Character and Fitness*, 373 U.S. 96 (1963).

gress has normally been silent, as it was, for example, in the Natural Environmental Policy Act of 1969.

Nevertheless, Congress is not likely to eliminate private intervention at this time even if, on balance, private intervention could be shown to be undesirable. The 1973 Alaska pipeline legislation creating an exception to NEPA, and other examples, illustrate the controversy that would surround any such effort. Indeed, the perception that the administration's nuclear-power-plant-siting legislation would somehow stand in the way of unfettered assertion of rights was probably an important reason why that proposed legislation was stalled in the 95th Congress.

Invocation of the due process terminology reflects a broad political consensus that citizens should have a right to have certain important issues, particularly in the environmental realm, decided in the courts or at least in an adjudicatory proceeding. That is, there should be a proceeding with oral testimony taken on a formal record, with rights of cross-examination, and with review in the courts not merely of the procedures followed but also of the adequacy of the evidence to support the agency decision. The use of the term "due process" thus conveys the notion that we value the ability of citizens to intervene in governmental decisionmaking. It is an illustration of the notion of participative democracy.

At the same time, we should be clear that raising intervention to the level of a quasi-constitutional right is not self-evidently consistent with fundamental notions of representative democracy. The courts are not designed to be a democratic institution. The regulatory agencies, though more amenable to legislative control, are intentionally shielded from the political process in which the legislative and executive branches must function. Intervention is being used to fight out fundamental social and political choices better resolved in the legislature.

Moreover, support for intervention reflects a widespread lack of confidence in administrative agencies in general, not just in the NRC. The common belief that issues bearing directly on the energy future of our country should be the subject of litigation is a signal that the Congress has seriously defaulted on its responsibility of making hard choices about trade-offs in the energy arena. That the courts have been so willing to fill the vacuum may surprise some lawyers and energy experts but would surprise no seasoned politician or bureaucrat. Legislatures, and particularly the Congress, have provided the framework for judicial activism by passing statutes that create administrative proceedings (and hence give rise to judicial review) but that fail to specify in detail the substantive standards to be applied in the hearing.

If constitutional concepts do not justify intervention, we must take a cost-benefit approach. Intervention has benefits. There are certain kinds of factual issues, especially those pertaining to a specific site or a specific plant component design, that may be better decided with the investment of the considerable time and expense associated with adjudicatory methods. An adjudicatory proceeding may uncover a great deal more about a specific narrow issue of fact than would be learned through ordinary bureaucratic processes. Let us call this benefit the "better decision effect."

However, this effect does not extend to decisions involving a ranking of values. There is no reason, for example, to believe that an adjudicatory process will lead to a better decision about an energy-environmental trade-off. Moreover, adjudication leads directly to judicial review, and it is an open question whether decisions by judges reflect society's values as well as decisions by agencies. Judges are not responsible to the electorate, but agency heads are at least subject to oversight and budgetary review by Congress. Some participants prefer adjudication because they value an adversary process in which spokesmen for conflicting interests articulate the strongest argument for their positions. But many of the advantages of an adversary process can be enjoyed in rulemaking hearings not involving trial type hearings with their attendant costs and confusions. Advocacy can help to clarify relationships between competing values in the rulemaking process. In sum, although there is a better decision effect, one cannot be sure what percentage of private interventions actually lead to better decisions, especially in view of the tendency of intervention to carry with it adjudicatory methods and detailed judicial review.

Another benefit is the sense of justice that the opportunity to intervene provides. Even if one is not an intervenor, one may feel better because one could intervene if one wanted to. Let us call this the "participation effect." Participation has its costs, and we will discuss them when we consider the cost side on the benefit-cost balance. More appropriate for consideration on the benefit side is whether most or all of the benefits associated with the participation effect can be achieved by notice and comment rulemaking proceedings. In such proceedings, private parties can make written and perhaps oral comments on the basis of a complete disclosure by the agency of the evidence it has already collected and of the theories of the decision it is contemplating. Although a judge's opinion does afford at least the appearance of deliberation and evenhandedness, adjudicatory processes themselves add little to the value of the participation effect. To be sure, those intent on stopping a particular nuclear plant and especially those resolved to stop nuclear power altogether will not be satisfied by rulemaking that grinds out an adverse rule, but then they would probably not be satisfied either by a full-scale trial ending in an adverse result.

What then are the costs to balance against the benefits of the better decision and participation effects? Let us put to one side legal fees, though these are not insubstantial. It can hardly be doubted that private intervention takes time. The intervention proceedings will not always be on the critical path, because some design work or actual construction can go on during the period of intervention. But when the intervention proceeding is on the critical path, the costs of delay are considerable. Putting aside economic accounting issues involving the nature of costs previously alluded to, one can in principle nonetheless make estimates of the cost of delay at different points in the plant construction cycle. A few such estimates have been set forth above.

Although in principle the costs of delay are knowable, it is much more difficult to determine when a particular delay is attributable to private intervention and when to other causes. In many instances this will be a subjective judgment on which reasonable men will differ, often requiring assumptions as to the motives of participants in the licensing process.

Assuming that a cost-benefit czar could determine the time period of delay attributable to private intervention, one could calculate the direct cost of delay for each project in the country, which could then be summed across all plant projects to arrive at some global direct cost of delay. This figure could be adjusted to give annual national direct delay costs or average delay costs per plant or per megawatt.

But there are also indirect costs. Even if a utility decisionmaker knows the average period of delay per plant, there is always the risk that the period for any particular plant will be greater. Utility representatives assert that many of the financing difficulties that led to delay or canceling of power plant projects were in fact attributable to the risk that the plant would be seriously delayed or even blocked entirely (after large expenditures were already made) as a result of private intervention and subsequent appellate litigation. It is a truism that uncertainty is costly, but it is unclear how to put a cost figure, for cost-benefit purposes, on what we may term the "uncertainty effect."

An entirely different kind of indirect cost of private intervention is analogous to the costs of defensive medicine engendered by the medical malpractice tort system. Let us call this the "defensive regulation effect." Some of the requirements may clearly be desirable, and the costs involved may be more than offset by the better decision effect. Others may be abstractly desirable, but their benefits may not be worth the costs. Still other requirements may be of no actual use or may even be undesirable; these are unmitigated costs of private intervention. The defensive regulation effect is extremely hard to value, not because the cost of additional safety and health measures cannot be calculated, but because it is extremely difficult to know what the regulators would have done if the possibility of private intervention had not been lurking in their future.

Our analysis emerges with two principal benefits—the better decision effect and the participation effect. On the other side of the balance, we have both the direct costs of delay and the indirect costs, including the uncertainty effect and the defensive regulation effect.

The benefits are difficult to value. This is especially true because the value of the participation effect tends to lie in the eyes of the beholder, and there is no market in citizen participation. More abstractly put, the participation effect is incommensurate with the other benefits and costs. This factor, of course, does not make it less important.

On the cost side, the direct costs of delay—notwithstanding certain problems outlined above—are relatively easy to measure. So too, in principle, is the uncertainty effect, though the actual calculation would be rather speculative. And the defensive regulation effect is very difficult to measure because we cannot know, with even modest confidence, what the regulators would have done in a hypothetical situation.

Some will say that analyzing private intervention within a cost-benefit framework is less problematic than so analyzing other public policy issues. That may be true. But some of these considerations may explain why the utility industry has not tried to cost out or even to collect statistics on private intervention in a form that could be useful to an outside evaluator. The costs of private intervention might turn out to be small, and the calculation would certainly be controversial. Therefore, the industry has chosen to debate the issue not in the cost-benefit terms of the economist but rather in anecdotal terms.

What balance should one then draw as to the desirability of private intervention? The evidence on which to base such a judgment is unsatisfactory. Although the overwhelming majority of nuclear plant proposals are challenged before the NRC at either the construction permit or operating license stage, the extent to which the typical proceeding is delayed by private intervention is not at all clear. Nonetheless, particular cases of private intervention clearly had a substantial delaying effect. Seabrook and Midland (Consumers Power) are two widely publicized examples, though even those instances involved what many would consider special circumstances.

The application to the NRC for Seabrook was filed in early 1973, and although the NRC issued a construction permit in 1976, construction was halted on several occasions as a result of EPA administrative proceedings and subsequent court appeals involving the plant's cooling system. The Midland NRC application was filed at the beginning of 1969, but construction permits were not issued until almost four years later. Although subsequent litigation culminating in a Supreme Court decision (discussed above) kept the lawyers busy, construction was never actually halted by agency or court order. This is not to say, however, that the uncertainties involved in the litigation as well as the ratcheting of requirements may not have slowed construction.

HARNESSING AND CHANNELING PRIVATE INTERVENTION

Even a massive research study would be unlikely to lead to any solid conclusion beyond the obvious fact that private intervention sometimes has delay costs. But proponents of private intervention have never denied that litigation sometimes causes delays. Inability to draw up a definitive balance on private intervention does not mean, however, that certain abuses of private intervention cannot be brought under control by better procedures.

One reform would be to make much more generous use of notice and comment rulemaking and much less use of adjudicatory techniques. For many kinds of issues, adjudication is simply not worth the delay costs. This is particularly the case for "generic" issues. For such issues, notice and comment rulemaking should be used, and the issues should thereafter be barred in site-specific permit litigation.

For an issue to be considered generic, it should be unrelated to the particular site and plant design. Two kinds of issues meet that requirement. The first, essentially procedural, involves the methodology and criteria to be used in particular license decisions with regard to the need for power, the relative costs and desirability of alternative energy sources, the comparison of the proposed site with alternative sites, and the weighing of costs and benefits in NEPA decisions (particularly at the operating license stage when issues of principle arise such as the proper treatment of sunk costs). A second kind of generic issue involves recurring issues such as the future availability and price of uranium. Although any such judgment must be reviewed frequently, it makes little sense to analyze the issue independently and fully in each licensing proceeding.

For many such issues, rulemaking techniques have an advantage over adjudicatory methods. The District of Columbia Court of Appeals has in the past put practical roadblocks in the way of the notice and comment approach, with its unpredictable and sporadic requirement of "hybrid" rulemaking, involving adjudicatory fact-finding methods for contested factual issues within the overall framework of notice and comment rulemaking. However, the Supreme Court's *Vermont Yankee* decision (discussed above) seems to have vindicated and protected the notice and comment approach where it is permitted by the underlying regulatory statute.

The second general recommendation would be to impose rigid limitations upon the time when private intervention would have to occur. Some intervention occurs quite late in the construction cycle and, in particular, later than one could reasonably require of intervenors. Nothing in the Constitution says that intervenors should be able to sit on thier hands until construction is far along and then intervene with the hope of causing the sponsors to abort the project.

To be sure, health and safety should come first, even if the intervention comes late; but if important health and safety risks are at stake, then potential intervenors should be given a strong incentive to intervene earlier. The concern that the health or safety risk is one that could not have been discovered or properly appreciated earlier (because of, say, new knowledge) can be met by a provision along these lines: If the late intervenor can document to the satisfaction of the agency or court that the intervention could not have been made within the specified period on the basis of knowledge then publicly available, the agency or court may waive the barrier to late intervention.

Rigorous time limits on intervention would deal with difficult procedural problems raised by the phenomenon of multiple intervention where one or more intervenors come in late. The usual legal rule is that, assuming a general statutory right of intervention, an unsuccessful intervention by one public intervenor cannot constitutionally bar a later intervenor unless the first intervenor can be found to "represent" the second intervenor. In this sense there is something to the appellation "due process." Once a statutory right of intervention is created by Congress, one citizen's recourse to that right cannot under the due process clause bar another citizen's recourse to it. It is true, of course, that the second intervention can be dismissed simply by referring to the precedent of the first decision. But if there are rigorous time limits on intervention, then all intervenors will have to come quite early into the regulatory or court proceeding. Once all intervenors are before the same tribunal at the same time, the costs and risks of multiple intervention are more manageable.

Time limit provisions would have to be coupled with measures to assure wide public awareness of new power plant applications. Some late interventions may have been attributable to utility attempts to keep construction plans out of the public consciousness until the latest possible time. Such a strategy raises the social cost of intervention and should be discouraged.

Several trends have come together to raise the issue of decisionmaking to one of critical importance: the emergence of highly vocal, well-organized citizens' groups that do not easily compromise and that have learned to utilize the courts and the media to project their views; the proliferation of aspirations far beyond "life, liberty, and the pursuit of happiness"; the inadequacy of institutions for reconciling conflict—the marketplace of ideas is largely for display, not for trading; a rising role for government to tackle a variety of problems that individuals cannot handle; a widespread distrust of both business and central government, often coupled with neoromantic notions of simple forms of social structure where problems are assumed to be more easily solved; the sheer size of industrial and political undertakings; the long lead times required to introduce innovation and make changes; the heavy risks and the possibility of catastrophes on a national or global scale. In the face of such an array of forces, ambitions, uncertainties, complexities, and hazards, the nation's traditional decisionmaking apparatus is hard pressed to process and deliver what is required from it. Observers are equally hard pressed to propose reforms. As the gulf between technical and social inventiveness continues to widen, perhaps all we can hope for is to improve the efficiency of the processes and to reduce the time for decisions to be reached while safeguarding the right of citizens to a voice in decisionmaking. That would be a task modest only in the language in which it is cast.

The Role of Science and Technology

15

Creating new facilities to supply energy, adapting existing technologies to changes in costs or regulation, and introducing new processes that incorporate advances in materials and technology are each aspects of a process that we call research, development, demonstration, and deployment (RDD&D). We will often refer to portions of this process as "R&D," meaning the research and development phases; or as "RD&D" for research, deployment, and demonstration; or as "D&D" for the demonstration and deployment phases. As this chapter points out, there are significant differences in the purposes of these phases; we will give examples of the goals and techniques appropriate to each.

This chapter emphasizes that the R&D process is essential not only to hardware development but also to planning and other nonhardware activities; that the process is characterized by and must accommodate uncertainty; and that there is a critical difference between a government's role in the energy RDD&D process and its more conventional role in the development of large systems for government use.

The discussion concentrates on RDD&D efforts on technologies of special importance in the next two decades, but not necessarily on those that will produce or save energy in this period. Some important technologies require at least twenty years of RD&D to be ready for large-scale deployment. Thus, it may be urgent to work during this entire period on certain programs that will not contribute to energy supply at all during the next two decades. In other cases, it will be important during this period to determine the merits of some technology but not to take a decision to deploy until the relative economics become favorable.

NATURE OF ENERGY RDD&D

Energy RDD&D is intended largely to facilitate the production of effective hardware to produce or conserve energy. Because much of the funding for this purpose comes from government, policymakers are inclined to see it in the same light as other types of RDD&D programs that use government funds and produce hardware for government use. But energy RDD&D is not like programs in space or defense; in those programs the government is essentially the only customer for the hardware developed. By contrast, energy R&D is more similar to research in the fields of health, environment, and agriculture; its immediate result is not hardware, but knowledge about hardware—its effectiveness, costs, side effects, and so on.

Because we feel that, with all its imperfections, the market system is a better way to operate the energy industry than government operation would be, we view the primary purpose of government-sponsored energy RDD&D to be the development and use of energy technology in the private sector. Further, energy RDD&D should seek to deploy useful technology wherever it has been developed. In some cases, major government programs are required to design and develop new technology. Fusion technology is an obvious example because of the large size, the long time to the marketplace, and the likelihood that alternative technologies such as nuclear breeder reactors will provide lower cost power. In other cases, the problem may simply be to accelerate the deployment of commonplace items, such as storm windows. In these instances, research may be needed to find ways to remove institutional obstacles to full deployment or to *find out* how much uranium or natural gas is producible up to a given cost (rather than actually *finding* the material).

These fundamental characteristics of energy RDD&D have important implications for every step of the process. In the balance of this section, we go into these implications in some detail, because they form the basis for the recommendations that follow. We first discuss aspects of the R&D process and the need for competitive ideas in it. Then we review the D&D process and develop our observations on the selection and management of energy RDD&D generally. Finally, because RDD&D is so often regarded as a way of producing energy production hardware, we emphasize the importance of other functions performed by energy research.

THE NEED FOR COMPETITIVE IDEAS

Precisely because new energy technology must compete successfully in private markets, energy RDD&D should place a large premium on the generation of possible competitors—that is, on new ideas and improved understanding likely to yield cheaper and better technology.

To the extent that new ideas and better understanding are to be generated by government programs, rather than by the market, however, the RDD&D process is a planned rather than an economically competitive one. Because government programs lack the competitive edge induced by the market, there are tendencies both to generate fewer competitors than the market might and not to drop ideas that do not appear economically attractive after a little investigation. Because of these tendencies, the managers of government RDD&D programs must pay special attention to generating new ideas and better understanding in government-sponsored energy RDD&D.

Unfortunately, temptation lies the other way. In a planned rather than a competitive system, some uncertainties connected with RDD&D may appear to be removed; it may be imagined that careful planning will produce a simple linear RDD&D process, as shown in Figure 15-1, but this notion is in large part illusory.

Because the burden for economic competitiveness is transferred from the market to a planning function, decisionmakers need to ensure that they have plenty of options from which to choose. If the decisionmaker is, in the end, required to select a single "best choice," the process leading to that technology must have a more reliable outcome than the process leading to any one of several competitors. The linear process of Figure 15-1 is simply not redundant enough to be that reliable.

It is thus important to rely on parallel R&D activities, as illustrated in Figure 15-2. Parallelism in R&D has several benefits. First, it avoids the problem of a chain, which is only "as strong as its weakest link." Two parallel programs, each like that of Figure 15-1, both of which produce an item to perform a given function, may have a considerably smaller risk of failure than a single chain. Actually, a woven fabric may be a better structural member than even two chains in parallel, and if there are horizontal connections between two chains of the type shown in Figure 15-1, at least one competitive idea will almost certainly emerge. The second benefit from parallel ("competitive") R&D arises from the simple possibility that one may choose the better product. The third benefit of parallel R&D is the competi-

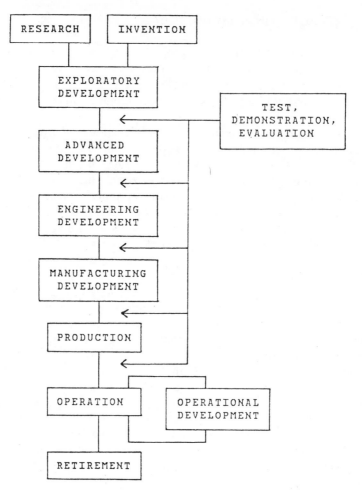

Figure 15-1. One View of the Involvement of R&D in Making Available an Operational Capability.

tive spur it provides, especially if one has the horizontal interconnections (woven fabric) type of parallelism. Periodic evaluation may thus lead to the termination of one or another element of the R&D program. Done right, this motivates the individuals to better and faster work, although it may also force an unfortunate emphasis on short-term measurable goals.

The premium on generating new ideas and better understanding also suggests that energy RDD&D should resist the rush to demonstration. Large-scale development, prototyping, and "first of a kind"

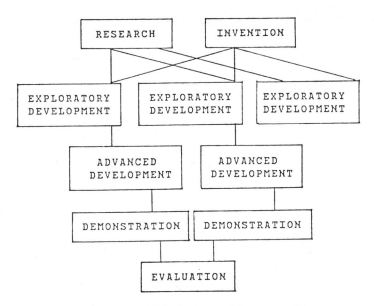

Figure 15-2. R&D in Support of Decisionmaking.

commercial plants are the expensive parts of the RDD&D program. There is, by definition, no room for novel ideas at these later stages. As a result, the later stages are avoided in private development, and should be avoided by government, until an idea emerges that appears reasonably competitive. On the other hand, substantial exploratory funds should be invested in a continual search for approaches that could supply energy at a real cost less than that of the widely available alternatives in the same area.

THE DEMONSTRATION AND DEPLOYMENT PROCESS

Precisely because demonstration and deployment have a chilling effect on research and development, the move to the D&D phase is not to be taken lightly. But there is another reason for caution. It is at this stage that the need to deploy technology in the private sector dominates, and the decisions on whether to put a technology into commercial use cannot be made by government nearly as well as by the private sector.

In the competitive commercial world, the early stages of R&D are dominated by the question of whether a potential product can be introduced and sold either because it will have a lower price or because it will perform better than other products. Thus the commercial question is not only whether the product will repay its cost to the customer, but also whether its cost of delivery (including borrowing costs, normal profit, and the like) is less than that of existing products and of those that competitors might put on the market some few years after the product is introduced.

In fact, new products will be bought only if their cost is sufficiently lower that the buyer will still save money after paying the inevitable transition costs of moving from old to new products. Transition costs are both real and uncertain. The introduction of even a successful competitive technology involves retraining, relocation, unemployment, and reemployment. But the exact costs of these items are uncertain, and the perception of uncertain transition costs is often a substantial impediment to innovation and progress in private organizations as well as in government. To the extent that the actual customer for the technology or the producer bears these transition costs, he or she takes them into account in deciding whether to use the new technology instead of the old or instead of alternatives. To the extent that these transition costs are borne by others than the users, they are typical of "externalities." That is, they are real costs (and they may be large), but they are not directly reckoned by the private decisionmaker.

In such cases, it is appropriate for society to enter the decision process, preferably by using a tax or other similar means to ensure that the externalities are reflected in the cost of introducing the technology. Whether those who bear the cost of the externalities can be or are compensated for them is another question. Indeed, choosing the best way to account for externalities in the deployment process is a matter for research and experimentation.

Externalities related to energy may take the form of environmental uncertainties, difficulties in predicting oil prices, or a national interest in accelerating the introduction of a new technology. The urgency of dealing with the energy problem pressures government policymakers to attempt a frontal attack on these externalities by spending substantial government funds in direct support of D&D. It is important to resist such proposals or at least to evaluate them soberly. The desire to get on with the job must not lead to D&D on unpromising technologies.

The objective of the D&D process must rather be to stimulate the private sector to deploy attractive technologies with minimum government interference. In particular, we do not regard the gathering of information or the creation of infrastructure benefits as sufficient justification for government intrustion at this large scale into an inherently commercial process. If information is really needed to decide whether to deploy a technology, then the technology is not ready for D&D, and the information can most cheaply be gathered by building a prototype plant—perhaps a fairly large one—rather than a full-scale facility.

The infrastructure argument is equally dangerous. It is, of course, good to train manpower, develop a general knowledge base, build awareness on the part of the public and private sector participants, and build implementation capability. But these benefits are real only if the technology is a winner; if the technology is a loser, the side effects can make it more difficult to stop the deployment of an uneconomical or otherwise disadvantageous approach.

SELECTION AND MANAGEMENT OF ENERGY RDD&D

In considering policy toward energy RDD&D, it is important to recognize that the producers of the materials and equipment embodying new technologies will be private, profit-making firms and that their customers will be dominantly private parties. This fact seriously complicates the problem of deciding who should select and manage energy RDD&D. In traditional government RDD&D programs, where government is both the customer and the sponsor, this role falls naturally to government. In energy RDD&D, where government is often the sponsor but rarely the user, placing selection and management responsibility is not so easy. The nature of the problem, moreover, differs depending on the industry and the technology in question.

For those industries and technologies where private firms have considerable technical sophistication and possess and support R&D programs, we may expect technological innovation to occur even without any government efforts to spur it and certainly without government direction on what the R&D menu should be. Private competitive enterprise has a number of advantages as a system for generating R&D support and allocating R&D resources. Firms have motivation, and sometimes are under severe pressure, to reduce costs or to keep them from rising and to produce products that will keep or enhance their share of the market. They possess detailed knowledge of the strengths and weaknesses of the technologies that they are using and of the needs and wants of their customers. Different firms are likely to place their R&D bets on different horses, and thus the industry as a whole is likely to support diverse projects. Because R&D proceeds within the same organization as production and marketing, the road from the laboratory to implementation is not barred by interorganizational barriers.

It should be stressed that these advantages are pragmatic and relative to other systems of R&D support and direction. There is no grand theorem from economics to the effect that private enterprise allocates resources in the best possible way. Indeed, the problems and limitations of free enterprise organization and support of R&D are relatively blatant and obvious. What firms see as profitable R&D ventures may match only loosely what would be socially fruitful. In some cases, firms may see little reason to spend research money on a technique that competitors could easily imitate, because then the firm would be able to capture on the market only a small fraction of its innovation's worth (and perhaps even fall short of recapturing its outlay). Conversely, once one firm has established a patent position, other firms may have an incentive to more or less duplicate technologies. Industrial secrecy may limit the knowledge that a firm has of R&D alternatives and their promise. A competitive industry will duplicate some projects and omit others that might be proposed by an expert committee.

However, it is far easier to point to these limitations and problems than it is to correct them. A central difficulty is that because R&D is an important instrument of competitive policy, firms are not likely to cooperate when government programs are proposed that might upset a firm's carefully crafted R&D strategy. Government or other outside interests may conjecture that the risks and limited capturability of certain technological ventures are deterring private investments, but they are likely to have great difficulty in finding out exactly what private firms are spending on these endeavors. Proposed government programs to push in these directions may meet a hostile reception. Proposals that companies share their technological knowledge are likely to unify the companies themselves and the Antitrust Division of the Justice Department in resistance. It is hard for public policy to fill in the holes in the portfolio when there is no solid information as to what that portfolio actually is and there are many good reasons for not finding out.

So far, we have been discussing industries and technologies where private firms are mounting an aggressive, competitive R&D effort. For industries and technologies where this is not so, the situation is quite different. The government cannot easily mimic the effects of a competitive R&D market. In public programs, there is no sharp profit lure or competitive threat to motivate and sharpen attention. The government's knowledge of technology in use and of market demands is not likely to be particularly detailed or accurate. If publicly supported applied R&D programs are mounted outside of private firms, transfer from R&D to application requires the crossing of organizational boundaries.

Again, the point is not that private R&D efforts are optimal or even satisfactory, but that it is not easy for public policies to improve upon the situation. The history of attempts by governments, here and abroad, to invigorate technically sluggish industries shows few success stories. The only notable exception has been agricultural research, and much of this research was concerned not with hardware but with procedure—with the development of knowledge to aid the farmer in choosing and using available materials. There is room for support of this type in the energy RD&D field, particularly as regards conservation, choice of appropriate processes, fuel substitution, and the like.

The problem is thus not that the private market does a bad job in selecting promising R&D projects, but that it may choose not to sponsor R&D in some areas at all. And in these areas, there is often a public benefit in having R&D done. Especially in the case of exploratory work, it is generally recognized that society can be richer, healthier, and more secure if more R&D is performed than would ordinarily be provided by the market system. Firms can recapture sales and production costs, but they cannot always fully recapture funds invested in R&D. The point is not that there is only some chance that R&D will "pay off." There is always a risk that a sales campaign will fail or that the market will not exist for a new plant. The difference is that if the sales campaign succeeds and the plant product is bought, the investors receive a return on their investment. But R&D on new ways of refining oil, or new means of exploration, or flue gas desulfurization may provide not a product but valuable knowledge. That knowledge can be used by any producer in the United States or abroad, among them competitors of the organization that paid for the R&D.

In many cases also, the actual sales and income from R&D are to be found at an uncertain time that is likely to be far in the future. Further uncertainty exists as to which companies will constitute the industry at that time, whether the organization presently contemplating R&D will still be in that business in the future, whether environmental or commercial regulation will permit the product to be sold, and like questions. Finally, there is knowledge to be obtained from R&D that in the short term may appear not to benefit the industry at all, but only the public—the identification of damaging pollutants, a better assessment of the overall availability of uranium resources as a function of price, the identification of alternatives to a large firm's existing commercial products, and the acquisition of knowledge that will permit regulation.

Government is thus presented with a delicate problem. There are good reasons for spending public funds on energy RDD&D. But there are also good reasons to conclude that government is unlikely to improve on private decisionmaking in the selection of what energy RDD&D is to do. How then to spend public funds responsibly if the public (represented by government) is not well equipped to decide what to do or how to do it?

The answer to this question is not that the government should do R&D or even select and fund it. It may be more effective and efficient to partially subsidize R&D through, for example, favorable tax treatment of private grants to universities for the performance of R&D. Or the government may simply sign a fixed price contract for delivery some years hence of a product resulting from R&D. Examples might include oil from shale or a turnkey coal-burning total energy package for a large government building. If the time is long enough and the contract large enough, there can be substantial incentive to perform the R&D that will bring innovation and improved efficiency.

When the government is dealing with a technologically aggressive industry, the range of government policies tends in practice to be limited to those that the industry considers as generally supportive and evenhanded and not threatening or disequilibrating. The United States and other OECD countries have commonly supported exploratory research and pilot development on certain technologies at universities, nonprofit institutions, and government laboratories. The United States has done this in atomic energy, coal, and more recently in a variety of other energy-related technologies. Evenhanded tax credits and formula cost-sharing arrangements have long been advocated by industry and are techniques employed in other OECD countries. In many cases the government can "make a market" for a new technology and draw private R&D efforts to meeting that market. In the case of atomic energy, this has been done by subsidizing utilities (the potential demanders of reactors) if they agree to build certain facilities that would test or demonstrate various reactor concepts. In smaller countries, where there is one or at most a very few firms in an industry, government support or cost sharing on particular R&D projects undertaken by industry has often been employed, but this is made feasible only because the companies in the country in question do not view themselves as competing against each other, but rather against foreign firms. In the United States the government is more limited in what it can do in support of particular industrial R&D projects.

From the foregoing, it should be clear that we believe that government should not try to decide the details of the energy RDD&D portfolio, even if public funds are paying for the work. Rather, it must find ways to harness economically motivated private decisions to sort out what ideas are most likely to be commercially useful in the end.

Similar caution should attend the decision on who actually conducts the RDD&D. Of course, the sponsor must ensure that the work is done on time and in the right way, so that all pieces fit together. But the government must avoid awarding R&D funds in a noncompetitive and inefficient fashion to organizations that are more interested in obtaining money for R&D over many years than in solving problems. Too often, such organizations fail to explore critical aspects of the technology that might reveal insuperable problems to commercialization, while aspects susceptible to development and analysis receive continuous but fruitless funding. Some of the very best and most capable organizations are "in the business of obtaining money for R&D." We simply caution that such an organization has no necessary spur (or for that matter, leverage) to see its R&D program terminated by commercial success.

Nor need the RDD&D be done in the United States. Manufacturers of equipment and plants such as coal gasifiers can and do perform RDD&D for their foreign markets where energy costs are already high. Some of these organizations are based abroad, and government has a real role in facilitating availability of knowledge of the state of technology and in encouraging import of actual hardware for government and commercial purposes. For example, Brazil has a large "gasohol" economy, and the French breeder reactor program is well advanced. Too often in the past, domestic industry has forced the United States to waste both resources and time by duplicating foreign efforts. In many cases it would be cheaper and quicker to import technology on an experimental or commercial basis.

However, we should be wary of joint programs involving several governments. These tend to lose sight of real goals and to reflect the greater inefficiency of the partners. Some truly multinational pure research efforts are quite successful, such as the European nuclear research center (CERN), but these are primarily shared *support* of a body that then chooses its program and procures equipment essentially as if it were in a single country. There are a few successful examples of joint *development* programs. Technically successful developments like the Concorde passenger SST by Britain and France can be severe economic burdens. Technologists often find their counterparts in another country, and such close relationships may serve as the instigation for a joint program that has little merit in comparison with alternatives for achieving the stated goals. Such seems to be the case in the U.S.-USSR collaboration on magnetohydrodynamic power and might become the case if fusion programs actually became joint efforts rather than a matter of sharing information.

NONHARDWARE ENERGY RDD&D

Our discussion of energy RDD&D to this point is of particular relevance to the development of hardware and especially of hardware to produce energy. This point of view both simplifies the discussion and produces results of importance to other kinds of energy RDD&D. But these other kinds of RDD&D, which we shall call nonhardware RDD&D for lack of a better term, are both important and different from the development of production hardware.

Research into environmental effects, energy economics, and energy policy are examples of nonhardware R&D. They are crucially important and, as we point out elsewhere in this report, are often underfunded or done poorly. And they are different because government, in these cases, usually is the customer for the RDD&D work. As a result, such programs are likely to be cast in a more traditional government R&D mold than most energy RDD&D.

Two other areas of nonhardware RDD&D require more extensive discussion—consumer RDD&D (for which see also Chapter 3) and RDD&D in support of planning. Both have special characteristics and present special management problems.

RDD&D on the Consumer Side

Businesses and industries will invest in new energy-saving capital stock in just the same way as they invest in new plant providing improved productivity, reduced emissions, reduced cost, and the like. The general health of the economy, investment credits, and other tax incentives have a major influence on the rate of installation of new plant, but we believe that competition and business awareness of costs and a desire to maximize profits will nonetheless result in substantially reduced energy consumption. That energy savings are delayed by imperfect information and competition among developers and manufacturers is not, in our view, worse for the economy and for the social good than the delay of productivity improvements or cost reduction for the same reason. Attention to competitiveness of industrial plant should be motivated by all of these benefits, not simply by reduced energy consumption.

However, the individual householder or automobile purchaser cannot afford the time and effort that a large corporation invests in making decisions on whether to buy equipment or buildings. For this reason, it may be efficient for the government to do the preliminary work in such a decision procedure, by making available information on products, their energy consumption, longevity, and the like. Mandatory appliance labeling for energy consumption is just being introduced, with labels showing annual energy costs as a function of usage and fuel price. The market response to the combination of initial purchase cost and operating cost will probably shift sales toward some appliances with reduced energy consumption.

A more difficult question is who does the RD&D that will make available these energy-efficient, cost-effective appliances? In general, the federal government has maintained that there are enough suppliers with enough research and development capability so the government has no role in this field. Government involvement has been limited to mandatory labeling, with some attention to the possibility of mandatory energy efficiency standards for appliances. But as we have noted, from a national standpoint, most firms invest too little in research and exploratory development because they cannot capture all of the benefits. If the benefit is simply reduced energy costs, such R&D may be underfunded by the industry as a whole, even in cases in which there is an industrywide R&D organization. This is particularly true in cases in which the results of R&D would involve changes to the individual product lines of the manufacturers and so would have different effects on manufacturers of different sizes and with different product cycles. Thus, while pricing energy at marginal cost would in general provide a major spur to improved energy efficiency, some subsidy or other stimulation of R&D (in contrast to deployment) into consumer products may be desirable. If energy is priced at average cost, much more stimulation or conduct or R&D by government is in order.

But if more government involvement in R&D for consumer products is required, it would be to produce knowledge, not products themselves. Automobiles use some 14 percent of U.S. energy, and it would be useful to know that fuel consumption could be cut in half by a new automotive engine. Similarly, it would be helpful to know that minor changes in household refrigerators would cut energy usage by a factor of two or three. But it would not be wise to suppose that government can actually design better cars or refrigerators.

RD&D in Support of Planning

Knowledge is valuable in decisionmaking and may be obtained through RD&D. This benefit of RD&D can arise even if the knowledge is never actually used to produce a physical product. One example in a competitive fuel supply market is the great influence on present ("scarcity") fuel prices of demonstrating that thirty to forty years from now an alternative fuel or energy source will be available, even at some substantially higher price than the fuel commands at present. Under certain plausible circumstances, the present efficient scarcity price of a depleting fuel will depend on the timing and cost of this alternative and nondepleting fuel.

Let us say, for example, that we know today that in twenty-five years we could make a synthetic crude oil from coal at a full price in today's dollars of $30 per barrel, or $5 per million Btu, and hence that natural oil cannot sell for more than $30 at that time. (All dollars here are adjusted for general inflation.) If we know that natural oil will last that long and will cost $3 per barrel to produce, then the present competitive "economic rent" (technically, the competitive price minus the production cost) of oil will be $27 per barrel in twenty-five years. In a competitive market, producers should be willing to sell their oil today if they can get an economic rent today at least as great as the discounted value of that future economic rent. Assuming a 7 percent per year inflation-free discount rate and a current extraction cost of $3 per barrel, the current competition price should be about $8 per barrel.[1] Because this is below today's market price of a barrel of oil, there would be a real benefit in demonstrating (but not deploying now) such a presently noncompetitive supply technology. Of course, the oil market is much more complex than this simple example suggests. It is not clear, for example, that the major oil producers can earn anything like a 7 percent per year inflation-free return on investments. Nonetheless, demonstrating alternatives—even without actually using them—might have significant effects on world oil prices.

As another example, we show how knowledge of what it would cost to deploy a breeder reactor in the future could help hold down today's cost of generating electricity from light water reactors (LWRs). Suppose that as a result of RD&D completed during the past decade, we knew that a 1,000 megawatt electric (MWe) breeder reactor could be built for $600 million more than it would cost to build a new LWR of the same size and also that uranium ore (U_3O_8) will be available until 2020 at a production cost of $30 per pound or

1. The $8 per barrel figure is the approximate present value of $27 twenty-five years from now, assuming a 7 percent discount rate, plus the $3 extraction cost today. The exact figure is:

$$\$3 + \frac{\$27}{(1.07)^{25}} = \$7.97.$$

less. (All monetary figures are in 1979 dollars. We assume, in this example, that safety risks and other externalities associated with both LWRs and breeders are small enough to be ignored.) The $600 million capital cost penalty means that the LWR would remain competitive with the breeder in 2020 provided that the price of U_3O_8 did not rise above about $150 per pound.[2] The present scarcity rent on U_3O_8 would therefore be this $150 per pound price in the year 2020 minus the $30 per pound production cost, discounted to 1979. Assuming a discount rate of 7 percent per year, the present scarcity rent of U_3O_8 would be about $7.50 per pound.[3] Added to the present production cost of $30 per pound, this would suggest that the present market price of U_3O_8 should be no more than about $37.50 per pound. If uranium producers were to raise their prices above this level, electric utility companies would presumably stop building LWRs and start ordering breeder reactors, if they were available to be ordered.

Even if breeders are not *presently* available, the fact of their future availability would tell uranium producers that they cannot expect

2. The $150 per pound U_3O_8 price is an approximate figure, determined as follows: Assuming a 7 percent discount rate and that both LWRs and breeders would have operating lives of thirty years, the contribution of a $600 million capital cost penalty to the levelized annual cost of electricity from a breeder is:

$$\$600 \times 10^6 \ \frac{0.07 \ (1.07)^{30}}{(1.07)^{30} - 1} = \$48 \text{ million.}$$

A 1,000 MWe LWR requires about 200 short tons of U_3O_8 each year (this corresponds to a burnup of 30,000 megawatt days per metric ton of uranium, LWR capacity factor of 65 percent, LWR thermal efficiency of 32 percent, 3 percent enrichment of uranium-235, 0.2 percent tails assay of uranium-235, and 15 percent uranium conversion and fabrication losses). The amount by which the price of uranium could rise without making LWRs uncompetitive with breeders is therefore the levelized annual cost penalty of the breeder divided by the annual fuel requirement of an LWR, or $48 million divided by 200 tons U_3O_8, which equals $120 per pound U_3O_8. If this is added to the assumed production cost of $30 per pound, the price of uranium ore at which electric utility companies would be indifferent between ordering an LWR or a breeder becomes $150 per pound U_3O_8. A more precise calculation of the relative economic value of breeders versus LWRs would require recognition of the fact that both uranium prices and electricity prices will vary over the life of the reactor.

3. The exact present value of $120 forty-one years in the future, assuming a 7 percent discount rate, is:

$$\frac{\$120}{(1.07)^{41}} = \$7.49.$$

very high future U_3O_8 prices, and hence they would compete more vigorously for present markets. Either way, U_3O_8 prices should fall. Whether or not the breeder is ever actually used on a large scale will depend on many things, including the costs of other sources that may prove cheaper. But just knowing that it is there as an option can lower current energy costs.

There are other examples of R&D in support of planning. For example, knowing how much uranium is available at low production cost is of crucial importance to the decision on breeder deployment. The private sector has little economic incentive to identify and "prove up" reserves much larger than ten or twelve times current annual production. And for planning purposes, government does not need to prove up reserves either, much less develop them; but it does need reasonably reliable projections to make long-term policy. In these and other cases, R&D in support of planning is often cheap insurance against errors in very expensive decisions and deserves substantial support.

Without present knowledge of future alternatives, we might now be paying prices for depleting resources very much higher than their true economic value. These uneconomically high prices would clearly not be in the interests of the users and may not even be in the long-term interest of the producers, if they result in premature adoption of costly alternatives and uneconomic delay in using the natural, low cost sources. Note, however, that present scarcity prices are sensitive to the assumed (inflation-free) interest rate available to resource owners, and note also the great uncertainty of actual costs projected far into the future. If research demonstrates that alternatives are even more costly than now believed, the result can be higher prices for depleting resources in the short run. Even in this case, however, the sooner the world knows the cruel facts, the better in the long run.

RECOMMENDATIONS FOR IMPROVING THE RDD&D PROCESS

Our analysis of the energy RDD&D process leads us to three important conclusions:

1. Government action is required to foster research, invention, and exploratory development in the energy field for the same reasons that such work is generally underfunded by the market system in many areas of research.
2. Large costs are incurred in demonstration and deployment, and programs at this stage not only rarely generate new ideas but tend

to repress new competitive concepts. The government must resist pressures for D&D if technical success will show only that the technology is too expensive to be bought. However, the government can properly support D&D to demonstrate an alternative for purposes of insurance; on occasion and on a modest scale to attain knowledge in support of economic or environmental planning; or for similar purposes.

3. Economic choices should guide the selection of RDD&D programs, and generally the private sector is in the best position to decide the details of such choices. Especially in nonhardware research, however, the government must play a larger role.

These conclusions form the basis for our recommendations on both the RDD&D process—presented in the present section—and on the content of the RDD&D program—developed in the subsequent section.

Improving the R&D Process

Government properly can play a major role in energy R&D to generate new ideas and improve understanding of basic energy processes. To perform this role, the government should:

1. *Encourage more basic research by industry* through a system of generous (perhaps 90 percent) tax credits on research grants to universities, individuals, or industry and by permitting industrial scientists to receive competitive research grants or contracts. In the latter case there should be mechamisms to ensure that the research results will be as generally available as if the work were done in a university.

2. *Create centers of excellence* in disciplines important to energy such as combustion, catalysis, and corrosion. In some cases (where special equipment is required, for example) a single site may be required. In others, researchers may be geographically dispersed, but woven into an integrated program. In any case, it is important to involve nongovernment scientists to the maximum.

3. *Insist on parallel development* in the stages of R&D preceding the commitment to large-scale hardware.

4. *Resist the commitment to demonstration* or large prototypes, unless the competitive strength of a technically successful technology is clear. This should not be solely a government decision, but one in which the private sector plays a large role. (We discuss the nature of the private sector role in the next section.) In any case, demonstration should be avoided if it is clear that if everything works the system will not be worth having. The exception, as noted above, is for insurance, but that decision should be taken explicity.

5. *Fund efficiently* those programs that are funded.[4] Too low a funding level delays the benefits, while too high a funding level may reduce the need for sharp thinking during the R&D program and ensures excessive costs in the event of even modest delay.

Targeting the D&D Process

The energy D&D process is different in kind from more familiar demonstration and deployment activities and also varies from technology to technology. Thus, it is important both to shed approaches that do not work for energy D&D and to design the D&D approach for the specific problem at hand. The function of demonstration and deployment contrasts with our advocacy of generally applicable research and exploratory development. D&D must be justifiable for the particular expected benefits; there is hardly a place for new concepts to arise in a D&D program, and if the product is to be readily copied and to be illustrative of realistic costs, it must be characteristic of a commercial program.

It is here that the relative roles of government and industry are most confusing. On the one hand, D&D is expensive; if the public benefits are great enough to warrant spending public funds, the sheer expense requires government to ensure that they are spent wisely. On the other hand, the more government is involved, the less scope the private sector has to ensure that the project is a commercial success.

In general, we believe that this problem can be sorted out in the following way. Government should limit its decisionmaking to determining how much it is worth to undertake the D&D project and then leave the details of technology selection and project management to the private sector. In this context, "worth" can be established in two ways.

First, worth can be the difference between the total benefit of a new technology and that portion of the total benefit that a firm can capture in the private market. Such noncapturable public benefits might include reduction in oil imports or environmental enhancement.

A second measure of worth is insurance against serious, but less than likely, future events. In such cases, if all goes well, there would be enough oil or other conventional fuel at acceptable prices. Under these circumstances, the D&D investment might pay off. But just as an individual buys insurance against major loss (and finds it worth-

4. By "efficient funding" we mean a funding schedule that will maximize the discounted present value of the algebraic stream of expected costs and benefits.

while, even though the expected return on that investment is negative), so in a detailed comparison of the probability of outcome, the
availability of technology for use in the less likely eventuality may
still be worthwhile as a kind of social insurance.

We regard these concepts as useful ways of thinking about how
much it would be worth for government to pay to deploy a new
technology, but we also recognize that they are subject to abuse.
Reducing oil imports, for example, is not an umbrella under which
many D&D projects should be afforded shelter. If a new technology is ready for deployment, the private sector will do the job.
Spending public funds on D&D is not a substitute for decontrolling
oil prices, developing efficient environmental regulation, or otherwise rationalizing the framework for private sector decisions. Un-
capturable public benefits and insurance as justifications of D&D
projects should be exceptions and subject to the closest scrutiny.

Assuming, however, that government is convinced that it is willing
to pay up to a specified amount to conduct a D&D project, then the
next step in sorting out government and industry roles is to leave the
details of technology selection and management up to the private
sector. This step requires finding a way to let industry bid competitively for government funding (up to the specified maximum that
the project is worth) and selecting the lowest bidder. Of course, if
no one bids, government's response should not be to raise the price
but to return to the search for better ideas.

This general principle must be adapted to a variety of different
D&D projects. In what follows, we suggest how the principle might
work in six types of projects. For each type, we describe the nature
of the project, the character of the decision government must make
to proceed, and the role it should play. The list is not exhaustive,
but it serves to illustrate our conviction that government can limit
itself to the assessment of public benefits while leaving the commercial details to industry. In each case, we assume that government
has decided to pay something to have the D&D done.

1. The true demonstration
   a. *Nature.* The technology carries a substantial technical risk that
   can only be resolved by demonstration at commercial scale.
   However, if the technical risk is resolved, commercial use will
   follow easily (with production by many suppliers), because
   returns are high, the technology is affordable, and other risks
   are low. New industrial unit processes are an example.
   b. *Commitment Decision.* If discussion with prospective users

shows that the problem is truly of the nature described above, the commitment should be made.

   c. *Role.* Fund the demonstration efficiently for the low bidder.

2. Selection among technology alternatives

   a. *Nature.* Government or industry has developed and tested on a small scale several technologies that produce the same product—alternative coal gasification technologies, for example. However, the technical and other risks of building a commercial unit prevent deployment in the private sector. Typically, the technology is expensive.

   b. *Commitment Decision.* Selecting the most promising technology is basically a commercial decision and should, to the extent possible, be left to the private sector. The government's decision should be limited to whether the product (e.g., coal gas) is needed. If it is needed, the government should offer incentives in the form of taxes, government purchases, and so on designed to call forth the commercial technology.

   c. *Role.* To permit a largely private sector decision, the government should offer to pay a maximum sum for a fixed amount of the product. Industry would bid fixed sums at or below the government ceiling. The ceiling would be higher or lower depending on the government's desire to have the product available. If no one bids, it would be a reasonable inference that the technology is a loser.

3. Transfer of government technology

   a. *Nature.* Government researchers have developed much of the technology to be deployed in the private sector. Deployment requires technology transfer, as well as other activities. (This is not a happy situation, but it occurs especially in nuclear technology.)

   b. *Commitment Decision.* Because of the technology transfer problem, two decisions are involved:

     (1) *What to Demonstrate.* A separate demonstration step is usually needed because industry is not schooled in the technology. To avoid problems like those of the Clinch River breeder reactor, whose design was largely specified by the federal government, the industries that will use and build the technology should decide the design to be demonstrated. The prototype large breeder reactor design studies are good examples.

     (2) *Who Should Deploy?* Once the technology is demonstrated in a commercial setting, the government must decide

whether to support deployment and whether one or more plants should be supported. Support of one deployed plant could lead to an advantage for one or a few companies; support of several projects is more costly. The lesser competitiveness of industry in producing large plants suggests that supposed economies of scale should be reviewed carefully. In cases where there is an absolute necessity to support a single large plant, the mechanism to avoid monopoly control deserves attention.

c. *Role.* Again there are two phases:

(1) *Demonstration.* This phase requires major government funding and direction of the program, much like weapon system demonstration. However, the program should be open to all prospective participants in commercial use of the technology.

(2) *Deployment.* The government must not foster or force deployment of uneconomic technologies simply to make the government program look good—that is, to justify money that has already been spent. To permit a largely private sector decision, the government should merely offer to pay a maximum sum for a fixed amount of the product. Industry would bid fixed sums at or below the government ceiling. The ceiling would be higher or lower depending on the government's desire to have the product available. If no one bids, it would (again) be a reasonable inference that the technology is a loser.

4. Mass production of expensive technology

a. *Nature.* The device will be produced in large numbers and has large front end costs. The production costs of initial units is much higher than the acceptable competitive price, although production costs ultimately fall below the selling price. However, the discounted return on the device is unattractive if (1) front end R&D costs are so high that they cannot be amortized by a single firm introducing the technology, because it cannot capture the entire market; (2) the initial production cost is uncertain and may be too high; or (3) there are serious uncertainties of market penetration at the desired rate. Small fuel cells are an example.

b. *Commitment Decision.* Deployment will require a subsidy of some kind, even if technology-forcing regulation is used (as in the automotive catalyst). That is, the public—as a consumer or taxpayer—winds up paying a premium to accelerate introduction of the technology. The commitment decision thus rests on whether the premium is reasonable. Actual demonstration and commercialization is in the hands of the private sector.

   c. *Role.* Depending on the nature of the uncertainties, the government (having decided that it is worthwhile to buy the insurance provided) can adopt one or more of the following roles: (1) pay all or part of the front end costs; (2) pay (perhaps through direct purchase) for initial production units; or (3) subsidize some part of losses in the early years.

5. Market development for cheap technology

   a. *Nature.* The technology is largely in hand, is relatively inexpensive to produce, and will be sold as the market develops. However, nontechnical obstacles slow market development. Many residential conservation devices fall into this category.

   b. *Commitment Decision.* By its nature, this case requires fairly small expenditures and so carries the great danger that government will intervene because it looks easy and cheap to do so. The key to the commitment decision is thus to be sure that government action will without doubt be effective and will supplement normal market activity. The decision to provide tax credits to a booming insulation market illustrates the potential problem. In this case, insulation was so attractive to consumers in comparison with continued expenditures on fuel that the demand exceeded production capacity. This situation led to quotas, price increases, and so on. It is possible that tax credits simply added to the demand without increasing the amount of insulation available or installed in the immediate future. A more careful phasing of tax credits could have provided the benefits at lower costs.

   c. *Role.* Possible government roles are numerous. They include education, tax credits, and promotional advertising.

6. Making know-how available

   a. *Nature.* The government has supported a research or exploratory development program (or otherwise learned how to do something)—for example, in reducing the energy consumption of existing housing stock without diminishing comfort. No nationwide industry exists to provide such services to homeowners. Utilities are prevented by federal law from making any profit by providing the service, and no individual piece of hardware can be sold at a high enough competitive price (considering modest production costs) to repay the cost of a private educational campaign.

   b. *Commitment Decision.* If analysis shows a net social benefit from an advertising and educational campaign to transfer this know-how, a program of appropriate scale should be mounted. Economic analysis should underlie the nature and magnitude of the campaign—some individuals are more readily reachable (as in the case of apartment buildings or centrally heated complexes), and reaching them will result in a greater social benefit. The campaign should be broad enough to provide the same marginal return as other investments.

c. *Role.* As for "market development for cheap technology," possible government roles include education, tax credits, promotional advertising, and the like. Naturally, the government should not itself run an advertising campaign. Rather, a government agency should specify evaluative criteria, make a preliminary analysis of the social return on programs of various sizes, and put the project out for competitive bid. In such activities the government must not only provide the funds but also set the standards for integrity. It is not a matter of energy saving whatever the cost—in money, integrity, and good taste—and it should be recognized that socially desirable governmental action may encounter not only industry indifference but even opposition.

RECOMMENDATIONS ON THE RDD&D PROGRAM

Over a twenty year span, improvement in the way RDD&D is conducted will likely have more impact on the RDD&D program than near-term revisions to the details of the energy budget. Although we have reviewed the energy RDD&D program and budget since 1975, we will not here attempt a point-by-point evaluation of it. What we will do in this section is to:

- Suggest points of emphasis for R&D activities,
- Indicate with some specificity the direction we see for the D&D program, and
- Present our fairly general observations on the relative value of R&D and D&D in a number of individual energy technologies.

The first two topics are discussed in this section, and the third is found in the appendix to this chapter.

The R&D Program

Our recommendations for the R&D program derive primarily from our discussion of what is important in R&D generally and from our experience that there is a tendency even in government to underfund projects in support of basic knowledge, planning, and the like. We therefore recommend that the government should do the following:

1. *Increase its relative funding of basic energy sciences,* especially including environmental research, work on energy usage patterns, and energy economics.

2. *Emphasize generally applicable energy technologies,* such as improved heat engines, energy storage systems, corrosion resistant coatings, combustion research, and the like. Improvements in components and increases in knowledge can then be used in a timely fashion in varied programs, rather than being forced in the context of a particular wind energy project, fluidized bed combustor, or the like.

3. *Fund more nonhardware RD&D.* Here are a few of the many possible areas of nonhardware research:

- Study of the prospects for land reclamation in different regions, including evaluation of experience;
- Assessment of potential consumer response to electric rate reform (e.g., prospects for greater use of point of end use energy storage, more energy-efficient home appliances, and the like);
- Empirical estimates and sampling in each economic sector to narrow the range in energy demand model parameters;
- Innovation in energy-conserving procedures applicable to existing capital stocks (e.g., lower thermostats, new driving habits, changes of industrial processes, and the like);
- Better understanding of the size and distribution of the deep natural gas resource;
- Better understanding of the environmental health patterns associated with coal liquefaction;
- Improved estimates of the health and environmental effects of air pollution associated with energy production and use;
- Assessment of policy implications of huge fossil fuel resources (e.g., Venezuelan tar sands);
- Better understanding of the cause of the decline in worker productivity in surface and underground coal mining;
- Quasi-hardware research (e.g., control and feedback systems, the effect of siting of houses);
- Legal and regulatory research (e.g., federal development of a model state waste oil recycling statute);
- The effect of tax laws on energy consumption patterns;
- Interfacing decentralized energy technologies with utility systems.

The D&D Program

We are more concerned about the current state of the D&D program than about the R&D program. We sense both that there is more uncertainty about the proper direction for the D&D program and that there is some likelihood that the program will demonstrate that many technologies cannot be deployed. For these reasons, our recommendations here are somewhat more pointed.

As a general rule, D&D should be initiated only when the technology involved can reasonably be expected to be economically competitive at about the time that it is in first commercial use. Because early D&D both is economically unwise and retards the R&D process, the government should resist pressures for D&D if technical success will show only that the technology is too expensive. If the technology does look competitive, the best reason for government involvement in D&D is to pay for clearly specified public benefits not capturable in the market.

On the other hand, D&D can provide insurance by bringing about progress on larger scale operation of a technology that may be required to cope with less than likely events. Unlike D&D of economically attractive technologies—which may be accelerated by a simple, one time subsidy—insurance requires more significant government involvement and funding. Particular care should be taken to minimize the cost of insurance. Insurance can often be bought most economically by building a modest technological prototype rather than a major demonstration facility.

With these principles in mind, we believe most of the D&D that government is undertaking should be abandoned in favor of R&D. However, four areas of D&D deserve more discriminating comment.

First, the use of coal is limited by demand, and technology can help remove this limitation. Technology to enable utilities to use coal serves a clear need, and government should err on the side of demonstrating promising pollution control devices, solvent-refined coal solids, and combined cycle plants. Technology would serve a similarly clear need in industry, but we believe that current government programs (e.g., small fluidized bed boilers) are on target and that other technologies (e.g., cogeneration, low and medium Btu gas) are sufficiently advanced that industry needs no D&D help to deploy them economically.

Second, high Btu gas from coal could serve residential and commercial markets, but we believe that sufficient sources of conventional and unconventional gas are available below the price of high Btu gas, so there is no need now for constructing a commercial plant with government support. Some insurance against lower levels of gas production is warranted, however, and we support construction of pilot plants (300 to 600 tons per day) of the one or two best new technologies in this field.

Third, the desirability of increasing domestic liquids production is clear. D&D of enhanced oil recovery, which is both useful and relatively inexpensive, deserves aggressive support. Conversion process costs are uncertain, many environmental and health hazards unresolved. Prototype plant construction seems timely for costly coal liquefaction. Rising pressure for "action" is best met by government requests for bids for specified amounts of liquid product from various sources at a modest price premium over crude. This would test the state of the art with minimum financial exposure and commitment.

Finally, we need a vigorous U.S. R&D program focused on providing candidates for a decision to build one or two large-scale (but not commercial) steam-generating breeder reactors. The target decision date, given the possibility that a commercial breeder may need to be deployed by 2010 or so, should be between 1985 and 1990. The decision criteria should include low capital cost (including low inventory of fissionable fuel) and the nature and cost of the fuel cycle facilities. Of course, in accord with the principles we have proposed to guide RDD&D in general, if the results of the U.S. R&D program at the time do not promise a breeder economically superior to the liquid metal fast breeder reactor (LMFBR), which will have been demonstrated abroad as Super Phénix and others, the foreign LMFBR technology should be adopted if a decision is made to deploy breeder technology.

As a footnote, it is useful to point out that these exceptions to our general belief that government should abandon much of its D&D work still add up to perhaps $5 to $7 billion of expenditures over the next five or six years. Thus, even modest D&D programs are expensive, and a more expansive view of D&D than ours could easily lead to the result of compressing the R&D program below tolerable limits.

APPENDIX: OBSERVATIONS ON ENERGY TECHNOLOGY

In the body of our chapter on energy RDD&D, we express support for R&D and some reluctance about D&D. In large part, these views derive from our understanding of the general nature of R&D and D&D programs in both government and industry—an understanding we share broadly in our study group. Our collective view is confirmed by familiarity, differing in degree from person to person, with most of the major energy technologies now in various stages of conceptualization or development. It would be both pretentious and misleading for our study group to offer itself as the source of a full-blown technology assessment; but neither do we wish to abstain from comments. Thus, the observations on selected energy technologies offered in this appendix serve as illustrations of the principles and criteria we have proposed for the evaluation of RDD&D activities and choices.

The list will make no one happy. Those espousing a specific technology will find their favored approach evaluated only in competition with others and only as it contributes to the availability of an energy option at a price. Our purpose is to indicate which technologies are likely to be more costly than competitors (and so least attractive for early demonstration) and to suggest where ideas are more important than hardware. We begin with a series of vignettes—by no means exhaustive—in areas other than fossil fuel. Fossil fuels are then represented by three examples—coal gasification, shale oil, and natural gas. We conclude with some notes on energy conservation and air pollution.

Fusion

Magnetic Confinement Fusion (MCF). As usually discussed, MCF consists of a heat source, steam turbine, and electric generator. In this regard, it can be compared side by side with fossil fuel plants, with light water reactors (LWR), or with advanced fission generators. At least, if it is technically successful it can be compared side by side in this way. Under such a comparison, the heat source in the existing plant is simply replaced by a fusion source. Since fusion fuel is cheap, comparison is most straightforward with the alternative technologies with lowest fuel costs—LWRs with almost zero present fuel costs and other fission reactors. It seems unlikely that MCFs can provide a heat source that is less costly than the essentially passive source represented by a fission reactor. With this view, some hundred years will be required for uranium supplies even to approach a price level to cover any substantial increment in capital cost between fission technology and MCF.

This comparison of "costs" of alternative technologies has assumed low fuel costs so that the comparison is largely on the basis of capital costs. We have implicitly assumed either low or equal costs of externalities for these alternatives, a point much challenged by opponents of the individual technologies. Although neither government nor industry has adequately demonstrated disposal of fission products or transuranic wastes, and although mine tailings pose a problem of radioactivity in the neighborhood and in the atmosphere, the cost of reducing these hazards by a substantial factor is small relative to the cost of building a large number of reactors and does not affect significantly the choice among the alternatives compared in this section. Major externalities such as the possibility of a "carbon dioxide catastrophe" can affect the choice between coal fossil fuel and nuclear power on the one hand or between fossil fuel and solar power on the other; and the study of the likely (and unlikely) climatic effects of carbon dioxide and what might be done about them is identified in Chapter 10 as an important research need.

Given this view of expected costs, we should "efficiently fund" the Tokamak fusion test reactor (TFTR) and operate it for some years as a potent source of thermonuclear reactors and of experimental results on plasma in the state of thermonuclear burn. We should also fully fund conceptual analyses, diagnostic tools, and particularly the exploration of concepts that have a chance of being much cheaper than the kind of MCF that has thus far been proposed. We should not build large plants that are "next in line" between TFTR and a commercial-sized fusion reactor unless there is a very good forecast that such plants will be cheaper than alternative sources such as the breeder reactor.

Inertial Confinement Fusion (ICF). About ten years ago it was proposed to obtain substantial conversion of deuterium, or deuterium and tritium, into thermonuclear energy not by confining the nuclei for minutes in a magnetic field but by using very high densities for the very brief confinement time of a billionth of a second available even in the high velocity explosion resulting from the burning of a substantial fraction of the initial material. This is the mechanism that has been used (on a much larger scale) since 1952 in hydrogen bombs, but the idea here is to use small pellets of material compressed and heated by lasers, ions, or even by energy supplied by pellets accelerated by magnetic guns.

If technically successful, ICF appears far from competitive with breeder reactors. Unlike MCF, where small amounts of recirculating power are needed for superconducting magnets, a large fraction of the electrical energy produced would have to be supplied to the accelerator in a very expensive form in order to provide the conditions for the next explosion (which might take place ten times per second for a 1,000 MWe plant). Thus, this approach may demonstrate technical achievability, but it has substantial problems in making the transition to something with a substantial energy surplus. Finally, it has great difficulty overcoming the barrier of lower cost alternatives such as the LWR and the breeder reactor. Unless and until its competition potential becomes more evident, inertial confinement fusion should be funded as indicated by its mulitary usefulness and not as a primary energy technology.

Foreign Work in Fusion. Because fusion is a technically exciting program, it is worked on worldwide. The United States can maximize the return on its research dollar by drawing in large part on this foreign work while keeping up a program of technical sophistication that allows us to contribute our own part of this total.

Fission

In this category should be substantial R&D on licensing techniques, cost reduction of construction, instrumentation, operator training, safety analyses and improvement, research and proof demonstrations of waste disposal forms, advanced isotope separation, and the like. Obviously, the results of such R&D would benefit energy consumers in the United States if the real cost of production of nuclear energy were thereby reduced, if the uncertainties associated with nuclear energy were reduced so that a trade-off could be made between excessive conservatism and reduced costs, and if the confidence in our ability to store nuclear waste safely for an interim and to dispose of it permanently in either irretrievable or retrievable form were increased.

Advanced isotope separation holds the promise of reducing the cost of enriched uranium for light water reactors to one-fifth or even one-tenth its present level, potentially increasing the availability of uranium by about 50 percent—in particular, allowing us to "mine the tails" of the existing depleted uranium stockpile. The effect of advanced isotope separation is not simply to expand the supply of uranium economically, but also to change the relative economics of different reactor systems. In particular, low cost isotope separation such as might be provided by a laser isotope separation process would give us increased flexibility in starting a breeder reactor economy with enriched uranium, thus decoupling the breeder deployment schedule from the history of light water reactor operations (and from availability of a plutonium stockpile).

A potential issue of major importance is whether advanced isotope separation techniques facilitate the proliferation of nuclear weapons to states that are now nuclear weapons powers. Both the technology and the pricing policy for uranium enrichment using such techniques may significantly affect the number and nature of foreign entrants into this field. For instance, deployment of an evidently low cost process by the United States, with enrichment services selling at a price equal to that for gaseous diffusion, will surely entice others into the field. Market forces will then reduce the price, but is it in the interests of world security to obtain low enrichment prices in this way rather than by foregoing temporary profits to the U.S. government? U.S. government pricing policy (for government-operated enrichment facilities) should consider such externalities.

Another broad issue requiring rethinking is the National Uranium Resource Evaluation (NURE) program. Fundamental to the scheduling of R&D on fission and alternative modes is the understanding of the future resource cost of uranium. NURE was a good idea. Its aim was to understand the availability and cost of extraction of uranium in the United States. However, it soon entered a bureaucratic niche where it did little toward its goal. NURE should concentrate less on development of tools and more on the determination of uranium availability, firming up new prospects and then progressing to delineating the more speculative resource categories. Random drilling, little used so far, should be a useful tool.

Research on recycling uranium fuel should proceed so that better choices can be made among the alternatives and to reduce the potential cost of extracting plutonium from LWRs (in the event that commercial plutonium reprocessing is some day permitted). That even the nuclear industry's knowledge of such processes is rather primitive is illustrated best by the commercial failure of the General Electric Morristown commercial fuel-reprocessing facility before it was even put into operation.

Advanced Converters and Breeders.   The benefit of advanced converters over the ordinary LWR may lie in lower cost, but mainly arises from the lower uranium consumption expected from advanced cycles. Lower consumption compensates higher capital costs only when high cost uranium is resorted to. If the investment in the advanced converter is in fact greater, then its deployment will follow substantial exhaustion of uranium resources—significantly past the turn of the century. But a program of research now may promote flexibility. For example, low in-core fissile inventory, rapid on-site reprocessing, and low out of core fissile inventory would allow substantial reduction in breeder reactor (or advanced converter) fuel cycle cost. The molten salt reactor (MSR) has a low in-core inventory, continuous reprocessing, and no fissile material out of core. It would be a good candidate for further exploration into reactor design and especially into the materials problems.

If such advanced reactors could have lower investment costs than LWRs (perhaps because they operate near atmospheric pressure rather than the pressures associated with hot steam), then they might be deployed earlier, simply as a matter of economics. More work is required on concepts, but not at present on demonstration plants, first of a kind commercial plants, and the like.

Fission and Fusion. Most of the fission and fusion plants are of large scale, require big organizations, and have a large overhead for safety and health concerns. These are real costs. But the usual method of improvement by competition, through either research by manufacturers on alternatives to their present nuclear plants or research funded by the government with these same manufacturers, has a serious problem. Commercial firms are reluctant to develop new concepts that will make their present product line obsolete. Thus, although limited amounts of research may be done on their own initiative by the large nuclear reactor manufacturers in this country and abroad, and some work may be funded by the U.S. government to these manufacturers, special care should be taken to provide R&D support and direction to those who have a real interest in the success of the R&D on which they are working—instrument manufacturers, component manufacturers, universities, and selected industry.

Geothermal Energy

Dry Steam. Dry steam is a commercial reality, profitably exploited at present in California and in several locations abroad. No research needs to be done. It would be useful, however, to have a somewhat better idea as to how much dry steam there is in the nation and in the world.

Hot Liquid. The problem with hot liquids is primarily corrosion. Disposal of the corrosive and polluting waste is also a serious problem. Government can usefully generate new ideas in these areas, but need not support D&D. Indeed, the private sector is doing D&D already.

Hot Dry Rock. This widely available thermal resource is in the early stages of investigation by means of a dual well natural circulation technique. The success of this method appears to rely on breaking up or "spalling" of the rock from which the heat has been extracted, in order to expose a fresh surface to the heat transfer medium. Because the hot dry rock is widely available and the system could operate in rather small sizes (10 MWe compared to 1,000 MWe for a fission reactor), this R&D work should be efficiently funded.

Solar Energy

Biomass. The growth of crops solely for the supply of fuel and on a continuing basis does not seem competitive economically with other use of the land. However, by-product or joint product agricultural materials can be used for crop drying or for local energy production. Some demonstrations of economically desirable use may be in order. The model of the agricultural extension service seems unusually applicable in this case. Naturally, the utility of by-product biomass for fuel may to some extent change the decision as to what to plant, although major changes in crop choice may result in only small changes in income or social good. Altogether, energy production from biomass is not primarily an energy but a land use problem, subject to many case-by-case considerations, and should be so evaluated.

Ocean Thermal Gradient Energy (OTE). This zero fuel cost approach would be effective at best only in very large sizes. It can be compared directly with the LWR or the breeder reactor in many ways. Side-by-side comparison appears to show no possibility of making OTE economically competitive with the fission energy approach, assuming that externalities are accommodated in each case with little or no cost. Therefore, support of ocean thermal gradient work should rely primarily on the exploration of new concepts. An example is the "salt fountain," the energy available from the osmotic pressure difference between salt water and fresh (equivalent to about 700 feet of hydrostatic head).

One point to note is that the technological advances required to make OTE competitive (low cost, corrosion-resistant heat exchangers and low cost vapor turbines) would also reduce the cost of its competitors such as solar ponds and bottoming cycles of conventional plants. New concepts are needed, and no large-scale work should be funded in this field unless a good probability of providing competitive electrical energy can be demonstrated.

Solar Photovoltaic Energy. The production of electrical power from the sun in the solar photovoltaic mode has a very small modular size. There is also a considerable range of application for which solar photovoltaic energy is economical now—powering navigational aids, telephone repeaters, satellites, emergency equipment, and the like. This market should be aggregated to provide a greater demand pull— that is, government should buy to meet real needs, thus creating a market that will open the door to competitive entry. However, the demand pull side of the technology acceleration may have a negative as well as a positive effect. If higher cost approaches can satisfy the customer (because they are available at a subsidized price), lower cost approaches may not be developed.

Thus, R&D is warranted on the supply side, but only R&D. Indeed, the conditions seem good for generating many competitive ideas, which is what is needed. Because of the small economic size of solar photovoltaic approaches, many firms can compete. Concepts include low-efficiency, spray-on films; increased efficiency of systems that are now too expensive; the hoped for substantial reduction in costs allocated to electricity production by sale of the waste heat; and the like.

Unfortunately, the technology at best works during the daytime and often not then because of clouds. Thus, an integral part of a solar photovoltaic approach must be the storage of the energy in electrical form. The costs of such a system must include that of the storage, which might include the production of some fuel and its recombination in a fuel cell.

Solar Thermal Electric (STE). The idea of focusing the sun's rays to provide very high temperature heat that can then be used in a normal heat engine to produce electricity is an old one. In large sizes, this zero fuel cost system clearly must compete with the light water reactor and advanced converters. In small sizes, it must compete with the solar photovoltaic approach, which has a much smaller overhead. The solar thermal electric approach produces energy only when the sun shines, and therefore can only be used as a peak shaver or with storage. Peak shaving is precisely what one does not want to do with a high capital cost technology. Storage of electrical energy produced requires a large capital investment for the electrical energy production equipment because all of the electrical energy is produced during the shorter period when the sun shines. Therefore, storage of the high temperature heat is likely to be the favored approach to extend electrical-generating capability of the STE plant through the day and perhaps for a day or so of clouds.

Once again, if such an economical (at best, zero cost) heat storage technology existed, its initial use could well be in association with LWR or fossil fuel base load plants, in order to provide a plant capable of handling peak loads with the smallest possible capital investment. Thus, an LWR plant could put heat into its thermal store during the night and during the day produce electrical power normally while also providing electrical power from the thermal store. This is an example of a generic technology development which at first sight would aid *one* approach—STE—but in fact render it more difficult to enter the field against the existing resource. Additional discussion of this and related subjects is provided in Chapter 13, especially those portions that deal with economics.

Solar Heating and Cooling. Much of the benefit of solar space heating and cooling can be obtained by design standard changes and by insulation and management that would allow more economical use of the solar heating and/or cooling. These approaches would be valuable even if one used electrical or fossil energy. The energy consumption of new commercial or industrial buildings can be reduced by those architects who are thoroughly abreast of such questions. An individual contemplating a new home can draw upon government publications and the knowledge of architects. After taking advantage of siting, insulation, sun lines, adaptive window shutters, and the like, it will make very little difference whether the remaining energy needs are supplied through solar energy or through fossil fuel, because either way their cost will be minimal. In either case, space heating and cooling costs will be only about one-fourth of what they would be otherwise. The key to such cost savings is cost knlowledge. The choice among architects and construction alternatives is simple if the construction costs are summed with present value costs of energy over the life of the building.

Although the application of solar heating and cooling to existing residences may not be promising, other markets are underemphasized in the present program. The neglected approaches are those of community scale. Heat can be stored, or a heat sink can be provided, by taking advantage of economies of scale associated with a large reservoir to handle the needs of hundreds of people, of a shopping center, or of a large commercial building.

Wind Energy. There are relatively few places in the United States where the wind blows intensely and steadily enough to make wind energy competitive with electricity from nuclear power or coal at the present state of our knowledge. If wind is to be a major contributor, new concepts are required, such as those that use the earth as a structural compression member, and the like. Of course, we will know more about wind energy twenty years from now than we do at present. Whether the outlook for wind energy is more favorable depends upon the relative increase of knowledge in this field compared with the alternatives, and the amount of RD&D funds to be spent in this field depends on our present understanding of the expected or potential benefit of funds expended in this field as compared with other fields. Thus one does not want to waste funds on large-scale and expensive demonstrations if, even when technically successful, such demonstrations show only that the cost of electrical energy from wind is too large to be incurred now and larger than that for other inexhaustible energy sources.

Solar Energy Technology for Export. There are nations or regions where the real cost of importing fossil fuel is felt to be very large. Those locales are potential customers for solar energy technologies even at presently foreseeable costs. Examples are wind power in northern windy locations, and solar heat engines in calm, sunny, southern climes. When cost is correctly calculated, however, it may even in such cases be as undesirable to import high capital cost equipment with zero continuing fuel cost as it is to import high cost fuel. After using all available by-products, biomass, or fuel, the next lowest cost energy may come from locally manufactured, high capital cost solar thermal approaches.

Fossil Fuels

Coal Gasification. Coal gasification typifies the need for basic research (and the lack thereof), the ability of private industry to provide energy supply on its own without government involvement when the price is right, and the need for demonstration of future supplies against the possibility of shortage and possible increase in cost.

Research in coal gasification involves tens of millions of dollars, compared with accumulated subsidies toward the production of oil (largely via tax provisions) in the tens of billions of dollars and compared with subsidies for hydro power and nuclear power in the billions or ten billion dollar range. It seems unlikely that the relative marginal benefits are so disproportionately against the expenditure of funds in coal research. Of course, other technologies that have little or no subsidy may be as promising as coal gasification. In any case, there is no necessity for society to fund only the most promising technology, because the products are not interchangeable. Major costs would be incurred in substituting electrical power for gasoline, even if the cost per energy unit were the same. Much better understanding of the basic processes involved in coal gasification would make it possible to explore alternatives in much smaller plants or without the building of pilot plants, saving years and hundreds of millions of dollars in direct cost and billions of dollars of lost or deferred product.

Coal can be transformed into low Btu gas of between 100 and 300 Btu/cu.ft., medium Btu gas of between 300 and 600 Btu/cu.ft., or high Btu (pipeline quality) gas in the range of 900 to 1,100 Btu/cu.ft. Although it is frequently asserted that only high Btu gas is suitable for transmission in existing pipelines, this is far from true. The increased cost of alternative fuels in the last five years makes it economical to transmit medium Btu gas by pipeline. Still, it may be preferable to use the expensive pipeline resource for the transmission of high Btu gas. Because many users cannot shift from high to low and back again with ease, it makes sense to consider the availability of high Btu gas from coal for distribution by the very large existing pipeline network in which tens of billions of dollars have already been sunk.

Still, individual large industries such as oil refineries, smelters, glass plants, and the like, as well as clusters of industries within ten to fifty miles of the source of such gas, can readily use low or medium Btu gas from coal as a convenient, clean fuel. No government subsidy seems necessary or desirable. Such activities are going forth to the profit of those involved. What this means is that a substantial fraction of the stationary fuel needs of the United States can be met by coal for the cost of shipping the coal. Low and medium Btu gas conversion is a way of burning coal. There are already known ways of mining coal, and although not very exciting, the combination can allow industry, including power plants, to operate within environmental regulations.

Centralized power plants are large enough that they need not convert the coal to gas (with an inevitable loss in fuel value), but can burn the coal itself either with flue gas desulfurization (FGD) or in a fluidized bed combuster (FBC) to remove sulfur and keep the production of nitrogen oxides sufficiently low. We have no doubt that a substantial conversion to low and medium Btu gas and to FGD and FBC in central power plants will occur as soon as the utilities are made to feel the price of imported oil, which we emphasize as desirable throughout this report.

A more difficult question is the availability of pipeline quality gas from coal. The same pipeline that could distribute this gas, however, is available for distributing natural gas if it is cheaper for the pipeline operator (or the ultimate user) to buy. Thus, the commercial success of a high Btu gasification plant may depend not so much on the price of imported alternative fuel (liquefied natural gas or oil) as it does on the cost and availability of natural gas within the United States. It appears that enough natural gas is or will be available from conventional and unconventional sources, at a price below high Btu coal gas, that little urgency attaches to deploying a high Btu gasification plant.

There is therefore time to select from among the three chief alternative methods of producing high Btu gas from coal. The operator will wish to minimize his cost (including the cost of meeting environmental regulations) and will have to have pilot plants of substantial scale to avoid excessive risk in the full-sized plant. Such pilot plants, large but still smaller than demonstration scale facilities, should be the concern of the program.

Shale Oil. Very large resources of shale oil exist in the United States and elsewhere, estimated in the range of trillions of barrels as contrasted with world reserves of crude oil in the range of 600 billion (10^9) barrels and resources of perhaps 2 trillion (10^{12}) barrels. This shale oil is not oil in the form of droplets in shale, but is rather a compound called kerogen in rock often known as marlstone. To mobilize this "oil," the material must be heated to about 800°-1,000° F or contacted with reagents to free the kerogen.

Two alternative methods have been tested. One is external retorting, in which marlstone is mined, brought to the surface, and then heated in conventional retorts. The second is in situ retorting, in which a portion of a cavity is mined out to allow for the expansion of the "shale" when it is heated. Most of the cavity is then made into rubble, and a flame front fed by air from the surface is induced to travel through the rubblized shale. This liberates kerogen, which drips to the bottom of the cavity and can be pumped to the surface. Although claims have been made for the better economics and lesser environmental impact of in situ retorting, it makes substantially less use of the resource, and there is no reason to believe at present that it is a better approach than the more conventional surface retorting, which has been in commercial use in various parts of the world, though on a modest scale only.

The problems with shale oil are that the expected cost of at least $25 per barrel exceeds that for alternative fuels and that it creates environmental and water problems. In fact, many believe that the environmental and water problems will prevent shale oil from ever providing an important share of U.S. liquid fuels. Still, it would be good to know how to make shale oil, how much it will cost to do so, and more about its environmental and water problems. But it is not clearly in the interest of the investor to spend the funds to obtain this knowledge.

Those in the industry have urged the creation of incentives such as a $3 per barrel tax credit for shale oil product. Under some circumstances such a tax credit could seem to be an acceptable approach, but we would opt for a firm contract by the federal government for a fixed delivery rate of shale oil at a competitively bid price, presumably above the market price for imported oil. The great uncertainty in the case of shale oil is not existence or size of the resource, nor the technical feasibility, nor so much the cost of production (though environmental and infrastructure costs may loom large). The problem is the cost of alternative fuels and (for the U.S. society as a whole) how much it is worth to displace a barrel of imported oil.

Natural Gas. Although the production of natural gas in the lower forty-eight states has been decreasing in recent years, the deregulation of gas price has called forth substantial increases in production, investment in pipelines, and so on in the intrastate market. As a result, the price of intrastate gas—although it has risen—has in some cases been below the price of gas available for sale to interstate pipelines. To what extent this same phenomenon will be repeated on a national scale as gas prices are deregulated is not known. More research results would certainly be valuable on this score.

In addition to conventional gas (that obtained from wells shallower than 15,000 feet in conventional sedimentary basins), new sources of unconventional gas are much discussed. These include tight gas sands in the West, Devonian shales, and abiogenic gas not associated with sedimentary rock and presumably not fossil at all (in that it has not been generated by the decay of once-living organisms). Estimates of the size of these unconventional sources vary widely. For example, the ratio between the "high" and "low" resource estimates for gas from Devonian shale at $3 per million cubic feet is more than a factor of twenty. Even conventional gas is subject to great uncertainties in estimates of resource and cost. The cost of gas at a cumulative production level of 400 trillion cubic feet ranges from $0.40 to $2.00 per million cubic feet.[5]

Estimates of the amount of gas available from western tight sands at a marginal cost of $3 per million cubic feet range from about 30 to about 220 trillion cubic feet. There are similar ranges of estimates of the availability of geopressured gas.

---

5. The energy in one trillion cubic feet of natural gas is equivalent to about one quad (10^{15} Btu). See also Chapter 9 on unconventional gas sources.

The Department of Energy's Market Oriented Program Planning Study (MOPPS) discusses costs of methane from coal. It cites quantities potentially available at $2 per million cubic feet ranging from about 80 to about 320 trillion cubic feet. "Conventional" gas cost does not appear to rise to $3 per million cubic feet until about 1,000 trillion cubic feet have been produced. This cost is less than that for coal gasification, for synthetic natural gas from coal, for geopressured methane, and perhaps less than that for gas from Devonian shale and western tight sands.

Given this situation, there is little incentive to produce a gas from higher cost resources. To guide planning, and perhaps to produce, it will therefore be necessary to provide artificial incentives to obtain the knowledge as to the ultimate cost of gas from these sources.

Energy Conservation

Every Btu produced will either be wasted or used, and energy adequacy can in principle be aided as readily by cutting consumption as by increasing production. As emphasized in Chapter 3, inexpensive and effective means for reducing the consumption of energy are known; in a list of measures to ensure the adequacy of supply at affordable costs, certain conservation measures come first.

Measures that can be deployed in tens of millions of homes, that can be offered or mandated in 10 million automobiles per year, or that may change a process in use by a large industry, deserve thought, improvement, and demonstration. We exemplify this large field by two applications, one to conservation in existing and new housing and the other in the automobile industry.

Energy Saving in Existing Households. The Twin Rivers experiment of Princeton University has demonstrated that use of energy for household space heating can be cut to one-fourth its normal level. This was achieved by relatively minor modifications of modern housing, performed without requiring the resident families to move out for even a single day. The actual modifications performed could cost-effectively be adapted to existing housing stock, but the thought and prescription for this experiment could not be replicated nationwide for lack of appropriate personnel and for reasons of cost. Some of the knowledge already won presumably will be made available to homeowners. However, it is not always easy for commercial enterprises to earn profits by disseminating such information, so additional effort is needed. Service contractors might market and install energy conservation packages, and the government can sponsor advertising.

Hardware can be perfected by private industry and offered for sale. Such socially desirable activity will be helped by government publications showing that there are real savings to be made and exhibiting examples of breakeven calculations or payback times. Among such hardware are inexpensive improved seals against convection at the top and bottom of floor-to-ceiling drapes, rigid foam shutters to close off windows at night, and flue closers for gas furnaces.

Automobile Fuel Efficiency. The 14 percent of the nation's energy supply that is now used in automobiles is scheduled to drop (at least on a per automobile basis) as improved fuel economy requirements are introduced, aiming toward 27.5 miles per gallon in 1985. We do not favor energy savings without regard to cost, but actual cost and performance are not known without RD&D. Improvements in fuel economy of the conventional automobile are clearly possible in the use of diesel engines, in improved transmissions and auxiliary equipment, and in the reduction of friction and drag. The demonstrated performance of nearly fifty miles per gallon in the diesel Volkswagen Rabbit is a milestone along this route.

Further improvement is available by the use of less conventional combined cycles, such as the installation of two engines of 10 horsepower and 70 horsepower or a scheme whereby individual cylinder valving and fuel are controlled according to the power demands.

An assortment of weight reduction measures should be demonstrated in prototype form to be ready for implementation depending upon the evolution of energy prices.

A real impediment to progress in saving energy in automobiles arises from the mandatory limits themselves. These "ceilings" on fuel consumption tend to become "floors," as the manufacturers do only enough to meet the requirements. A fuel consumption tax on new automobiles would provide an incentive for manufacturers continually to improve the fuel consumption of their automobiles. Such a tax could be arranged to have zero fiscal impact by the creation of a pool that could be refunded to manufacturers or buyers of automobiles with better than average fuel consumption. Unfortunately, major manufacturers are reluctant to demonstrate techniques for improved fuel economy under a mandatory fuel economy regime, fearing that such a demonstration will lead to the requirement to use this technique. Such reluctance would be countered by market forces in the case of the fuel economy tax. The tax scheme has lost out to the mandatory limits in the past, because it would take on the guise of a subsidy to foreign producers who are efficient producers of fuel-conserving vehicles. We note that these producers are also OECD nations with a common interest in reducing oil imports.

Nonhardware Energy RD&D

We have already discussed the need for nonhardware and quasi-hardware energy RD&D, which is of greater importance than its relatively low cost might suggest. We remind the reader of this important sector but do not repeat the list here. We discuss in some detail only one item.

Air Pollution. Numerous studies suggest an important effect of air pollution on human mortality and morbidity, but reliable quantitative estimates of the overall impact are almost entirely lacking. For example, estimates of the proportion of deaths attributable to air pollution range from 0.1 percent to 10 percent. Moreover, air pollution is a highly complex and dynamic component of the environment, incorporating physical, chemical, and biologic components. The individual contributions of these elements and their interactions with health effects have not been assessed. The Clean Air Act amendments of 1977 require decisions regarding regulation and control technology that carry major economic as well as health impacts. The scientific basis for making these decisions is woefully inadequate.

It is particularly urgent that knowledge be strengthened on the health effects of the by-products of fossil fuel use. Specific substances that deserve attention include sulfur oxides, nitrates, metals and metalloids, photochemical oxidants, organic nitrogen, and other organic compounds.

Air pollution within localized environments, such as homes, is increasingly relevant and often neglected. As homes are constructed to be better insulated, the concentrations of air pollutants will rise because of reduced ventilation and air circulation. The potential hazard to health of energy-conserving home construction has barely been addressed. Where it has been studied, results have shown that chemicals produced in the home or evolving from building materials, such as toluene from resins, may reach higher proportions in energy-conserving dwellings than in more conventional, less efficient buildings.

In spite of the difficulties of studying human populations in ways that will permit separation of the influences of individual environmental factors on health, epidemiologic research provides the most compelling evidence that polluted air damages health. Partly because of the difficulties of such studies and partly from lack of foresight by policymakers, important opportunities have been missed to study changes of health status when major changes were being made in environmental factors. Large changes took place in air quality in the United Kingdom starting in 1956 and in the United States about ten years later. During this critical period of changing air pollution, neither country mounted a planned and systematic study to determine whether there were any health benefits from cleaning the air. With the present prospect of future degradation of the air in this country at least on a regional basis, it is urgent that plans be made to monitor the possible concurrent effects on health.

The nation should continue to support prospective studies to assess the health effects of exposure to pollutants resulting from use of fossil fuels. These exposures come primarily through pollution of air, although the possibility of surface and ground water contamination must also be considered. Long-term commitment is required for these studies because the changes to be assessed may take several years to develop to a degree that is measurable.

Populations studied should include:

1. Subjects living in communities likely to experience significant changes in the level of exposure;
2. Subjects with unique occupational exposures;
3. Panels of subjects regularly followed who might be available for assessment during documented acute pollution episodes;
4. Subjects in whom other known or suspected risk factors for the development of chronic respiratory disease exist, such as cigarette smoking, occupational risks, or childhood history of bronchitis; and
5. Cohorts of children, since changes in lung function in young persons may become apparent after a shorter duration of exposure than is the case in adults and since young children are also less subject to the confounding factors of tobacco smoking and occupational risk.

Glossary

ACCELERATOR (PARTICLE ACCELERATOR): A device which imparts high velocities, and therefore high energies, to atoms, nuclei or subatomic particles.

ACID RAIN: Abnormally acidic rainfall, most often dilute sulfuric or nitric acid.

ACRE-FOOT: Amount of water covering 1 acre of surface to height of 1 foot; equals 326,000 gallons.

ACTINIDES: A group name for the series of radioactive elements from element 89 (actinium) through element 103 (lawrencium).

ADJUDICATORY HEARING OR PROCEEDING: A process by which administrative agencies decide on the application of a law or regulation to a specific individual, company, or fact situation. Usually cross examination and other features of a judicial trial are permitted.

AERODYNAMIC NOZZLE PROCESS: A method of uranium enrichment. A gaseous mixture of UF_6 and hydrogen is forced to flow at high velocity in a semicircular path, establishing centrifugal forces which tend to separate the heavier molecules from the lighter ones.

AEROSOL: A suspension of fine solid or liquid particles in gas; smoke, fog, and mist are examples.

ALLUVIAL BOTTOM LANDS: Lands where a former river bed has deposited sand, gravel, and earth.

ANAEROBIC: Living, active, or occurring in the absence of free oxygen.

ANTHROPOGENIC: Of, relating to, or influenced by the impact of humans or human society on nature.

AQUIFER: A water-bearing stratum of permeable rock, sand, or gravel.

BACK-END COSTS: The portion of nuclear electric generation costs incurred from the so-called back end of the fuel cycle linkages. The back end of the fuel cycle is that portion of the cycle occurring after in-core irradiation. Typically, this includes spent fuel cooling and interim storage, reprocessing, and ultimate disposal.

BASE LOAD GENERATING COSTS: The cost of electricity generated by an electric facility designed to operate at constant output with little hourly or daily fluctuation.

BATCH PROCESSING: Any process in which a quantity of material is handled or considered as a unit. Such processes involve intermittent, as contrasted with continuous, operation.

BIOMASS: In energy context, with potential for heat or energy generation.

BITUMINOUS COAL: An intermediate rank coal with low to high fixed carbon, intermediate to high heat content, a high percentage of volatile matter, and low percentage of moisture.

BLOWDOWN WATER: A side stream of water removed from the cooling system in order to maintain an acceptable level of dissolved solids.

BOTTOMING CYCLE: Generating power by exploiting the (otherwise wasted in atmospheric rejection) heat content of the engine working fluid. When that working fluid is a liquid boiling at a relatively low temperature, bottoming cycle generation is relatively easy.

BREEDER REACTOR: A nuclear reactor that produces more fissile material than it consumes.

BREEDING RATIO: The number of fissile atoms produced per atom fissioned within a nuclear reactor. If the ratio is greater than one, a reactor is called a "breeder."

BRITISH THERMAL UNIT (BTU): The amount of energy necessary to raise the temperature of one pound of water by one degree Fahrenheit, from 39.2 to 40.2 degrees Fahrenheit.

BURNER REACTOR: Any nuclear reactor that produces less fuel than it consumes.

BURNUP: The amount of thermal energy generated per unit mass of reactor core fuel load. Though burnup will vary with core management practice, it typically is one measure of reactor operating efficiency.

BUS-BAR COST: The cost of electric power as it leaves the generating facility.

CANDU: A nuclear reactor of Canadian design, which uses natural uranium as a fuel and heavy water as a moderator and coolant.

CAPACITY FACTOR: The ratio of energy actually produced to that which would have been produced in the same period (usually a year) had the unit operated continuously at rated capacity.

CAPITAL CHARGES: The annualized costs of borrowing plus the amortization of investment and allowance for taxes.

CARBON DIOXIDE PROBLEM OR CATASTROPHE: See "Greenhouse Effect."

CARRYING CHARGE: Generally the cost of carrying capital. In nuclear power context, frequently used to refer only to cost of carrying fuel inventory.

CATALYSIS: A modification—especially an increase—in the rate of a chemical reaction that is induced by material that remains unchanged chemically at the end of the reaction.

CENTRIFUGE (ISOTOPE SEPARATION): A new isotope enrichment process that separates lighter molecules containing uranium-235 from heavier molecules containing uranium-238 by means of ultra high speed centrifuges. The process is now being successfully exploited by the URENCO consortium in Western Europe.

CESIUM-137: A radioactive isotope of the element cesium that is a common fission product. Its half-life is 30 years.

CHAIN REACTION: A reaction in which one of the agents necessary to the reaction is itself produced by the reaction so as to cause similar reactions. In a nuclear reactor (or bomb) a neutron plus a fissionable atom cause a fission, resulting in a number of neutrons which in turn cause other fissions.

CLINCH RIVER BREEDER REACTOR (CRBR): A proposed demonstration of the liquid metal fast breeder reactor. The CRBR, which would operate on a plutonium fuel cycle and be cooled by molten sodium, would have an electrical output of about 350 MWe.

COAL CONVERSION: Now most commonly used to denote coal gasification and liquefaction, but strictly speaking, any process that transforms coal into a different from of energy (e.g., electricity).

COAL GAS: Low, intermediate, or high Btu content gases produced from coals.

COAL LIQUEFACTION: The conversion of coal into liquid hydrocarbons and related compounds by the addition of hydrogen.

COGENERATION: The generation of electricity with direct use of the residual heat for industrial process heat or for space heating.

COMBINED CYCLE: The combination, for instance, of a gas turbine followed by a steam turbine in an electrical generating plant.

COMPETITIVE PRICE: Price of a product or service established in a competitive market.

CONSTANT DOLLARS: Dollar estimates from which the effects of changes in the general price level have been removed, reported in terms of a base year value.

CONVECTION: The circulatory motion that occurs in a fluid not at a uniform temperature owing to the variation of the fluid's density and the action of gravity. The term may also denote the transfer of heat which accompanies this motion.

CONVERSION EFFICIENCY: The percentage of total thermal energy that is actually converted into electricity by an electric generating plant.

CONVERTER REACTOR: Any nuclear reactor that is not a breeder reactor, i.e., that consumes more fuel than it breeds.

CRITICALITY ACCIDENT: An unintended nuclear chain reaction which, depending upon the materials involved and their physical configuration, can release varying amounts of energy and radioactivity at various rates.

CURRENT ACCOUNT: In matters of foreign trade, the value of goods and services entering imports and/or exports; excludes capital transactions.

CURRENT DOLLARS: Dollar values that have not been corrected for changes in the general price level.

DAUGHTER PRODUCTS: Atoms formed as a result of the radioactive decay of other atoms. Daughter products may themselves be radioactive, continuing to decay into other daughter products.

DENATURE: The addition of a nonfissionable isotope to fissionable material to make it unsuitable for use in nuclear weapons without extensive processing.

DEUTERIUM: A hydrogen isotope of mass number two (one proton and one neutron).

DEVONIAN SHALE: Gas-bearing black or brown shale of the Devonian geologic age; underlies over 160,000 m^2 of the Appalachian Basin.

DIFFUSION PLANT: A type of plant for uranium enrichment which uses gaseous uranium hexafluoride as its feedstock, and in which isotopic separation occurs from the differential rates at which UF_6 atoms diffuse through membranes.

DISCOUNT RATE: A rate used to reflect the time value of money. The discount rate is used to adjust future costs and benefits to their present day value. The effect of the discount rate (r) on the present value of a cost or benefit at time (t) in the future (C_t) is given by the expression

$$C_{(t)} \left[\frac{1}{(1 + r)^t} \right].$$

The selection of discount rates appropriate to particular situations is a matter of debate among economists, although 10 percent has been used in government calculations.

DISCOUNTED PRESENT VALUE: See "Present Value."

DISTRICT HEATING: A system for providing heat in commercial and residential buildings in which heat or byproduct heat from an electrical generating unit is distributed via an underground pipeline system.

DOSE-RESPONSE CURVE: A graph or function showing the relationship between dose or stimulus and its effect on living tissue or organism(s).

DRAFT ENVIRONMENTAL IMPACT STATEMENT: A preliminary version of an environmental impact statement (EIS) which is made available for public review and comment prior to preparation of a final EIS.

ECONOMIES OF SCALE: A production process is characterized by economies of scale when costs per unit output fall as plant size is increased.

ECONOMIC RENT: Income accruing beyond that needed to bring a firm into, or keep it, in operation; arises from a variety of advantages enjoyed by the firm (e.g., location, quality of inputs, ability, changes in public policy or other external factors, often referred to as "windfalls").

ELECTRICAL GRID: A net or system that connects a number of power-generating stations and users of electricity (load centers) to permit interchanges and more economical utilization of participating facilities.

EMISSION CHARGE: A charge levied on emitters of pollutants per unit of pollutant or per unit of input of the emitting substance.

ENHANCED OIL RECOVERY: Technology for lifting oil from wells beyond amount recovered through natural reservoir pressure. Commonly defined to exclude simple devices such as injecting water as a driving force.

ENRICHED URANIUM: Uranium in which the percentage of the fissionable isotope uranium-235 has been increased above that contained in natural uranium.

ENRICHMENT: The process by which the percentage of the fissionable isotope uranium-235 is increased above that contained in natural uranium.

ENTITLEMENT: A device for allocating the higher cost of imported oil among refiners so that every refiner pays the price for crude he would pay if he purchased it in the proportion between domestic and imported oil that prevails for the industry as a whole.

ENVIRONMENTAL IMPACT STATEMENT (EIS): An analysis of the environmental effects of a proposed action. Under the U.S. National Environmental Policy Act of 1969, an EIS must be prepared in conjunction with any proposed action of the federal government that would significantly affect the quality of the environment.

EPIDEMIOLOGY: A branch of medical science that deals with the incidence, distribution, and control of disease in a population.

ETHANOL: A colorless, volatile, flammable liquid of the chemical formula C_2H_5OH, which is found in fermented and distilled liquors; also made from petroleum hydrocarbons.

EUROCHEMIC: Spent fuel reprocessing plant in Belgium.

EUROCURRENCY (XENOCURRENCY): A balance expressed in a currency held in a bank outside the territory of the nation issuing the currency; originally applied to deposits held in European banks, but now widely used in the generic sense.

EURODIF: A joint venture, involving France (52 percent), Italy, Belgium, Spain, and Iran, which is building a gaseous diffusion plant in France.

EXPECTED BENEFIT: In cost-benefit analysis, the expected benefit (EB) is, broadly speaking, the weighted average of the monetary benefits of the possible outcomes (B), the weight for each being its probability of occurrence (P): $EB = P_a B_a + P_b B_b + \ldots P_n B_n$.

EXPECTED COST: In cost-benefit analysis, the expected cost (EC) is, broadly speaking, the weighted average of the monetary costs of the possible outcomes (C), the weight for each being its probability of occurrence (P): $EC = E_a C_a + E_b C_b + \ldots E_n C_n$.

FAST BREEDER REACTOR: A breeder reactor that relies on very energetic ("fast") neutrons.

FEDERAL PRE-EMPTION: The right granted to the federal government by the Constitution to supersede state or local laws that are in conflict with federal laws or regulations.

FEEDSTOCK: Any raw or processed material input stream fed into some production process.

FISSILE MATERIAL: Atoms such as uranium-233, uranium-235, or plutonium-239 that fission upon the absorption of a low energy neutron.

FISSION: The splitting of an atomic nucleus with the release of energy.

FISSION PRODUCTS: The medium-weight, highly radioactive daughter products of U-235 fission.

FLUE GAS DESULFURIZATION (FGD): The removal of sulfur oxides from the gaseous combustion products of fuels before they are discharged into the atmosphere.

FLUIDIZED BED COMBUSTION: The process of burning finely divided coal in a sulfur-capturing bed of limestone-based particles, with both coal and limestone particles supported by a stream of air or other gas.

FOCUSING COLLECTOR: A solar energy collector that gathers and concentrates the incoming radiation.

FOOD CHAIN: An arrangement of the organisms of an ecological community according to the order of predation in which each uses the next, usually lower, member as a food source.

FORWARD COST: A concept of cost used in U.S. estimates of uranium reserves that omits cost incurred in early stages, such as exploration, land acquisition and the like.

FRONT-END COSTS: Not as clearly defined as back-end costs; ambiguously used to cover costs arising in nuclear power generation through in-core radiation.

FUEL BANK: A proposed international uranium stockpile from which nations could obtain uranium, enriched if necessary, in the event of an interruption of normal supply.

FUEL CELL: A device for combining hydrogen and oxygen in an electrochemical reaction to generate electricity. Chemical energy is converted directly into electric energy without combustion.

FUSION: The combining of certain light atomic nuclei to form heavier nuclei with the release of energy.

GAS CENTRIFUGE ENRICHMENT PLANT: See "Centrifuge."

GAS-COOLED REACTOR: See "High Temperature Gas-Cooled Reactor."

GAS CORE REACTOR: A nuclear reactor in which the fuel is in gaseous form (uranium hexafluoride).

GASIFIER: Equipment for converting coal to gas.

GASOHOL: Variously refers to either alcohol derived from an organic material (grain, sugar, waste, etc.) or a mixture of such alcohol with gasoline, both intended as motor fuel.

GENETIC DAMAGE: Injury to the genetic material of a cell. Often denotes specifically damage to an organism's germ cells which can be passed on to offspring.

GEOPRESSURED BRINES: Formation water contained in some sedimentary rocks under abnormally high pressure; unusually hot and may be saturated with methane.

GEOTHERMAL ENERGY: The heat energy in the earth's crust which derives from the earth's molten interior. Can be tapped as steam, or by injection of water to form steam.

GLOBAL INSOLATION: The total insolation on a horizontal surface on the earth, averaged over some specified period of time.

GREENHOUSE EFFECT: A possible increase in the average temperature of the earth's atmosphere that might be caused by the release of carbon dioxide from fossil fuel combustion.

GROSS DOMESTIC PRODUCT: Equals Gross National Product, diminished by net factor income originating in foreign enterprises and investment.

GROSS NATIONAL PRODUCT: Aggregate value of goods and services produced in a national economy.

GNP DEFLATOR: A price index that translates GNP in current dollars into GNP in constant dollars, i.e., corrects for change in GNP due to price changes.

HALF-LIFE: The period required for the disintegration of half of the atoms in a given amount of a specific radioactive substance.

HARD CURRENCY: A currency easily exchangeable anywhere at the official rate of exchange.

HEAT ENGINE: Any device that converts thermal energy (heat) into mechanical energy.

HEAT PUMP: A device for transferring heat from a substance at one temperature to a substance at a higher temperature, by alternately vaporizing and liquefying a fluid through the use of a compressor.

HEAVY OIL: Crude oil of such low viscosity (usually $15°$ or less) that it does not flow freely enough to be lifted from a reservoir reached by the drill.

HEAVY WATER REACTOR (HWR): A nuclear reactor that uses heavy water as its coolant and neutron moderator. Heavy water is water in which the hydrogen of the water molecule consists entirely of the heavy hydrogen isotope of mass number two (one proton and one neutron).

HIGH-BTU GAS: Gas that has a heating value between 900 and 1,100 Btu per cubic foot.

HIGH-ENRICHED URANIUM: Uranium sufficiently enriched in the fissionable isotope 235 to be usable as weapons material without further enrichment or processing.

HIGH-LEVEL WASTES: The actinide and fission product component of the waste stream from spent fuel reprocessing. The intense radioactivity of this waste component requires long-term isolation.

HIGH-TEMPERATURE GAS-COOLED REACTOR (HTGR): A graphite moderated, helium cooled reactor using highly enriched uranium as initial fuel and thorium as a source of new fuel.

HYBRID RULEMAKING: A more formal kind of rulemaking proceeding in which public participation is usually permitted through a hearing that includes opportunities for at least rudimentary cross examination. Hybrid rulemaking is often used where there are questions of both fact and policy to be decided upon.

HYDROLYSIS: A chemical decomposition in which a compound is broken up and resolved into other compounds by reaction with water.

HYDROSTATIC HEAD: The pressure created by the weight of a column of water.

INCOME ELASTICITY OF DEMAND: A measure of the change of consumer demand for a particular good with change in income. It is defined as the ratio of the percentage change in demand to the percentage change in consumer income. If income elasticity of demand for a commodity is low, there will be little change in consumer demand for that commodity in response to changes in income.

INDIFFERENCE COST: A cost at which either of two alternatives is equally attractive, i.e., the user is indifferent as to choice.

INFRASTRUCTURE: The underlying or associated foundation or basic framework required to utilize a given product or service (as of a system or organization).

INSOLATION: The solar energy per unit time (power) crossing a unit area.

INTERMEDIATION (FINANCIAL): Process facilitating the flow of funds from original sources to ultimate borrowers.

ION: An atom or group of molecularly bound atoms that carries a positive or negative electric charge as a result of having lost or gained one or more electrons.

IONIZING RADIATION: Radiation, including alpha, beta, or X-rays, sufficiently energetic to knock off the electrons—or "ionize"—the atoms of matter while in transit through that matter.

ISOTOPE: One of perhaps several different species of a given chemical element, distinguished by variations in the number of neutrons in the atomic nucleus but indistinguishable by chemical means.

ISOTOPE SEPARATION: The process of separating the normally

and naturally commingled, and chemically identical, isotopes of a given element.

KINETIC ENERGY: The energy associated with an object's motion; varies directly in proportion with the object's mass and with the square of its velocity.

LASER ISOTOPE SEPARATION: A new isotope enrichment process now in the development stage. Atoms of uranium-235, or molecules containing them, would be selectively ionized or excited by lasers, allowing physical or chemical separation of one of the isotopes.

LEVELIZED COST: The spreading out of a cost or charge in even payments over a period of time.

LIFE-CYCLE COSTING: Distributing one-time cost of acquisition of a piece of equipment over its estimated lifetime to calculate annual cost.

LIGHT-WATER REACTOR (LWR): A nuclear reactor that uses ordinary water as a coolant to transfer heat from the fissioning uranium to a steam turbine and employs slightly enriched uranium-235 as fuel.

LIGNITE: The lowest rank coal with low heat content and fixed carbon, high percentage of volatile matter and moisture; an early stage in the formation of coal.

LINEAR NO-THRESHOLD MODEL: A dose-response model which assumes that there is no threshold or "safe" dose level and that the effects are directly proportional to the size of the dose.

LOAD: The amount of energy or electric power required of an energy system during any specified period of time and at any specified location or locations.

LOAD FACTOR: The annual average output of a utility system divided by its maximum potential output.

LONG-WALL MINING: A mining method, widely used outside the United States, in which the entire coal seam is removed at a "long wall" and the roof allowed to cave behind the coal face.

LOW BTU GAS: Gas that has a heating value between 100 and 300 Btu per cubic foot.

LOW-ENRICHED URANIUM: Uranium that has been enriched in the fissionable isotope 235, but not sufficiently enriched for weapons use.

LOW SULFUR COAL: Coal with sulfur content generally less than 1.0 percent.

MAGNETOHYDRODYNAMIC POWER: Generation of electricity by moving hot, partially ionized gases through a magnetic field, where they are separated by electrical charge, generating an electric current that is then collected by electrodes lining the expansion chamber. Usable in combination with conventional power generating plants.

MARGINAL COST PRICING: Charging users for all units consumed at the rate that corresponds to the cost of the final unit that needs to be supplied to meet demand. Sometimes referred to as "incremental" pricing.

MASS NUMBER: The total number of neutrons and protons in the nucleus of an atom.

MEDIUM BTU GAS: Gas that has a heating value of between 300 and 600 Btu per cubic foot.

MEGAWATT: The unit by which the capacity of production of electricity is usually measured. A megawatt is a million watts or a thousand kilowatts.

MELTDOWN: A nuclear reactor accident in which the radioactive fuel overheats, melting the fuel cladding and threatening release of the core radioactive inventory to the environment.

METHANE: The lightest compound of the paraffin series (hydrocarbons) with the formula CH_4; occurs naturally in oil and gas wells and is the principal constituent of natural gas.

METRIC TON: 2,205 pounds.

MICRON: One millionth (10^{-6}) of a meter.

MILL OR MINE TAILINGS: See "Tailings."

MORBIDITY: Prevalence of a disease; or rate of illness.

MUTAGEN: A substance or agent that tends to increase the frequency or extent of genetic mutation.

NATURAL URANIUM: Uranium in which the naturally occurring proportions of the isotopes of mass numbers 234, 235, and 238 have not been artificially altered as, for example, through enrichment.

NEUTRON: An elementary particle with approximately the mass of a proton but without any electric charge. It is one of the constituents of the atomic nucleus. Frequently released during nuclear reactions and, on entering a nucleus, can cause nuclear reactions including nuclear fission.

NEUTRON CAPTURE: The absorption of a neutron by the nucleus of an atom. Depending on the kind of nucleus and upon the speed of the incident neutron, three outcomes are possible: The neutron can be absorbed, leaving a stable isotope of the original element; the new nucleus can decay to another of approximately the same size; or there can be fission—breakup into two smaller nuclei.

NEUTRON SPECTRUM: The velocity distribution of the free neutrons in a chain-reacting assembly—for example, a reactor core.

NOMINAL DEMAND: Demand expressed in current dollars.

NONTRANSURANIC WASTES: All of the constituents of nuclear waste other than the transuranic elements. The most highly radioactive of these are the so-called fission products—the approximately equal fragments into which U-235 splits.

NOTICE AND COMMENT RULEMAKING: A specific kind of rulemaking proceeding in which public participation is sought through the publication of a proposed rule in the *Federal Register* and written or oral comments are solicited prior to the agency decision.

NUCLEUS: The small, massive, positively charged core of an atom that comprises nearly all of the atomic mass but only a minute part of the atom's volume. The nucleus consists of neutrons and protons, in the common isotope of hydrogen, which consists of one proton only.

NUCLEAR FUEL CYCLE: The series of steps involved in supplying fuel for nuclear power reactors and disposing of it after use. It can include some or all of the following stages: mining, refining of uranium or thorium ore, enrichment, fabrication of fuel elements, their use in a nuclear reactor, spent fuel reprocessing, recycling, radioactive waste storage or disposal.

NUCLEAR WASTE: The radioactive products formed by fission and other nuclear processes in a reactor. Most nuclear waste is initially in the form of spent fuel. If this material is reprocessed, new categories or waste result: high-level, transuranic, low-level wastes and others.

OCEAN THERMAL GRADIENT ENERGY (OTE): Power generation by exploiting the temperature difference between surface waters and ocean depths.

OFF-PEAK: Period of low demand, usually during 24-hour period. When applied to capacity, refers to facilities preferably employed during such periods. When applied to customer rates, refers to rates charged during consumption at such periods.

OIL SHALE: Rock containing organic matter (kerogen) that upon being heated to $800°$-$1,000°F$ yields commercially useful oil and/or gas.

ONCE-THROUGH FUEL CYCLE: Any nuclear fuel cycle in which spent fuel is disposed of without reprocessing or recycling to recover fissionable materials from the reactor.

OSMOTIC PRESSURE: Pressure produced by or associated with diffusion of a solvent through a semipermeable membrane.

OVERBURDEN: Rock, soil, and other strata above a coal bed that is to be strip mined.

OZONE: A triatomic form of oxygen (O_3) that is a bluish irritating gas of pungent odor, is formed naturally in the upper atomosphere by a photochemical reaction with solar ultraviolet radiation, and that is a major agent in the formation of smog.

OZONE LAYER: A region in the earth's upper atmosphere containing ozone which helps shield living organisms from the sun's ultraviolet radiation.

PARTICULATES: Fine solid particles that remain individually dispersed in emissions from fossil fuel plants.

PASSIVE SOLAR ENERGY: Means of utilizing solar energy for heating and cooling, such as window shutters, eaves, or insulation, that do not consume electricity, pump fluids, or have powered mechanical components.

PAYBACK PERIOD: The length of time required for the cumulative net revenue from an investment to equal the original investment. Often used in connection with outlays for energy conservation.

PEAT: Name given to layers of dead vegetation that is sometimes regarded as the youngest member of the ranks of coal.

PERMEABILITY: The rate at which a liquid or gas can pass through a porous solid substance.

pH: A measure of acidity, calibrated logarithmically, whose values run from 0 to 14 with 7 representing neutrality, numbers less than 7 increasing acidity, and numbers greater than 7 increasing alkalinity.

PHOTOCHEMICAL: Of, relating to, or resulting from the chemical action of radiant energy, especially light.

PHOTOGALVANIC PROCESSES: Light-induced production of electrical current.

PHOTON: A quantum or discrete bundle of electromagnetic radiation. According to the quantum theory underlying modern physics, all matter comes in discrete bundles, or "quanta"; "electromagnetic radiation"—X-rays, radiowaves, and visible light are examples —is no exception.

PHOTOSYNTHESIS: The biologic process by which the chlorophyll-containing tissues of plants use water, carbon dioxide and light energy to synthesize carbohydrates.

PHOTOVOLTAIC ENERGY: Energy obtained from devices that convert sunlight directly into electricity.

PLASMA: A dilute gaseous mass of electrically charged particles.

PLUTONIUM: A radioactive metallic element that is formed in nuclear reactors. The plutonium isotope 239 has a half-life of 24,000 years and can be fissioned to yield energy within nuclear reactors or nuclear bombs.

PLUTONIUM SEPARATION: The process by which plutonium is

recovered from the uranium, transuranic elements, and fission products that are in spent reactor fuel.

POISON (OR NEUTRON POISON): Any material (other than the deliberately introduced moderator) which absorbs neutrons within a nuclear chain reaction, slowing down or halting the rate of nuclear fission. Because of the accumulation of neutron poisons in the fuel rods, these rods must be withdrawn from a reactor core after about one year.

POLYNUCLEAR ORGANIC MATERIALS: Aromatic hydrocarbons that are important as pollutants and possibly as carcinogens. Arise in coal conversion processes.

PRESENT VALUE: The current value of a future stream of costs or benefits calculated by discounting these costs or benefits to the present time. (See Discount Rate.)

PRICE ELASTICITY OF DEMAND: The responsiveness of demand to changes of price. It is defined as the ratio of the percentage change in the quantity demanded to the percentage change in the price of the commodity. If a given change in price results in a large change in demand, then demand is elastic; if the change in price has only a slight effect on demand, then demand is inelastic.

PRIMARY ENERGY: Energy before processing or conversion into different or more refined form.

PUMPED STORAGE: Method of generating hydroelectricity by using low cost (usually off-peak) power to pump water to an elevated storage area and releasing it at peak demand periods into a river or other body of water.

PYROLYSIS: Breaking down petroleum hydrocarbons by high temperatures into simpler and lighter compounds.

QUENCH WATER: Water used to cool hot gases or solids.

RADIATION ACTIVATED SPECIES: Radioactive nuclei which have been produced from other stable nuclei by neutron irradiation.

RADIATION DAMAGE: A general term for the adverse effects of radiation upon living and nonliving substances.

RADIOACTIVE HEATING: Heating caused by the decay of radioactive materials.

RADIOACTIVE WASTE: See "Nuclear Waste."

RADIUM-226: A naturally occurring radioactive isotope of radium with a half-life of 1,600 years.

RADON: A chemically inert, radioactive element which has 86 protons in its nucleus. The most common isotope is radon-222, which has a half-life of 3.8 days.

REACTOR CORE: The region within a nuclear reactor that contains the fissionable material.

REACTOR YEAR: The operation of a nuclear power reactor for one year. As the term has come to be used, unless otherwise stated, the capacity of the reactor is assumed to be 1,000 megawatts (electric).

RECYCLING: The reuse in a nuclear reactor of uranium, plutonium, or thorium that has been recovered from spent fuel.

REPROCESSING: The chemical and mechanical processes by which plutonium-239 and the unused uranium-235 are recovered from spent reactor fuel.

RESERVE CURRENCY: Currency in which central banks maintain foreign exchange reserves; role played by U.S. dollar for most of post-World War II period.

RESERVES: Resources well identified as to location and size and commercially producible at current prices and with current technology.

RESISTANCE HEAT: Heat generated by an electric current as it is conducted through a substance opposing the flow of the current.

RESOURCES: Sometimes used to denote the sum total of a given resource estimated to exist; at other times only those portions estimated to remain discoverable and recoverable in the future.

RETORT: To subject substances to heat for the purpose of distillation or decomposition; equipment for such a process.

ROLLED-IN PRICING: Injecting a high-cost product into the market by averaging its cost with that of the bulk of the same product produced at lower cost. Typical for public utility marketing of gas and electricity.

RULEMAKING HEARING OR PROCEEDING: A process by which administrative agencies make regulations (usually establishing policy affecting classes of individuals or companies) that implement legislative mandates.

SAFEGUARDS: Physical or institutional arrangements aimed at preventing the illegal diversion of weapons-usable materials from civilian nuclear programs.

SALT DOME: A subterranean geologic formation frequently associated with occurrence of oil and/or gas; after original hydrocarbon material has been withdrawn, can be used for storage of oil or gas.

SAUDI MARKER CRUDE: A quality of light crude oil, with 34° viscosity, used as the basic reference material for determining differential prices for other types of oil.

SCRUBBING: Technique for removing pollutants such as noxious gases and particulates from stack gas emissions.

SEMICONDUCTOR: Any of a class of solids, such as germanium and silicon, whose electrical conductivity is between that of a conductor and that of an insulator in being nearly metallic at high temperatures and nearly absent at low temperatures.

SEMIPERMEABLE MEMBRANE: A thin, soft, pliable sheet through

which some molecules (usually smaller ones) can pass while others (usually larger) cannot.

SENSITIVE FACILITIES: Any facilities employed in civilian nuclear research or nuclear power programs which could conceivably be converted to weapons material production uses. Typically these are research reactors, enrichment plants, and reprocessing facilities.

SEPARATIVE WORK UNITS (SWUs): A measure of the work required to separate uranium isotopes in the enrichment process. It is used to measure the capacity of an enrichment plant independent of a particular product and tails. To put this unit of measurement in perspective, it takes about 100,000 SWUs per year to keep a 1,000 MWe LWR operating and 2,500 SWUs to make a nuclear weapon.

SHORT TON: 2,000 pounds.

SOLAR ELECTRIC CELL: A device which converts sunlight directly into electricity.

SOLAR POND: An insulated pond used to store solar energy.

SOLVENT REFINED COAL: Product of a process in which coal is dissolved in oil, treated at moderate conditions with hydrogen and filtered to produce a solid with a high heat value (approximately 16,000 Btu/lb) and an ash and sulfur content much below that of the input coal.

SOMATIC DAMAGE: Injury to the normal functioning of a living organism, which may or may not result in death.

SPENT FUEL: The fuel elements removed from a reactor after several years of generating power. Spent fuel contains radioactive waste materials, unburned uranium and plutonium.

SPIKING: The addition of highly radioactive materials to nonradioactive weapons-usable nuclear materials. The intent is to discourage illegal diversion.

SPOT PRICE: Price formed in a market for one-time, usually quick, delivery of a specific cargo and destination. In times of stringency, spot market prices can greatly exceed prices for identical products sold under long-term contracts.

STACK GASES: Gases resulting from combustion of fuels and emitted through the stack.

STRIPPERS, STRIPPER WELLS: Oil wells producing less than 10 barrels per day.

STRONTIUM-90: A radioactive isotope of the element strontium. A common fission product, its half-life is 28 years.

SUBMICRON: Smaller than a micron.

SUNK COST: The unrecovered or unrecoverable balance of an investment. A cost already paid or committed.

SUPER PHÉNIX: French commercial-scale liquid metal fast breeder reactor under construction.

SUPERCONDUCTING MAGNET: A magnet which is at very low temperature (near absolute zero).

SYNERGISM: Combined or cooperative action of discrete agencies such that the total effect is greater than the sum of the effects taken independently.

SYNTHETIC FUELS: A term now commonly reserved for liquid and gaseous fuels that are the product of a conversion process rather than mining or drilling; most frequently applied to cover liquids or gases derived from coal, shale, tar sands, waste, biomass.

TAILINGS: Rejected material after uranium ore is processed. Since uranium ore contains less than 1 percent uranium, essentially all of the processed ore is left as tailings near uranium mills.

TAILS: Uranium, depleted in the isotope uranium-235, remaining after production of enriched uranium.

TAILS ASSAY: The percentage of uranium-235 in tails. Natural uranium contains 0.71 percent uranium-235; the tails assay may be 0.3 percent or lower.

TAR SANDS: Also known as bituminous sands; unconsolidated sands permeated with bitumen, subject to surface mining and refining, and being produced in commercial-sized installations in Canada.

TERMS OF TRADE: A measure of change over time in the ratio of export prices to import prices; used to judge changing advantage in foreign trade.

THERM: 100,000 British thermal units.

THERMODYNAMIC EFFICIENCY: The highest fraction of input energy that can be converted into useful work by a heat engine. The Second Law of Thermodynamics establishes that maximum.

THERMONUCLEAR EXPLOSIVE OR DEVICE: A nuclear bomb, such as a hydrogen bomb, that derives its energy from the fusion of light elements into slightly heavier ones.

THERMONUCLEAR REACTOR: A nuclear reactor that derives its energy from the controlled release of fusion energy.

THORIUM: A radioactive metallic element. When bombarded with neutrons, the thorium isotope of mass number 232 forms thorium-233, which in turn decays through protactinium to form the fissionable isotope uranium-233.

THRESHOLD: A dose level or concentration (as of a drug, toxin, or pollutant) below which there is, or it is assumed there is, no effect, especially no adverse effect, on a recipient tissue or organism.

THROWAWAY FUEL CYCLE: See "Once-Through Fuel Cycle."

TIGHT SANDS FORMATION (GAS): Gas-bearing geologic strata that hold gas too tightly for conventional extraction processes to bring it to the surface at economic rates without special stimulation.

TOKOMAK FUSION TEST REACTOR (TFTR): An experimental thermonuclear reactor, originally developed in the Soviet Union and now being developed also in the United States (for example, Princeton University). Many fusion experts believe that the Tokomak concept is the fusion reactor concept "most likely to succeed" as a commercial reactor.

TOLUENE: A liquid aromatic hydrocarbon ($C_7 H_8$) that is produced commercially from coke oven gas, light oils, coal tar, and petroleum.

TRANSURANIC ELEMENTS: All of the elements that have more protons in their nuclei than does uranium. Among them are neptunium, plutonium, americium, and curium. The transuranic elements are all radioactive.

TRITIUM: A radioactive isotope of hydrogen which has mass number three (one proton, two neutrons) and a half-life of 12.3 years.

TURNKEY: A ready-to-operate industrial facility sold at an agreed price.

ULTIMATE DISPOSAL: Final, usually irretrievable, disposal of hazardous materials, especially nuclear wastes.

URANIUM: A radioactive metallic element that exists naturally as a mixture of the isotopes 234, 235, and 238 in the proportions of 0.006 percent, 0.71 percent, and 99.28 percent, respectively. The isotope 235 is the principal fuel of light-water reactors. Occurs in various kinds of ore.

URANIUM-232: A relatively stable isotope of uranium with a half-life of 74 years.

URANIUM-233; A fissionable isotope of uranium that can be formed by bombarding thorium-232 with neutrons.

URANIUM-235 (U-235): An isotope of uranium with mass number 235 that fissions when bombarded with slow neutrons. The principal fuel for light-water reactors, U-235 can, if highly enriched, be used to produce nuclear explosions.

URANIUM-238 (U-238): An isotope of uranium with mass number 238 that absorbs energetic ("fast") neutrons, forming U-239 which then decays through neptunium to form plutonium-239.

URANIUM HEXAFLUORIDE (UF_6): A gaseous compound of uranium used in various isotope separation processes.

URANIUM OXIDE ($U_3 O_8$): The most common oxide of uranium that is found in typical ores. Also, the unit of measurement in which uranium resources and requirements are commonly calculated and reported.

URENCO: A joint British, Dutch, and West German organization that operates centrifuge isotope separation plants.

VISCOSITY: Characteristic of a material determining its ability to resist flow (the lower the viscosity, the less the flowing ability).

WELLHEAD TAX: A tax imposed on oil (or gas) at the top of the production well.

WEAPONS GRADE MATERIAL: Any mixture of uranium or plutonium isotopes that could be used to create a nuclear explosion.

WORKING FLUID: Fluid used in an electrical generator that is heated by the energy source and then expands through a turbine to produce electricity and then is recycled and reused.

Abbreviations and Acronyms

ACRS	Advisory Committee on Reactor Safeguards
AEC	U.S. Atomic Energy Commission
AIA/RC	American Institute of Architects Research Corporation
APS	American Physical Society
BACT	Best Available Control Technology
bbl	barrel
BEPS	Building Energy Performance Standards
Btu	British thermal unit(s)
CAA	Clean Air Act
CANDU	Canadian deuterium-uranium reactor
CHESS	Community Health and Environmental Surveillance System
CIA	U.S. Central Intelligence Agency
CO	carbon monoxide
CO_2	carbon dioxide
CONAES	Committee on Nuclear and Alternative Energy Systems (National Academy of Sciences)
CPI	Consumer Price Index
CRS	U.S. Congressional Research Service
D&D	demonstration and deployment
DOE	U.S. Department of Energy
ECPA	Energy Conservation and Production Act of 1976

EEC	European Economic Community
EIA	Energy Information Agency, U.S. Department of Energy
EIS	Environmental Impact Statement
EPA	U.S. Environmental Protection Agency
EPCA	Energy Policy and Conservation Act of 1975
ERDA	U.S. Energy Research and Development Administration
EURATOM	European Atomic Energy Community
F.A.S.	free alongside ship (price quotation)
FBC	fluidized bed combustion or fluidized bed combustor
FBR	fast breeder reactor
FEA	U.S. Federal Energy Administration
FGD	flue gas desulfurization
GDP	gross domestic product
GNP	gross national product
GWe	gigawatt(s) (electric) = 1,000 MWe
HP	horsepower
HTGR	high temperature gas-cooled reactor
HWR	heavy water reactor
IAEA	International Atomic Energy Agency
ICC	U.S. Interstate Commerce Commission
IDA	International Development Association
IEA	International Energy Agency
IFC	International Finance Corporation
kg	kilogram(s)
kw	kilowatt(s)
kWe or kwe	kilowatt(s) (electric)
kwh	kilowatt-hour(s)
LAER	lowest achievable emission rate
LMFBR	liquid metal fast breeder reactor
LWA	limited work authorization
LWR	light-water reactor
m^2	square meter(s)
m^3	cubic meter(s)
mbd	million barrels per day
Mcf	thousand cubic feet

MER	marketable emission right
Mg	microgram (10^{-6} grams)
MHD	magnetohydrodynamic
mil	$0.001
MSR	molten salt reactor
MW	megawatt(s) = 1,000 kw
MWd	megawatt day(s)
MWd (t)	megawatt day(s) (thermal)
MW (e)	megawatt(s) (electric) = 1,000 kwe

N-A Region	Non-Attainment Region
NAAQS	National Ambient Air Quality Standard(s)
NAS	National Academy of Sciences
NEPA	National Environmental Policy Act of 1969
NNPA	Nuclear Non-Proliferation Act of 1978
NODCs	non-OPEC developing countries
NO_2	nitrogen dioxide
NO_x	nitrogen oxide
NPDES	National Pollution Discharge Elimination System
NPT	Nuclear Nonproliferation Treaty
NRC	U.S. Nuclear Regulatory Commission
NSF	National Science Foundation
NSPS	New Source Performance Standard(s)

O and M	operation and maintenance
OECD	Organization for Economic Cooperation and Development
OPEC	Organization of Petroleum Exporting Countries
OTA	U.S. Office of Technology Assessment
OTE	ocean thermal gradient energy
OTEC	ocean thermal energy conversion (or converter)

PAN	peroxyacetyl nitrate
ppm	parts per million
PSD	Prevention of Significant Deterioration

quad	one quadrillion (10^{15}) Btu(s)

R&D	research and development
RD&D	research, development and demonstration
RDD&D	research, development, demonstration and deployment

SCRAP	Students Challenging Regulatory Agency Procedures

SDR	Special Drawing Rights
SEPA	State Environmental Policy Act
SIP	State Implementation Plan
SO_2	sulfur dioxide
SRC	solvent refined coal
STE	solar thermal electric
tcf	one trillion (10^{12}) cubic feet
TFTR	Tokomak fusion test reactor
Th	thorium
therm	100,000 British thermal units
TMI	Three Mile Island (nuclear reactor accident)
TPE	total primary energy consumption
UF_6	uranium hexafluoride
U_3O_8	uranium oxide
U-232	uranium-232
U-233	uranium-233
U-235	uranium-235
U-238	uranium-238
USDI	U.S. Department of the Interior
W	watt(s)

Appendix

Below is a listing of background papers commissioned by the Study Group to be published as a supplementary volume by Ballinger Publishing Company.

SELECTED STUDIES ON ENERGY:
BACKGROUND PAPERS FOR ENERGY:
THE NEXT TWENTY YEARS

Dimensions of Energy Demand—*William Hogan*
Energy Policies and Automobile Use of Gasoline—*James L. Sweeney*
The Electric Utilities Face the Next 20 Years—*Irwin M. Stelzer*
Price Regulation and Energy Policy—*Philip Mause*
Institutional Obstacles to Industrial Cogeneration—*Norman L. Dean*
Preparing for an Oil Crisis: Elements and Obstacles in Crisis Management—*Edward N. Krapels*
Energy Prospects in Western Europe and Japan—*Horst Mendershausen*
Prospects and Problems for an Increased Resort to Coal in Western Europe, 1980–2000—*Gerald Manners*
Prospects and Problems for an Increased Resort to Nuclear Power in Western Europe, 1980–2000—*Gerald Manners*
Energy Policy in the Soviet Union and China—*Marshall I. Goldman*
Energy in Non-OPEC Developing Countries—*Dennis W. Bakke*
OPEC and the Political Future of the World Oil Market—*Dankwart A. Rustow*

Index

and national ambient air quality
 standards, 377–78, 382, 384
NSPS, 378–81, 384, 392–96
and PSD, 382–84, 407
SIPs, 376–78
and threshold concept, 344–46
U.S. Clean Air Act Amendments of
 1977, 311, 323–24, 380, 382,
 386, 391, 408, 416
and coal-fired plants, 531–32
nonattainment regions, 396–99
and NRC, 519–20
NSPS, 392–96, 399
and PSD, 398–400
and SIPs, 397
U.S. Department of Energy (DOE),
 36–37, 40, 292, 515, 521
and conservation, 129–31
and entitlements, 188–89
and MOPPS, 581
and NSPS, 380–81
on nuclear energy, 421, 423, 465
U.S. Department of Housing and
 Urban Development, 128
U.S. Department of the Interior,
 291–92
U.S. Environmental Protection Agency
 (EPA), 344, 374–82, 391,
 393, 406
and CHESS program, 361–62
and coal-fired plants, 530
and health effects from air pollution,
 344, 347–48, 361–62, 364
NPDES, 519, 522, 530
on nuclear energy, 519–21, 540
U.S. Institute for Technological Cooperation, 269
U.S. Interstate Commerce Commission, 516
U.S. Nuclear Regulatory Commission
 (NRC), 437, 439, 445
ACRS, 518
and delay costs, 532–35
and environment, 519–21, 523, 527
and health and safety, 518–21, 527,
 529
licensing procedures, 517–19,
 527–29
and need for power, 523–25, 527
U.S. Supreme Court, 518–19, 541
USSR, 11
breeder reactors in, 427, 447
climate changes in, 329, 331
coal consumption in, 281
coal production in, 283
coal resources and reserves, 229, 255,
 280, 283
and enriched uranium, 59
natural gas in, 9, 255
NPT and, 442–44
as oil exporter, 226, 260, 262
and OPEC relations, 252
petroleum reserves in, 260–61, 269
uranium, 461

Uranium. *See also* Enriched uranium;
 Enrichment, uranium
and alternative fuel cycles, 425–32,
 572–73
for breeder reactors, 427–28, 447
consumption, reduction of, 425–31,
 449, 461–62
exports, 458–62
and LWRs, 426–30, 448–49
for nuclear energy, 414–15
and R&D, 558
resources and reserves of, 235–38,
 248, 461–62, 558
in U.S., 415, 422
in USSR, 461
for Western Europe and Japan, 255,
 415
URENCO, 459, 464

Venezuela, 15, 203, 240, 264
Vermont Yankee case, 517, 541

Wages, and price increase, 21, 25,
 158–60, 162–64
WASH-1400, 437
Waste disposal, 26
coal, 311–12
coal conversion, 314–16
nuclear, 411, 413, 415, 427, 436–
 42, 451–52, 456–58, 464, 513,
 525–27
Water pollution
and coal conversion, 314–16
and coal production, 296–98, 311
and legislation, 519–20
Water resources, in Western U.S.,
 316–23
Wellhead tax, 47, 214
Western coal, in U.S., 230, 278,
 285–86, 291–93, 295–96, 300–
 301, 378
and CAA, 323–24, 383
coal and energy development in,
 316–24
and cost, 304, 313
Western Europe
and coal imports, 289
coal production in, 325
coal use in, 226, 255–57, 281, 325
conservation, and demand reduction,
 255–60
energy policy in, 253–60
energy sources in, 226, 256–58
nuclear energy in, 257, 413–15,
 424–25
oil imports in, 10, 206, 225, 257–58,
 265
uranium in, 415
West Germany
breeder reactors in, 427
coal production in, 256, 325
coal reserves in, 255
LWRs in, 427